Engineered Transparency 2016

Engineered Transparency 2016

Glass in Architecture and Structural Engineering

Editors: Jens Schneider, Bernhard Weller
Editorial Office: Katharina Lohr, Silke Tasche

Cover: +e Kita, Marburg, opus ARCHITEKTEN, Darmstadt, Germany
Photograph: Eibe Sönnecken, Darmstadt, www.eibefotografie.de, Germany

Library of Congress Card No.:
applied for

British Library Cataloguing-in-Publication Data
A catalogue record for this book is available from the British Library.

Bibliographic information published by the Deutsche Nationalbibliothek
The Deutsche Nationalbibliothek lists this publication in the Deutsche Nationalbibliografie; detailed bibliographic data are available on the Internet at <http://dnb.d-nb.de>.

Coverdesign: Sophie Bleifuß, Berlin
Production management: pp030 – Produktionsbüro Heike Praetor, Berlin
Printing and Binding: CPI books GmbH, Leck

Printed in the Federal Republic of Germany.
Printed on acid-free paper.

Print ISBN: 978-3-433-03187-2

Foreword

The conference »Engineered Transparency« was founded in 2007 at the Columbia University in the City of New York and has accompanied the »glasstec« in Düsseldorf – the world's leading trade fair of the glass industry – since 2010. The unique liaison of research, novel developments and built examples always meets the demands of a broad expert audience.

The conference covers various subjects ranging from conceptual design, to planning and realization as well as to relevant research topics in glass and facade constructions. Additionally, the mini-symposia »embedded functions« and »glass technology« address current trends in the field of glass in building. Those special sessions were initiated in close collaboration with the professional associations Hüttentechnische Vereinigung der Deutschen Glasindustrie e.V./ Deutsche Glastechnische Gesellschaft e.V. (HVG/DGG) and Bundesverband Flachglas e.V. (BF) as well as Fachverband Konstruktiver Glasbau e.V. (FKG).

More than 60 peer-reviewed contributions received from authors more than 20 different nations create this exceptional spectrum of topics. We want to express our sincerest thanks to all contributing authors and speakers who share their ideas and knowledge with great commitment and often in addition to their day-to-day business. It is a great pleasure to welcome Werner Sobek, Werner Sobek Group, Stuttgart, Tom Minderhoud, UNStudio, Amsterdam, Juan Lucas Young, Sauerbruch Hutton, Berlin and Johann Sischka, Waagner-Biro, Wien as keynote speakers who lead to the conference with their inspiring ideas.

We would like to thank the members of the scientific committee for their valuable suggestions and the review of all papers. Particular thanks are due to Mrs. Stürmer and her team at Ernst & Sohn and Mrs. Horn and the staff at Messe Düsseldorf for their understanding and active support.

Prof. Dr.-Ing. Jens Schneider
Technische Universität Darmstadt

Prof. Dr.-Ing. Bernhard Weller
Technische Universität Dresden

Engineered Transparency 2016
Glass in Architecture and Structural Engineering

www.engineered-transparency.eu

Editors
Prof. Dr.-Ing. Jens Schneider, Technische Universität Darmstadt
Prof. Dr.-Ing. Bernhard Weller, Technische Universität Dresden

Editorial Office
Dipl.-Ing. Katharina Lohr, Technische Universität Dresden
Dr.-Ing. Silke Tasche, Technische Universität Dresden

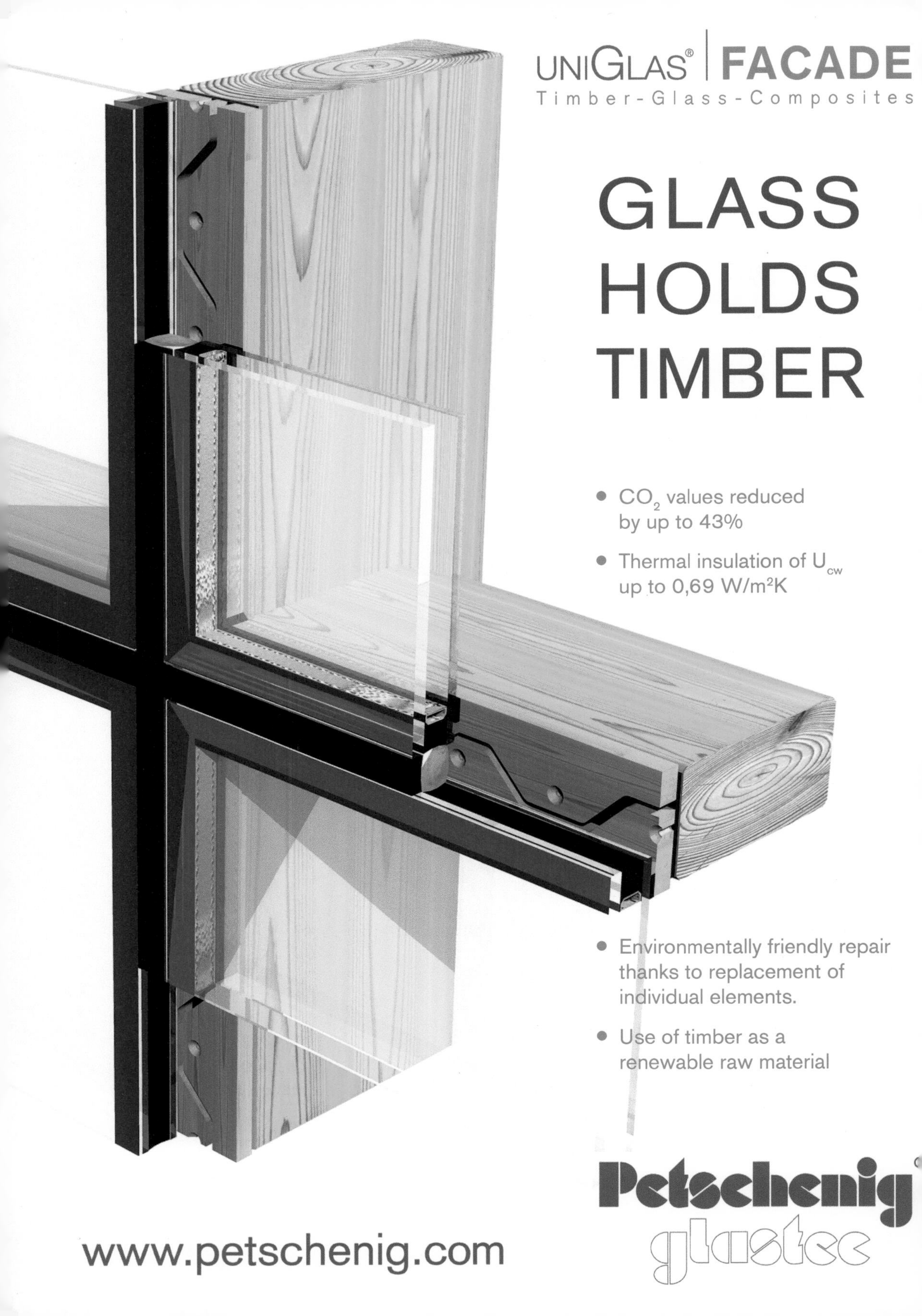

Inhaltsverzeichnis

Facade – Architectural Design

Facade – Structural Design

Facade – Projects

Solar Technologies

Glass – Structural Glass

Glass – Structural Design

Glass – Composites and Coatings

Glass – Projects

Laminated display based on printed electronics

Mikael Ludvigsson[1], Peter Leisner[2], Bo Schödt[2], Peter Dyreklev[3], David Nilsson[3], Björn Norberg[3], Lina Grund Bäck[1], Ingemar Malmros[1], Ulf Clausén[4], Peter Andersson Ersman[3]

1 Glafo, the Glass research institute, PG Vejdes väg 15 SE-35196 Växjö Sweden, Mikael.Ludvigsson@glafo.se, Lina.Grund.Back@glafo.se, Ingemar.Malmros@glafo.se

2 SP Technical research institute of Sweden, Box 857 SE-501 15 Borås Sweden, Peter.Leisner@sp.se, Bo.Schodt@sp.se

3 Acreo Swedish ICT, Box 787 SE-601 17 Norrköping Sweden, Peter.Dyreklev@acreo.se, David.Nilsson@acreo.se, Bjorn.Norberg@acreo.se, Peter.Andersson.Ersman@acreo.se

4 Forserum Safety Glass, Sörbyvägen 9 SE-571 78 Forserum Sweden, Ulf.Clausen@fsglass.se

Transparent intelligence, functional glass and smart windows are important areas of development for the global glass industry. Transparency is vital for quality of life. Functionalities such as solar control and low emissivity coatings are necessary constituents in the climate shell of an energy efficient building. Information technology has become a part of the modern infrastructure and internet of things will be incorporated in the glass solutions of tomorrow. Glafo the Swedish Glass research Institute, the electronics department at SP Technical Research Institute of Sweden and Acreo Swedish ICT investigated the possibility and challenges related to the lamination of printed electronic displays between two glass panes. The printed electronic display, which has been developed at Acreo Swedish ICT, was manufactured by using screen printing technology. The technology allows for both opaque and transparent displays; this study investigates an opaque version. The lamination of the display was performed at the Swedish company Forserum Safety Glass using two pieces of 4 mm float glass panes and EVA (ethyl vinyl acetate). The laminated displays were investigated in a climate chamber at different settings of temperature and humidity and compared with reference displays that were not laminated. It was obvious that the laminated displays performed better than the reference displays. Some challenges were identified during the lamination process, for example that the display moved and changed position with as much as one centimeter. Due to this, some samples lost contact with the external power source. It could also be observed how gases were evaporating from the display during the lamination process. These are challenges that will be investigated in further studies.

Keywords: printed electronic displays, glass, lamination, transparency, laminated displays

1 Introduction

Glass is becoming more and more important as a building block in architectural buildings, and this is also a strong motivation to add electronic functionalities to the glass itself. There are a large number of applications in which an electronic functionality could add value to the glass. Smart windows might be one of the most obvious examples, in which

Engineered Transparency 2016. Glass in Architecture and Structural Engineering. First Edition.
Edited by Jens Schneider, Bernhard Weller.

an electrochromic layer that can be switched between two different optical absorption states is incorporated into the window. The change in optical absorption changes the color of the window in many cases, but even more important, the absorption of the near-infrared wavelengths is also affected by the electrochromic switch. Hence, a combination of glare and heat control can be obtained in such smart windows, which in turn results in lowered energy consumption of the air conditioning system. Smart windows are manufactured either directly onto the glass surface, or by lamination of a separate electrochromic layer onto the window. Historically, smart windows have typically been based on inorganic electrochromic materials that are deposited in vacuum equipment by sputtering techniques (Avendano et al. [1]). However, a lot of currently ongoing research aims at the development of the large-area electrochromic systems based on organic electrochromic polymers. In such systems, the electrochromic material is processed from solution and the deposition is carried out by e.g. coating or printing techniques. Electrochromic smart windows are relying on large areas that are coated with an electrochromic material, an electrolyte layer and a counter electrode layer, hence, no patterning of the deposited layers is required.

However, electrochromic smart windows are just one type of application in which an electronic functionality is added to the glass, and as already mentioned, they are manufactured by either coating the layers directly onto the glass or by pre-manufacturing the electrochromic film on e.g. a plastic or thin glass that subsequently is laminated onto the glass window. The future will involve many other kinds of electronic components and systems incorporated onto glass, into glass, or by laminating the electronics between glass sheets. Examples of such components are solar cells and displays, which both can be manufactured by printing techniques and also require the transparency of the glass. Furthermore, many printed electronic components are relying on organic materials, which results in that the devices show a dependence, or even degradational behavior, upon exposure to e.g. humidity fluctuations or oxygen; two critical issues that can be prevented by the hermetic sealing property provided by utilizing glass as the carrying substrate. Therefore, in order to take advantage of both the transparency of the glass as well as the hermetic sealing property, it is demonstrated how printed electrochromic displays can be laminated between two glass sheets by using a standard lamination process for safety glass. It is then demonstrated that the display functionality is maintained also after the lamination process, and additionally, it is also shown how the inherent humidity dependence of the printed display is circumvented after the lamination procedure.

Acreo Swedish ICT has developed an electrochromic display technology in which the display devices are manufactured solely by printing techniques (Kawahara et al. [2, 3]). Screen printing, which is a very straightforward printing method, is used to deposit all printed layers of the display device, and the compatibility of the manufacturing process has been verified in sheet-to-sheet as well as in roll-to-roll production setups. The display architecture is relatively simple; a conducting polymer serves as the electrochromic material. Thanks to its intrinsic electronic conductivity, the very same conducting polymer serves also as the conducting electrode providing the electrons required to the color switch. This is a unique feature implying that no additional conducting material is required, such as indium tin oxide (ITO) or similar transparent conducting oxides (TCO), which typically is the case in other display technologies. Thus, the conducting polymer is deposited by screen printing directly on top of a plastic surface, e.g. polyethylene terephthalate (PET). Hence, the content of the display is observed through the plastic sheet.

The electrochromic color switch requires ions, and an electrolyte layer is therefore screen printed on top of the conducting polymer layer. There are many different electrolyte alternatives that can be used, e.g. polymer electrolytes, polyelectrolytes and ionic liquids. The resulting display device is vertically stacked, which therefore adds curability of the electrolyte as a requirement, such that the counter electrode layer can be subsequently deposited by screen printing. Hence, it is not only the printability that is of importance when developing functional inks, the compatibility with previously deposited (wetting property) as well as subsequently deposited (curability) layers are equally important. The electrochromic display devices presented here are operated in reflection mode, which adds another requirement on the electrolyte; it must be optically opaque in order to hide the underlying counter electrode material. The counter electrode material can consist of the same conducting polymer used in the color changing later, or other conducting polymers, or a printed layer of graphite, but there are a lot of other electrically conducting materials that can be used as an alternative counter electrode material as well. It should also be noted that it is possible to create electrochromic display devices that are operating in transmission mode. This would of course require a transparent electrolyte as well as that the electrochromic material and the counter electrode material must be as transparent as possible in order to provide maximum color contrast.

2 Experimental

The conductive polymer PEDOT:PSS (poly (3,4-ethylenedioxythiophene):poly(4-styrenesulfonate)) was used as the electrochromically active material, Figure 2-1 (left) shows the chemical structures of PEDOT (bottom) and PSS (top). PEDOT is chemically oxidized during manufacturing, and PSS is therefore required to maintain the charge neutrality of the polymer complex. The oxidized state of PEDOT is electronically conducting and almost transparent, while electrochemical reduction of the polymer results in lowering of the electronic conductivity by many orders of magnitude. Additionally, electrochemical reduction also implies an electrochromic switch by that the optical absorption is changed; the appearance of the polymer is changed from the almost transparent oxidized state to a dark blue color in the reduced state. Figure 2-1 (middle) shows the optical absorption characteristics of PEDOT:PSS. As already mentioned, electrochromic displays can be operated in either transmission or reflection mode; the displays reported herein are operated in reflection mode and thereby using an opaque electrolyte layer hiding the counter electrode layer. The images in Figure 2-1 (right) are photographs showing the actual appearance of the reduced blue-colored and oxidized white/gray-colored states. The white/gray color in the oxidized state is simply explained by that the PEDOT:PSS is switched to its most transparent state, such that the opaque white/gray-colored electrolyte becomes visible.

The PEDOT:PSS (commercial ink provided by Heraeus) was screen printed on top of a PET foil (Polifoil BI-AS provided by Policrom Screens, thickness 125 µm, tensile strength 22 daN/mm^2), the ink was thermally dried at 110 °C for a few minutes. Then the electrolyte (commercial ink provided by Acreo Swedish ICT) was screen printed on top of the PEDOT:PSS layer. UV-curing of the printed electrolyte layer provides mechanical robustness. The shape of the electrolyte that is brought in contact with the PEDOT:PSS defines the pattern of the display segment. Finally, the counter electrode material was deposited by screen printing. In this case the very same type of PEDOT:PSS ink was used also as the counter electrode material, and therefore also thermally dried at 110 °C for a few minutes. Additional layers of carbon ink and silver ink (both commercially available inks provided by DuPont) were also screen printed in order to form robust contact pads with high electronic conductivity. The carbon and silver inks were both thermally dried at 110 °C for a few minutes. It should be noted that PEDOT:PSS could serve as the contact pad material as well, but it is more sensitive to the mechanical scratching that occurs upon contacting the display device during measurements, as compared to the carbon and silver layers.

Before laminating the printed display between the glass panes, the functionality was checked as well as that long gold wires were glued onto the screen printed silver contact pads. The assembly of the gold wires was carried out by SP Technical Research Institute of Sweden, and the purpose of the gold wires is simply to enable electronic contact with the display device also after lamination. The lamination of the display was performed at the Swedish company Forserum Safety Glass. The display was placed between two pieces of 4 mm thick float glass panes, and the display was circumvented by ethyl vinyl acetate (EVA, EVASAFE001 provided by Bridgestone Corporation, thickness 400 µm, tensile strength > 15 MPa) covering the remainder of the glass area. The thickness of the display device is approximately 200 µm, including the 125 µm thick PET substrate, and the thickness of the EVA was approximately 400 µm in order to smoothen the pressure across the complete area of the glass surface, without adding too much pressure on the area covered by the display device, during the lamination process carried out under vacuum and 110 °C. To complete the encapsulation, the edges of the glass sheets were sealed by the deposition of rubber after the lamination process.

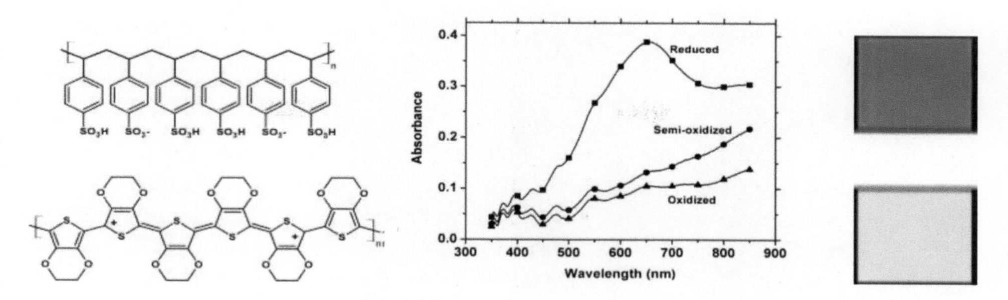

Figure 2-1 Left: The chemical structures of the counter ion PSS (top) and the oxidized form of PEDOT (bottom) are illustrated. Middle: The graphs show the optical absorption characteristics of the different oxidation states of PEDOT:PSS. The reduced form is blue-colored and electronically semiconducting, while the oxidized form is almost transparent and electronically conducting. The curve denoted semi-oxidized represents the intrinsic oxidation state of PEDOT:PSS, which is relatively transparent and electronically conducting. Right: The images represent the color states of a printed electrochromic display that is operated in reflection mode.

3 Results

Some challenges were identified during the lamination process. The display moved, and thereby changed its position with as much as one centimeter, during the vacuum process. Due to this, some of the samples lost their contact with the external power source. In Figure 3-1, it can also be seen how gases are evaporated from the display during the lamination process, which in turn creates bubbles circumventing the display. This might not be crucial for the proof of concept demonstrated in this report, but are clearly challenges related to the lamination process as well as the display design that both need further investigation and optimization in future studies, for example by adjusting the thickness of the EVA material circumventing the display device between the glass sheets.

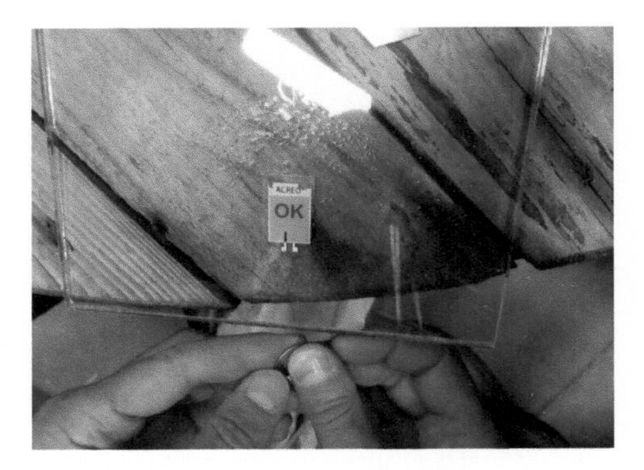

Figure 3-1 The photograph shows a printed electrochromic display device laminated between two sheets of glass. The display segment 'OK' is switched to its blue-colored on-state by a 3 V battery. Some bubbles were formed during the lamination process, they can especially be observed above the display device in the image.

The laminated displays were investigated in a climate chamber at different settings of temperature and humidity, and the results were compared with reference displays that were not laminated. In dry (15 % RH) and relatively cold (20 °C) environment, it was obvious that the displays that were laminated between glass sheets performed much better than the reference displays without encapsulation after four days of storage in this environment, see the charging current vs. time characteristics of the laminated and non-laminated display in Figure 3-2 and Figure 3-3, respectively.

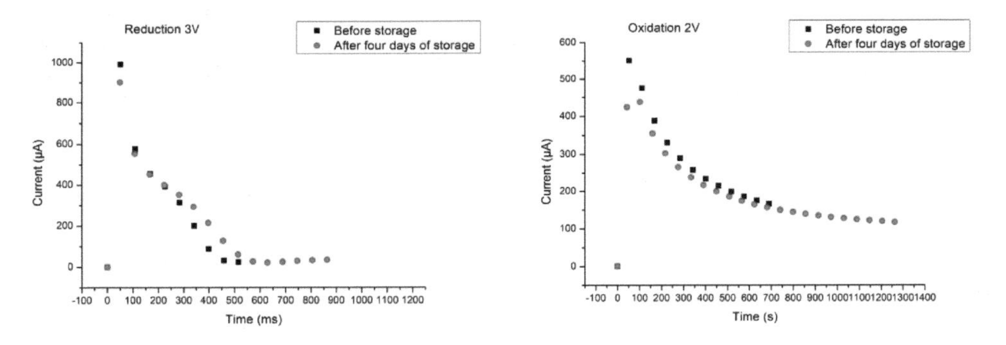

Figure 3-2 The switching behavior is shown after storing a laminated display device at 20 °C and 15 % RH. The black curves represent the initial switching behavior and the red curves represent the switching behavior after storing the display for four days in this environment, i.e. the display is more or less unaffected by the storage in such dry environment. Left: the curves correspond to the switching behavior upon reducing the display segments to their blue-colored states by applying 3 V. Right: the curves correspond to the switching behavior upon oxidizing the display segments to their white-colored states by applying 2 V of the opposite voltage polarity.

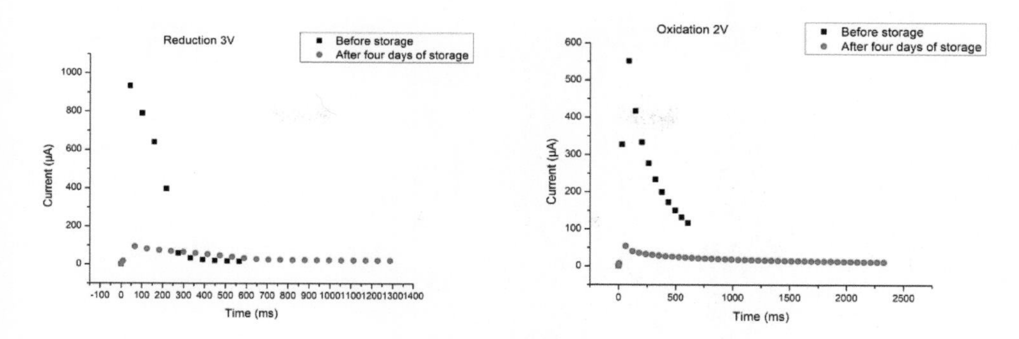

Figure 3-3 The switching behavior is shown after storing a non-laminated display device at 20 °C and 15 % RH. Left: the curves correspond to the white-to-blue color switching event, while the curves on the right hand side correspond to the opposite switching event. The black curves represent the initial switching behavior, while the red curves represent the switching behavior after four days of storage in this dry environment. 3 V was applied during the white-to-blue switching event, while 2 V of the opposite voltage polarity was applied during the opposite switching direction. The low current levels in the red curves indicate that the electrolyte has become too dry during the storage time, which in turn results in very long switching time of the electrochromic reaction. Hence, the large difference when comparing the current levels between the black and red curves proves that the non-laminated display is severely affected by the dry environment.

The settings of the climate chamber was then adjusted towards a more humid (93 % RH) and relatively hot (40 °C) environment. The same samples were used as in the previous storage test in dry and cold environment, and a storage time of four days was used again. The results are not as obvious as in the case of storage in dry and cold environment, even though the characteristics of the charging current vs. time indicate that the laminated display behaves similarly after four days of storage, see Figure 3-4. On the contrary, the display without encapsulation shows elevated current levels, which in turn is a strong indication of parasitic electrochemical reactions caused by increased water content inside the display device, see Figure 3-5.

In addition to the measurements of the charging current levels vs. time, photos of displays with and without glass encapsulation were also recorded in order to further prove the possibility to prevent the humidity dependence by laminating the displays between glass sheets, see Figure 3-6. Elevated humidity levels typically results in that the color spreads outside the display segment area predefined by the patterned electrolyte layer, i.e. like a halo, in displays without encapsulation. This creates a blurry appearance of the display segment. On the contrary, displays that are encapsulated between glass sheets do not suffer from this effect; the color spreading is prevented as observed by the crisp appearance of an encapsulated display segment.

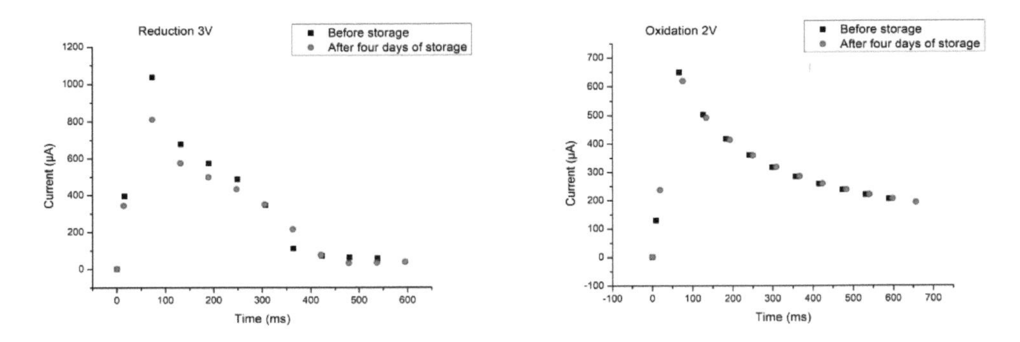

Figure 3-4 The graphs indicate that the laminated display is more or less unaffected by a storage time of four days in 40 °C and 93 % RH. Left: the curves correspond to the white-to-blue color switching event at 3 V, while the curves on the right hand side correspond to the blue-to-white color switching event performed at 2 V and reversed polarity.

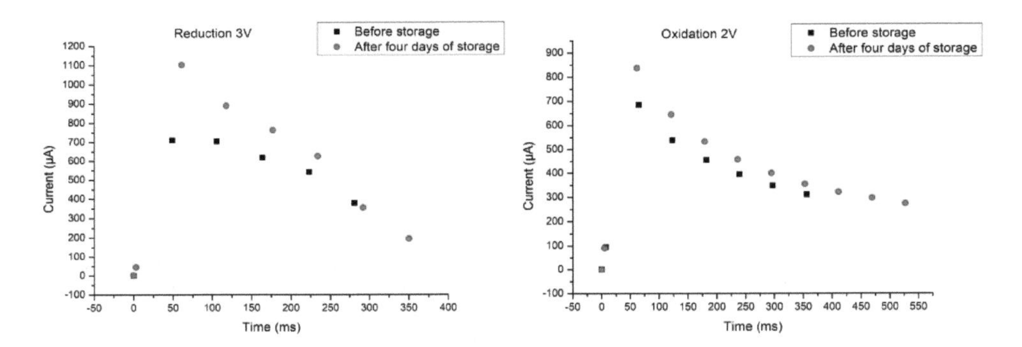

Figure 3-5 The graphs indicate that the display without glass encapsulation is affected by a storage time of four days in 40 °C and 93 % RH. This is concluded by the elevated current levels after four days of storage, which is an indication of increased water content inside the display device. Left: the curves correspond to the white-to-blue color switching event at 3 V, while the curves on the right hand side correspond to the blue-to-white color switching event performed at 2 V and reversed polarity.

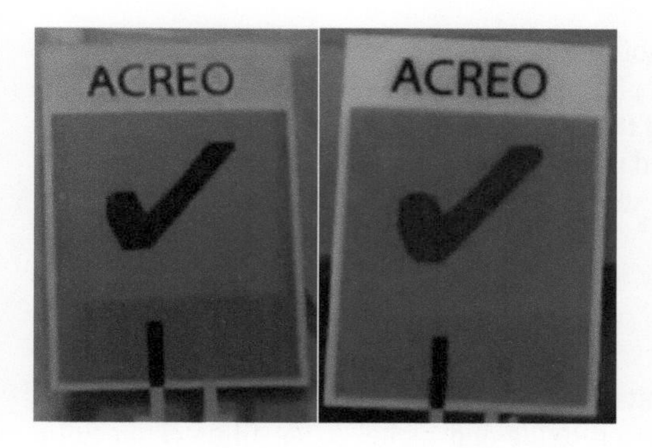

Figure 3-6 The displays shown in the upper images and the lower right image have been stored in 40 °C and 93 % RH during four days. The 'tick mark' segment in the upper right image, which shows a blurry appearance, corresponds to a display without encapsulation, while the color spreading is prevented in the encapsulated display shown in the upper left image. The display shown in the lower right image has no encapsulation, which can be easily observed by the blurry edges of the 'OK' segment. The display shown in the lower left image is also non-encapsulated, but it has instead been stored in ambient atmosphere during four days, such that the crisp appearance of the 'OK' segment is maintained.

4 Conclusions

It has been demonstrated that printed electrochromic displays based on conducting polymers can be laminated between glass sheets. The display functionality is maintained, and even improved, after the lamination process performed in vacuum. The improvement in display performance is evidenced by that the glass sheets encapsulate the display and thereby circumvents the inherent humidity dependence of the display device. This enables the possibility to incorporate electronic functionality in various applications that already are relying on the transparent and encapsulating properties of glass.

Additionally, conducting polymers are more prone to degradation upon exposure to the combination of UV light and oxygen, i.e. photooxidation. Hence, the lifetime of the PEDOT:PSS serving as the active material of the electrochromic display should be extended by that the amount of oxygen is minimized upon encapsulating the display between glass sheets; this will be investigated in further studies.

5 References

[1] Avendano, E., et al., Electrochromic materials and devices: Brief survey and new data on optical absorption in tungsten oxide and nickel oxide films. Thin Solid Films, 2006. 496(1): p. 30-36.

[2] Kawahara, J., et al., Improving the color switch contrast in PEDOT:PSS-based electrochromic displays. Organic Electronics, 2012. 13(3): p. 469-474.

[3] Kawahara, J., et al., Flexible active matrix addressed displays manufactured by printing and coating techniques. Journal of Polymer Science Part B-Polymer Physics, 2013. 51(4): p. 265-271.

Regulated transparency. A brief review of the existing and the most promising technologies

Marcin Brzezicki[1]

1 Wroclaw University of Technology, Wroclaw, Poland, marcin.brzezicki@pwr.edu.pl

The possibility to change light transmission properties of glass has been of interest to architects and engineers not only because it can be used for microclimate control, but also because it helps to achieve a variety of visual, formal and architectural expression. In this process, the optical transparency can be altered qualitatively or quantitatively: by changing the quality of transmitted light (e.g. by scattering, wavelength selection/discrimination), or by changing the amount of transmitted light (e.g. by modifying the luminous flux transmitted through the envelope in all wavelengths). Although various technologies of transparency alteration are known, in this paper I will briefly describe the existing „analogue" solutions (e.g. shutters, blinds and rollers) and focus on emerging technologies, which allow for the modification of transparency using physical/chemical/electrical/mechanical processes at the microscopic level (e.g. at the scale of the coating – thermochromic, electrocromic, LCD, etc.).

Keywords: smart glazing, smart glass, switchable glazing, regulated transparency, privacy glass, sun valve

1 Introduction

Humans need daylight to live a healthy life. Vitamin D is synthesized only in the presence of daylight and the diurnal cycle of rest and activity is governed by the day-night cycle. Windows have a positive impact „as long as they do not cause glare, thermal discomfort or a loss of privacy" (Hellinga [1]). Smart glazing – the general term describing the technologies capable of regulating light transmission through glass – greatly influences the building envelope performance in: *(i)* thermal management; *(ii)* daylight harvesting and regulation; *(iii)* maintenance of views; *(iv)* power capture and finally also *(v)* activating the envelope as information display (Andow et al [2]). The possibility to change light transmission properties of glass has been of interest to architects and engineers not only because it can be used for the above mentioned reasons, but also because it helps to achieve a variety of visual, formal and architectural expressions. The optical properties of a transparent envelope play an important role for both the interior and exterior of a building. This paper addresses the visual/optical issues of the so-called „smart glazing" and briefly presents the wide range of effects that can be achieved by the application of various technologies. The reviewed technologies are presented in a matrix (table), with indication areas in which technologies are currently being developed, as well as „empty" areas with potential for development. Graphical representation/visualization of data might accelerate the development of technologies based on previously unknown mechanisms.

Engineered Transparency 2016. Glass in Architecture and Structural Engineering. First Edition.
Edited by Jens Schneider, Bernhard Weller.

2 Typology. The qualitative and quantitative approach.

Optical transparency can be altered qualitatively or quantitatively: by changing the quality of transmitted light (e.g. by scattering, wavelength selection/discrimination), or by changing the amount of transmitted light (e.g. by modifying the luminous flux transmitted through the envelope in all wavelengths). Although the techniques used to achieve such effects have evolved over time, they generally fall into one of the two categories mentioned above (e.g. according to the approach presented in this paper, a venetian blind is a device that controls the light quantitatively, whereas a light-scattering LCD screen is a device that simply filters the light and is thus regarded as a qualitative device). The wide range of available solutions also calls for the formulation of an intermediate category, where both the qualitative and quantitative change could be observed, e.g. in electrochromic devices or SPDs (suspended particles devices), where a dark-blue tint simultaneously limits the amount of light and changes its color. The above categories might be further differentiated according to a temporal scale, which determines if the change between extreme states (from clear to translucent, or from clear to tinted and then to dark) occurs gradually or rapidly (in an on/off manner or by dimming/tinting). Such a typological approach allows light regulation techniques to be grouped according to the dynamics of the visual effect that is produced. The temporal scale might be of similar interest to architects and interior designers as is the quality of vision. Another equally important factor is the spatial scale distinction which, in simple terms, allows to classify the available technologies as „perceptible" (macro-scale technologies), or „imperceptible" (micro- or nano-scale technologies). This differentiation can help to determine whether the change in a pane's light-transmitting features is heterogeneous, i.e. if the change occurs only locally (as in the case of venetian blinds), or homogenous (the change affects the whole pane as in the case of photochromic glazing). Micro-scale systems can change the characteristics and properties of the light-transmitting material (see Fig 3-1).

3 The quantitative approach

The amount of light flowing through a building's envelope can be regulated quantitatively at the macro- and micro- and nano-scale (i.e. inside the „imperceptible" coating).

3.1 Macro-scale solutions

In the classical approach, light flow is regulated before it reaches the light-transmitting envelope. This includes external venetian blinds, rollers and shutters. The first attempts to regulate the flow of light into a building were made long before the introduction of glass windows, and can be traced back to the Middle Ages when elements of Arabic architecture, such as *mashrabiya*, were used. In more temperate climates, where glazing is a necessity, sunlight was managed mainly through the use of venetian blinds (horizontal slats), rollers (rolled fabric) and shutters (pairs of exterior window coverings) – which serve as an additional functional layer mounted inside or outside the building.

The amount of sunlight that entered the building was controlled by *local* shading of the interior. The classical solutions are still successfully used in a more advanced technological form and serve as a functional layer of the so-called „climate adaptive building shell" (Lonnen et al [3]). In contemporary applications, such systems are managed by rather simple, yet highly effective, mechanical actuators. e.g. recent studies show that venetian blinds „can reduce solar heat gains up to 35 %" (Para et al [4]) when used in double facades.

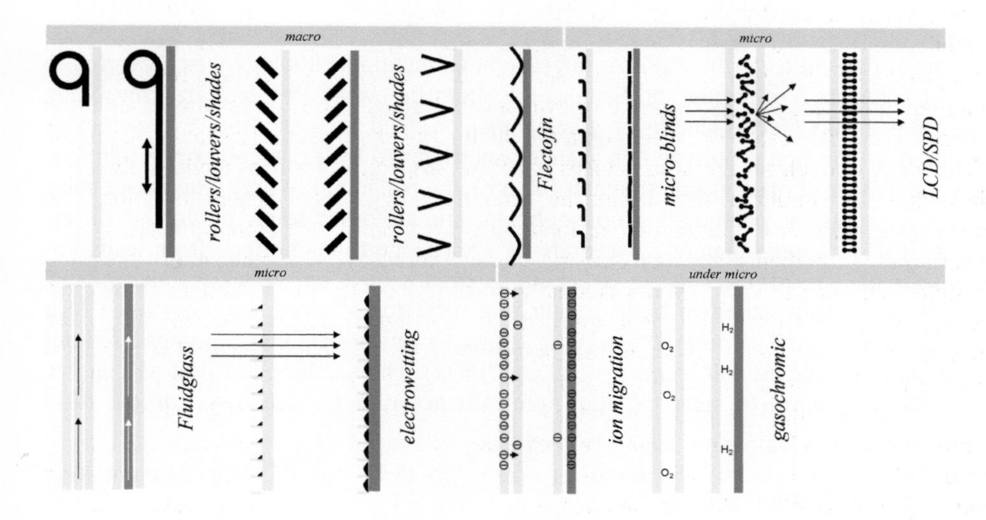

Figure 3-1 Schematic illustration of various smart glass mechanisms.

Regulation devices can also be placed on the interior of the glazed envelope, but this only improves daylight control while heat gains remain almost the same. This is why the vast majority of regulated transparency systems are developed to reduce the heat gains through the light-transmitting envelope.

3.2 Micro- and nano-scale management

Micro-blinds were developed at the National Research Council (Canada). Micro-blinds are composed of trapezoid-shaped curling micro-thin metal blinds on a transparent conductive layer. In the absence of voltage the blinds are curled and light passes through. Once voltage is applied, the difference of potential is created and the electrostatic force stretches the blinds so that light is blocked. The advantage of this solution lies in its instant reaction time (milliseconds) and UV resistance (Lamontagne [5]). The technology is at prototype stage.

Another effective and „instantaneous" form of quantitative regulation is the application of *LCD display* in the so-called transmissive mode. The TIM-LCD element

(TIM = Transparent Imaging Matrix) patented by AEG MIS is „clearly transparent in the *off state* and black in the *on state*" (US Patent no. 20070018943 [6]). This result is achieved with the use of a liquid crystal layer, transparent electrodes and two perpendicular polarizing filters. This technology allows for rapid on/off regulation of the light flow simply by blocking the light passage. The change of the state from transparent to black is instant (within milliseconds). This soultion can also filter light, which is manifested in several levels of „tint". This technology is currently used in information-displays or, less frequently, in partition walls for offices. The high-resolution LCD backlit matrixes operate on a similar principle, i.e. the LCD light-valve regulates the amount if transmitted light. Another novel light extinction technology was presented in 2007 by Kazuki Yoshimura and Shanhu Bao from the Japan Research Institute. They presented a 70 x 60 cm sample of glazing that changed its state from transparent to mirrored. In the first prototype this change was triggered by a reaction between hydrogen and a layer of magnesium, titanium (40 nm) and palladium (4 nm). The newer solutions do not use the potentially problematic gas and are based solely on voltage differences. It takes approximately 15 seconds for the change to take effect. Both technologies were advertised (the latter in 2012) but apparently have not been developed (AIST [7]). A previously unknown optical switching behavior of glass was developed in a laboratory using the *nanocrystal-in-glass* approach by introducing „tin-doped indium oxide nanocrystals into niobium oxide glass (NbOx). The resulting material will enable the dynamic control of solar radiation transmittance through windows (Llordés et al [8]).

4 The qualitative approach

The flow of light can be regulated qualitatively across the entire visible spectrum (the regulation device works as the so-called „gray filter") or for selected wavelengths, thus creating a color tint. In most cases such a color effect is not intentional, but rather a byproduct of the reaction between certain chemical compounds. Another method of qualitative regulation of light flow is light scattering which significantly affects the direction of light rays, but does not influence the amount of light transmitted through the envelope. Qualitative regulation is also possible with the use of „classical" solutions – according to the aforementioned typology, rollers made of translucent or translucent-colored fabric should also fall into the category of qualitative regulation devices.

4.1 Wavelength discrimination

Wavelength discrimination is usually the consequence of changes taking place on the atomic, or sub-atomic level (e.g. ion migration in the substrate due to the application of electrical current). The currently available technologies include extrinsic (requiring external regulation e.g. the flow of electrical current and potential difference) or intrinsic (auto-reactive – self regulated) solutions, the former being user-controlled while the latter are user-independent. *Electrochromic* smart glass is an example of the first type. It „utilizes an electrochromic film with an ion storage layer and ion conductor placed between two transparent plates. The electrochromic film is usually made of tungsten oxide, owing to the electrochromic nature of transition metals. An electric potential initiates a redox reaction of the electrochromic film transitioning the color and the transparency of the smart glass“ (Wong and Chan [9]). The result is a change from transparent (or almost transparent) to a brown or blue tint. The commercially available systems are „able to modulate the transmittance by up to 68 % of the total solar spectrum” (Baetensa and Jelle [10]). Another commercially available extrinsic technology with a similar visual output (i.e. from transparent to a gray or blue tint) is the *SPD glazing*. This technology is based on the principle of aligning the suspended particles to allow light to pass through the device when an electric field is applied. At present, this technology is commercially available in products such as a thin foil that can be cut to size and laminated between or glued onto sheets of transparent panes (glass, acrylic or polycarbonate). The roof of the USA Pavilion at the 2015 World's Fair in Milan was made of SPD-SmartGlass and consisted of 312 panels that made up over 900 m^2 surface. Wavelength discrimination can also be accomplished with *thermochromic* and *photochromic* coatings, both of which are auto-reactive. The change of transparency is triggered either by heat (in thermochromic coatings) or by light (in photochromic coatings). While thermochromic technology has recently reached the stage of commercially available product (e.g. a 8 - 54 % change in light transmission and a solar heat gain coefficient of 0.16 - 0.36), the photochromic coatings are currently available only on a small scale (e.g. as lens coatings in consumer products). Large-scale products are not available yet.

4.2 Scattering of light

The most advanced light-scattering technique is the *LCD dispersed filter* which operates on the principle of aligning LCD crystals that are suspended in a crystal-bearing polymer (PDLC – polymer dispersed liquid crystals). The difference of potentials between two transparent electrodes causes the crystal dipoles either to align in a light-transmitting mode (under current), or arranges them randomly in a light-scattering mode (no current). The visual output is outstanding and allows the haze to be modulated from 7.5 % to almost 90 %. Since this technique changes only the direction of light rays, it can hardly be used for shading (only 1 % difference of L_t is recorded), but it is frequently used for privacy control or in exhibition design (a translucent surface can be used as a projection screen – e.g. in Jewish Museum in Berlin). The change of optical parameters is instant (in milliseconds) and the technology is commercially available (SGG Privalite), which makes LCD dispersed filter technology one of the most popular types of smart glass (the newest SGG Privalite XL glass allows for L_t of 77 - 50 %).

Thermodynamic glazing is another type of glazing that creates a light-scattering effect as a byproduct. It occurs as a result of physical phenomena connected with the change of the material’s phase. This technology uses small containers (approx. 5 - 10 cm) filled with a chemical solution (salt solutions called hydrates, or certain types of paraffin) that changes its phase from solid to liquid in a relatively low temperature of approx. 23 -

25 deg. C. This thermodynamic mechanism is primarily used to store heat in energy-aware architecture (so-called latent heat storage) but due to its phase-changing property, this solution simultaneously changes its optical parameters: from translucent in solid state to transparent in liquid state. The resulting effect is spectacular, as the crystallization process resembles frost forming ice flowers. Because the change is produced by heat (as in the case of thermochromic glazing), this technology is intrinsic.

5 Emerging technologies

In response to a considerable demand for affordable and simple-to-use smart glazing in responsive building skins, new technologies and solutions are being developed (see Fig 5-1). Some of them are still at the conceptual stage while others are already in prototype stage. Recent studies (de Hass [11]) show that one of the most promising technologies is based on the physical phenomenon of *electrowetting*, i.e. the ability to change the contact angle of a droplet on a hydrophobic surface by applying voltage (electrical current). This opens the doors to micro lenses of regulated shape and focal length. „In result, it is possible to change the direction and intensity of incoming light into a building and create optimal light surroundings for every task you are doing" (de Hass [11]). A similar technology of light-reflecting displays that is currently being researched is the *electro-fluidic display*, where the pixel color change is produced by an electrostatic force which „sucks out" the colored pigment from a small reservoir (5 - 10 % of the visible pixel area). Both technologies can be used to regulate the local light-transmitting properties of materials, thus allowing the exploration of new technologies of selective/personal shading or anti-glare protection (e.g. using the so-called dead pixels to create a local tint on a transparent pane that produces a local shadow e.g. to shade only the workspace). Other emerging liquid technology is Fluidglass®, where liquid is circulated in spaces of transparent window pane, transforming passive façade into active heat collector. System provides the external (outer) and internal (inner) cavity for liquid circulation. Liquid can be clear (transparent), or dyed (thus producing the impression of tinted glass). The management of the energy transfer results in energy savings potential of 50 % - 70 %. The system is currently under development and the project is due 31/08/2017. Other technologies have recently been developed as well and are in their prototype stage with potential for further development. Among the macro-scale solutions is the *RSS Rotating Sun Screen* by Marcel Bilow (two checkered transparent disks are placed on top of one another and spun so that their patterns overlap) or the *pneumatic blind*, recently developed at MIT. The latter solution is based on a simple phenomenon: a balloon (elastic diaphragm) is coated with black paint and pressurized. As the pressure is building up, the balloon inflates and the distance between the particles of paint is increased, allowing the light to pass. Inspired by a mechanism observed in nature *Flectofin®* developed a natural hingeless system that allows two elastic surfaces to „flap". A minimal displacement of an elastic longitudinal spine causes a 90 deg. rotation of perpendicular fins attached to it. In collabration with Clauss Markisen, Flectofin® used this concept to build a façade shading, which was awarded the International Bionic-Award in 2012 (Lienhard et al [12]).

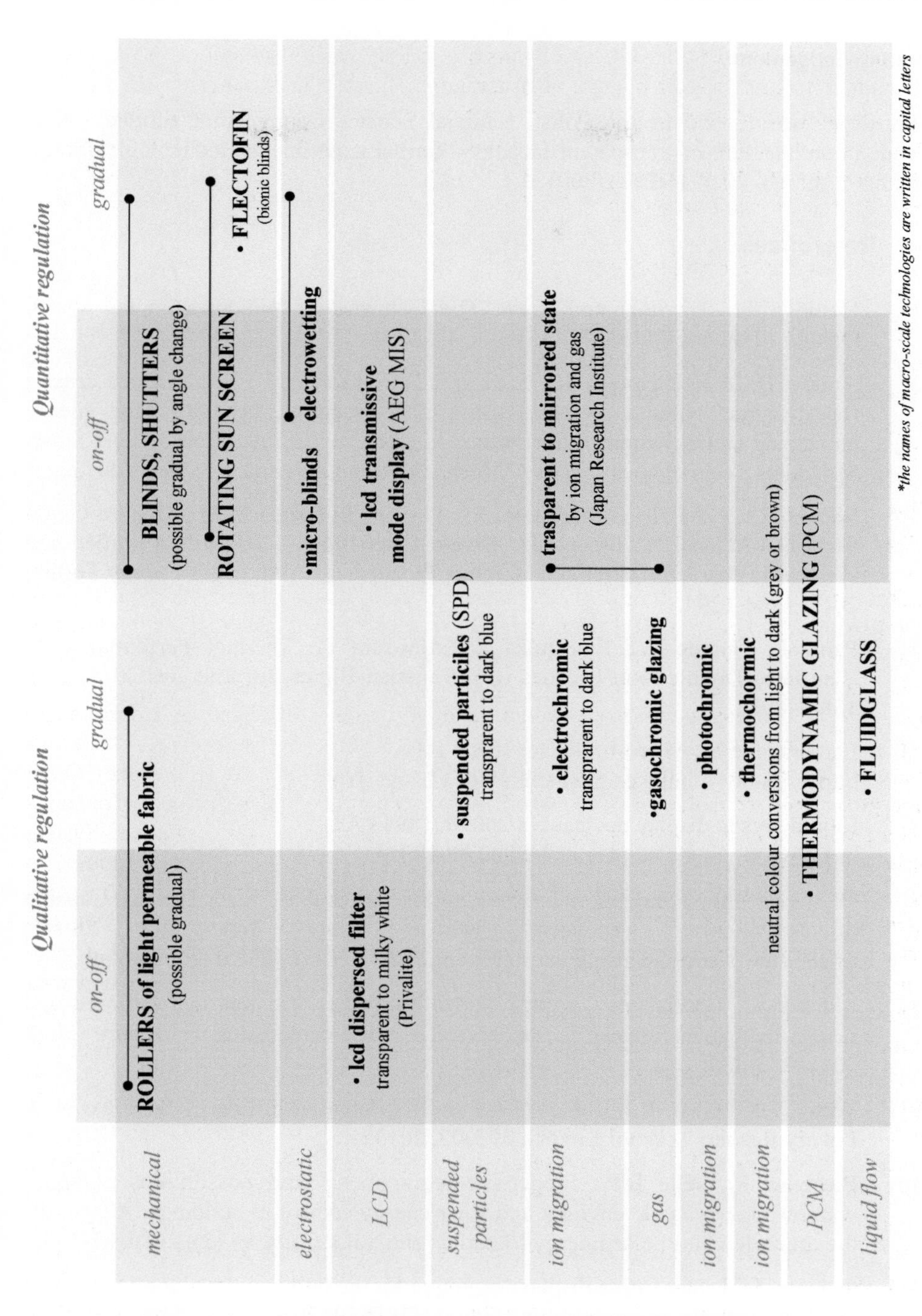

Figure 5-1 The matrix representing the qualitative and quantitative regulation and the type of control.

Acknowledgements

This paper was funded by the Polish National Science Centre grant entitled: „New trends in architecture of transparent facades – formal experiments, technological innovations ", ref. no. 2014/15/B/ST8/00191.

6 References

[1] Hellinga, H.: Daylight and View. The Influence of Windows on the Visual Quality of Indoor Spaces, PhD Thesis, TU Delft, 2013.

[2] Andow B.C.; Krietemeyer B.; Stark P.R.H.; Dyson AH..; Performance criteria for dynamic window systems using nanostructured behaviors for energy harvesting and environmental comfort, In: Proc. SPIE 8692, Sensors and Smart Structures Technologies for Civil, Mechanical, and Aerospace Systems 2013.

[3] Loonen R.C.G.M.; Rico-Martinez J.M.; Favoino F; Brzezicki M.; Menezo C.; La Ferla G.; Aelenei L.: Design for façade adaptability – Towards a unified and systematic characterization, Conference Proceedings of the 10th Energy Forum, 2015, 1284-1294.

[4] Parra J.; Guardo A.; Egusquiza E.; Alavedra P.: Thermal Performance of Ventilated Double Skin Façades with Venetian Blinds. In: Energies 2015, 8(6), 4882-4898.

[5] Lamontagne B.; Micro-blinds for smart glass https://www.youtube.com/watch?v=RqwL2egaqYY.

[6] Liquid crystal display device, US 20070018943 A1 http://www.google.ch/patents/US20070018943.

[7] High-efficiency Fabrication Technology for Switchable Mirror Devices Capable of Switching between Mirror and Transparent States http://www.aist.go.jp/aist_e/latest_research/2012/20120409/20120409.html.

[8] Llordés A.; Garcia G.; Gazquez J.; Milliron D.J.: Tunable near-infrared and visible-light transmittance in nanocrystal-in-glass composites, In: Nature (500), 2013, 323-326.

[9] Wong V.K.; Chan R.J.: Smart Glass and Its Potential in Energy Savings, In. J. Energy Resour. Technol 136 (1), 012002, 2013.

[10] Baetensa R.; Jelle B.P.: Properties, requirements and possibilities of smart windows for dynamic daylight and solar energy control in buildings: A state-of-the-art review. In: Solar Energy Materials and Solar Cells, 94 (2), 2010, 87-105.

[11] de Hass, M. I. Innovative display technique for the building envelope. A research on electrowetting as new glazing method, TU Delft, 2015.

[12] Lienhard J.; Schleicher S.; Poppinga S.; Masselter T.; Müller L.; Sartori J.: Flectofin® A Hinge-less Flapping Mechanism Inspired by Nature, ITKE, University of Stuttgart, 2012.

Modern media facades – symbiosis of architectural design and lighting information

Ralf Krüßel[1]

1 ONLYGLASS GmbH, Siemensstraße 15-17, 27283 Verden, Germany, Kruessel@onlyglass.de

This document would like to introduce the idea of modern media façades. The call for light, integrated solutions that, optimally, are practically invisible when switched off is, for this reason, absolutely understandable from an architectural as well as urban-planning point of view. This in turn means that the glass element takes on a double function; on the one hand it is the thermally-insulation skin and on the other hand that of a modern media façade. Due to the fact that modern facades are increasingly being constructed using glass in order to create greater transparency and brighter spaces, "integrated" in this context can really only mean; integrated into the façade. Only when reduced down to this one single level can it be assured that the demands for high transparency, integration and reduced static stress are met.

Keywords: modern media façade, architectural design, lighting information

1 Evolution of lighting & facades

The staging of architecturally well-designed buildings through illumination is an idea, which has existed for centuries. Architecture combined with lighting adds prestige to a building. The idea of combining architecture and lighting is pretty much the same today; however, the only difference really from the past to today is that lighting technology has advanced. Today we use lighting as an opportunity to transform buildings into communicating objects. One way to do that is by using projectors to display images onto a building. The company URBANSCREEN for example has completed several projects with this technique. However, with this technology the façade of the building acts as a "passive" element and does not have an "active" role in the sense of media screen. Due to the regulatory environment regarding projections (space in front of the building, usage only at dawn or at night, sensitivity to extraneous light) the application of this technology is limited to special events. As always, the development of media façades has been influenced and characterized by their economic value, which can be explored through selling advertising content for example. However, media facades changed the role of the façade from "passive" to"active"- transforming the building into a communicating object.

Engineered Transparency 2016. Glass in Architecture and Structural Engineering. First Edition.
Edited by Jens Schneider, Bernhard Weller.

Figure 1-1 Eiffel Tower as advertising medium.

In 1925, André Citroen has been the first person to transform the Eiffel Tower into an advertising medium by installing 250,000 light bulbs on the beams of the tower. This campaign lasted for ten years and was visible to up to 30 km. The next generation of lighting has been known in Germany as “neon lamps”. This kind of advertising technique used light bulbs as well as light tubes to display advertising content.

A simple animated effect was possible by switching the “neon lamps” on and off. Starting at the end of the last century “neon lamps” have been increasingly replaced by LEDs enabling large LED screens. The development of LEDs and therefore LED screens has been a significant step. Information could be presented to the spectator through text and images and changed at any time directly on the LED screen. This technological development opened up new opportunities for artists and media agencies.

However all of these technologies described above have the following disadvantages:

- In general, they are installed in front of the façade and have a substantial influence on the architectural design of the building.
- Due to the missing transparency the application is limited to buildings with a glass façade as they block daylight from entering the rooms behind the screen.
- Their weight in addition to supporting structures has a significant influence on the static of the façade. Without additional static requirements the installation is unlikely to be realized.
- They are difficult to clean.

With the beginning of the 21st century the requirements for media facades have changed and became more demanding – especially the need for transparent solutions. Today, modern architecture often combines media screens with the glass façade. The so-called "mesh system" solutions are somewhat transparent as the LEDs are placed onto steel mesh, rope mesh or self-supporting wire lead. The "mesh system" only solves the issue of transparency partially; however, all other issues (architectural design, weight, ability to clean) remain unchanged. Especially the substantial influence on the architectural design is often criticised. The "mesh system" may suit a few buildings; however, it is not a revolutionary approach to media facades in the sense of a direct integration into the facade.

Therefore architects would much prefer a solution, which is integrated directly into the façade of a building, causes no additional static requirements and is almost invisible when switched off. To integrate an LED screen directly into a façade can therefore only be possible by integrating the LEDs directly into the insulating glass unit (IGU). Using this technique the IGU does not only keep its thermal insulating glass function but is at the same time a modern media façade. Only this revolutionary solution can assure the combination of high transparency, integration, no static issues and easy to clean. ONLYGLASS MEDIAFACADE is one of the first products in the market to fulfill these characteristics.

2 MEDIATECTURE – What is that all about – Hamburg Reeperbahn – a living example

Figure 2-1 Klubhaus St. Pauli, Reeperbahn Hamburg – living example for MEDIATECTURE.

If one looks at the development which architecture has taken over the past number of decades, it has been one of constant change. This has not only been down to the creativity of the architects who have continuously integrated new designs into their constructions. With the technical development that has taken place within the construction industry and the desire to new forms have been continuously realised create ever more dynamic façades.

Figure 2-2 MEDIAFACADE while it is switched off (left) and during operation (right).

Life has also become far more fast-paced: what is cutting-edge today will be obsolete tomorrow – everything is fluid. Simply having a roof over one's head to protect us from the elements is certainly no longer enough. In the best case scenario, the buildings and façades are able to adapt over time to the changing needs of the people. The development of the new media has seen not only a new technology based on constructional physics become an option, but also a technology that can only able to be integrated into urban planning with great difficulty at the moment. The use of the new media has changed the communication needs of people and society. A living example is the Klubhaus St. Pauli/Hamburg which was opened in September 2015 and makes ever since this idea visible to everybody in the public space around the Reeperbahn. Apart from 16:9 formats the ONLYGLASS MEDIAFACADE will attract much more people through its special design and contents- this does not only apply to advertising contents. The LEDs are embedded within double glazed units and are therefore a part of the façade, which does not disturb the architecture at all. The technical parameters and the principle of integration of the individual elements a shown on the following page.

Technical datas

- Total amount of integrated Pixels: 51.000
- Pixel pitch: 30 mm x 30 mm
- Brightness: 5.300 nit (cd/m^2)
- Transparency: 73 %
- Connected Load: 35 kW
- Power Consumption: 12-14 kW
- Pane size: 2.45 m (W) x 3.45 m (H) m
- Control: DVI

Figure 2-3 Mediafacade element before mounting into the façade (left); Mediafacade element on its way to the façade (right).

Figure 2-4 Mediafacade element integrated into the sub construction (inside view) (left); Mediafacade double glazed units after installation (right).

This development means that modern society is approaching a new phenomenon: MEDIATECTURE. MEDIATECTURE is everything that modern architecture is not allowed to be – conditioned, temporary and decorative. The term MEDIATECTURE goes beyond the utilisation of buildings for media purposes, meaning that the term must also fulfil other tasks in addition to the new communication role which it has been assigned. The challenge arises of creatively combining the potential offered by the new media with the technical systems of a building. According to a definition from Christoph Kronhagel, MEDIATECTURE describes "the interface between real, concrete places and the virtual world. Through the direct integration of new media into the structural environment, connections between spaces that can be physically experienced and ones of an immaterial intellectual nature can be created."

The coinage "MEDIATECTURE" is made up of the words media and tecture (derived from the word architecture). This accurately implies that architecture continues to play a very important role because the façades are part of the architecture, which need to be designed in a mediatectural way. Mediatectural constructions are designed so that they can variably change, for example, with a variably programmable external as well as internal skin that can be designed using the electronic media. It is in this way that the mediatectural space becomes a medium for communicative processes. And it is not just that we as people should adapt to the space, the space should adapt to us.

Figure 2-5 Example for billboard installation at Picadilly Circus London, Zero transparency and in front of the façade.

This makes it clear that the boundary between architecture and MEDIATECTURE is fluid. It is for this reason that it is no longer enough to just decorate building facades with billboards or LED Boards, because the architectural element is completely ignored as a result. Modern architecture in the meantime has developed a far more facades through its analysis of the new media and their meaning for buildings and the urban space. Up to now, media facades had the same functions as a screen on which advertising or art was displayed. With MEDIATECTURE, we have to leave the world of 16:9 behind us and enter into different, completely new and unexpected dimensions.

Many urban planners now also follow a similar argumentation in the meantime. Urban development has become a significant challenge because it is not yet clear what tasks the urban environment of tomorrow will even have. One thing is fact though and that is that the world lives from communication. This not only holds true on the political stage, but also in our personal lives and interpersonal relationships in which the solitude of the urban space becomes more prevalent. Knowing things about each other, results in pan-societal intimacy, and in an increased sense responsibility for each other. In the modern technical world, there are no more meeting points like the village well or the local pub. This is how new markets must be created. A market characterised by modernity does not necessarily need MEDIATECTURE, but it can help to fulfil the challenges in such places much quicker. Public spaces in cities and communities are practically never suitable to create the village well effect because they are seldom places where residents come together for the purposes of social interaction. Media facades, when not being used for advertising purposes or for reporting that day's headlines, could replace the village well and offer the reason "to want to be there" through intelligent content. Intelligent content first-and-foremost means – in addition to integration offers and the Aha!-moments - everything that fuels local communication.

Figure 2-6 Totally integrated. MEDIAFACADE – Embedded in double glazed units.

The call for light, integrated solutions that, optimally, are practically invisible when witched off, is, for this reason, absolutely understandable from an architectural as well as an urban-planning point of view. Due to the fact that modern facades are increasingly being constructed using glass in order to create greater transparency and brighter spaces, "integrated" in this context can really only mean; integrated into the facade. This in turn means that the glass element takes on a double function; on the one hand it is the thermally-insulating skin and on the other hand that of a modern media facade. Only when reduced down to this one single level can it be assured that the demands for high transparency, integration and reduced static stress are met. An ONLYGLASS MEDIAFACADE, like that on offer from ONLYGLASS is a media-tectonic construction element that meets all the demands mentioned. When the media facade is turned-off, it "disappears" into the outer-shell of the building and can no longer be perceived as a display. When it is turned-on, all possible content imaginable can be displayed on the facade - the only limit is human creativity. This also emphasises the variable and versatile character of the facade: advertising, art, illumination, films and interactive communication with the social environment can be freely adapted to meet the needs of the time and the people. Again, the Klubhaus St. Pauli at the Reeperbahn has been proofing for many months after its opening, that MEDIATECTURE is feasible nowadays, provided one decides for the right concept from the very beginning.

...towards the layer it actually occurs – Integral facades vs. symptomatic building management

Christian Wiegel[1]

1 Delft University of Technology, Architecture and the Build Environment, Architectural Engineering + Technology, Solarlux GmbH, Osnabrück Münsterstraße 144, 46397 Bocholt, Germany, c.wiegel@solarlux.de

This paper exemplifies solutions for primary energy saving in predominantly glazed buildings, which is one of the most challenging concurrent engineering subjects. Both aims, energy saving and ensuring appropriate interior thermal comfort, seem hardly to be consolidated the more the transparency ratio of building skins increases. Besides, global warming pervasively exceeds and climate change by means of four of nine ecological outrun limits (Potsdam Institute for climate impact research) still irreversibly establishes, in particular the building sector takes a key part in permutation in developing a sustainable society. Whereas thermal resistance of building envelopes encounters its physical limits and thereby tends to turn over into unsatisfactory interior daylight contribution, fossil fuel substitution by renewable energy institutes as cardinal virtue of modern architecture. Moreover, research evidences that multi-functional facades originate comparable satisfactory thermal comfort in entirely glazed spaces with less energy consumption than building services might be able to accomplish. Correspondingly, research and development focus rather on implementing building services and smart functions in facade structures than simply on eclectic attachment to building envelopes. Thus, facades concurrently discontinue to provide simply envelopes, but rather former "material layers develop to devices", ancient functions like load bearing and weather shell contemporarily "change into building automation". Consequently, such developments result in "liberation of the floor plan" and into an offset of building management away "from the functional core towards the functional periphery". Integral building skins aim on considerable fossil fuel substitution by e.g. solar thermal collectors and even on autarkic building operation whereupon users should not encounter any forfeit in interior thermal comfort. Hence, for example, experiments with perennial empirical measurements on sun space envelopes have been elaborated, which were integrally equipped by diverse solar thermal collector technologies. Glazing layers and structural parts as well have been technically enhanced in order to absorb and convert irradiation in cascaded heating energy for seasonal storing and exergetically optimized floor heating. However, performance optimization of annual yield, temperature level and façade layer integration may induce undesired negative impact on thermal comfort. Developments contradict presumed research objectives. But finally, exactly this experimental and numerically simulated investigation illustrates that entirely glazed spaces are eligible to contribute satisfying thermal comfort by exclusively envelope generated thermal renewable energy.

Keywords: multi-functional facades, energy saving, thermal comfort, building skin integrated renewable energy collector, visual comfort, fossil fuel substitution

Engineered Transparency 2016. Glass in Architecture and Structural Engineering. First Edition.
Edited by Jens Schneider, Bernhard Weller.

1 First cut Net-Zero approaches

Since the late 1980s architects and engineers have designed and developed building skins (Disch, 1994, Wurm, 2013) that provide more than simply protection (Figure 1-1) against local climate impacts. Building skins regulate energy exchange between spaces and the environment. Building skin performance intensively determines both interior thermal comfort and energy consumption for maintenance [1]. Low building skin performance is responsible for a 30 % fraction of annual total end energy demand related to the building sector and comparable proportions of CO_2 emission in Germany. In research and application as well, the next two exemplified parallel tendencies concentrate on optimum building façade performances regarding additional utilization of clean environmental energy.

Figure 1-1 Integral facades: Heliotrop by R. Disch, Algae façade by Jan Wurm, passive solar gains: Manitoba Hydro headquarter, Canada, T. Auer.

Particularly, Knaack [2] puts emphasis on research into integral facades. Integral facades include several domestic service techniques to enhance running costs or smart materials which enable to adapt subtly to changes in local climate. Thus, Knaack entitles this emerging field of energy and building science ´performance oriented envelopes´, which constitutes an innovative branch in modern architecture and building engineering.

On the contrary, Auer [3] shows with the Mannitoba Hydro headquarter a huge profanatory building complex (Figure 1-1) that sufficiently works exclusively with multi-story glass-alumina recreation sun spaces. The sun spaces are detached from working areas and provide heating the entire year. In the same conceptual line, the R 128 house designed by Sobeck in 1999 simply utilizes spacial solar gains to feed sensible buffer tanks of domestic heating. However, research and practical experiences show, that profanatory, office buildings or residential homes as well located in certain European climatic regions, which spaces are oriented unfavorably, cannot be solely run by renewable energy, i.e., solar gains. In the long run, bivalent driven heating energy systems are required, which conventionally predominantly base on fossil fuels.

Logically, as a rule of thumb, the more dominant the glazing proportion (transparency of building envelope) of a space is, or the more the space is off-oriented from south, the lower is the opportunity to run the space exclusively by solar gains. On the contrary, spaces with building systems exclusively based on solar gains can suffer from overheating. Hence, maximum solar wins are restricted by limits of acceptable indoor thermal comfort as Bluyssen [4] (2009) explains, thus, need to be controlled.

2 Research & Developments and state-of-the-art façade technologies

Different approaches of integrated facades have been evaluated and developed in science and applied sciences and even by manufacturer, which meet several functions of weather shielding, perusal, sun protection and heating energy generation. Durst [5] illustrates in 2013 a development of a high rise post&beam façade being developed by a joint venture of façade and tube collector manufacturers. Heatpipes, as is known maintenance friendly since being fully hermetic heat pump systems, have externally been embedded into the post structure (Figure 2-1), which also provide certain sun protection qualities.

This idea is been followed up by researchers of the Technical University of Munich, in order to replace heatpipes by CPC-collector types, which principally achieve higher efficiency due to parabolic reflectors (Figure 2-1). Indeed, this concept provides further synergetic benefits like additional sun and glare protection by perforated reflector sheets. Similarly, the unique vacuum tube technology prevents any energy losses to the environment, what improves system performance to the max and prevents undesired internal radiation as impact on the space thermal comfort. Despite this, as being directly flooded collector types, CPC-tubes require higher maintenance effort compared to heatipes and considerably reduce perusal.

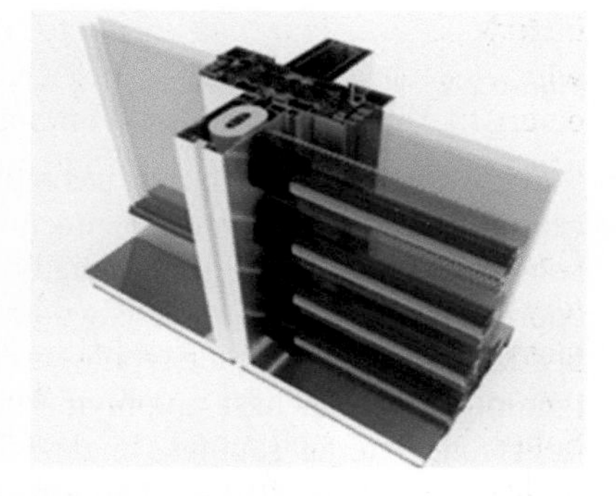

Figure 2-1 Vacuum tube collectors, sun shading; CPC collectors, enhanced sun shading.

Opposed to that concept, Wurm (2013) invented and still evaluates external sun protection lamellas for both residential and high rise buildings to meet perusal quality requirements.

The concept (Figure 1-1) bases on tempered double glazing that is flooded by a liquid algae solution. Circulated and exposed to the sun, this algae solution firstly heats up sensibly and secondly produces biomass by increase of algae population. The sensible heat is immediately free for energetic utilization in buildings and energy storages, additionally heating energy from biomass can be derived heating energy at a later state, i.e. in winter time. If required, the lamellas can track daily sun path. Conceptually, the free adjustable external lamellas provide 100 % perusal and sufficient sun protection. Characteristically, a several inch distance between lamellas and glazing prevents the façade cavity from overheating. In spite of this, the drawbacks of this concept are massive, inasmuch the effort for algae solution circulation and piping is considerable high and what is more, a biomass conversion plant is necessary to utilize the latent energy from algae biomass.

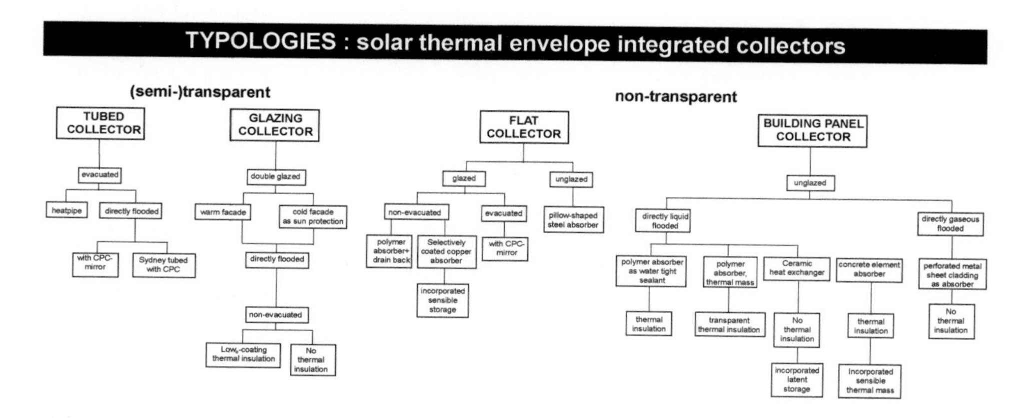

Figure 2-2 Typology survey.

Previously mentioned approaches represent technologies that are integrated in facades to certain extend. However, architectural engineering also aims on distinct integration, namely towards the layer it actually occurs. On the one hand, since the 1990s various approaches cover well established and experienced flat collector technologies to be embedded in different façade typologies (Figure 2-2). On the other hand, contemporary system improvements base on polymeric absorbers [6,7] which opposed to metal absorbers target on material savings, unglazed, oversize and free-form integration into modern building envelopes [8]. Furthermore, recent research [9,10] parallel develops and evaluates opaque ceramic solar thermal collectors and vice versa free cooler as well in order to better balance energy fluxes among the facades especially adapted to specific local climate conditions (Figure 2-3).

Obviously, besides different other drawbacks, all these concepts do not meet the requirements of perusal qualities of space users at all, but deliver perfect sun protection comparable to solid opaque external wall typologies. Those flat collector concepts except vacuum flat collectors namely suffer from simply moderate feeding temperatures (< 80 °C), which solely allow to feed daily water buffer tanks of domestic heating systems.

In spite of them, highly capacitive sorptive storage technologies for decentralized seasonal energy storing are focus of recent research [11,12]. In fact, systematically those flat collector technologies do not match temperature level requirements for sorption storages.

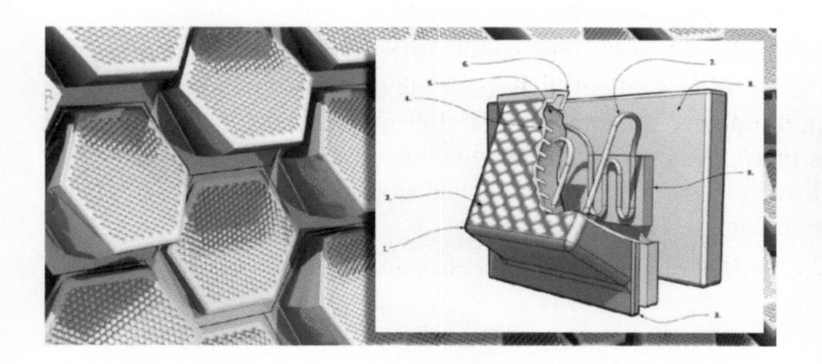

Figure 2-3 Advection based ceramic façade tile, in-situ embedded solar thermal glass.

Although modern collector systems provide thermal insulation material in order to minimize energy losses to the environment, researchers [13] have identified considerable negative impact on space thermal comfort by undesired radiation from collectors. In order to prevent a significant raise of internal wall surface temperatures, either cost intensive aerogel insulation has to be applied [14] or external wall and collector need to be separated by incorporated cavities to enhance rear ventilation. This indeed, also limits the freedom of application in regard to slim high rise building façade structures, which principally do not remain any space for cavities. The field of conflict leads to a completely different typology of building skin integrated collectors, which basically distinguish itself by incomparable high ratio of façade integration.

Whereas, the above mentioned typologies mainly are placed in front of the actual required transparent glazing, and hence, represent a further façade layer, double glazing collectors have been developed. Cavities of these glazing collectors characteristically contain either serpentine copper tubes (Figure 2-3) with attached highly selectively coated absorber sheets or completely different integrated vertical sun protection lamellas, which beneficially function as solar absorber areas. Both approaches offer a certain degree of perusal and provide a persistent quality of sun protection, either fix or adjustable.

In 2010 a variant, entirely flooded by highly capacitive transport fluid excluding any internal piping was scientifically evaluated by the author in context of a master thesis, verifying considerable cooling and heating performance qualities. Thus, these typologies represent highly efficient and smart multi-functional elements, which meet different contradictive requirements in an integral matter: sun protection, perusal, single layer structure, heating energy collection along with incomparable architectural qualities.

Additionally, this concept of double glazing and track-wise piping offers opportunity to utilize the element for space heating in winter time, what – if provided by renewable energy – anyhow lowers heating losses.

In any case, even these typologies encounter some difficulties with distributing the transport fluid through the absorber element, inasmuch the conventional glazing sealing needs to be penetrated and piping requires space parallel to the element in the glazing rebate area. But, what is more, besides feeding of sensible daily buffer tanks, transport fluid characterizations allow feeding of thermo-chemical storages according to comparable less convection losses in the inert gas filled cavities.

Notwithstanding, undesired negative radiation caused by high absorber temperatures can have negative impact on the thermal comfort of the adjacent space related to double-glazing principle. If internal surface temperatures of layer #4 exceeds appropriate room mean temperature (21 °C) for more than 23 degree (DIN EN ISO 7730-2007), thermal comfort quality lowers. In this context, intensive research has been done by Solarlux on approximately 100 % transparent spaces, namely sun spaces, on both fossil fuel substitution by solely passive solar gains and on integral facades as well, regarding interior thermal comfort qualities.

Moreover, two different test set-ups running for one to three years and aiming on diversified measurements firstly reveal how effective irradiation utilized in sun spaces can be for appropriate thermal comfort. Secondly they help determining how embedded functions in preferred facades layers converting irradiation to renewable heating energy can additionally contribute and enhance non-sufficient thermal comfort.

3 THE TEST SET UP – Boundary Conditions

In summer 2013 a solitary sun space has been erected at Solarlux headquarter site in Bissendorf, Osnabrück, Niedersachsen. As a solitary squared building with edge lengths of 6.00 m, an eaves height of 2.2 m and a roof ridge height of 3.4 m, it encloses a space volume of approximately 72 m^3, while providing living space of approx. 27 m^2 (Figure 3-1). While one quarter of the ground floor area is separated and thermally segregated to be utilized as a building services and monitoring room, the sun space offers a south and a west directed space partition as well.

The space formulates facades of four different directions as well as 30° declined south and west directed roofs by thermally disconnected Solarlux alumina profiles. A water driven floor heating is fed by a sensible buffer tank. Internal loads can be neglected, since artificial lightning is based on LED, and since the space dispenses with human occupation or any other equipment.

Figure 3-1 Isometry and functions, in-situ test set-up, attached heatpipe arrangement.

Vertical sun space construction elements incorporate triple-and double glazings with a U_g-value of 0.8, respectively 1.2 W/m²K. In the roof construction embedded double-glazings assess a U_g-value of 1.40 W/m²K regarding roof inclination. Thus, the vertical elements provide an overall U_w-value of 1.40 W/m²K, while the roof elements generally define an energetic quality with a U_w-value of 1.55 W/m²K. The space flanks are sun shaded by each an automatic internal canvas blind. Diverse sensors have been installed both externally and internally in order to monitor cause and effect of climatic changes as internal impacts. Especially, irradiation, external and internal ambient air, relative humidity and surface temperatures of three different heights on glazing surfaces, the floor, and main alumina profiles allow to determine dry bulb temperature, resultant temperature, radiation asymmetry, horizontal temperature gradients, vertical striation, relative humidity and internal solar gains as well.

A computational software has logged the data so far and controlled all actions as to be mentioned the floor heating, mechanical ventilation, internal sun shading activation and artificial lightning. Typical for the site is a maximum irradiation of 880 W/m², an annual solar yield of 960 kWh/m², a minimum external temperature of -12.2 °C and maximum external temperature of 34.2 °C.

The most promising since being most integral concurrent solar thermal façade collector was implemented in the south facing façade of the sun space test set-up. Glazing was replaced and solar thermal glass was installed while ducts and piping was managed to be distributed invisible within the SL80 80 mm profiles. Changes in space interior surface temperature were monitored by sensors that were placed in three different heights (0.10 m; 1.70 m; 2.70 m above floor).

3.1 STEP I – performance evaluation

Sun spaces are the most eligible construction principle to provide maximum living space quality by maximum perusal and , thus, intermediate contact to the environment. Hence, transparent parts need to be kept free from collector devices what reduce perusal. Consequently, the idea is to integrate highly efficient solar thermal collectors rather in rafter axis than in transparent areas. In this line, the test sun space was equipped by vacuum tube collector (heatpipes) units in rafter axis. Each a rafter and the hip rafter was equipped by two double heat exchangers carrying two tubes each. In the first step the tubes were mounted to the rafter surface with a distance of 80 mm, thou not actually being integrated but being attached to the façade. In total 44 vacuum tubes were installed (Figure 3-1), which provide heating power of 2.2 kW on the south roof and 1.5 kW on the west roof, in total 3.7 kW_p. Ducts were installed along the roof attics and lead to the maintenance service room. Both heatpipes circles (south & west) worked independently from each other which circuit pumps were activated by a temperature difference of buffer tank-to-heatpipe of 10 K.

In particular, heatpipes have been chosen, as being principally closed collectors and thus changing heat to a secondary circle. This, exchange in case of maintenance is considerably easier than compared to directly flooded CPC collectors or Sydney tubes.

3.2 STEP II – product development HEATBEAM© I

A different test assembly (Figure 3-2) was erected in the close near to the test sun space. This assembly represents a further development of sun space rafter integrated vacuum tube collectors that provides an enhanced quality of façade embedded functions. In particular, this assembly physically simulates sun space rafters exaggerated by an alumina U-shaped profile (Figure 3-2). This profile carries another all sided closed U-profile covered by a a.) 3 mm low-iron tempered, or b.) 6 mm low-iron single non-tempered, c.) 2 x 4 mm tempered laminated glass sheet.

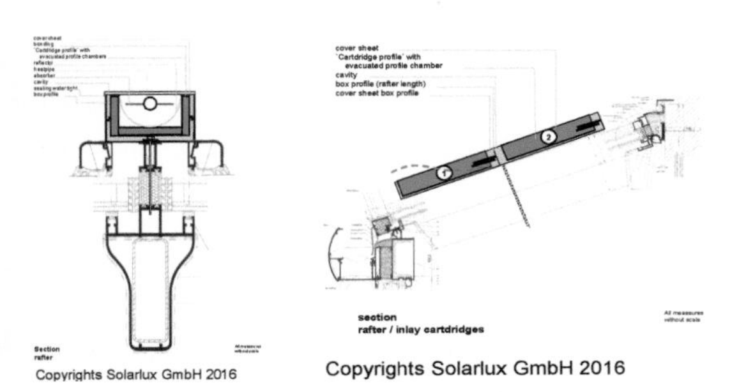

Figure 3-2 HEATBEAM©I test set-up, section and longitudinal section scheme.

This all-sided closed and glass sheet covered profile contains a heatpipe (closed copper tube with selectively coated copper absorber fins, $\alpha = 95$; $\varepsilon = 5$), that is inserted into the carrying U-profile like a cartridge (Figure 3-2) regarding easy removal. The cartridge profile is evacuated to 30 mbar remaining pressure. Innovatively, a linearly designed heat exchanger is inserted in the top, respectively ridge related area of the carrying U-profile. Test variants were equipped by each a temperature sensor monitoring the alumina cartridge surface temperature and the "presumed space oriented" carrying U-profile surface temperature to evaluate possible heat radiation to the environment or, to what is more of concern, the ambient space air. Temperature sensors plugged into the transport medium in each a heat exchanger monitored actual feeding temperature of each a heatpipe.

3.3 STEP III – product development HEATBEAM© II

Product development and HEATBEAM© heatpipe design have been continued so far inasmuch simplicity in manufacturing, vacuum tightness, transparent appearance and functionality have been enhanced. HEATBEAM© II bases on a transparent polymer case (Figure 3-3) that allows a flatter (h = 30 mm) construction with an additional function to clamp (Figure 3-3) the roof glazing and making conventional glazing bars obsolete. Contemporary research [6-8] in polymer technology allows to provide UV stable, heat resistant (> 200 °C), environmental chemically inert and highly light (> 90 %) and as well UV light transmitting materials.

Moreover, these polymers also allows thermal insulation by evacuation. HEATBEAM© II, compared to the alloy/glass approach I, is considerably light weight and easier to customize by extrusion process.

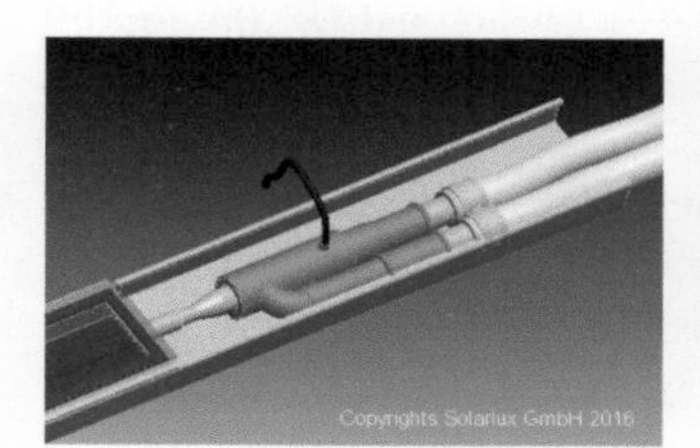

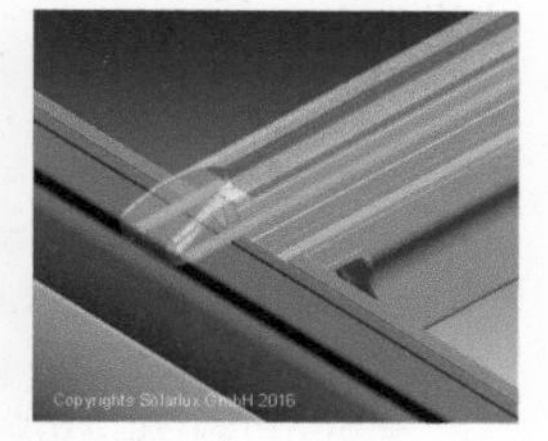

Figure 3-3 HEATBEAM©I/II linear heat exchanger, rafter integrated polymer evacuated tube.

4 Research Outline

In order to evaluate the quality of complex integral façade technologies three major parameters have to be taken into account : a.) cumulative frequencies of hours with resultant temperature T_{res} > 20 °C from 17 to 22 o´clock and from 22 to 17 o´clock a1.) solely by solar gains, a2.) explicitly by renewable energy. Additionally, two more parameters help to identify the potential of such a technology development: b.) the amount of heating energy generated a year (required: approx. 1.600 kWh) and c.) interior surface temperatures. Before determined three parameters allow to conclude about a feasibility of 100 % autarky in respect to sufficient interior thermal comfort.

Consequently, for the research have been defined the following investigational steps:

1. Run: thermal comfort by explicitly solar gains

2. Run: thermal comfort by solar gains and heatpipes with buffer tank

3. Run: energetic and comfortable performance of HEATBEAM©

5 Results

The histogram below shows cumulative frequencies of hours with resultant temperatures > 20 °C measured for one year in the test set-up sun space. In total 3587 hours (of 8760 a year) with temperature of 20 °C or higher could have been observed. Forty one percent of annual hours regarding interior temperature in this sun space are sufficient in respect of heating. This temperature level could not have been achieved in December and January.

A time span between 17 and 22 o´clock has been presumed to be predominant for preferred sun space occupation in general. In 950 hours a year (Figure 5-1) the resultant temperatures in the sun space was 20 °C or higher between 17 and 22 o´clock. Another 2672 hours a year with temperature of > 20 °C could have been measured between 22 and 17 o´clock.

The entire cumulative frequency of hours with > 20 °C between 17 and 22 o´clock was observed to be exaggerated by façade embedded solar thermal collectors by 67 hours (Figure 5-2). Integral facades in combination with a sensible buffer tank were able to elongate predominated occupation time for another 7 % and more if technical fall outs can be eliminated. Moreover, multi-functional sun space envelope components manage to prolong cumulative frequencies of hours with sufficient thermal comfort beyond the predominant occupation time between 22 to 17 o´clock for additional 200 hours. This results in additional 7.5 % time regarding sensible energy storages.

Thus, the test set-up incorporating integral sun space facades illustrates the potential of in total 267 hours and more a year surplus of sufficient thermal comfort in 100 % transparent spaces. Significant comfort time prolongations could have been observed for September and March. Nonetheless, sun space user benefit from integral facades also in October, November and April in respect of additional fossil fuel-free renewable heating energy.

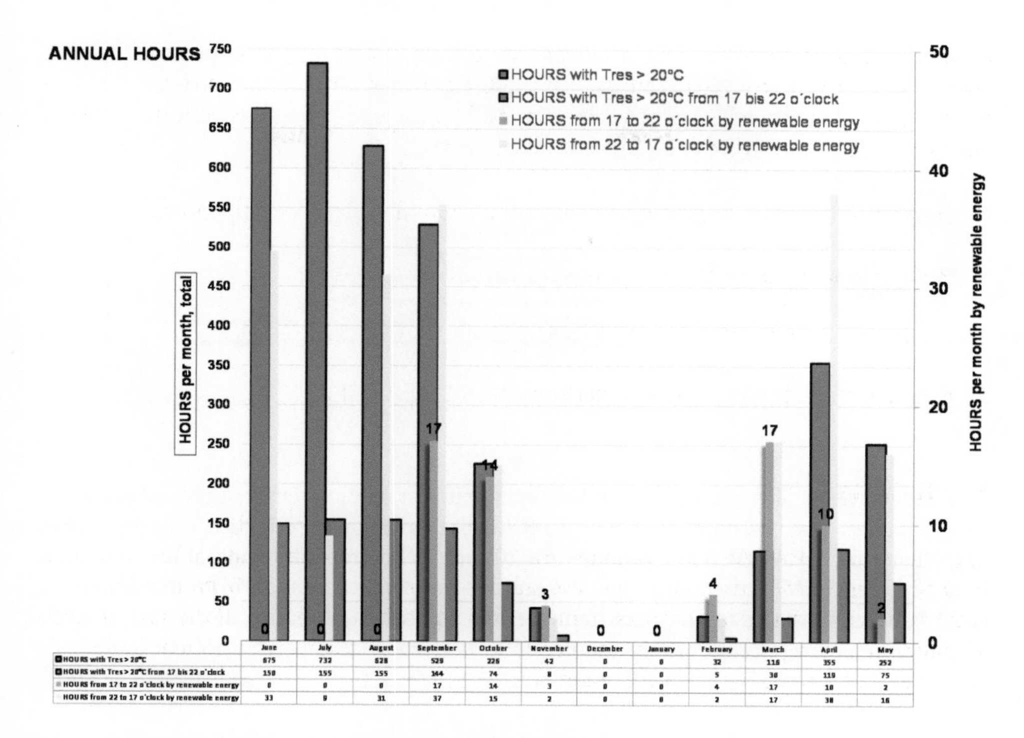

	June	July	August	September	October	November	December	January	February	March	April	May
HOURS with Tres > 20°C	675	732	628	529	226	42	0	0	32	116	355	252
HOURS with Tres > 20°C from 17 bis 22 o'clock	150	155	155	144	74	8	0	0	5	30	119	75
HOURS from 17 to 22 o'clock by renewable energy	0	0	0	17	14	3	0	0	4	17	10	2
HOURS from 22 to 17 o'clock by renewable energy	33	9	31	37	15	2	0	0	2	17	38	16

Figure 5-1 Cumulative frequencies of hours with T_{res} > 20 °C.

A double-glazing solar thermal collector mounted in a south directed vertical facade and two HEATBEAM© variants are compared. Figure 5-3 shows maximum feeding temperatures of a glazing collector in March, July and October. All three graphs impressively illustrate that feeding temperatures of more than 100 °C and 110 °C is feasible during normal operation and stagnation as well, actually in mid of October. This results in a buffer tank temperature increase from 22 °C to 48 °C in case of July. Moreover, space adjacent collector surface temperature rises to 38 °C and even 51 °C. Related to measured irradiation the collector surface temperature ascends significantly compared to east and west orientated reference glazings. This results in a raise of the resultant temperature, in case of March and October. The resultant temperature develops paralleled shifted to surface temperature curve. The figures 5-3 demonstrate daily temperature changes of a vacuum tube collector system related to irradiation changes for representative examples in April, August and end of October.

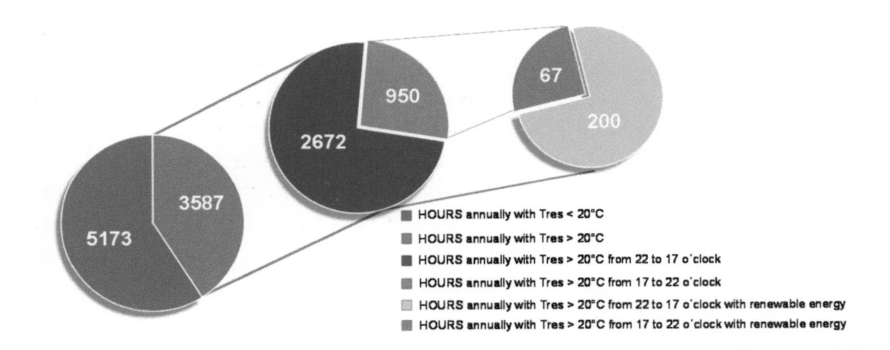

Figure 5-2 Cumulative hours with T_{res} > 20°C for different time periods in a bivalent operation mode.

In observed cases of August and October the maximum feeding temperatures of alumina-low iron glass case heatpipes are solely 12 K lower than a mass production reference tube. Significant for HEATBEAM© variants is a 30 min. delay in peak temperature compared to the reference. Measured alumina case surface temperatures range from maximum 25 to 50 °C what is related to non-persistent vacuum and solar up-heating of the case as well.

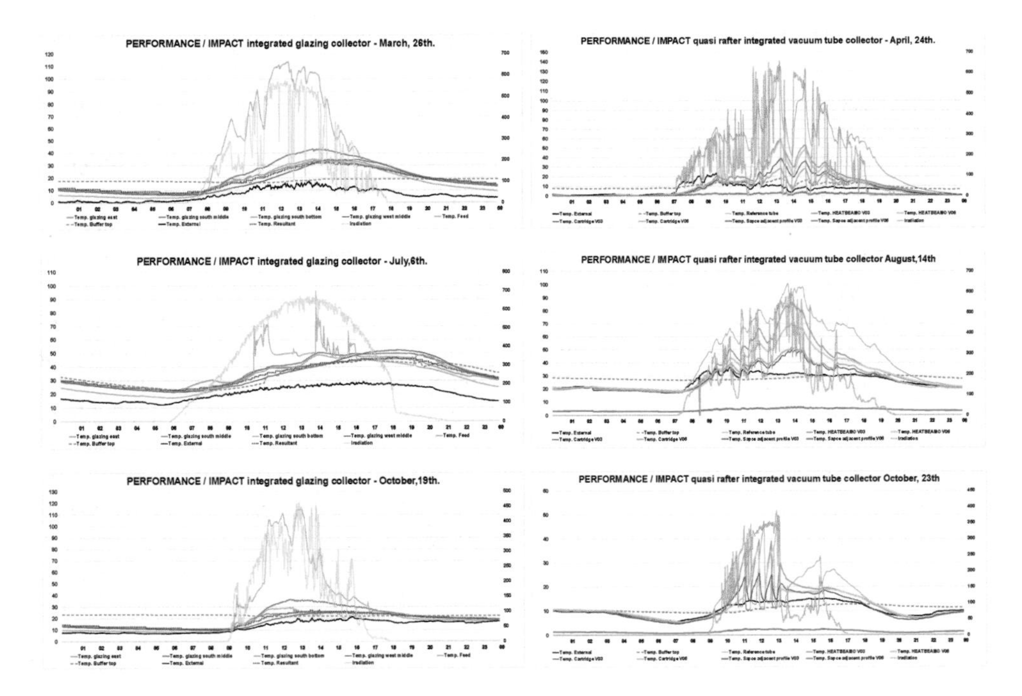

Figure 5-3 Cumulative hours with T_{res} > 20 °C for different periods and energy source (glazing collector or heatpipes).

According to the construction principle these surfaces are in contact to the environment and are disconnected to the cartridge profile and further load bearing parts by a 5 mm cavity. Cartridge surface temperatures that according to design are oriented to the ambient space solely rises to maximum 5 to 8 °C what is assumed to have no worth mentioning impact to internal thermal comfort. The heating effect on the sensible buffer tank of the 8-heatpipe assembly in all three cases is considerable low since thermal insulation of the tank was removed to provide an effective heat sink. That method alongside generates low temperature of recirculation media for optimum energy transfer.

6 Summary

In research and applied sciences have been established several tracks of developing, running and monitoring different façade embedded renewable energy collectors. Collectors and functions in general embedded in facades offer additional space and area for installation if space according to building typology is limited. Furthermore, facades equipped by diverse functions operating as multi-functional facades help managing to establish completely different building management systems. Such systems keep the core area free and lowers the distance between actual occupation area of space users and the facades, where user manually and actively adapt to climatic impacts. Thermal comfort becomes individual and decentralized by integral facades. Technologies mainly differ in form (case, tube or glazing), absorbing material and qualities and the façade layer, where the technology actually is located. Within the last ten years collector technologies have focused on diverse variants of energy and exergy efficient vacuum tube collectors and on glazing collectors since renewable energy collection is no longer contradictive to perusal and sun protection, whereas conventional flat collectors or any other opaque derivative do not serve this façade quality. Hence, these focused collector technologies are eligible solutions for optimum façade integration, respectively high integration ratio. However, focusing on development on layers where integration actually occurs reveals further demand for research in science. Since façade integrated thermal collectors – if not completely evacuated – suffer from heat losses and causes negative radiation towards the actual living or working space, consequently, the effective efficiency of renewable heating energy generation needs to be lowered while being corrected by the energy amount for mechanical cooling. This paper shows a rough survey of meanwhile available technology and illustrates the effective utilization of passive solar gains and additional façade generated renewable energy for 100 % transparent space heating. The test results demonstrate the close interrelationship between maximum energy yield per area and simultaneous possible negative impact on interior thermal comfort. Consequently, research needs to concentrate beside innovative polymeric absorber materials on highly effective thermal insulation systems which both enhances energy yield and eliminate negative thermal impact.

7 References

[1] Cody, B., 2008, Fassaden – Blick in die Zukunft, VDI Wissensforum GmbH, VDI-Berichte Nr. 2034, 2008.

[2] Knaack, U., Grenzen des Machbaren – Ein Versuch, Fassaden neu zu interpretieren, 2008, VDI Berichte 203.

[3] Auer, T., Lauter, M., Olsen, E., High Comfort – Low Impact, Transsolar Climate Engineering, 2009, Transsolar Energietechnik.

[4] Bluyssen, P., The Indoor Environment Handbook – How to make buildings healthy and comfortable, 2009, earthscan, U.K.

[5] Durst, Anna, 2013, Fassadenkollektoren mit Durchblick – Fenster mit integrierten Vakuumröhren gewinnen solare Wärme, spenden Schatten und leuchten Räume gleichmäßig aus, bine projektinfo 07/2013.

[6] Köhl, M., Meir, M., Fischer, S., Wallner, G., 2014, Polymeric Material for Solar Thermal Applications – 2006-2014, Final Presentation.

[7] Lang. R., 2013, Polymerwerkstoffe und die Transformation des Energiesystems, Innviertler VDI-Dialog: Energie Effizienz Denken, Technologische An-sätze zur industriellen Nachhaltigkeit.

[8] Singh, G., Mantell, S., Davidson, J., 2014, Polymere Durability for Solar Thermal Applications, Mechanical Engineering University of Minnesota.

[9] Hausladen, G., de Saldanha, M., Liedl, P., ClimateSkin – Building-skin concepts that can do more with less energy, 2006, Birkhäuser Verlag, Berlin.

[10] Vollen, J., Winn, K., 2014, Advection based adaptive building envelopes: Component surface morphology and entropy management of a ceramic building façade, Center for Architecture, Science and Ecology.

[11] Fink, C., Preiß, D., 2014, Roadmap "Solarwärme 2025" – Eine Technologie und Marktanalyse mit Handlungsempfehlungen, AEE-Institut für Nachhaltige Technologien.

[12] Abedin, A., Rosen, M., 2011, A Critical Review of Thermochemical Energy. Storage Systems, The Open Renewable Energy Journal, 2011, 4, 42-46.

[13] Matuska, T, Sourek, B., 2006, Facade solar collectors, Solar energy No. 80, S. 1443-1452, sciencedirect.

[14] Zauner, C., Sterrer, R., Gosztonyi, S., (2014), Verhalten innovativer Solarthermischer Fassadenkollektoren im Rahmen des Forschungsprojektes MPPF, Austrian Institute of Technology, Austria.

Flexible transparent OLED lighting in safety glass composites

Jacqueline Hauptmann[1], Stefan Mogck[1]

1 Fraunhofer Institute for Organic Electronics, Electron Beam and Plasma Technology FEP, Winterbergstraße 28, 01277 Dresden, Germany, jacqueline.hauptmann@fep.fraunhofer.de

Flexible OLED devices on flexible substrates for large area homogenous lighting have unique features in contrast to inorganic LED lighting devices. Nowadays, OLED technology will be applied in the pilot production for OLED lighting on rigid glass panels and mass production for AMOLED displays in smart phones. Recent developments of flexible transparent OLED lighting and OLED displays show the unique feature for this technology. However, the costs has to be brought significantly down by simplifying the fabrication. Flexible and transparent OLED devices into glass laminates for customized lighting solution could be a great opportunity to functionalize glass façade and interior furnishing. However, several challenges for the integration in glass laminates need to be evaluated: The high temperature load during the autoclave process and reliable electrical connection can be critical for the flexible OLED devices. Finally, a smart power supply to drive the OLED lighting devices need to be developed for proper designs. In the present paper, the fabrication chain concepts for OLED lighting integration into glass laminates with open issues and opportunities will be discussed.

Keywords: OLED, transparent, integration, autoclave, glass laminates

1 Overview of available technologies and flexible OLED production capacities

The OLED technologies for OLED display and OLED lighting is predominately a vacuum based coating process of evaporations of small molecules. Only this technology meets the needed requirements for brightness, efficacy and device lifetime in contrast to polymer based printable OLED materials coated under atmospheric conditions at present.

OLED display mass production is well established since 2012 and OLED displays are well established in the market. The lead manufacturers are LG Display and Samsung Display and they have made commitments for developing flexible OLED displays. This is the next market entry for flexible OLED display devices in the medium term. OLED lighting panels are currently under pilot production scale on rigid glass panels. LG Display has been started to provide flexible OLED lighting panels on plastic substrate. Konica Minolta is putting a lot of efforts to bring roll-to-roll OLED mass production into operation. In Europe Osram OLED and OLEDWorks have upgraded their lines and have supplied many luminaire manufacturers with prototype products [1].

Sheet processing is sufficient for initial OLED lighting volumes. Roll-to-roll processing capability will help drive down cost at high volume [1].

Engineered Transparency 2016. Glass in Architecture and Structural Engineering. First Edition.
Edited by Jens Schneider, Bernhard Weller.

Fraunhofer FEP establish flexible OLED devices using different process types and equipments (see Table 1-1.). Thin glass foil, plastic barrier films and metal foils are suitable substrate materials for flexible OLED devices.

Table 1-1 Fraunhofer FEP OLED processing technologies.

equipment	substrate size	substrate thickness	deposition
sheet-to-sheet processing	max. 200 x 200 mm²	50 µm ... 100 µm	inorganic and organic thin-films
roll-to-roll processing	max. 300 mm width	50 µm ... 200 µm	inorganic and organic thin-films

2 Sheet to sheet production on flexible substrates at FEP

The sheet-to-sheet (S2S) process flow of flexible OLED fabrication is described in Figure 2-1. An “Attach and Release” process is necessary for deposition and encapsulation. Resulting in large area lighting modules on flexible materials as thin glass, plastic and metal foil with life time of several thousands of hours.

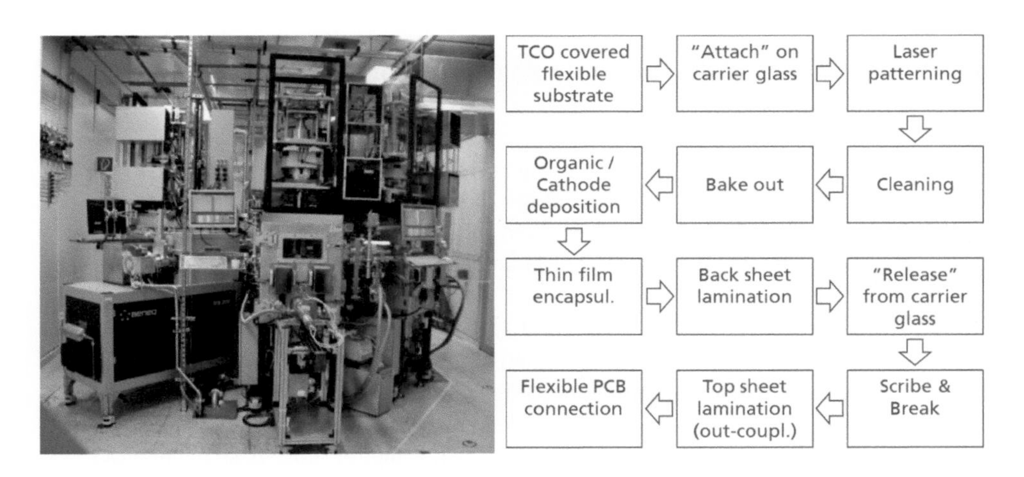

Figure 2-1 Sunicel 200 vacuum cluster tool (left) and the S2S process flow for flexible substrate processing.

Flexible OLED devices can be produced in small batch series, as shown in Figure 2-2.

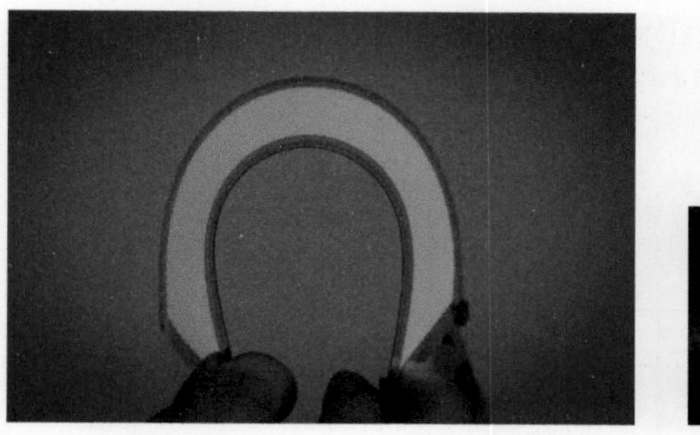

Figure 2-2 Flexible OLED lighting modules produced at Fraunhofer FEP using sheet to sheet process (size about 175 mm x 50 mm).

3 Roll-to-roll production on flexible substrates at FEP

The roll-to-roll (R2R) OLED fabrication line (Figure 3-1) covers the whole process chain including the optical web quality inspection, the customized web patterning with a passivation layer, the OLED deposition in vacuum coater and the OLED encapsulation with a barrier film.

Figure 3-1 The process flow used in R2R OLED fabrication.

Flexible OLED devices can also be produced in small batch series. The active lighting area can be easily changed depending on the customer needs. The lighting areas are limited in width of 10 cm and length of about 50 cm. Figure 3-2 shows an OLED a customized layout of different OLED lighting shapes.

Figure 3-2 Orange OLED lighting modules not separated from the roll with 1 x 1 cm² (test OLEDs), 10 x 10 cm² (Fraunhofer logo) and 10 x 25 cm² (Dresden sky line).

4 OLED application fields and cost roadmap

The biggest challenge in OLEDs fabrication is to develop acceptable, cost-effective manufacturing processes and identifying lighting applications that play to the strengths of OLED technology. OLED devices can be flexible and a homogeneous large area illumination in contrast to the inorganic LED lighting.

Substantial cost reduction will be needed for commercial success in OLED TV and OLED lighting markets. Since 55" LCD TVs illuminated by LED backlights can now be purchased at retail for below $800, broad market penetration of OLED TVs will require manufacturing costs for OLED panels to be around $250/m². If this will be achieved, the corresponding cost for OLED lighting panels with a luminous emittance of 10,000 lm/m² would be near $25/klm. The long-term target of the DOE SSL (Solid State Lighting) Program for OLED panels is $10/klm [1].

5 OLED lighting integration into glass laminates

5.1 First results of OLED integration tests

The OLED integration into glass-glass laminates seems to be a promising possibility using the natural barrier properties of glass. Opaque or transparent lighting modules can be integrated by autoclave or lamination processes. Laminated glass-OLED-glass sandwich with a hot melt adhesive film, typically of polyvinylbutyral (PVB), Thermoplastic polyurethane (TPU) or ethylene vinyl acetate (EVA) are realizable.

Proof of concept was carried out at Fraunhofer FEP to integrate flexible OLEDs plastic substrates in glass-glass sandwiches:

Figure 5-1 Lighting OLED in glass-glass laminate (10 x 10 cm²).

The hot lamination process was carried out at Fraunhofer FEP using TN glass (0.7 mm, 20 x 20 cm²) and commercial available EVA at 150 °C, 0.3 m/min and about 5 bar (Figure 5-1). OLED was integrated on commercial barrier PET film and dried under nitrogen atmosphere.

Figure 5-2 Green OLED produced in roll-to-roll equipment (left) and with degradation area because of residual water inside the barrier film (right).

Autoclave processes for laminated glass composites use temperatures in the range of 100-150 °C and 1-15 bar pressure. Feasibility autoclave tests were carried out, but degradation and short cuts needs to be further minimized. The enclosed moisture, shrinkage, mechanical and thermal stress destroyed the OLED system and would result as shown in Figure 5-2, right.

The real mechanism and required parameters are still not clear understood and needs more studies and tests. However further optimization needed to minimize the enclosed moisture, shrinkage, mechanical and thermal stress.

5.2 Study of process parameter for OLED integration

PET and PEN based barrier film are nowadays in flexible OLED production the basic materials because of efficient production cost. Critical issues for PET/PEN based barrier film for OLED application are different parameters:

- shrinkage (Figure 5-3) and the natural expansion of the film,
- the glass transition and upper operating temperature (Figure 5-4),
- the residual water content of the PET or PEN barrier film.

Figure 5-3 illustrates how the shrinkage of heat-stabilized PET (Melinex®) and PEN (Teonex®Q65) change with temperature [2]. Mechanical stress can damage the OLED laminate generated by cracks inside the barrier film.

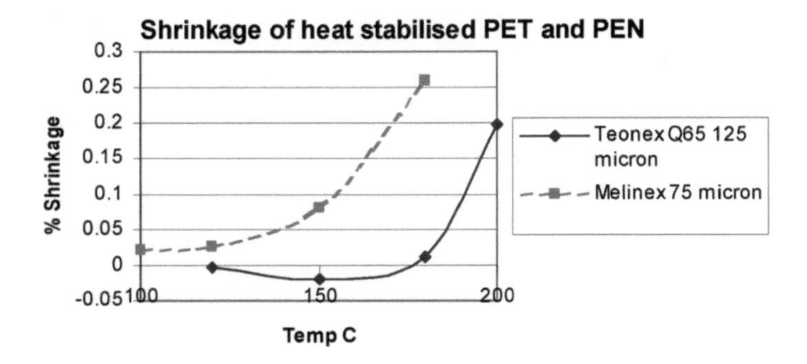

Figure 5-3 Shrinkage of heat-stabilized PET and PEN film vs. temperature [2].

The upper processing temperature for heat stabilized materials is another critical parameter during integration. Crystallization of the film material starts with time at temperatures above 100 °C. Planarization films are necessary for smoothing the growing cyclic oligomers at the film surface. In Figure 5-4 the upper processing temperature of substrates in dependence of operating temperature for different materials are shown.

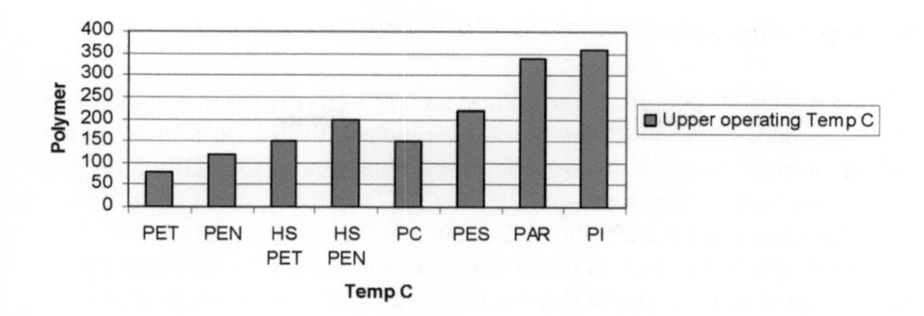

Figure 5-4 Upper processing temperature of film substrates of interest for flexible-electronics applications [2].

For reduction the residual water content of the OLED laminate an optimal drying process and storage concept was established. Measurements of the water content were carried out at Sartorius WDS400. Figure 5-5 shows drying and storage results at PET film. The water content of the material were measured in air (black line), after storage in nitrogen (green line) and in vacuum (orange line). The effective drying condition to reach 1000 ppm was obtained after drying in vacuum at 80 °C. Taking out the material after drying and bring to air the water content increase drastically and very fast.

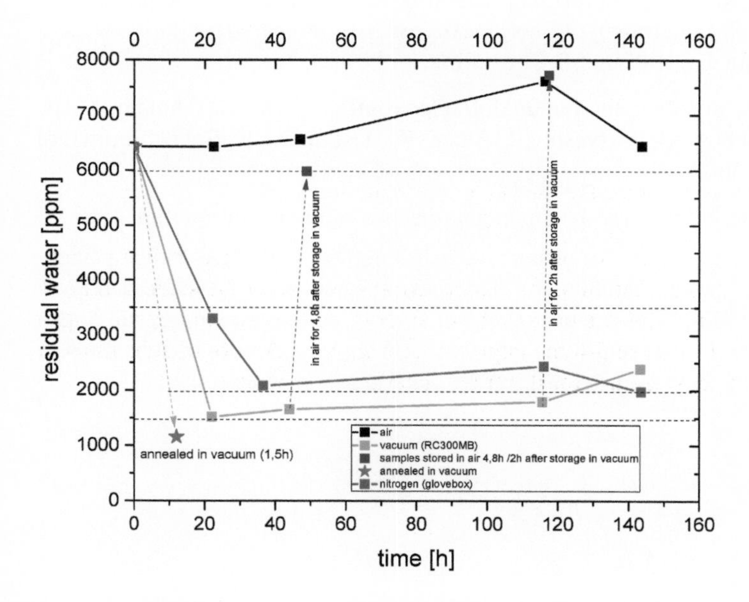

Figure 5-5 Water content of PET in dependence of drying conditions.

Storage in nitrogen and drying in vacuum atmosphere in front of the organic deposition is the best concept to keep moisture out of the OLED laminate.

6 Summary and Outlook

First OLED lamination trials into glass-glass laminates showed encouraging results after establishment of a proper substrate drying process for the roll-to-roll OLED fabrication. This open the opportunity to integrate cost effective and large area OLED lighting into glass-glass laminates. Further developments for the OLED stability at high temperature > 120 °C is necessary to allow OLED lighting film integration in existing production lines for glass-glass laminates. In parallel, material development of the adhesive interlayer is needed to bring the autoclave temperature, pressure and process time down. Furthermore, a cost effective storage concept under dry atmosphere for the adhesive interlayer and flexible OLED film needs to be established.

The hybrid integration of transparent OLED- and photovoltaic modules in glass-glass laminates on large areas is a great opportunity to create energy self-sustaining lighting systems. A first example for solar active car roof was already realized from Heliatek in cooperation with Webasto. “Roof integrated solar systems for a better CO_2 balance” [3].

7 References

[1] Manufacturing Roadmap, Solid-State Lighting Research and Development Prepared for the U.S. Department of Energy, August 2014, Prepared by Bardsley Consulting, Navigant Consulting, SB Consulting, and SSLS, Inc.

[2] Latest advances in substrates for flexible electronics, W. A. MacDonald, M. K. Looney, D. MacKerron, R. Eveson, R. Adam, K. Hashimoto, K. Rakos, Journal of the SID 15/12, 2007.

[3] Heliatek website, http://www.heliatek.com/de/anwendungen/automotive.

Insulating glazing with integrated functions – case study of the ETA-factory

Dr.-Ing. Johannes Franz[1], *Dr.-Ing. Frank Schneider*[2], *Dipl.-Ing. Andreas Maier*[3]

1 OKALUX GmbH, Am Jösperhecklein 1, 97828 Marktheidenfeld, jfranz@okalux.de

2 OKALUX GmbH, Am Jösperhecklein 1, 97828 Marktheidenfeld, fschneider@okalux.de

3 TU Darmstadt, Institut für Statik und Konstruktion, Franziska-Braun-Str. 3, 64287 Darmstadt, maier@ismd.tu-darmstadt.de

This article deals with the energy-efficient technologies of the research project ETA-Factory; in particular the insulating glass units (IGU) with integrated functions installed in the façade of the ETA-Factory. This includes IGU with three different types of inlays in the cavity: sun control louvres, translucent insulating material (TIM) or vacuum insulation panel (VIP). Depending on the geographical orientation of the building and its utilization, the choice of the inlay has a significant influence on the incidental light in the building and the thermal transmission. The sun control louvres mainly affect the selective light transmission and the total solar transmittance. In contrast, the vacuum insulation panel is an outstanding thermal insulator and has no properties regarding the selective light transmission, because the panel is opaque. The capillary slab is a translucent insulating material which influences both the selective light transmission and the thermal transmission. This article describes the properties of the different types of IGU used in the façade of the ETA-Factory in Darmstadt. The case study reveals that integrated functions in an IGU such as sun control louvres, capillary slabs or vacuum insulation panels not only improve the comfort in the building but also the costs of the building life cycle when all aspects are included in the planning.

Keywords: IGU, Louvres, TIM, VIP, energy, transmission

1 Research Project ETA-Factory

Reducing the energy consumption and the CO_2-emission of a building during its building life cycle is becoming increasingly important in light of global warming and rising energy costs. In 2009, the European Emission Trading set a reduction of the CO_2-emissions of buildings to about 20 % in the year 2020 in comparison to 1990. One way to achieve this purpose is to build with the goal of net „zero-energy-buildings“. In 2010, the directive of the European Parliament on the energy performance of buildings passed this objective for all new buildings from 2021.

This energy efficiency should also be applied in factories. However, with regard to factories, diverse subsystems like building structure, technical infrastructure and machine structure, have until now been considered separately as far as energy is concerned. The increase of energy efficiency in industrial production will be investigated with the

Engineered Transparency 2016. Glass in Architecture and Structural Engineering. First Edition.
Edited by Jens Schneider, Bernhard Weller.

help of the ETA-Factory research project funded by BMWi (TU Darmstadt [1]). The objective is not to consider the identified subsystems individually, but the energetic interaction of all subsystems: with the aim of saving 40 % of the energy consumption through the production and building structure with the building automation (Maier [2]). To verify the results achieved in the theoretical research approaches developed, a full-size model factory has been built on the campus of the TU Darmstadt, which might be able to put exactly these research approaches into practice (see Fig. 2-1).

The building envelope needs to be considered in a holistic context to achieve the potential savings. For this reason, the article deals with the installed façade of the ETA-Factory in the context of energetic and light transmission aspects.

2 Building Envelope of the Energy-Efficient ETA-Factory

2.1 Floor Plan and Overview of the Building Envelope

The ETA-Factory is located on the Campus of the TU Darmstadt in the south-east of Darmstadt. The factory and its geographical orientation are shown in Fig. 2-1. The factory has a width of 20 m and length of 40 m. The building height is about 12 m. The machine hall itself is 20 m wide and 30 m long. The machine hall is room-high and reaches from the ground floor up to the roof. It is located in the south of the factory and has no basement. In the north of the factory lecture rooms, entrance hall, washing rooms and stairs are located along the extent of the building and along a 10 m wide strip. In the basement of this area, building services and store rooms are planned. Furthermore, offices and meeting rooms are on both upper floors.

The building envelope of the ETA-Factory on the roof and longitudinal to the building are constructed to ensure that the building envelope fulfills the functions of the brick partition, the required structural demand, thermal insulation and weather protection. To achieve the structural demand, the building envelope in this area consists of double-web concrete slabs (Maier [2]). Inside the concrete slabs, a capillary tube network is installed, in which water serves as a climate control of the building. The thermal insulation and weather protection have been realized with an insulation layer consisting of mineral foam in combination of hardened pre-fabricated components with micro-reinforced, ultra-high-performance concrete.

Table 2-1 Overview of the installed types of IGUs including surface area, thermal resistance U_g, light transmission L_t and total energy transmittance coefficient g.

Façade	Product	IGU type	Surface	U_g	L_t	g
North	OKATHERM	Standard	118 m²	1.1 W/m²K	0.69	0.37
	OKALUX HPI	High Performance Insulation Glazing	70 m²	0.23 W/m²K	-	-
East / West	OKALUX +	Capillary Slab	30 m²	0.9 W/m²K	0.33	0.21
	OKATHERM	Standard	5 m²	1.1 W/m²K	0.5	0.27
South	OKASOLAR F	Integral Sun Control Louvres	111 m²	1.7 W/m²K	0.02 - 0.35	0.13 - 0.28
	Schüco Parametric System	Standard with silk-screen printing	78 m²	1.1 W/m²K	0.40	0.24

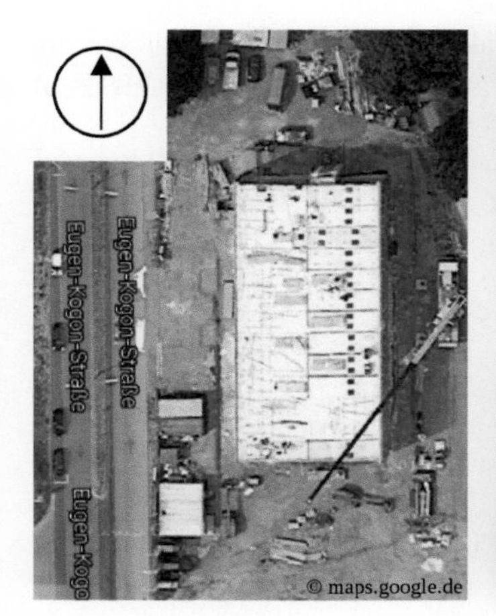

Figure 2-1 Orientation of the ETA-Factory: top view (left) and view from south (right).

2.2 Description of the Installed Insulating Glazing in the ETA Factory

According to the utilization concept explained above and the geographic orientation of the ETA-Factory, the façade of the factory was designed with insulating glass units (IGUs) of diverse functionalities. An overview of the installed IGUs is given in Tab. 2-1. To fulfill the required thermal insulation of the north façade, opaque façade elements with high performance insulation (HPI) modules were installed in addition to the standard IGU. Thus, no sun shading systems like exterior or interior sun control louvers, which might cause glare and are much more expensive, need to be considered. In addition, the architects wanted a stain pattern. This could be realized with the variety of opaque and transparent façade elements. The mode of action of the HPI is clarified in Fig. 2.2: due to the very low resistance coefficient of the HPI modules, condensation, which might freeze in the winter, builds up on the external glass pane. The reason is that the external glass panes of the IGUs with HPI are colder than the external glass panes of those without HPI. The façades in the east and in the west of the ETA-Factory were predominantly constructed with micro-reinforced, ultra-high-performance concrete plates (see Fig. 2-2). The openings intended in the façade are located in the upper third of the façade. The main reason for these openings was not to have an unobstructed view, but rather to get diffuse daylight deep into the machine hall. In addition, direct glare due to solar irradiation was to be avoided. IGUs with capillary slabs were installed to achieve these requirements. The machine hall adjacent to the façade in the south has lower requirements on the thermal coefficient (room-nominal temperature 12-19 °C) in comparison to the offices adjacent to the façade in the north (room-nominal temperature > 19 °C). Due to the cubature of the factory, the architects wanted the façade in the south to be transparent. To avoid direct glare due to solar irradiation while achieving a maximum of visual light transmission, IGUs with sun control louvres were used in the upper two-thirds of the façade in the south. The lower third of the façade is a parametric system which loosens up the monotony of the façade, makes the design more interesting and enables an intervisibility between the ETA-Factory and the surrounding Campus of the University.

Figure 2-2 Upper left: east/west façade; upper right: south façade; lower left: north façade; lower left: north façade; lower right: functionality of the opaque HPI modules.

3 Details of the Installed Insulating Glazing with Integrated Functions

3.1 General

In the following, the installed IGUs in the ETA-Factory will be explained in more detail regarding their specifications. The fact that the main focus of this article is on IGUs with integrated functions, the standard IGU will not be dealt with. We will concentrate solely on IGU with high performance insulation, IGU with capillary slabs and IGU with integrated sun control louvers.

3.2 High Performance Insulation Glazing

The OKALUX HPI product used in the ETA-Factory is an innovative high performance insulation glazing consisting of a vacuum insulation panel (VIP) which is installed in the cavity between the panes of an IGU (see Fig. 3-1). Thus, VIP permits very slender set-ups with outstanding thermal insulation.

The VIP can be used in façade areas where no incidental light is required (e.g. spandrel area), without interfering with the façade system. In general, OKALUX HPI is available in numerous design variants that are provided by a variety of design inlays which can have a significant influence on the appearance of the façade.

For the ETA-Factory, a double-glazing IGU (8+20+12 mm) with an 18 mm thick VIP in the cavity was chosen which results in a thermal transmittance coefficient of U_g = 0.23 W/m²K and almost no solar transmission. The product is opaque and there is no light transmission at all. In comparison to the standard IGUs (U_g = 1.1 W/m²K) used, there is a significant improvement regarding thermal insulation and solar transmittance. To emphasize the stain pattern of the façade, a black fibre tissue was applied on the inner and external side of the VIP.

3.3 Light Diffusing Insulating Glass

In the east and west façade Light Diffusing Insulating Glass, the product OKALUX+, was installed to direct a maximum of diffuse daylight into the machine hall. The general design of Light Diffusing Insulating Glass is given in Fig. 2-2. The capillaries are a translucent insulating material and achieve an optimum, uniform transmission of light into the building irrespective of irradiation conditions. Beyond that, the product has good colour rendering, visibility and glare protection. The light transmission and total solar energy transmittance are controlled as required through the thickness of the capillaries, as well as through the quantity and type of additional glass fibre tissues. As a result of the variability of the capillaries and additional glass fibre tissues, the thermal insulation, the UV protection and the sound insulation can also be controlled as needed. The capillaries reduce the heat transfer in the cavity between the panes in terms of convection and heat radiation. The set-up of the installed double-glazing IGU with capillary slabs is 6+18+6 mm. In the cavity, there is a 10 mm thick capillary slab covered from both sides with glass fibre tissues. That set-up of OKALUX+ achieves a thermal transmittance coefficient of $U_g = 0.9$ W/m^{2}K, a visible transmittance value of about $L_t = 33$ %, and a total energy transmittance coefficient of about $g = 21$ %.

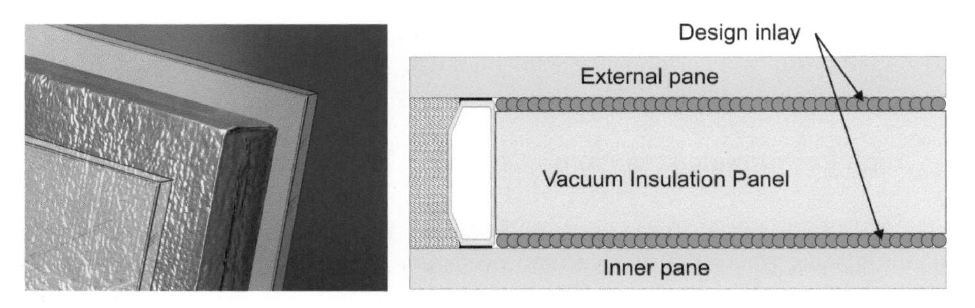

Figure 3-1 Components of IGU with vacuum insulation panel.

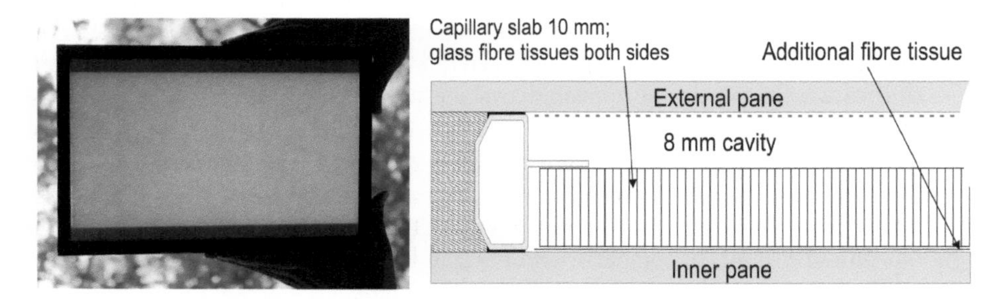

Figure 3-2 Components of Light Diffusing Insulating Glass with capillaries.

3.4 Glazing with Integral Sun Control Louvres

As mentioned above, glazing with integral sun control louvres, the product OKASOLAR F, was installed in the south façade of the ETA-Factory to control solar irradiation. OKASOLAR F provides two different types of louvres, which are shown in Fig. 3-3. The figure shows that the louvre type O provide light redirection into the building. In contrast, the louvre type U function as retroreflection. The louvre types are three-dimensionally shaped and highly reflective. Both louvre types can be combined in one IGU. In spite of the efficent directionally selective solar control, a partial through-vision is always given. In addition, the total energy transmittance is efficiently controllable.

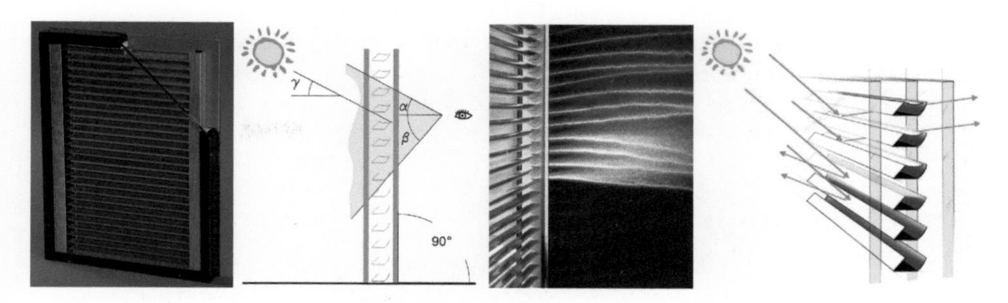

Figure 3-3 Typical examples of Okasolar F; right: louvre types O (upper half) and U (lower half) and their function.

In the case of the ETA-Factory, the louvre type U was not used because standard IGU with a silk-screen printing was installed in the lower part of the south facade. The louvre type O is sufficient for the upper two-thirds of the façade as the solar diagram, which was especially designed for the ETA-Factory, verifies according to Fig. 3-4: Glare protection is always given except from December to March. According to the solar diagram, glare protection is given when there is no intersection between the black line and the dotted lines. Beyond this, the daily and monthly solarization under consideration of the louvre types, the location of the factory as well as the solar irradiation in Darmstadt are decisive according to the solar radiation database PVGIS-CMSAF. Fig. 3-5 shows that integral sun control louvres significantly reduces the solarization in comparison to a 100 % solar transmission. Thus, OKASOLAR F with the set-up 10+16+10 mm was installed in the ETA-Factory. That IGU has a thermal transmittance coefficient of $U_g = 1.7\ \mathrm{W/m^2K}$, a visibility transmittance value L_t in the range of 2-35 %, and a total energy transmittance coefficient g in the range of 13-28 %.

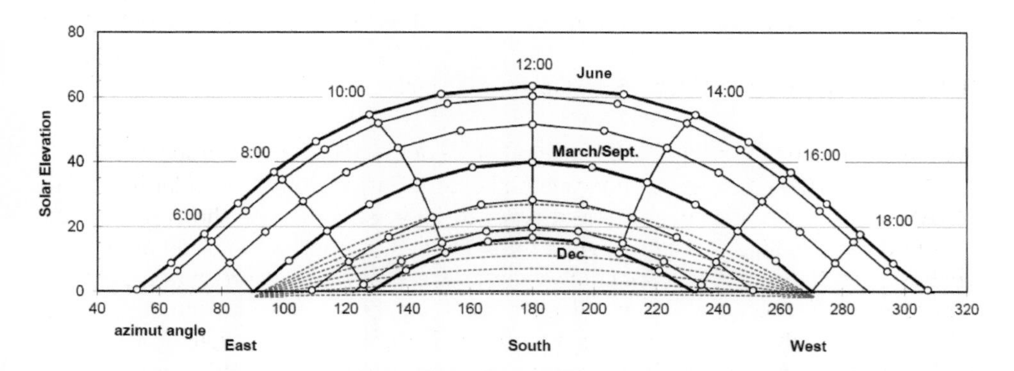

Figure 3-4 Solar diagram of the installed OKALUX F type O in the ETA-Factory in Darmstadt. Existing through-vision: intersection of area of through-vision (dotted lines) and solar orbit (black lines).

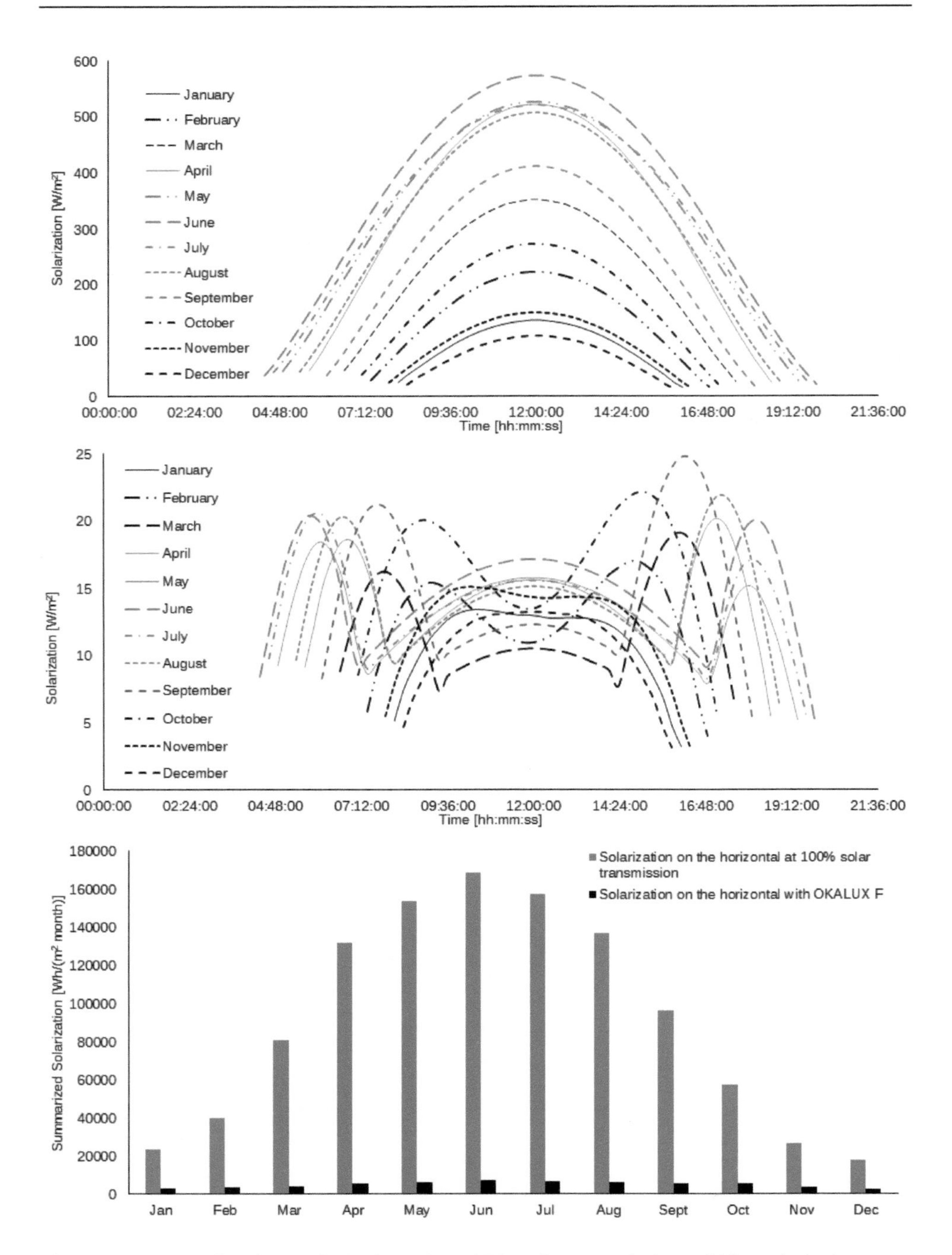

Figure 3-5 Top: solaration on the horizontal at 100 % solar transmission; middle: solarization on the horizontal with OKASOLAR F; bottom: comparison of monthly summarized solarization between 100 % solar transmission and OKALUX F.

4 Conclusion

In Tab. 2-1 the major structural-physical properties of the IGU installed in the ETA-Factory are given. The comparison of the thermal transmittance coefficient U_g, visibility transmittance L_t and total energy transmittance coefficient g reveals that IGUs with integrated functions can control the selective light transmission, the thermal transmission as well as the solar properties. The research project ETA-Factory shows that a smart choice of suitable IGUs not only enhances the subjective comfort in the building, but also improves the energetic requirements significantly. These factors consequently optimize the building life cycle costs.

5 References

[1] TU Darmstadt: www.eta-fabrik.de.

[2] Maier, A.; Kleuderlein, J.; Schneider, J.: Einfluss von Bauwerksverformung auf Fassadenkonstruktionen aus Glas am Beispiel der ETA-Fabrik In: Glasbau, 2016.

Review of optical properties of switchable glazing applications in the automotive industry

Aline Desjean[1]

1 Daimler AG, Mercedes-Benz Cars, RD/KEG, 71059 Sindelfingen, Germany,
aline.desjean@daimler.com

In the automotive industry, integrating shading functions to glazing displays an attractive feature to the standard glass roofs. However, the ageing of such a device must be investigated before any application. Hereby our investigations focus on the optical characterization of various dimming devices.

Keywords: automotive, smart windows, suspended particle devices, electrochromic, liquid crystal devices

1 Introduction

Vehicles are equipped with larger and larger glass surface windows to improve the comfort and spatial atmosphere for the customers. Drawbacks of this trend are the overheating of the interior especially with the current big panorama glass roofs and the glaring effects. An attractive solution to avoid these problems is the use of diverse dimming functions integrated to glazing in the vehicle. Different technologies for a dimming function are available on the market according to the desired optical effects. In this study, three technologies will be described with their applications in the automotive industry: the suspended particle, polymer dispersed liquid crystals and electrochromic devices. The different optical properties of these devices are investigated.

2 Switchable glazing

With suspended particle devices, polymer dispersed liquid crystals and electrochromic devices, the transmission or reflection of the glazing is changed by applying a voltage.

2.1 Suspended particle device (SPD)

This technology is used for panorama roofs with switchable transparency under the commercial name Magic Sky Control® by Mercedes-Benz Cars. A functional SPD foil is laminated between two glass substrates.

Engineered Transparency 2016. Glass in Architecture and Structural Engineering. First Edition.
Edited by Jens Schneider, Bernhard Weller.

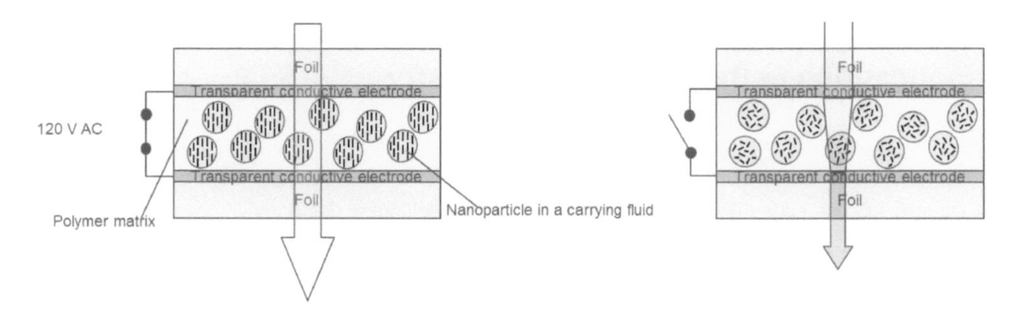

Figure 2-1 Working principle of SPD.

A transparent conductive electrode is coated on the foils, framing a polymer matrix which contains droplets in an emulsion. The droplets enclose rod shaped nanoparticles in a carrying liquid. By means of an electrical field (120 V AC) the nanoparticles align themselves in the direction of the field enabling the light to pass through the layer. Without the presence of an electric field, the particles are in a disordered state and the light is absorbed.

2.2 Polymer-dispersed liquid crystal (PDLC)

This technology is used as a roof in Maybach 62 or as a privacy window in buildings. The same way than SPD the PDLC foil is laminated between two glass windows. In this technology, liquid crystals are dispersed into a polymer matrix. With no voltage the crystals are randomly oriented and the window appears milky. By applying an electrical field (100 AC), the crystals are aligned and the light passes through the layer.

2.3 Electrochromic Device (EC)

Electrochromic devices are currently used in auto-dimming rear-view mirrors. This technology can also be used on a roof (e.g. Ferrari Superamerica 575) or as a window (e.g. Boeing 787 Dreamliner). Electrochromic devices use the redox-reaction like a rechargeable electrochemical cell in which a color change occurs during charging and discharging of the cell.

Electrochromic devices can have different designs generally from 5 to 7 layers (Monk et al. [1]). For a window application, the functional layers are located between the transparent conductive electrodes coated on a substrate (for instance glass). The functional layers are the electrochromic layer, an electrolyte insuring the ionic transfer and an ion storage layer.

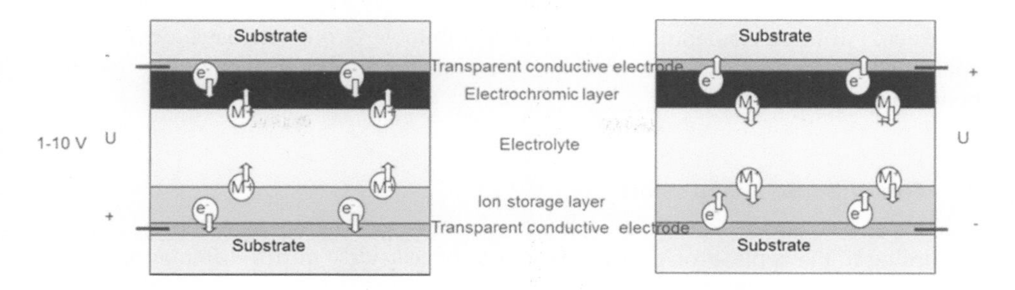

Figure 2-2 Working principle of EC.

By applying a voltage (low < 5 V), reduction for instance (respectively oxidation) of the electrochromic layer occurs, enabling a color change. Oxidation (respectively reduction) occurs at the same time in the ion storage layer. By reversing the applied potential the electrochromic layer will be oxidized (respectively reduced), the device bleaches. The colored or bleached state can be maintained without applying any voltage for a short time.

In current rear-view mirrors the functional layers are in a fluidic form, reducing the number of layers. A reductive layer is also required for this application and can be substituted to one of the conductive electrode or coated between the electrode and the substrate (Tonar et al. [2]).

3 Experimental

A panorama roof using suspended particle devices (SPD), laminated between 2 glasses and commercialized electrochromic outside rear-view mirrors will be characterized (Moll, [3]). A SGS Priva lite® sample is used as a PDLC device. All measurements are performed under controlled atmosphere at 21 °C.

Transmission or respectively reflection measurements are carried out using a Perkin Elmer Lambda 950. Haze measurements in transmission are carried out using a Haze-Guard plus, BYK Gardner. For electrochromic mirrors, the electrochemical properties of the mirrors are evaluated using a potenstiotat SP200 from BioLogic Science Instruments, with a 2 electrodes set-up. Kinetics performances of the switchable devices are measured using the spectrometer at 550 nm and by applying a simultaneous voltage (Desjean et al. [4]). For the SPD technology, a Pacific power source 108 AMX UPC12 is used. For the EC technology a chronocoulometry measurement is performed, enabling the observation of charge transfer (Desjean et al. [4, 5]).

The reflective properties of the electrochromic rear-view mirrors are also estimated by measuring the double image using a laser set-up and a screen. This set up includes a helium–neon laser (632.8 nm, 2 mW) as source a camera Nikon D-700 with a graph paper as screen. All these elements are on a honeycomb table in order to have a defined and controlled distance between the screen and the mirror, enabling to calculate the angle between the glasses.

4 Results and Discussion

According to the application as a window, roof or rear-view mirror different optical requirements are to be fulfilled.

4.1 Roof application

For a roof application like Magic Sky Control® the difference between bleached and colored state shall be at once noticeable for a customer. The switching speed has to be fast. The haze requirements are nevertheless less important that for a window.

4.1.1 Reflection/Transmission

The reflection of the magic sky control roof without voltage, in the dark state in the visible range (380 - 780 nm) is relatively high at 41 %. The transmission in this state in the visible range is 0.3 %. In the bleached state the transmission reaches 26.4 % in this range. The switching effect is clearly visible.

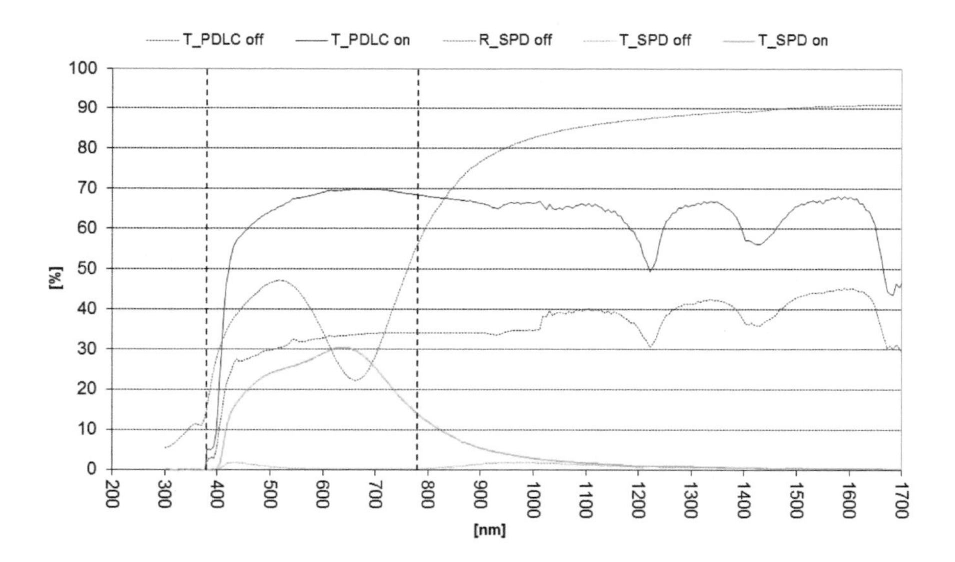

Figure 4-1 Transmittance of SPD and PDLC.

The PDLC device has a transmission of 29.5 % in the milky-state and of 61.9 % in the bleached state. The reflection of PDLC in visible range in the milky state is at 7 % and 12 % in the clear state.

These devices can also reduce the overheating of the vehicle like for Magic Sky Control does. With the reflection and transmission values, the TTS (transmission of total solar energy) value can be calculated. It determines the amount of energy from solar radiation staying in the vehicle. The lower the TTS value, the more energy-efficient the vehicle will be.

4.1.2 Switching time

The switching time is defined as the time from 10 to 90 % of the transmission stroke. For the SPD technology the switching time is 2.6 s by bleaching and 7.5 s by darkening, which is really fast. PDLC has a faster switching time than SPD being less than 1 s in both directions.

4.1.3 Haze

The haze can only be measured in transmission.

For the SPD device, the haze in the dark state is 29.3 % and 3.6 % in the bleached state. For PDLC, the haze is 94 % in the diffuse state and 6 % in the bleached state. As a comparison, a standard car window has a haze from 0.2 %

This represents a challenge for the switching technology.

4.2 Rear-view mirror application

For a mirror application also the difference between the dark and normally reflective state shall be noticeable at once. The switching shall be fast to be useful in dazzling situations. Another challenge for the rear-view mirrors is the multiple reflections inside the mirror itself responsible for double images. These can be easily perceived especially for headlights of the following cars.

4.2.1 Reflection

The rear-view mirrors have a reflection in the visible range of 52 % in the bleached state and of 9.5 % in the colored state. In the colored state a peak at 493 nm corresponding to blue/green and at 716 nm corresponding to red (but where the standard eye has no sensibility) are observed. It corroborates the blue/green color of the mirrors.

Figure 4-2 Electrochromic rear-view mirror.

4.2.2 Switching time

The switching time is defined as the time from 10 to 90 % of the reflection stroke. The rearview mirror darkens in 5.4 s and bleaches in 10.3 s, which is actually slower than the SPD or PDLC technology taking into account the surface difference.

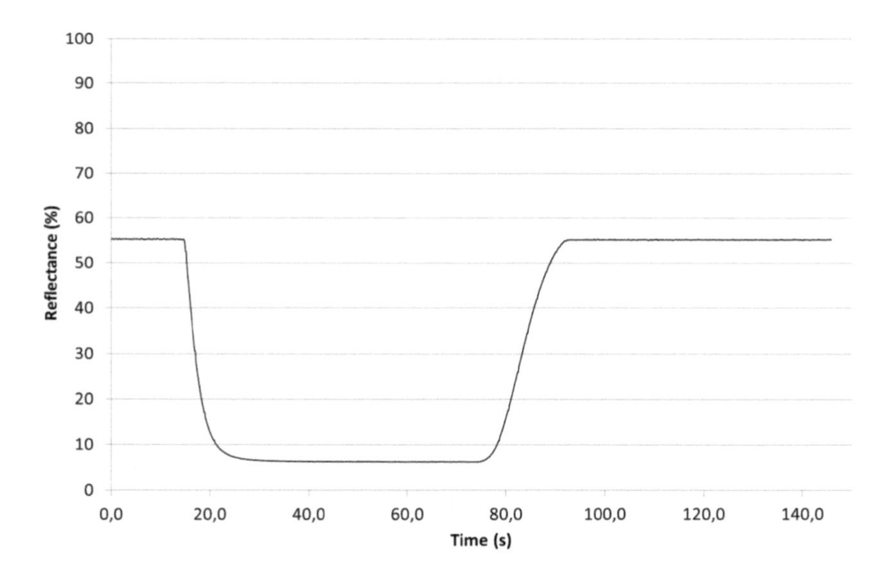

Figure 4-3 Switching time of an electrochromic mirror.

This technology is faster in the coloring phase contrary to the 2 other technologies.

4.2.3 Double Image

The double image from EC mirrors is due to the reflections on the first transparent glass and on the reflective layer. Eventually a third reflection due to the internal reflection between the two glasses can also be visible as shown in figure 6. In a good mirror only the central spot would be visible.

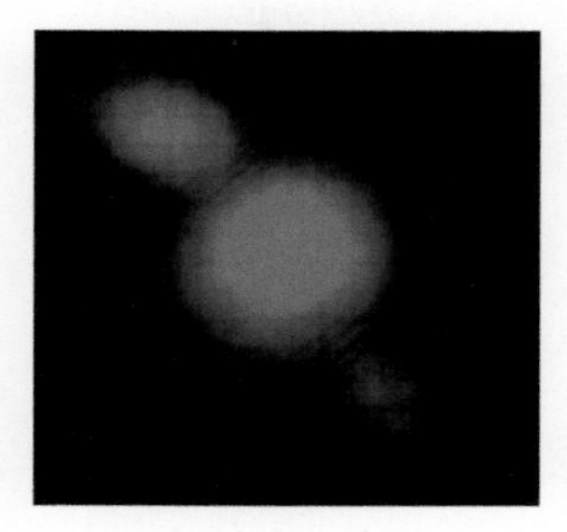

Figure 4-4 Double image phenomenon.

This phenomenon appears when the glasses are not strictly parallel and the angle between both glasses is high enough to be disturbing for the driver.

This angle can be measured and calculated to insure the quality of the mirrors.

5 References

[1] Monk, R.M.S.; Mortimer, R. J.; Rosseinsky, D.R.: Electrochromism: Fundamentals and applications, VCH, 1995.

[2] Tonar, W. L.; Forgette, J.A.; Anderson, J. S.; Bechtel J.H.; Carter, J. W.; Stam J.S.: Patent US 6.512.624, 2003.

[3] Moll, J.: Development of Magic Sky Control with Focus on Aging Stability. In: Glass performance days, Tampere, Finland, 2011.

[4] Desjean, A; Heim, H.-P.; Gövert, S.: characterization of electrical ageing of polymer-based electrochromic devices. In IME-10, Holland, MI, USA, 2012.

[5] Desjean, A.: Charakterisierung, Beurteilung und Modellierung des Langzeitverhaltens einer elektrochromen Kunststoffverscheibung, Kassel university press, 2014.

Prediction of the performance of IGU's with a thermoplastic warm edge spacer

Christian Scherer[1], Ernst Semar[1], Harald Becker[1], Wolfgang Wittwer[1], Jens Wolthaus[1], Thomas Scherer[1]

1 Kömmerling Chemische Fabrik GmbH, Zweibrücker Str. 200, D-66954 Pirmasens, christian.scherer@koe-chemie.de

Modern structural glazing façades with gas filling and warm edge are highly demanding on the edge seal. This is especially true with silicone secondary sealing, which has no gas retention capability and is mandatory due to its UV stability. Especially for the use in structural glazing (SSG) applications a new generation warm edge systems has been developed. Ködispace 4SG is an integrated polymer matrix incorporating the desiccant. This matrix meets the high requirements regarding long term stability and in particular the demands for noble gas tightness of insulating glass units (IGUs) with silicone secondary sealant. In contrast to rigid spacer frames, this thermoplastic spacer utilizes the whole inner gap size of the IGU to absorb movements caused by environmental stresses and allows full flexibility in shape of IGU. Furthermore Ködispace 4SG provides the opportunity to cold bend the whole IGU. The flexibility of a thermoplastic spacer provides a maximum of gas tightness unlike a metallic spacer, where the tightness is limited to the bendability of the spacer and the ability of the butyl layer to compensate relative displacements between spacer and glass. Excellent durability of the edge seal is insured by chemical bonding of the spacer matrix to glass and silicone secondary sealants. Due to the computer controlled application and the low permeability of the spacer, IGUs fulfil the requirements of EN1279-3 (gas tightness) even under standard mass production process conditions. This allows for an easy production even of large formats of triple IGU's for SSG façades with best Ug-values with reliable long-term performance. It's furthermore possible to predict the long term behaviour of these edge compounds. In this work state of the art analysis techniques are used, combined with a new material model, to calculate the behaviour of the IGU in different load cases. It is shown, that Ködispace 4SG does not only have excellent mechanical characteristics but its behaviour is also predictable and therefore reliable, making it an excellent choice for innovative designs. The innovative reactive thermoplastic spacer is a milestone in IGU technology with excellent durable energy efficiency for façades and contributes a significant step towards energy sustainability in building envelopes.

Keywords: long term material behavior, warm edge, thermoplastic spacer, structural silicone glazing (SSG), energy efficiency, flexible shape

1 Introduction

The 2010/31/EU [3] policy for example stipulates for new buildings the demanding "almost zero-energy building" specifications for government buildings beginning in 2019 and in 2021 for all other buildings. To decrease the demand for primary energy sustainability, an optimized thermal insulation – especially in the façade and window surface of

Engineered Transparency 2016. Glass in Architecture and Structural Engineering. First Edition.
Edited by Jens Schneider, Bernhard Weller.

buildings – has to be ensured. As a result of highly effective glass coatings as well as energetically optimized façade/frame profiles, gas filled double and even more triple pane IGUs have developed to highly thermal insulating assembly parts. Therefore, one of the last options to optimize thermal insulation performance of IGUs, the edge seal, has become the focus of attention. The traditional aluminum spacer systems create a significant thermal bridge in the edge seal and are therefore affecting the thermal properties of IGUs and reducing the energy efficiency of the building envelope. On this background, a wide variety of spacer systems -so called warm edge solutions- have been developed to reduce this thermal loss. Metal free thermoplastic spacers represent such high-end warm edge systems. They offer the lowest level ψ-values, which lead to improvements in the U_w values of the window and the U_{cw} values of structural glazing elements [3]. Compared to the ψ-values of the edge seal, the contribution of gas filling plays an even more important role for superior thermal transmission coefficients of double or triple glazing. To maintain the low U_g-values for the whole life time of a façade IGU, only the lowest gas loss rates are acceptable. According to the European product standard for insulating glass EN 1279-3 [2], the maximum limit for the gas loss rate is one per cent per year after a specified climatic load cycle. Some warm-edge systems may have proven gas tightness to this standard even with silicone sealant. But with many of those new warm edge spacer systems, large façade units are not easy to produce. Furthermore, there is a growing discussion about the durability of edge seals, especially in more challenging climates and edge loads. Modern structural glazing façades with gas filling and warm edge are highly demanding on the edge seal. Additionally a secondary sealing with silicone, which has inherently no gas retention capability, is often mandatory due to its UV stability. It is somewhat difficult to produce durable gas filled IGUs with silicone secondary sealing. However, there is a growing demand for such units for structural glazing façades. The gas will only stay inside the IGU with an absolutely tight primary butyl seal – for conventional edge bond with spacer profiles this is a nearly impossible demand. With focus on the use in structural glazing applications, Kömmerling has developed Ködispace 4SG, the new generation of thermoplastic warm edge system. Besides developing a new it is furthermore important to be able to describe its mechanical behavior. Since these thermoplastic systems have a strong non-linear behavior the spacer has not been considered in static analyses so far. To integrate the stiffness of a thermoplastic material into a structural analysis it is necessary to describe the mechanical behavior with a proper material model. This works presents a possibility to model the viscoelastic behavior of a thermoplastic spacer.

2 The reactive thermoplastic spacer system

Ködispace 4SG is a warm edge spacer system based on polyisobutylene, especially designed for silicone sealed units, which replaces the conventional edge seal components: metal or plastic spacer, desiccant and primary sealant (see Figure 2-1).

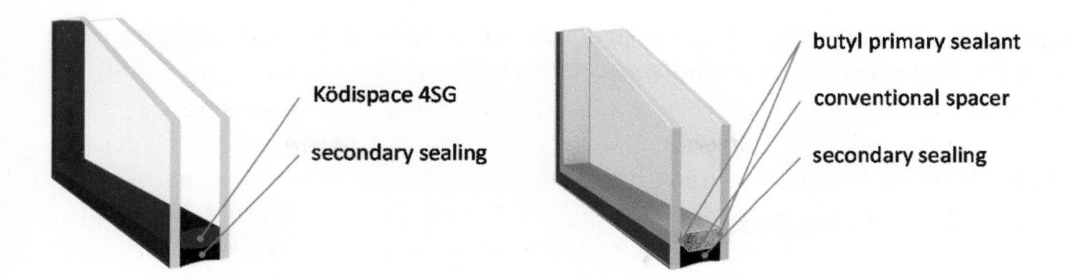

Figure 2-1 Comparison IGU with Ködispace 4SG (left) and aluminum spacer (right).

The reactive thermoplastic spacer is, in contrast to these components, an integrated polymer matrix incorporating the desiccant, which meets the high requirements regarding long term stability and in particular the demands for gas tightness of IGUs with silicone secondary sealant. Different to conventional TPS-types, this reactive spacer bonds chemically to both the glass surface and the silicone sealant (see Figure 2-2). As a result the whole edge seal melts to one integrated system and any dislocation of the spacer is impossible.

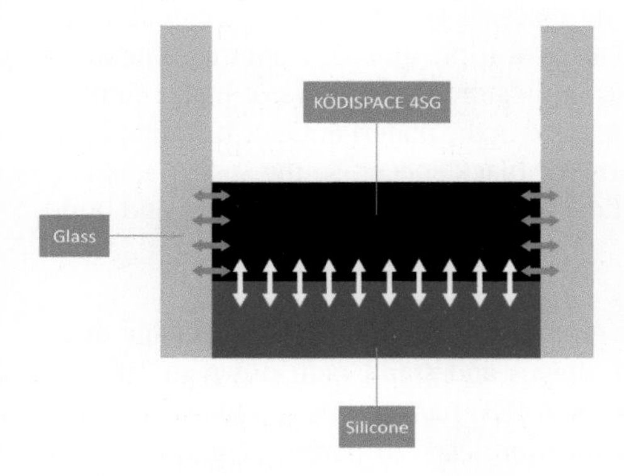

Figure 2-2 Sketch of the chemical bonding of Ködispace 4SG to glass and silicone.

In addition this material has an extended service temperature range up to +90 °C coming along with excellent UV and weathering resistance. Gas leakage and moisture penetration are critical factors, which have to be considered when working with silicone – which is almost exclusively used for structural glazing applications – as secondary sealing. The reactive thermoplastic spacer utilizes the well-known very low gas permeation and moisture vapor transmission rates of butyls. Therefore, even with silicone sealing, which has no gas retention capability, it is possible to produce absolutely tight IG units using Ködispace 4SG. Due to the chemical bond between the reactive thermoplastic spacer and the silicone additional stability is achieved, compared to a nonreactive TPS product which does not bond chemically to silicone. In addition automatic application options for this flexible, high-end warm-edge spacer system can dramatically improve efficiencies while offering the best available long-term sustainable performance.

3 Application

3.1 CNC-Controlled Production

Thermoplastic spacers are applied automatically by a CNC controlled robot directly onto the glass without any manual steps in between. For the application, the material is pumped at a temperature of approx. 130 °C directly from drums to the robot head of the applicator, which is controlled by a computer with a precision of a tenth of a millimeter. The joint between start and stop is perfectly sealed by special patented closing technique. As the application robot is following the glass edge, the spacer material is always precisely positioned, even for large sized spacer frames. This is why a thermoplastic warm edge system is so convenient for large façade units.

Because of its special composition, the reactive thermoplastic spacer adheres immediately to the surface of the panes and can be applied on glass with an inner space of up to 20 mm. Due to the easy application, the thickness of the spacer is very flexible and can be varied from 3 mm to 20 mm in any combination and also for triple glazing from the same drum. The gas filling press is perfectly aligned to the applicator. Corner keys or linear connectors are not required and desiccant is already included in the material. Once installed inside a window frame, the frame color is reflected by the black spacer, so the spacer adapts its color to its background improving the aesthetics – an advantage for architects and building owners.

In addition a thermoplastic spacer allows the precise adjustment of glass package thicknesses as well as the production of irregular shapes and small radii down to 100 mm. Because of this precision, the thickness of the final IGU package fits perfectly to frames or design variations. The fully automated application leads to perfect alignment of the two spacers in triple pane units since the robot applies the bead perfectly to the glass unit for unit and therefore there is no misalignment of the two frames (see Figure 3-1) – a fact that is often the cause of complaints with bent frames made of warm edge spacer profiles. Due to the CNC application various design possibilities referring to the shape of the units are feasible, which are very difficult or even impossible to achieve with standard spacers.

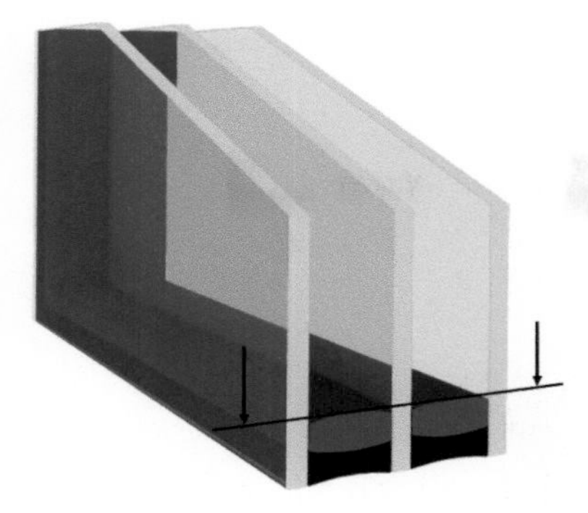

Figure 3-1 Exact Alignment of the two frames with Ködispace 4SG.

3.2 Shape Independent Production

The flexibility and the shape independent application of a thermoplastic spacer offer a wide range of opportunities regarding the shape of IGUs. For instance, it is possible to apply a thermoplastic spacer on various plane geometries (see Figure 3-2).

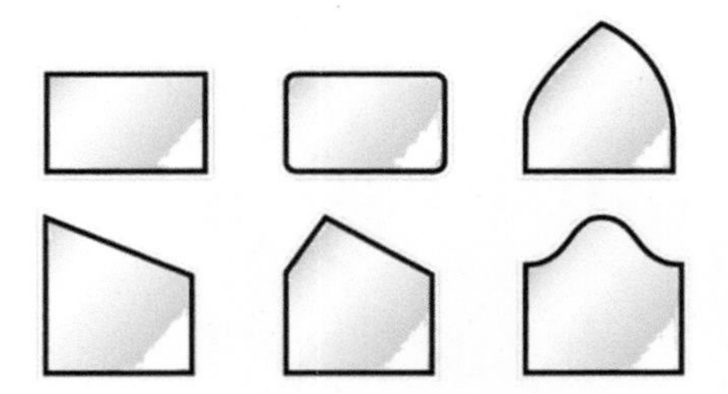

Figure 3-2 Examples for realizable planar glass shapes.

Furthermore a thermoplastic spacer provides the opportunity to either assemble IGUs out of warm bent glass or even cold bend the whole IGU on site. In case of a bent IGU the flexibility of a thermoplastic spacer provides a maximum of gas tightness unlike a metallic spacer, where the tightness is limited to the bendability of the spacer and the ability of the butyl layer to compensate the relative displacements between the spacer and the glass. Figure 3-3 shows a variety of bent shapes.

Cold bending of the whole IGU leads to deformation of the edge compound. Because of the high difference in stiffness between a conventional spacer and the butyl gaskets, the butyl will see very high strains that might cause gas leakages. A reactive thermoplastic spacer is much more flexible and capable of assuring gas tightness even at high deformations. This makes it a much better choice for these applications in comparison to conventional spacer systems.

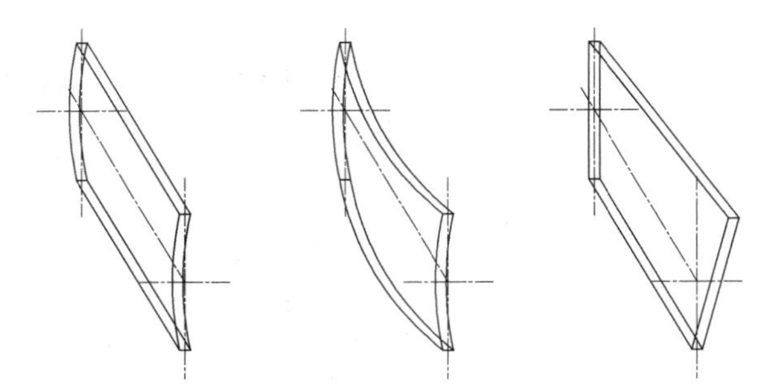

Figure 3-3 Examples for possible bent glass shapes, 1-axial bending (right), 2-axial bending (central), Torsion (right).

4 Prediction of the long term behaviour

4.1 Determination of a viscoelastic material model

The material behaviour of thermoplastic spacers is time and temperature dependent. To develop a material model that describes the long term behaviour of Ködispace 4SG the time and temperature dependent performance was measured using dynamic mechanical thermal analysis (DMTA). The used test specimens and the test assembly are depicted in Figure 4-1.

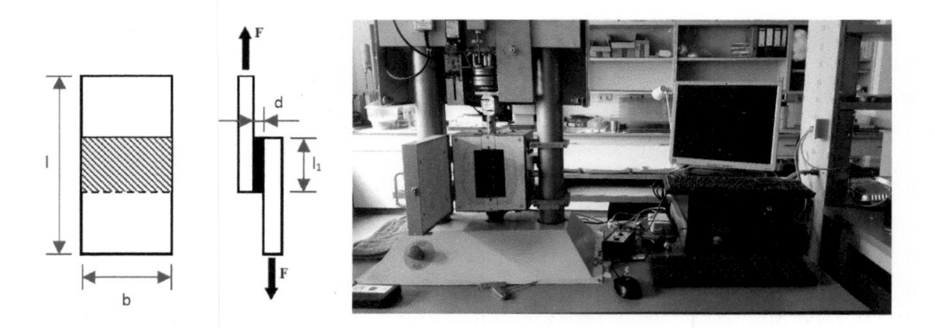

Figure 4-1 DMTA test assembly with shear test specimens.

The tests have been performed in shear mode at a temperature between 183 K and 303 K (-90 °C and + 30 °C). The tests have been repeated with four different load frequencies (1 Hz, 3 Hz, 10 Hz and 30 Hz). The results of these measurements are shown in Figure 4-.

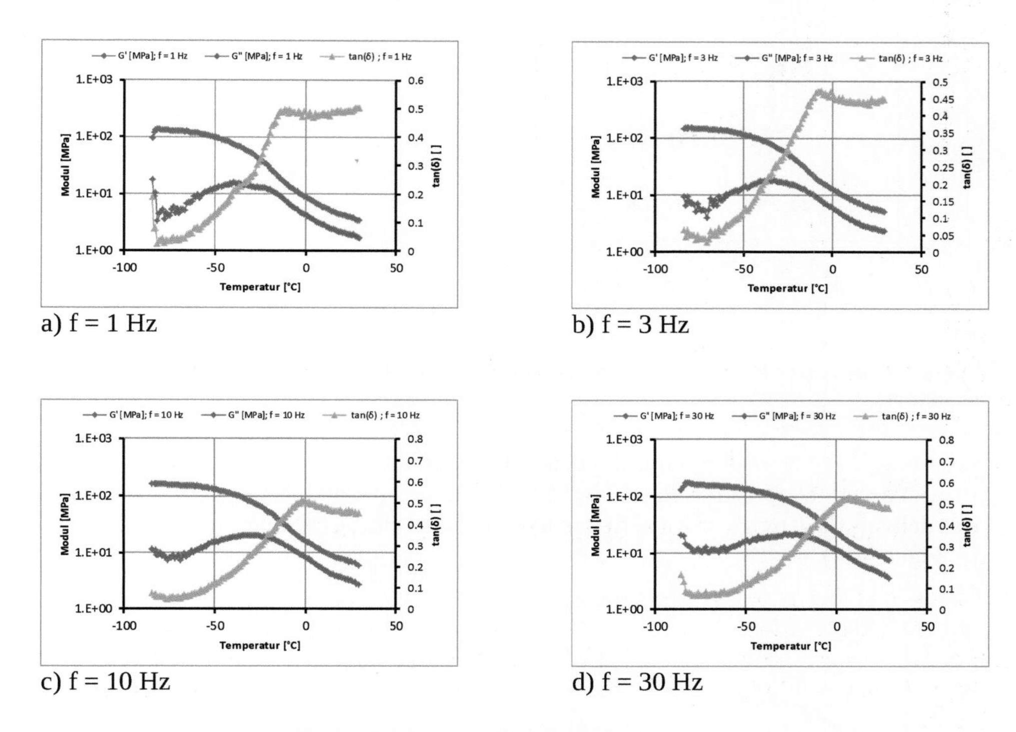

Figure 4-2 Results of the DMTA tests at different load frequencies.

The measured results have been used to determine the glass transition temperature by means of the max(tan(δ))-method. The glass transition temperatures can then be used in an Arrhenius-Plot to derive the activation energy. The determined glass transition temperatures for the different frequencies are shown in Table 4-1 and the Arrhenius-Plot is depicted in Figure 4-3.

Table 4-1 Technical data of the IGUs used for the 1-axial bending tests.

Load Frequency	Glass Transition Temperature
1 Hz	263.8 K
3 Hz	269.6 K
10 Hz	275.1 K

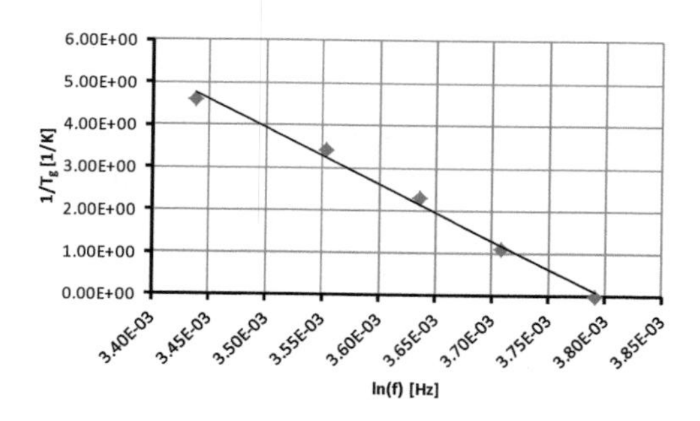

Figure 4-3 Arrhenius-Plot with the glass transition temperatures and corresponding load frequencies.

Assuming linear viscoelasticity, the determined activation energy can then be used to perform a time-temperature-shift based on the time-temperature superposition principle. This delivers the temperature and frequency dependent elastic modulus of Ködispace 4SG (cf. Figure 4-4).

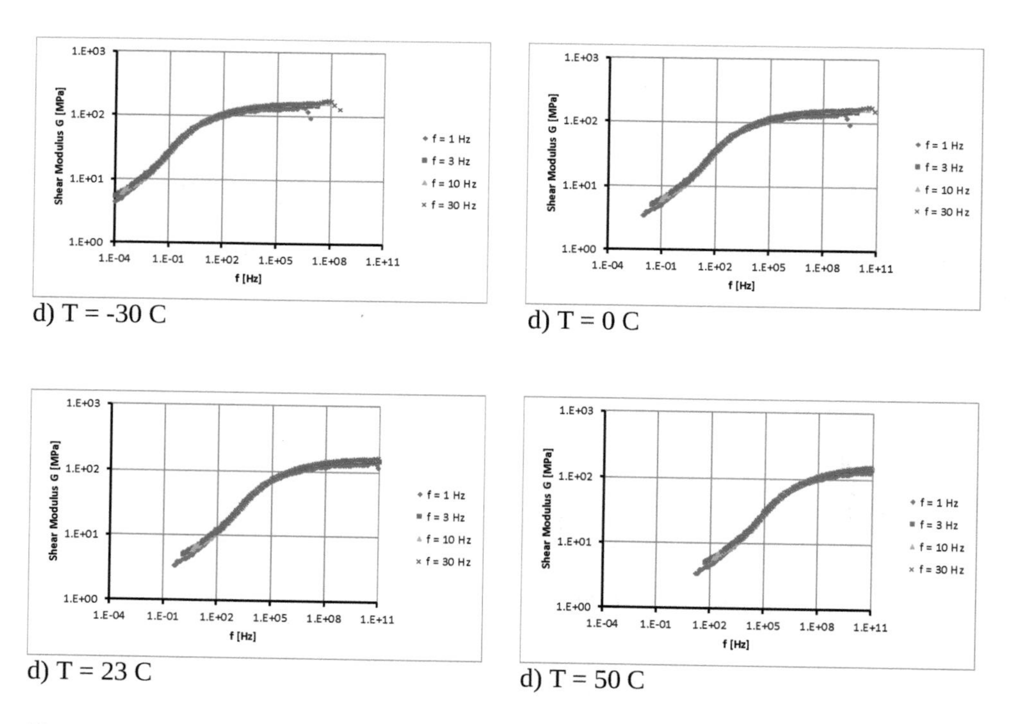

d) T = -30 C

d) T = 0 C

d) T = 23 C

d) T = 50 C

Figure 4-4 Frequency dependent shear modulus at different temperatures.

4.2 Validation of the material model

The results described in chapter 4.1 can be used in common FE-Tools. Therefore the modulus over load frequency curves are fitted with a prony series [1]. In this work the material model is used in Abaqus to calculate the creep behaviour of a H-test-specimen with an applied constant creep load. The used test specimen is shown in Figure 4-5 (left) and the test assembly is shown in Figure 4-5 (right).

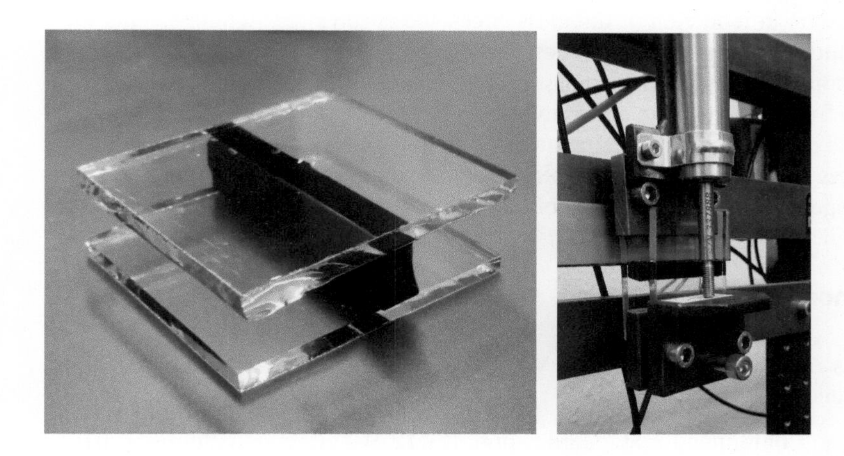

Figure 4-5 H-test-specimen (left) used for the creep tests and corresponding test assembly (right).

In the shown test assembly the deformation over time is measured by a linear potentiometer. The compared results of the test with the calculated FEA results are shown in Figure 4-6.

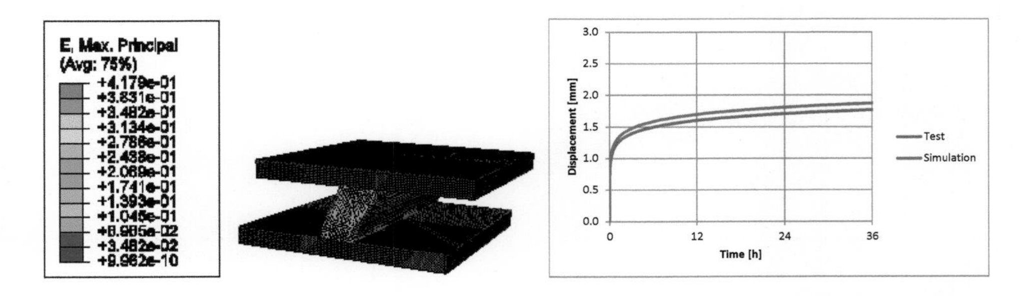

Figure 4-6 Comparison of measured and calculated results for creep tests of H-test-specimens.

The comparison shows a good match between measured and calculated results. This proofs that the shown material model is capable to predict the long term mechanical behaviour of Ködispace 4SG.

5 Conclusion

Ködispace 4SG allows the standard production of gas filled IGUs for façades with a silicone secondary sealant creating a new standard for durability in high-end thermal insulation. Structural glazing façades with best Ψ- and U_g-values are now available with reliable long-term performance. Furthermore a reactive thermoplastic edge compound guarantees a better mechanical performance as shown in the mentioned measurements and calculations. The high elasticity of the reactive thermoplastic spacer combined with a maximum of adhesion results in reduction of the glass stress and ensures gas tightness. This work shows furthermore that it is possible to describe the long term mechanical behaviour of a reactive thermoplastic spacer with common FEA–tools. All these advantages provide a greater design freedom and make more advanced constructions feasible. Those who want to be at the fore of these market developments should put the appropriate fenestration and façade products in place. This will enable you to match every increasing requirements and more stringent energy requirements of tomorrow.

6 References

[1] Abaqus Analysis User's Manual: Version 6.13, Dassault Systems Simulia Corp., Providence, RI, USA 2012.

[2] European Committee for Standardization: EN 1279 – Glass in buildings. (2010).

[3] Glassmagazine.com: Defining warm edge in the commercial market; Glass Magazine, http://glassmagazine.com/article/commercial/defining-warm-edge-commercial-market (2015).

[4] The European Parliament and The Council: DIRECTIVE 2010/31/EU of the European Parliament and of the Council on the energy performance of buildings. (2010).

Enamelled glass – new research studies

Michael Elstner [1], Iris Maniatis[2]

1 Michael Elstner, AGC Interpane, Germany

2 Iris Maniatis, Laboratory and Institute for Structural Engineering, University of the Armed Forces Munich, Germany

Enamelled glass has been in use in architecture for a long time. The enamel coating on the glass surface of thermally-toughened glass panes reduces their characteristic strength. This is taken into account in European product standards which specify a lower minimum tensile strength for such panes. Whereas in the United States for enameled glass the strength values for uncoated, un-enameled, glass are usually used. The tensile strength values mentioned in the European standards have been in use for many years and are valid regardless of the color and manufacturing process of the enamel. A current research project examined different parameters such as glass type, color and manufacturing process of the enamel by performing four point bending tests in accordance with the European standard EN 1288-3. An influence of the color and the manufacturing process on the glass strength was observed.

Keywords: enamelled glass, tensile bending strength, application process, color

1 Production Process

1.1 Color Components

Essentially the dye powder consists of the following main components: flow system (70 - 95 %) and pigment system (5 - 30 %). After firing the transparent and soft glass-flow surrounds the color pigments contained in the dye and binds these together permanently. The adhesion to the glass results from ionic bonding's of the SiO groups (Krampe 2013). Table 1-1 lists the most important components for the colors examined in the research project (black, red and white).

Table 1-1 Components of the colors black, red and white as per manufacturer's specifications (Technical Information Ferro 2012).

	Black	Red	White
Flow system	$ZnO\text{-}B_2O_3\text{-}SiO_2$	$ZnO\text{-}B_2O_3\text{-}SiO_2$	$ZnO\text{-}B_2O_3\text{-}SiO_2$
Pigment system	$CuO\text{-}Cr_2O_3$	Cd-S-Se	TiO_2
Particle size	9 µm	7 µm	9 µm
Density	3.0 kg/dm^3	2.9 kg/dm^3	3.2 kg/dm^3
Linear thermal expansion coefficient	$93*10^{-7}$ 1/K	$93*10^{-7}$ 1/K	$93*10^{-7}$ 1/K

Engineered Transparency 2016. Glass in Architecture and Structural Engineering. First Edition.
Edited by Jens Schneider, Bernhard Weller.

1.2 Color Application Process

There are several different processes used for the application of the color. These are, essentially: roller coating, screen printing, digital printing or "curtain" coating. Usually the enamel coating is applied on the "air surface" of the float glass (that is to say, on the opposite side to the tin bath surface). During the thermal toughening process of thermally toughened safety glass or heat strengthened glass the ceramic coating is baked and firmly bound together with the glass. These processes are described in detail in (BF-Bulletin 015 2014).

Depending on the chosen application process the colors need to have different viscosities. For roller coating, for example, more liquid colors are to be chosen (i.e. with a higher organic part). Correspondingly each different application procedure will require a different drying time. After the application of the color the printed pane has to be dried (at a temperature of approximately 170 °C to 190 °C). Then the pane is thermally toughened and at the same time the enamel coating is baked onto the surface of the glass.

Depending on the chosen manufacturing process different thicknesses of the enamel coating will result (see Fig. 1-1).

Method	Symbol	Thickness [μm]
Digital Print	-	15 – 20
Screen Print	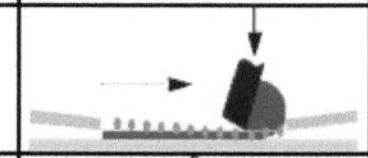	25 – 35
Spraying	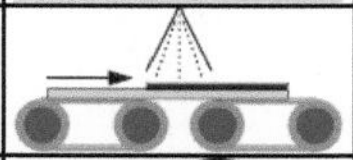	10 – 250
Roller Coating	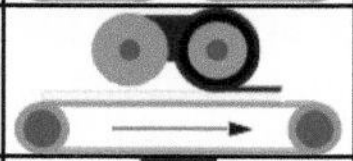	50 – 150
Curtain Coating	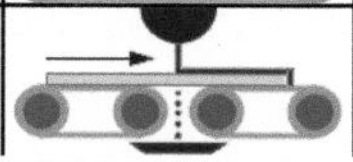	150 – 350

Figure 1-1 Different enamel application processes with average coating thicknesses (after baking process) (Krampe 2013).

The color systems currently available are capable to represent a large number of colors, like for example those of RAL and NCS chart of color range. In order to achieve a specific color the manufacturers of such color systems provide recommendations regarding which products need to be applied in which percentages.

At a furnace temperature of approximately 620 °C to 720 °C (corresponding to glass temperatures of approximately 620 °C to 660 °C) the ceramic color is baked into the glass. The total baking cycle required amounts to approximately 40 seconds per millimeter of glass thickness. For proper baking an oxidizing atmosphere in the furnace care has to be ensured.

Should the enamel prove to be porous after the baking process for example because of unfavorable baking conditions this can lead to a negative influence on the weather resistance, the chemical stability and the mechanical strength of the enameled glass. In (Technical Information Ferro 2012) several possible ways of avoiding this are listed:

- Improving the drying of the enamel coating before baking to optimize the burning of the organic medium
- Increasing the quantity of energy available for the fusing of the color into the glass by optimizing the baking conditions (raising of the temperature) or
- Lowering the fusing temperature of the glass enamel by admixture of some transparent flow-system

The enamel produced in this way will be scratch-proof as well as weather, UV and solvent resistant.

To check the quality of the enamel the following, among others, test methods are recommended according to (Technical Information Ferro 2012):

- Gloss value as according to EN ISO 2813
- Surface scratch resistance as according to EN 15771 or ISO 1518
- Un-dried thickness of the enamel coating or the degree of porosity of the baked ceramic coating

The evaluation of the visual quality of glass panes enameled across the whole or part of their surface can be carried out in accordance with the method described in (BF-Bulletin 015 2014).

2 Strength

From an architectural and manufacturer's perspective important considerations for enameled glass are: aesthetics, function (i.e. use as solar-control glass with an additional functional coating added to the enameled surface), manufacturing process, visual assessment and application.

However for the proper dimensioning of the glazing the characteristic strength values of the glass types must be known. The values given in EN 12150-1 (EN 12150-1:2000-11) and EN 1863-1 (EN 1863-1:2012-02) for enameled thermally toughened safety glass (fully toughened glass - FT) with or without heat soak test (HST) and heat strengthened glass (HS) have been used for many years (see Table 2-1). The reduced values compared with those of non enameled glass show that such ceramic frits have a weakening effect on the mechanical strength of thermally toughened glass. The values are valid independently of the color tint and the manufacturing process of the enamel coating.

The experimental data for the current values were determined decades ago and the scientific background of these investigations is unknown. Up to now there have been only a few systematic investigations examining the reasons for the reduction in strength (see Krampe 2013). In (Krampe 2013) following feasible causes for the stress reduction are given:

- Differentials in thermal expansion coefficient
- Fineness of grind
- Firing penetration and porosity of the enamel
- Leaching processes within the boundary layer of the glass to the enamel
- Changes in thermal conditions during the firing process

Table 2-1 Bending tensile strength for FT and HS glass.

Glass type	Minimum values for bending tensile strength [N/mm²]	
	FT glass according to EN 12150-1	HS glass according to EN 1863-1
Float: clear, tinted, coated	120	70
Enamelled float glass (enamelled surface under tensile stress)	75	45

3 Current Research Results

3.1 General Remarks

Based on the scientific literature explaining the reduction in mechanical strength of enamelled glass in the current research project (Siebert and Maniatis 2015) further investigation is carried out.

Samples of different glass types (HS, FT, FT with HST), different colours (red, black and white) and different application processes (screen printing and roller coating) were subjected to four point bending tests according to EN 1288-3 (EN 1288-3: 2000-09). The average coating thickness for screen printed glass is 45 µm and for roller coated glass 75 µm.

The results were compared with un-enamelled panes of respectively each of the glass-types used. In order to make possible a comparison between the glass-types and between enameled and un-enameled glass, float glass of a single batch was used. For each series 10 samples were tested and the glass thickness of all samples was 8 mm. All specimen were sprinkled with corundum before testing to reduce the dispersion of failure stresses.

In the following sections the results obtained from the individual series are presented in stress-deformation diagrams. These are divided up according to application processes and glass-types. Statistical evaluation was performed using standardised normal distribution. Evaluation using Weibull distribution produced somewhat lower results and requires more spot checks. The Anderson-Darling test was used in order to determine which of the two statistical evaluations is more suitable. The result was that both statistical methods are more or less equally suitable.

3.2 Four point bending test results

3.2.1 Heat strengthened (HS) glass series

In Figure 3-1 and Figure 3-2 the failure stress values are given for each manufacturing process. Also shown are the failure stress values for the un-enamelled reference samples. The statistical evaluation of all series is listed in Table 3-1.

The results can be summed up as follows:

- The reference samples show a high dispersion of failure stresses despite the sprinkling with corundum.
- For the reference samples the mean value of the tensile bending strength is 70.4 MPa and the fractile value 51.9 MPa (confidence level 95 %, fractile 5 %).
- Enamel on the side exposed to tensile bending stress: Except in the case of the "roller coating, black" series, the minimum bending strength of 45 MPa was reached in the case of all other colors and manufacturing processes.

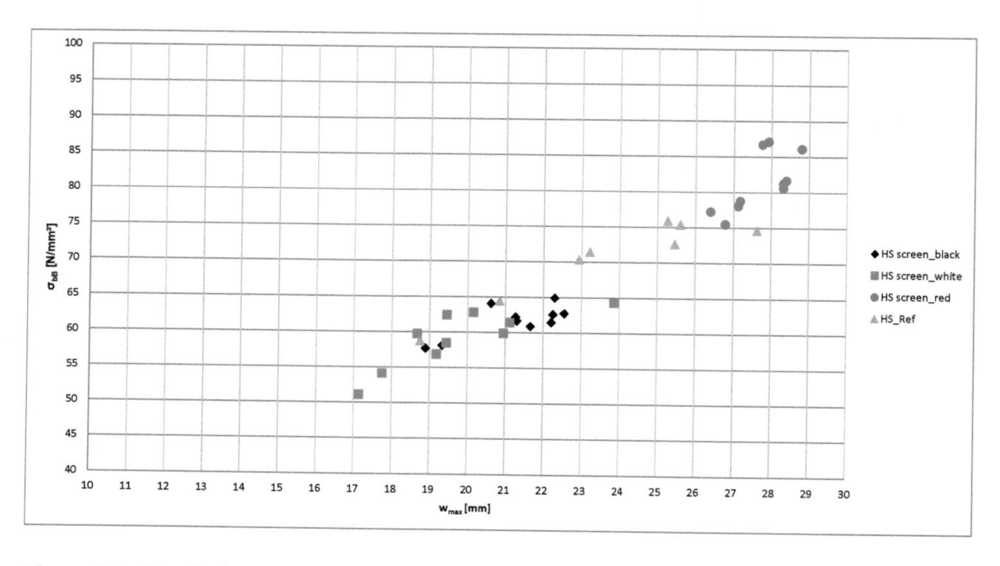

Figure 3-1 Results for series 'HS glass with screen printing on side exposed to tensile bending stress'.

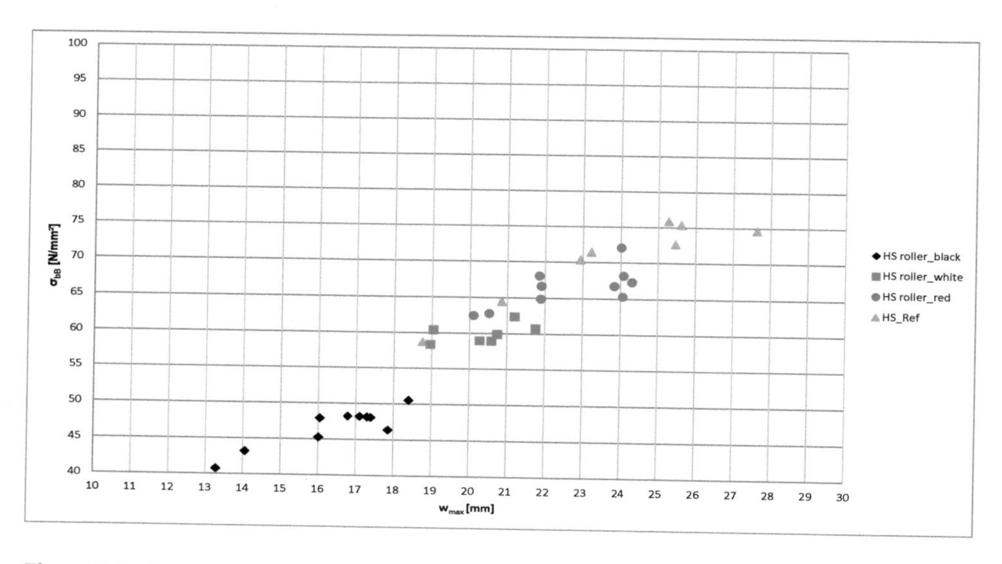

Figure 3-2 Results for series 'HS glass with roller coating on side exposed to tensile bending stress'.

Table 3-1 Statistical evaluation of the HS glass series.

Manufacturing process	Location of enamel		Black	Red	White
Screen printing	Surface under tension	Confidence level 95 %, 5 % fractile [MPa]	54.99	69.38	47.27
		Mean value [MPa]	61.63	81.29	59.10
		Standard deviation	2.31	4.14	4.12
Roller coating	Surface under tension	Confidence level 95 %, 5 % fractile [MPa]	38.39	58.36	55.32
		Mean value [MPa]	46.66	66.43	59.95
		Standard deviation	2.88	2.81	1.38

3.2.2 Thermally Toughened (Fully Toughened – FT) Glass Series

In Figure 3-3 and Figure 3-4 the failure stress values are given for each manufacturing process. Also shown are the failure stress values for the un-enamelled reference samples. The statistical evaluation of all series is listed in Table 3-2.

The results can be summed up as follows:

- For the reference samples the mean value of the tensile bending strength is 145.7 MPa and the fractile value 127.2 MPa (confidence level 95 %, fractile 5 %). The minimum bending strength of 120 MPa for FT was reached.
- Enamel on the side exposed to tensile bending stress: For all colors and manufacturing processes a minimum bending strength higher than 75 MPa was reached.

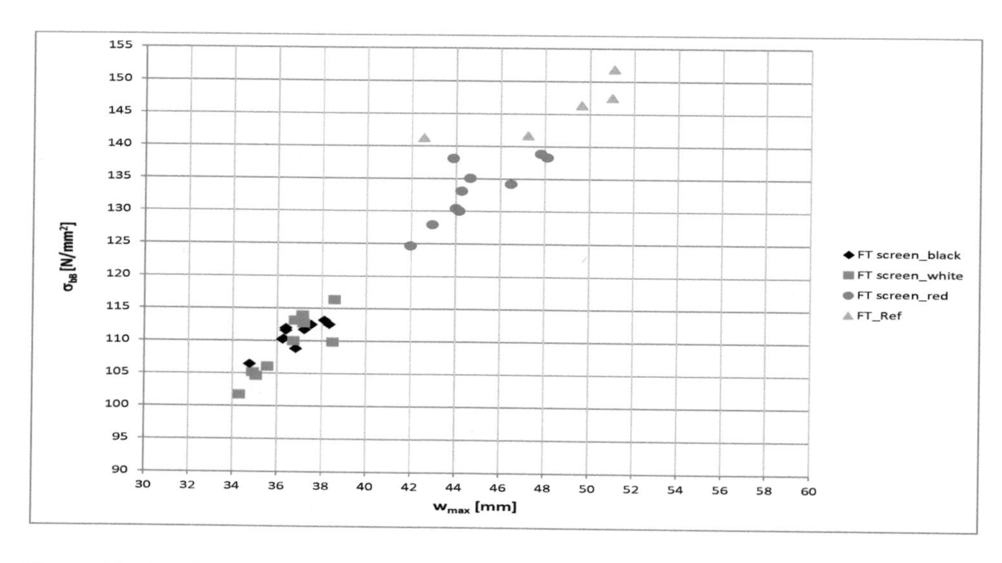

Figure 3-3 Results for series 'FT glass with screen printing on side exposed to tensile bending stress'.

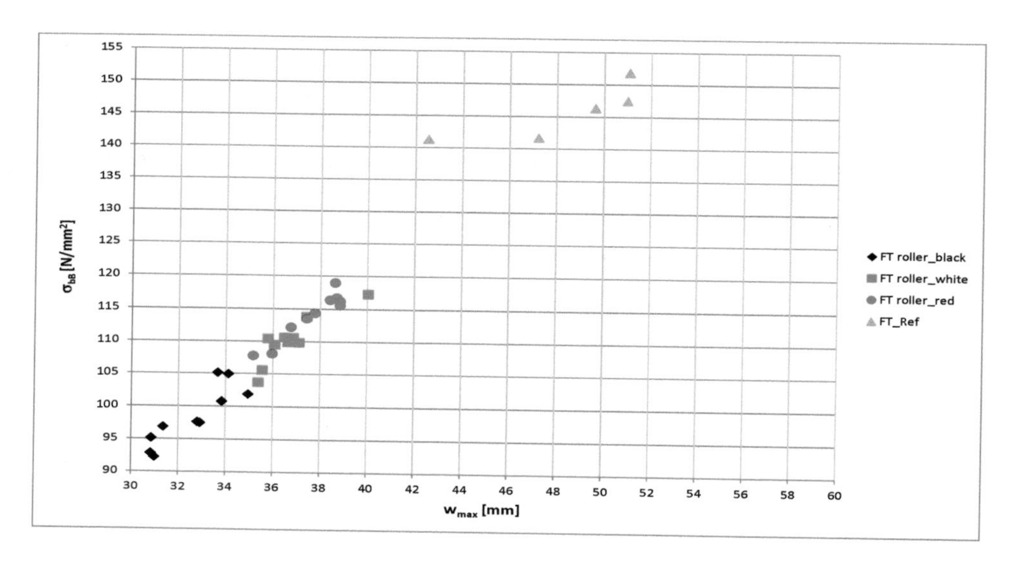

Figure 3-4 Results for series 'FT glass with roller coating on side exposed to tensile bending stress'.

Table 3-2 Statistical evaluation of the FT glass series.

Manufacturing process	Location of enamel		Black	Red	White
Screen printing	Surface under tension	Confidence level 95 %, 5 % fractile [MPa]	105.02	119.34	95.63
		Mean value [MPa]	110.89	133.05	109.34
		Standard deviation	2.04	4.77	4.77
Roller coating	Surface under tension	Confidence level 95 %, 5 % fractile [MPa]	85.28	103.36	99.24
		Mean value [MPa]	98.46	113.96	110.08
		Standard deviation	4.59	3.69	3.77

3.2.3 Thermally Toughened and Heat Soaked (FT-HST) Glass Series

In Figure 3-3 and Figure 3-6 the failure stress values are given for each manufacturing process. Also shown are the failure stress values for the un-enamelled reference samples. The statistical evaluation of all series is listed in Table 3-3.

The results can be summed up as follows:

- For the reference samples the mean value of the tensile bending strength is 134.4 MPa and the fractile value 126.5 MPa (confidence level 95 %, fractile 5 %). The minimum bending strength of 120 MPa for FT was reached.
- Enamel on the side exposed to tensile bending stress: For all colours and manufacturing processes a minimum bending strength higher than 75 MPa was reached.

Table 3-3 Statistical evaluation of the FT-HST glass series.

Manufacturing process	Location of enamel		Black	Red	White
Screen printing	Surface under tension	Confidence level 95 %, 5 % fractile [MPa]	100.33	113.59	98.08
		Mean value [MPa]	105.33	121.58	104.94
		Standard deviation	1.74	2.78	2.38
Roller coating	Surface under tension	Confidence level 95 %, 5 % fractile [MPa]	79.06	81.23	88.11
		Mean value [MPa]	93.01	94.92	98.01
		Standard deviation	4.85	4.76	3.44

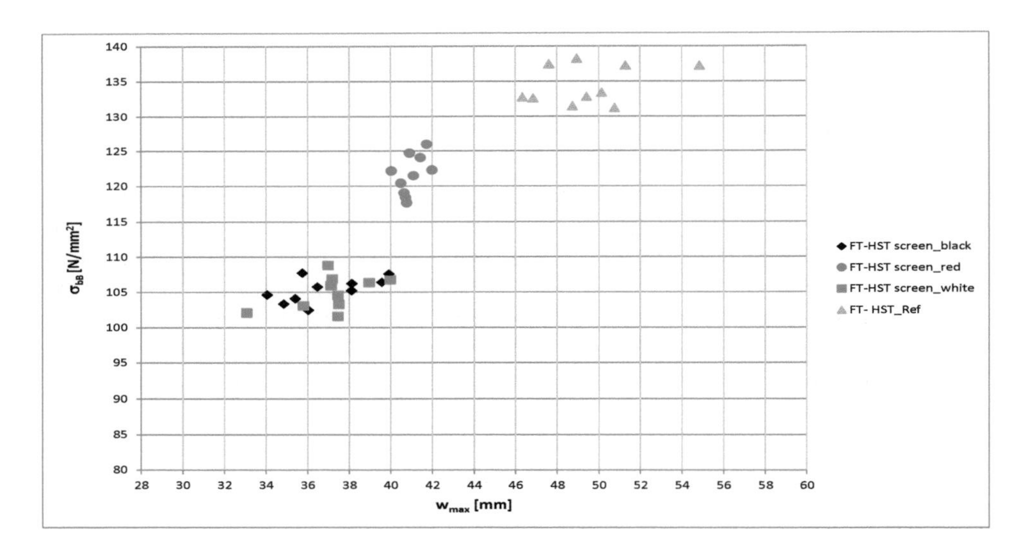

Figure 3-5 Results for series 'FT-HST glass with screen printing on side exposed to tensile bending stress'.

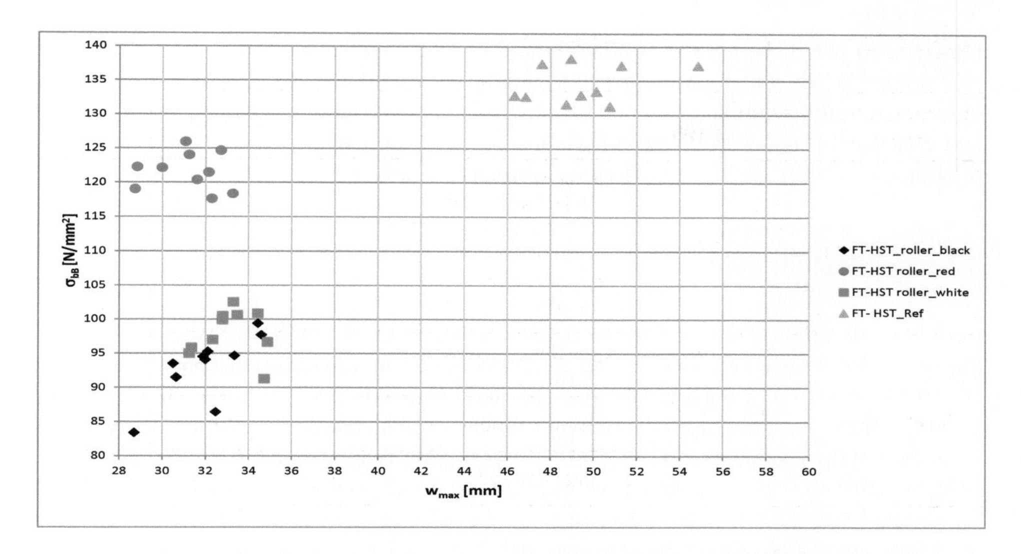

Figure 3-6 Results for series 'FT-HST glass with roller coating on side exposed to tensile bending stress'.

3.2.4 Summary of Results

The colours investigated in the present research project (black, red, white) and the application processes (screen printing, rolling coating) lead to different tensile bending strengths. A result which is to be traced back essentially to the different colour compositions of the enamels (different size of the colour grains, viscosity of the dyes (percentage of flow system, which vary depending on the manufacturing process) as well as the different thicknesses of their coatings. From these factors there result different parameters for the baking process (duration, temperature) which can then lead to different bending strengths. In order to determine the influence of the individual parameters, the colour compositions and the baking parameters would need to be known. On this basis it would be possible to numerically simulate and optimise the thermal processes.

There was noted a tendency for samples with thinner enamel coatings (screen printing, approximately 45 µm) to display greater strengths than the coatings manufactured by means of roller coating (coating thickness of approximately 75 µm). To give a more precise statement, coatings of greater thickness would need to be tested.

The tensile bending strengths ascertained for enamelled glass panes are, in some cases, significantly greater than the minimum bending strengths specified in the product standards (enamel on the side exposed to tensile bending stress), especially the enamel coatings produced by screen printing.

However to use these greater bending strengths as characteristic strength values for dimensioning of the glazing more precise parameters would need to be established regarding, for example, manufacturing and baking processes, colour tint and colour composition and coating thickness. But reproducibility of these parameters would also have to be proven, e.g. within the framework of a National or European Technical Approval.

4 Conclusion

In light of (Krampe 2013) and of the results of (Siebert and Maniatis 2015) it is obvious that several different factors can lead to the reduction in strength that concerns us here. Before these different tensile strengths that tend to result from different coating thicknesses, color types, color application processes etc. can be taken up into the specifications contained in the product standards, further investigations are needed. It has become clear, however, that the values which have been adopted and applied, for decades already, in the product standards for enameled fully toughened glass and enameled heat soak tested fully toughened glass lie on the safe side and thus, can be used for the design.

The samples of the heat strengthened glass series showed a high dispersion of failure stresses. To get reliable statements the number of samples would need to be increased.

5 References

[1] BF-Bulletin 015/2013 "Guideline for assessing the visual quality of enamelled glass".

[2] EN 1288-3: 2000-09 Glass in Building – Determination of the Tensile Strength of Glass – Part 3: Test With Sample Supported on Two Sides (Four-Point Bending Procedure).

[3] EN 1863-1: 2012-02 Glass in Building – Heat-Strengthened Soda-Lime-Silicate Glass – Part 1: Definition and Description.

[4] EN 12150-1: 2000-11 Glass in Building – Thermally-Toughened Soda-Lime-Silicate Safety Glass – Part 1: Definition and Description.

[5] Krampe, P.; Zur Festigkeit emaillierter Gläser. Dissertation, Technische Universität Dresden, 2013.

[6] Shand, E. B.: Strength of Glass – The Griffith Method Revised, Journal of American Ceramic Society 48 1965.

[7] Siebert, G., Maniatis, I.: Untersuchungen zur Biegezugfestigkeiten von emaillierten Gläsern, Forschungsbericht – Nr. b-01-14-24, Universität der Bundeswehr, 2015.

[8] Maniatis, I, Elstner, M.: Investigations on the Mechanical Strength of Enamelled Glass, in "Glass Structures and Engineering", in preparation.

[9] Technical Information "Flat Glass" from the Ferro Corporation, 09/2012.

[10] Technische Regel RAL-RG 529 A3:2007-07: Email(le) und emaillierte Erzeugnisse – Begriffsbestimmungen/Bezeichnungsvorschriften, Deutsches Institut für Gütesicherung und Kennzeichnung e.V.

[11] Wagner, Ekkehard: Glasschäden, Oberflächenbeschädigungen, Glasbrüche in Theorie und Praxis, Fraunhofer IRB Verlag 2012.

[12] Weller, B., Krampe, P., Reich, S.: Glasbau-Praxis – Band 1: Grundlagen. Konstruktion und Bemessung. 3. überarbeitete und erweiterte Auflage, Beuth Verlag 2013.

Design of structural silicone joints in unitized curtain walling exposed to earthquake

Viviana Nardini[1], Florian Doebbel[2]

1 Sika Services AG – Building Systems & Industry, Tueffenwies 16, 8048 Zurich, Switzerland, nardini.viviana@ch.sika.com

2 Sika Services AG – Building Systems & Industry, Tueffenwies 16, 8048 Zurich, Switzerland, doebbel.florian@ch.sika.com

This paper provides a brief overview of the approach adopted by Standards of different markets (Europe, U.S. and Japan) to design curtain walling exposed to earthquake; seismic forces and inter-storey drifts result key factors to consider in controlling the seismic performances of façade components. By comparison to capped curtain walling, the benefits offered by SSG systems to reduce damages due to earthquake are pointed out; although these are well recognized by regulations and scientific literature, no existing standard currently provides precise criteria to design SSG joints subjected to seismic impacts. In line with the performance-based engineering approach defined by Japanese Standard JASS 14, a concept and calculation procedure for seismic design of SSG joints is proposed. Three performance levels associated to different design requirements are defined, with the intent of balancing costs and risks with no compromise on safety. The concept is based on results from small-scale tests performed on sealant H-specimens and it is validated by results from full-scale racking tests on mock up panels.

Keywords: structural silicone, earthquake, inter-storey drift, seismic design, SSG system

1 Introduction

Past earthquakes have focused the attention on the performance of facades and architectural glazing, revealing their vulnerability. Two major concerns related to their performance during and after a seismic event exist (Beher [1]):

- Hazard to people: injuries and deaths at street level from shattered storefront and elevated glazed systems are recognized threads.
- Building downtime and costs to repair: bringing operations and services "back to normal" can be impeded by a breached building envelope due to damages to glazing systems.

As a result, interest in the design of buildings and façades to resist seismic loads and displacements has increased. In this context, the benefits offered by Structural Sealant Glazing (SSG) systems in curtain walling exposed to earthquake are widely recognized,

Engineered Transparency 2016. Glass in Architecture and Structural Engineering. First Edition.
Edited by Jens Schneider, Bernhard Weller.

but still limited guidelines are available to assess the seismic behavior of the façade systems and to size the structural silicone joints, whose correct design is crucial to ensure transfer of seismic loads and accommodation of seismic inter-storey drifts.

2 Overview of international standards for seismic design of façades

In European markets, EN 1998-1 [2] establishes guidelines for design of structures for earthquake resistance and partially deals with curtain walling and partitions elements, considered as non-structural elements. Seismic design of façade components basically focuses on a force-based approach: elements need to resist seismic actions if their failure can cause risks to people, affect the main building structure or services of critical facilities. No requirements are specified by EN 1998-1 [2] about the capability of the façade elements to accommodate the displacements that the main building structure experiences during the earthquake.

In U.S. markets, ASCE 7-10 [3] specifies that seismic demands for curtain walling components need to focus on both transfer of equivalent static forces and accommodation of relative displacements due to seismic inter-storey drifts, which do represent a key factor in controlling the seismic performance of a façade system.

As exterior wall panels can pose a life-safety hazard, they have to be designed to accommodate the differential displacements D_p caused by the earthquake (Section 13.3.2 of ASCE 7-10 [3]); additionally, glass in glazed curtain walls, storefronts and partitions have to be designed and installed to accommodate the relative displacement due to the building inter-storey drift $\Delta_{fallout}$, which causes glass fallout from frame. $\Delta_{fallout}$ has to be determined by engineering analysis or in accordance with AAMA 501.6 [4], which provides an experimental method for determining under controlled lab conditions and by dynamic motion simulation the seismic drift amplitude $\Delta_{fallout}$.

It is worth to mention that the dynamic test of AAMA 501.6 [4] substantially differs from the static test of AAMA 501.4 [5], which describes a test method to evaluate the performance of curtain wall systems subjected to smaller inter-storey drifts induced either by low-scale earthquake or by wind loads. Indeed, while AAMA 501.4 [5] test method focuses primarily on the seismic serviceability limit state behavior of a wall system, AAMA 501.6 [4] focuses on the seismic ultimate limit state of its glass. The two different test methods introduce the concept of calibrating the performance requirement to the magnitude of the seismic input, as per design philosophy promoted by NEHRP (National Earthquake Hazard Reduction Program) defining four seismic design performance levels, still at a conceptual stage:

- Operational Level, with essential no damage to cladding elements.
- Immediate Occupancy Level, with moderate damage to non-structural elements and light damage to structural elements in the primary structural system of the building.

- Life Safety Level, with moderate damage to structural and non-structural elements.
- Near Collapse Level.

In Japanese market, JASS 14 [6] is specifically dedicated to façades and curtain walling and provides design criteria for their seismic design. The energy released by the earthquake occurs in the forms of P-waves and S-waves acting in vertical and horizontal direction respectively; façade components need to be verified against equivalent static forces to be applied at their mass center.

Specific focus is given by JASS 14 [6] to the effect of the seismic inter-storey drifts of the main structure, which can introduce deformations into the façade system to be properly accommodated. Based on the building inter-storey height H, JASS 14 [6] sets three different seismic levels diversified by potential hazard and probability of occurrence:

- LEVEL 1 – Maximum inter-storey drift: H/300
  No damages to internal and external components have to occur.
  This seismic grade is related to earthquakes frequently occurring in Japan.
- LEVEL 2 – Maximum inter-storey drift: H/200
  The stress in all external components has not to exceed the allowable standard limits; after the seismic event, the full functionality of the façade is ensured with sealing repairing works admitted.
  This seismic grade is related to the largest scale earthquake happened in the past.
- LEVEL 3 – Maximum inter-storey drift: H/100
  Neither the damage of the glass pane nor drop-out of any component is allowed.
  This seismic grade is related to the greatest earthquake to happen in the next 100 years.

Please refer to Nardini et al. [7] and relevant codes for a detailed overview about design recommendations provided by mentioned standards.

3 Inter-storey drift effects on SSG systems

Typical SSG-systems for unitized curtain wall elements consist of glazed panels bonded to a main aluminum frame by SSG joints (Figure 3-1). Along mullions and transoms, the panels are provided with stack joints of adequate clearance (Figure 3-1) designed to ensure free accommodation of any movement that the building structure can experience during its service life; the panel frame is usually hanged to the slab of the main structure by hinge brackets (Figure 3-2).

In this section, the typical design concept adopted for standard SSG unitized panels to accommodate seismic slab movements is briefly described.

- Vertical differential movements between slabs
  The upward and downward differential seismic movements of the slabs are usually accommodated by the vertical stack joint along the transoms; adequate vertical clearance should be designed to avoid clashing of the panels (Figure 3-1). Therefore, no displacement is imposed to the SSG-joints by such movements.
- Horizontal differential movements between slabs (out-of-plane component)
  The out-of-plane differential seismic movements of the slabs due to inter-storey drift are usually accommodated by the brackets, which should allow for free rotation of the panel at the supports (Figure 3-2). Therefore, no displacement is imposed to the SSG-joints by such movements.
- Horizontal differential movements between slabs (in-plane component)
  The in-plane differential seismic movements of the slabs due to inter-storey drift produce a racking motion of the unit characterized by rigid translation and rotation of the glass panel within the frame, which can deform (Figure 3-2) (Beher [1], Memari et al. [8], Galli [9]).
  As a consequence, differential displacements between glass and frame occur and stress is introduced into the SSG-joints due to such inter-storey movements.

It should be noted that in-plane movements of the slabs represents the most critical ones for the integrity of the system and it is often demanding to predict their effect accurately. Depending on the design solutions adopted, different rotation points can be identified in the panel racking motion and any component which prevents free rotations can have significant impact in the behavior of the whole system.

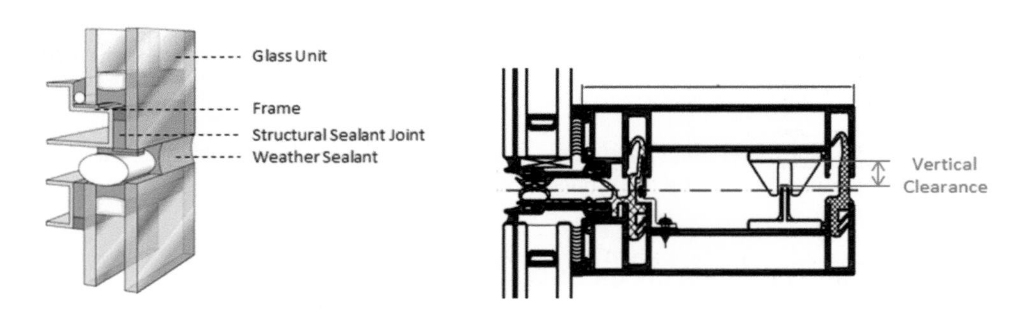

Figure 3-1 Typical SSG system detail (left) and vertical stack joint between transoms (right).

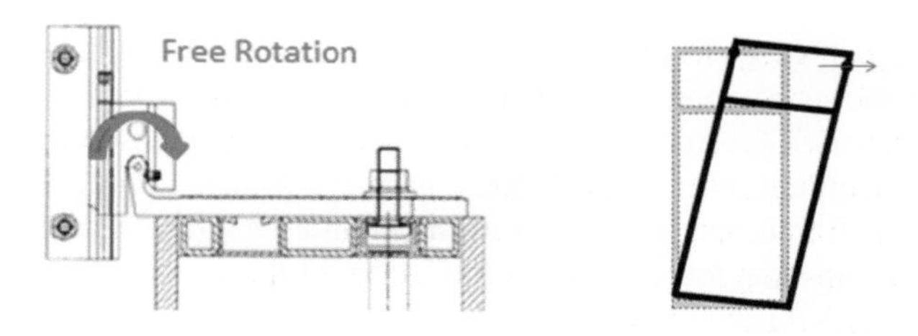

Figure 3-2 Typical hinge bracket (left) (Galli [9]); frame deformations due to panel racking (right).

4 Benefits offered by SSG systems in areas prone to earthquake

Captured glazed systems in unitized curtain walling typically consist of glass panels retained to a main frame by mechanical means able to transfer the required loads (Figure 4-1). Such systems highly differ from conventional SSG systems when seismic performances are analyzed.

Figure 4-1 Captured Glazed system (Beher [1]) (left); stick system after earthquake in New Mexico, 1985 (right) (Galli [9]).

The benefits offered by SSG systems compared to captured systems in areas prone to earthquake are widely recognized:

- The resilient attachment of the glass panel to the supporting framework by the structural sealant joint has proven to be beneficial in controlling and in some case eliminating breakage normally experienced during a small to moderate earthquake. Since the glass panel is not captured in metal glazing pocket, the opportunity for it to impact the metal surfaces during lateral displacements is minimized, eliminating a primary cause of breakage (ASTM C 1401 [10]).

- Experimental studies (Beher [1]) on glass panels retained by mechanical caps have shown that in-plane displacements of slabs produce at first a rigid racking motion of the glass panel as per typical SSG-systems, but mainly limited by the available clearance between glass and capping profiles. Additional inter-storey drifts produce high contact stresses between frame and glass, making it prone to fracture and to fallout under the in-plane compression forces (buckling effect) which are introduced into the capped system but avoided in the SSG one.
- When a glass lite break does occur, the SSG system can retain much if not all of the broken glass due to its continuous attachment along the edges, if a laminated glass panel is used and provided that the structural joints retain sufficient integrity (ASTM C 1401 [10]).
- Although conventional SSG systems can perform well in an earthquake, consideration could be given to isolate the lite from building frame movements. One method to consider is to structurally bond the glass panel to a sub-frame and then attach the sub-frame to the primary curtain wall frame with mechanical fasteners in slotted holes, dimensioned to accommodate the required seismic displacements (ASTM C 1401 [10]).

Correct dimensioning of SSG-joints results crucial to exploit the benefits offered by the system, to properly transfer seismic forces and accommodate imposed movements; seismic performance requirements associated to different damage levels can be satisfied depending on adhesive properties and joint dimensions.

5 Concept for seismic design of SSG joints

5.1 Performance-based concept

Existing regulations do not provide clear design criteria to evaluate the performance of SSG joints in façade exposed to earthquake, even if benefits they offer in this regard are quite well recognized. This section proposes a design concept to assess their seismic performances.

In line with the design philosophy adopted by JASS 14 [6], the utilization limit for the joints is defined depending on the seismic requirements set for the façade. Three different performance levels associated to corresponding allowable strengths and deformation rates of the adhesive joints are proposed.

- LEVEL 1 – Damages to the façade components must not occur and the full functionality of the façade system must not be compromised after the seismic event.
  During and after the seismic event, a minimum global safety level of 6 has to be ensured for the SSG-joints; the stress on the joints is limited by the dynamic tensile and shear strengths $\sigma_{des,1}$ and $\tau_{des,1}$ (Figure 5-1) defined by EOTA ETAG002 [11] for typical wind design.

- LEVEL2 – The full functionality of the façade is ensured; some sealing repair works might be needed and inspection of the SSG-joints is required.
  During the seismic event, the movement capability certified for the adhesive (ASTM C 719 [12]) is exploited and the allowable strengths $\sigma_{des,2}$ and $\tau_{des,2}$ (Figure 5-1) are set to correspond to a joint movement capability of 12.5 %.
  The strength values are defined based on statistical analysis of results obtained on a population of minimum 10 sam ples 12 mm x 12 mm x 50 mm tested in tension and shear; Figure 5-2 shows the stress vs. strain average curves obtained by the tests and representative of the tensile and shear behavior of structural silicone Sikasil® SG-500; $\sigma_{des,2}$ and $\tau_{des,2}$ (Figure 5-1) represent the characteristic strengths giving 75 % confidence that 95 % of the test results will be higher than the values adopted at the tensile deformation rate of 12.5 % (equivalent to shear deformation rate of 51.5 %).
  After the seismic event, the SSG-joints shall be able to withstand the loads occurring in the future service life of the façade and therefore a minimum safety level of 6 has to be restored.
  Figure 5-3 left shows the behavior of joints 12 mm x 12 mm x 50 mm (structural silicone Sikasil® SG-500) after Hockman Cycles representing an accelerated life cycle simulation consisting of (a) immersion in water for 7 days (b) exposure in an oven at 70 °C for 7 days while under compression (c) automatic compression and extension cycling to 12.5 % elongation rate at room temperature and (d) alternate compression and extension up to 12.5 % elongation rate at high (70 ±2 °C) and low temperatures (-26 ±2 °C) respectively under conditions described by ASTM C 719 [12].
  If compared to Figure 5-1, the graph proves that the final strength of the joint is not reduced after it has repeatedly experienced stress levels corresponding to 12.5 % elongation.
  Therefore, the earthquake associated to Level 2 will not compromise the future performance of the structural joints and a minimum design safety level of 6 will be ensured under future loads.
- LEVEL 3 – Drop-out of any component is not allowed.
  During such unique and extreme event, design focuses mainly on life safety and the demand for the structural joints is to be earthquake-resistant: the allowable strengths $\sigma_{des,3}$ and $\tau_{des,3}$ (Figure 5-1) are set to correspond to joint deformations of 25 %.
  As per level 2, the strength values are defined based on statistical analysis of results obtained on a population of minimum 10 samples 12 mm x 12 mm x 50 mm tested in tension and shear and $\sigma_{des,3}$ and $\tau_{des,3}$ (Figure 5-1) represent the characteristic strengths giving 75 % confidence that 95 % of the test results will be higher than the values adopted at the tensile deformation rate of 25 % (equivalent to shear deformation rate of 75 %).
  After the seismic event, a minimum residual strength must be ensured by the structural joints as the façade could be seriously damaged and substantial repair works are to be accounted for.
  Figure 5-3 right shows the behavior of joints 12 mm x 12 mm x 50 mm (structural silicone Sikasil® SG-500) after the Hockman Cycles (ASTM C 719 [12]) described

for Level 2, but associated to compression/elongation rate of 25 %. If compared to Figure 5-1, the graph highlights that the final strength of the joint is reduced by repeated stress levels corresponding to 25 % elongation; however, after this extreme event a minimum safety level of 2.5 is still ensured.

A proper inter-storey drift Δ associated to each performance level should be set by project specifications or local standard based on risk assessments.

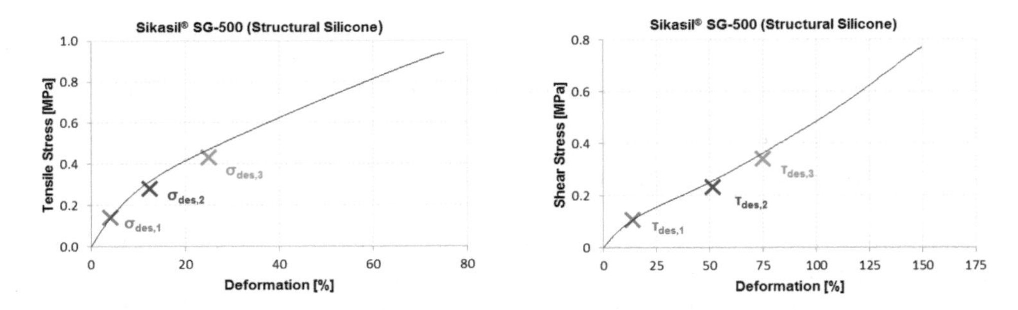

Figure 5-1 Strength limits: tensile (left) and shear (right) stress vs. strain at performance level *i*.

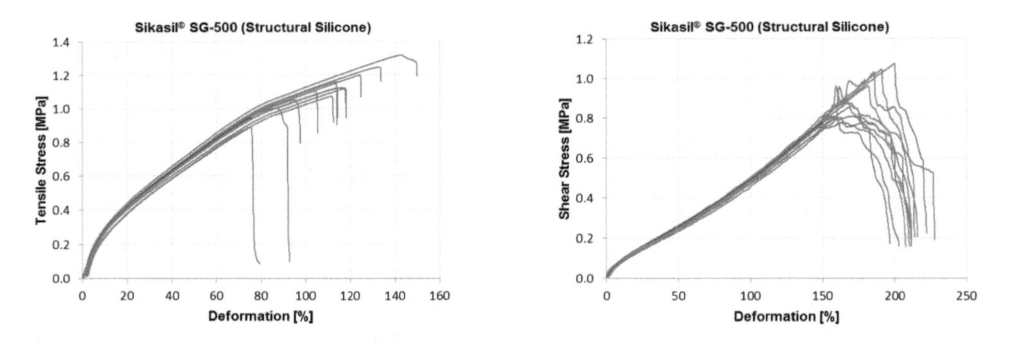

Figure 5-2 Average tensile (left) and shear (right) stress vs. strain.

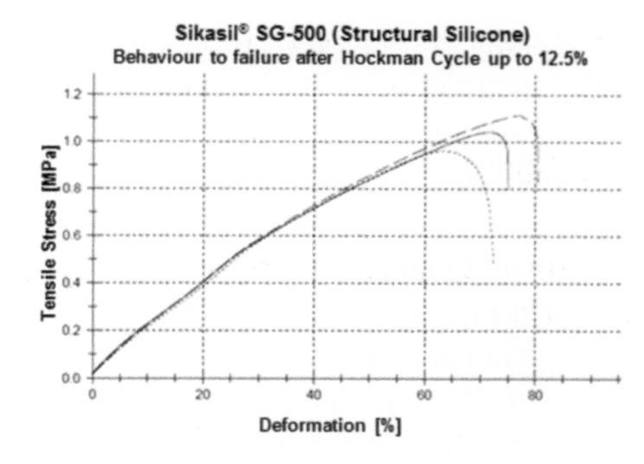

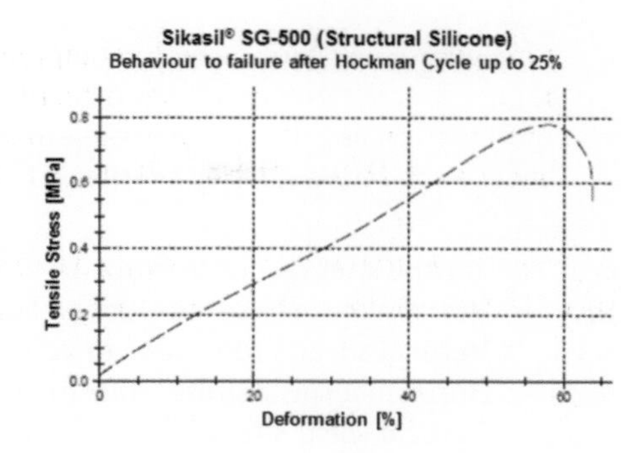

Figure 5-3 Residual tensile strength after Hockman cycles at deformation rates of 12.5 % (left) and 25 % (right).

5.2 Calculation procedure

The design concept proposed allows to evaluate the adequacy of the joint thickness to accommodate the movements imposed by seismic racking; for each performance level, the maximum shear stress in the joint can be calculated in line with ETAG002 [11]:

$$\tau_{S,i} = \frac{S_i G_i}{e} \leq \tau_{des,i} \tag{5.1}$$

S_i	displacement imposed to the joint for the seismic performance level i
G_i	adhesive shear modulus for the performance level i
e	joint thickness
$\tau_{S,i}$	shear stress due to S_i

For performance level 1, G_1 is defined according to ETAG002 [11].
For performance level 2, G_2 is defined as the secant modulus between the deformation boundary limits [0; $\varepsilon_{des,2}$] covering the shear deformation range 0 % < $\varepsilon \leq$ 51.5 % (equivalent to tensile deformation range 0 % < $\varepsilon \leq$ 12.5 %).
For performance level 3, G_3 is defined as the secant modulus between the deformation boundary limits [0; $\varepsilon_{des,3}$] covering the shear deformation range 0 % < $\varepsilon \leq$ 75 % (equivalent to tensile deformation range 0 % < $\varepsilon \leq$ 25 %).
For Sikasil® SG-500 G_1 = 0.50 MPa, G_2 = 0.49 MPa, G_3 = 0.48 MPa.

The global stress level in the SSG joint due to both seismic forces and imposed displacements needs to be calculated for each performance level i; as further step, the global utilization level of the joint needs to be checked.

$$\mu_{tens,i} = \sigma_{H,i} / \sigma_{des,i} \leq 1.0 \tag{5.2}$$

$$\mu_{shear,i} = \left[\left(\tau_{V,i} + \tau_{SV,i}\right)^2 + \left(\tau_{H,i} + \tau_{SH,i}\right)^2 \right]^{0.5} / \tau_{des,i} \leq 1.0 \tag{5.3}$$

$$\mu_i = 0.5\mu_{tens,i} + \sqrt{\left(0.5\mu_{tens,i}\right)^2 + \mu_{shear,i}^2} \leq 1.0 \tag{5.4}$$

$\sigma_{H,i}$	Tensile stress – due to horiz. force (out-of-plane) at seismic level *i*
$\tau_{H,i}$	Horizontal shear stress – due to horiz. force (in-plane) at seismic level *i*
$\tau_{V,i}$	Vertical shear stress – due to vert. force (in-plane) at seismic level *i*
$\tau_{SH,i}$	Horizontal shear stress – due to horiz. displacement imposed at seismic level *i*
$\tau_{SV,i}$	Vertical shear stress – due to vert. displacement imposed at seismic level *i*
$\sigma_{des,i}$	Tensile strength at seismic level *i*
$\tau_{des,i}$	Shear strength at seismic level *i*

Combination of relevant shear and tensile stress has to be defined depending on simultaneity of forces and displacements.

6 Mock Up Test

Seismic mock up tests performed by Permasteelisa Group on four unitized façade panels are here used to validate the performance-based concept proposed in Section 5. Test procedure, system configuration and experimental results are comprehensively provided by Galli [9]. Tests mock up consisted of four unitized panels composed by a single monolithic glass 1452 mm x 3752 mm bonded to its main aluminum frame by structural sealants; Sikasil® SG-500 joints 10 mm x 6 mm were used to bond the glass elements of two panels, while Sikasil® SG-550 joints 6 mm x 6 mm were applied on the other two panels in order to compare the behavior of the two structural sealants. The following test sequence was implemented, aiming at investigating the seismic behavior of the systems based on the performance requirements set by JASS 14 [6]:

- Air leakage test (EN 12153 [13]).
- Racking test: an inter-storey drift of H/300 ($\Delta 1$ = 12.5 mm) was imposed (20 cycles), as per performance Level 1 of JASS 14 [6].
- Air leakage test (EN 12153 [13]).
- Racking test: an inter-storey drift of H/200 ($\Delta 2$ = 18.75 mm) was imposed (10 cycles), as per performance Level 2 of JASS 14 [6].
- Racking test: an inter-storey drift of H/100 ($\Delta 3$ = 37.5 mm) was imposed (5 cycles), as per performance Level 3 of JASS 14 [6].

The following test results were obtained:

- Racking test representative of seismic Level 1 did not cause any damage to the façade panels; the air leakage tests before and after the imposed storey drift $\Delta 1$ proved that the functionality of the façade was not altered.
- Racking test representative of seismic Level 2 did not cause any damage to the façade panels; air leakage tests after this test was not repeated as performance level 2 by JASS 14 [6] allows for repair works on sealing joints to restore the tightness efficiency of the system.
- Racking test representative of seismic Level 3 did not cause any damage to the glass panes and no fallout of any component occurred. Failure of the screws located in the transoms and used for panel alignments occurred.

Test results listed above are mainly provided with focus on behavior of the structural joints; please refer to Galli [9] for complete information about the tests and their results. Control transducers were properly applied to measure the vertical and horizontal displacements of glass and frame in each test phase. The maximum differential displacements recorded during each racking phase are here used as inputs to calculate the joint deformation produced by the inter-storey drifts. Figure 6-1 summarizes the results obtained in each racking phase (and seismic level) with specific focus on panels bonded by structural silicone Sikasil® SG-500. The results show that a preliminary design based on the deformation limits set for the adhesive could ensure the resistance of the joint to the seismic inter-storey drifts specified and compliance with the performance requirements set for the façade elements.

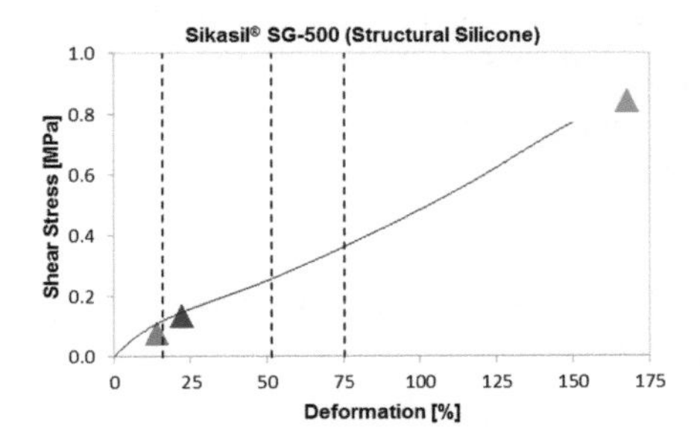

Figure 6-1 Mock up test results at the different performance levels.

7 Conclusions

Compared to capped systems, SSG systems in unitized curtain walling can provide an effective solution to minimize damages due to earthquake. Although this is well recognized, no standard currently provides clear criteria for the design of SSG joints subjected to seismic impacts. Depending on adhesive properties and joint dimensions, performance requirements associated to different damage levels can be satisfied. In line with the performance-based engineering approach proposed by JASS 14, a concept to design SSG joints affected by seismic forces and displacement due to inter-storey drifts is proposed in this paper. Three performance levels associated to different design requirements are defined, with the final intent of balancing costs and risks with no compromise on safety and of not affecting the appearance of the façade for a unique and extreme event. The concept is based on results obtained from small-scale tests on sealant H-specimens and it is validated by full-scale racking tests performed on mock up panels.

8 References

[1] Beher, R.A.: Architectural Glass to Resist Seismic and Extreme Climatic Events, Woodhead Publishing Limited and CRC Press LLC, 2009.

[2] EN1998-1, Eurocode 8: Design of Structures for Earthquake Resistance – Part 1: General Rules, Seismic Actions and Rules for Buildings, 2004.

[3] ASCE 7-10, Minimum Design Loads for Buildings and Other Structures, 2010.

[4] AAMA 501.6-09, Recommended Dynamic Test Method for Determining the Seismic Drift Causing Glass Fallout from a Wall System, 2009.

[5] AAMA 501.4-09, Recommended Static Testing Method for Evaluating Curtain Walling and Storefront Systems Subjected to Seismic and Wind Induced InterStorey Drift, 2009.

[6] JASS 14, Japanese Architectural Standard Specification – Curtain Wall, AIJ, 1996.

[7] Nardini, V.; Doebbel F.: Performance-Based Concept for Design of Structural Silicone Joints in Façades Exposed to Earthquake, Challenging Glass 5, Conference on Architectural and Structural Applications of Glass, Ghent, 2016.

[8] Memari, A.M.; Shirazi, A.: Development of a Seismic Rating System for Architectural Glass in Existing Curtain Walls, Storefronts and Windows, 13th World Conference on Earthquake Engineering, Vancouver, Canada, 2004.

[9] Galli, U: Seismic Behavior of Curtain Wall Facades – A Comparison Between Experimental Mock Up Test and Finite Element Method Analysis, Politecnico di Milano, 2011.

[10] ASTM C 1401-09, Standard Guide for Structural Sealant Glazing, 2009.

[11] EOTA ETA002-1, Structural Sealant Glazing Systems – Part 1, 2012.

[12] ASTM C 719, Standard Test Method for Adhesion and Cohesion of Elastomeric Joint Sealants Under Cyclic Movements (Hockman Cycles), 1998.

[13] EN 12153, Curtain Walling – Air Permeability – Test Method, 2000.

Dimensioning of spherical bent insulated glazing units for the Kazakhstan – Pavilion, Expo 2017

Lisa Heinze[1]*, Mascha Baitinger*[2]*, Christian Wolkowicz*[3]

1 Verrotec GmbH, Lerchenstraße 28a, 22767 Hamburg

2 Verrotec GmbH, Romano-Guardini-Platz 1, 55116 Mainz

3 Lindner Steel & Glass, Lange Länge 5, 97337 Dettelbach

As part of the Expo 2017 in Kazakhstan, the country's pavilion, called *The Sphere*, will be built at the center of the Expo area. The pavilion will be a steel glass structure in shape of a sphere with a diameter of about 80 m. Due to its shape, spherical bent, trapezoidal insulated glazing units (IGUs) will be used. In order to satisfy the architectural requirements, the IGUs will be used as *Structural Glazing Façade* with mechanical fasteners (toggles) in the outer edge sealing. The challenge of the glass dimensioning was to define a suitable design concept, which on one hand considered the complex boundary conditions in a sufficient way and on the other hand lead to an economic and statically realizable solution.

Keywords: spherical bent insulated glazing, structural glazing, climatic loads, outer edge sealing

1 Introduction

For the first time, since the first World's Fair in London 1851, the EXPO will take place in Kazakhstan's capital Astana in 2017. Entitled "Future Energy", the EXPO addresses energetic supply as well as ecological handling of energy in both production and use. The Setup of the EXPO area already indicates the title: The area has the shape of a drop of water; the pavilions are arranged similar to a wind turbine and at the center, the Kazakhstan-Pavilion towers over the area (Figure 1-1).

Engineered Transparency 2016. Glass in Architecture and Structural Engineering. First Edition.
Edited by Jens Schneider, Bernhard Weller.

Figure 1-1 View on the sphere (© Adrian Smith + Gordon Gill Architecture [Design Architecs]).

Lindner Steel & Glass tasked the engineering office Verrotec GmbH with the static proof of the spherically bent IGUs considering the given boundary conditions.

The challenge in the case at hand was, that a normative analysis usually with conservative assumptions, regarding both (climatic) loads and calculational boundary conditions (hinged support, rigid outer edge sealing), did not lead to economically acceptable outcomes and, accordingly, more precise methods of calculation had to be, project-based, developed.

Architectural requirements had to be considered as well as manufacturing and mounting limitations (glass and outer edge sealing, installation).

2 Design

2.1 Supporting structure

The Sphere is a self-supporting steel-glass sphere with an outer diameter of about 80 m. The main supporting structure of the Sphere is a steel skeleton, which can be subdivided into primary and secondary supporting structures. Based on the terrestrial globe, the structure is segmented into longitude and latitude. The supporting structure and glass plate's geometries are rotationally symmetrical around the vertical axis of The Sphere. The main supporting structure consists of 20 identical longitudes so that statical analysis of the spherical bent insulated glazing can be reduced to a minimal amount of relevant unit positions (Figure 2-1).

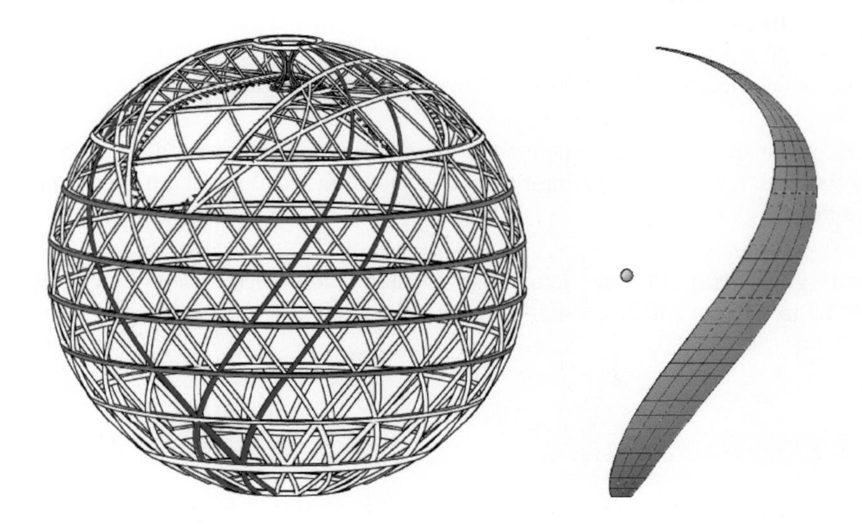

Figure 2-1 Primary structure with longitudes (blue) and latitudes (red), © Werner Sobek Stuttgart (left); analyzed longitude, © Lindner Steel & Glass (right).

Due to the geometry depicted in Figure 2-1, parallelogram-shaped supporting steel framework elements emerge. A parallelogram normally comprises sixteen glazing elements (Figure 2-2) which are, analog to the spherical shape, biaxial bent. The two top and bottom parallelograms, next to the 'poles', comprise eight IGUs each.

Figure 2-2 Segment of the parallelogram with secondary support and glazing (© Lindner Steel & Glass).

2.2 Glazing

The design of the glazing is the same in all positions (Figure 2-3, left)

Top layer:	106.4 laminated glass made of heat strengthened glass
	20 mm internal gap
Bottom layer:	88.4 laminated glass made of heat strengthened glass

The strengths of the spherically bent laminated glass – made of tempered glass – were provided by the Italian glass manufacturer Sunglass (bending tensile strength according to heat strengthened glass (TVG))

The relevant geometries of the laminated glazing have been determined as b_{max} x h_{max} ≈ 3.2 m x 2.7 m and b_{min} x h_{min}≈ 1.33 m x 3.9 m (Figure 2-3, right).

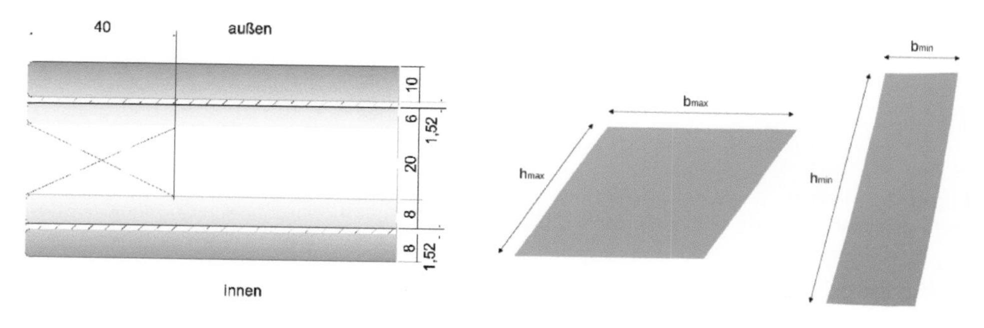

Figure 2-3 Glass assembly with enamelling (left), dimensions of the relevant plates (right).

2.3 Support

The glazing is a Structural Glazing Façade whose top layer is connected to the bottom layer only through the outer edge sealing and therefore with the substructure. This applies to both the northern hemisphere and the southern hemisphere.

Edgetech Europe GmbH's Super Spacer TriSeal Premium Plus will be used in combination with Dow Cornings's silicone sealant DC 3362. Because of the statical assumptions, the outer edge sealing's width could be reduced to an, architecturally speaking, appealing size. The bottom layer is connected to the substructure at regular intervals by use of punctual glass fixings (toggles). The toggle fits an aluminum u-profile which is integrated into the outer edge sealing by width of 19 mm (Figure 2-4).

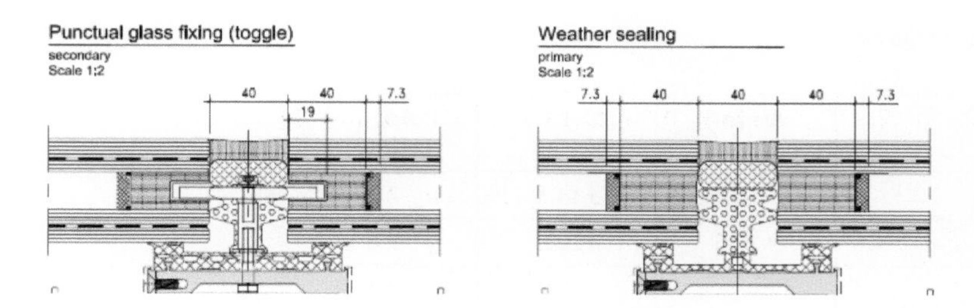

Figure 2-4 Detailed outer edge sealing with toggle (left), without toggle (right) (© Lindner Steel & Glass).

The toggles normally are mounted at 150 mm from the edges and less than 400 mm from each other. Additionally, glazing supports are used for the dead load (Figure 2-5).

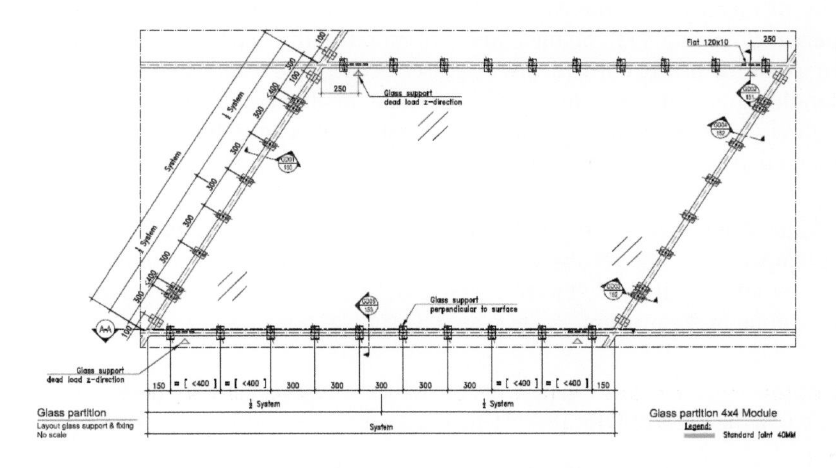

Figure 2-5 Position of the toggles und glazing blocks (© Lindner Steel & Glass).

3 Loads

As part of the structural analysis, loads like dead loads, climatic loads (wind loads, snow loads, thermal loads, etc.), live loads (only in the areas surrounded by the northern hemisphere) and, moreover, loads by deformation of the substructure have to be taken into consideration. Particularly in terms of climatic loads, formulated by structural analysis, boundary case calculations have to be done through which possible vacancies will also be considered ('in use' and 'not in use').

Since both manufacturing and installation locations are known, the actual difference in altitude ΔH can be formulated for internal pressure calculations (Table 3-1).

Table 3-1 Climatic load factors.

Condition	**ΔH [m]**	**ΔT [°C]**	**Δp_{met} [kN/m²]**
Summer ‚in use'	+ 316	+14.1	-2.0
Summer ‚not in use'	+ 316	+19.9	-2.0
Winter ‚in use'	-	-30.0	+4.0
Winter ‚not in use'	-	-61.1	+4.0

4 Dimensioning concept

Conventional calculation of the IGUs with hinged supported edges is, in this case, to be avoided in order to increase economical and technical realizability. As is well known from various publications (e.g. [1]), on the corners of monoaxially bent IGUs, because of the rigid hindrance of the lifting corners, peak stresses occur. This effect amplifies when trapezoidal / parallelogram-shaped spherical, thus biaxial, bent glass plates are used. Because the Structural Glazing's material is known, a realistic, yet safe, formulation of rigidity in the outer edge sealing can be considered.

As part of M. Minasyan's matser's thesis [2], rigidities of similar products were determined at room temperature. The results were used to safely estimate the anticipated rigidities of the outer edge sealing. To cover climatic and structural loads, boundary case calculations were done in the process.

Dimensioning of the Sphere's IGUs was, with the consent of everyone involved in the project, conducted on the basis of DIN 18008 [3]. The loads were combined in accordance with Eurocode.

5 Structural Analysis

5.1 Determination of loads on the glazing

Structural analysis of the glazing was done applying the finite element method (FEM) by use of the program Dlubal RFEM 5.05. With the help of the module RF-Glas, glazing can be, considering climatic loads, statically determined. Here, in standard cases occurring climatic loads are, with regard to geometry and rigidity of the glazing, determined program internally. In special cases, e.g. for twice-bent IGUs, loads in the interlayer cannot be determined without further intermediate calculation steps. Thus, climatic loads cannot be determined using RF-Glas. Therefore, to determine the climatic loads, the actual rigidity of the biaxial bent glazing had to be taken into account. The climatic loads result from possible changes in volume of the glazing. Curvature, setup of layers and rigidity of the outer edge sealing plays a crucial role here. The smaller/more rigid the IGU is the higher the climatic loads are; which makes it necessary to determine also climatic loads with unfavorable sealing effects. The outer edge sealing's rigidity was modeled as a spring support. By considering these factors, a climatic load, working as surface load, could be determined. Through extensive comparative calculations, it was made sure that this approach is safe.

5.2 Glazing

The relevant glazing positions were examined along the selected longitude (Figure 5-1). Consequently, a certain position in sector (level) 8 turned out to be of great significance, which will be further examined below.

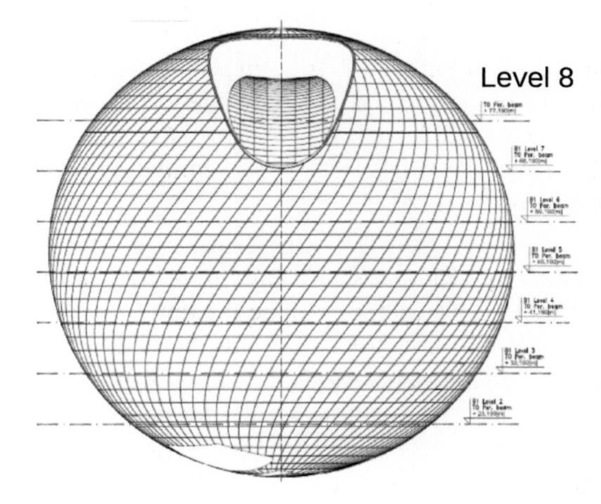

Figure 5-1 Overview of the Sphere's glazing (© Lindner Steel & Glass).

Structural dimensioning of the IGUs was conducted in consideration of the individual supports. For the outer layer the appropriate outer edge sealing's rigidity was considered. Support of the bottom layer, in the direction of wind suction, was done using toggles; in the direction of wind pressure using linear support. The sealing effect of the PVB-Foil was considered where it was relevant for dimensioning (mV). By having identified the environmental load as a surface load, the layers of the IGUs could be treated as isolated systems, once again under consideration of the appropriate factors of the sealing effect as well as the outer edge sealing's rigidity.

Figure 5-2 to 5-5 exemplarily illustrate stresses in the outer layer for the loading case: wind suction and climatic load: winter 'not in use'. The effect of the climatic loads on the outer edge sealing's rigidity is particularly well noticeable.

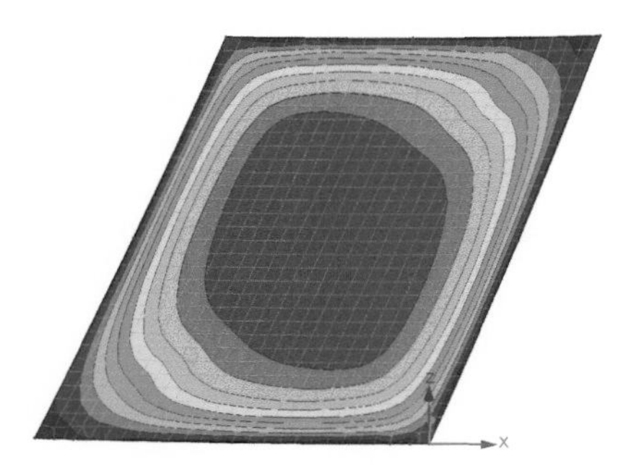

Figure 5-2 Principal tensile stress caused by wind suction, σ_1 = 7.75 N/mm², RVmax mV.

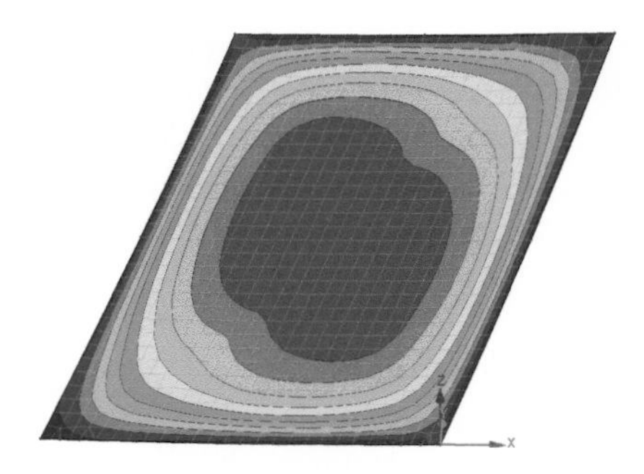

Figure 5-3 Principal tensile stress caused by wind suction, σ_1 = 8.45 N/mm², RVmin mV.

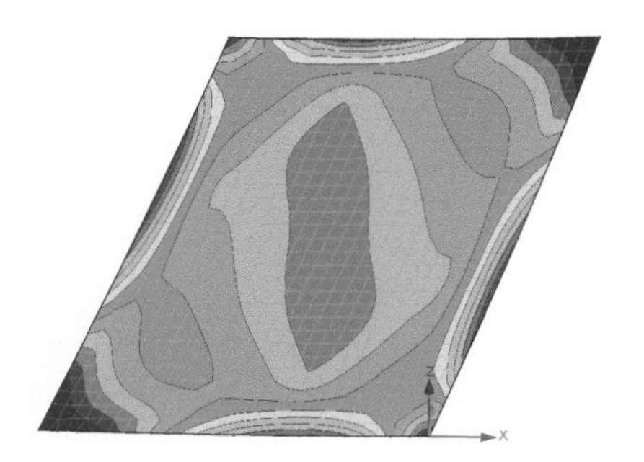

Figure 5-4 Principal tensile stress caused by climatic load case winter 'not in use', σ_1 = 8.45 N/mm², RVmin mV.

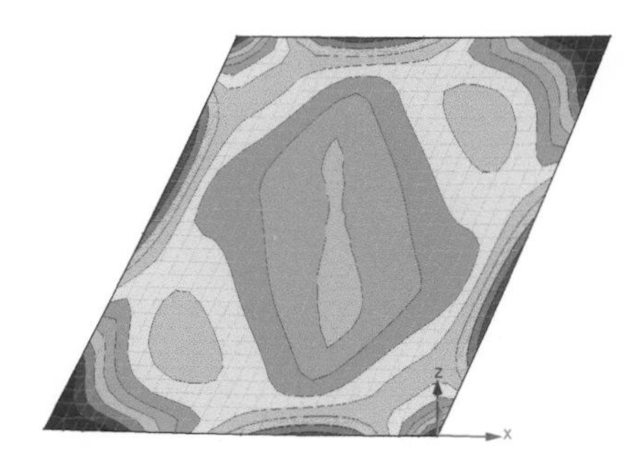

Figure 5-5 Principal tensile stress caused by climatic load case winter 'not in use', σ_1 = 17.64 N/mm², RVmin mV.

In order to analyze the bottom layer, stresses were overlaid conservatively. The relevant principal tensile stress occurs once again, considering loads by normal use, at maximum calculative outer edge sealing rigidity. Figure 5-6 illustrates the effect the toggle has on the glazing.

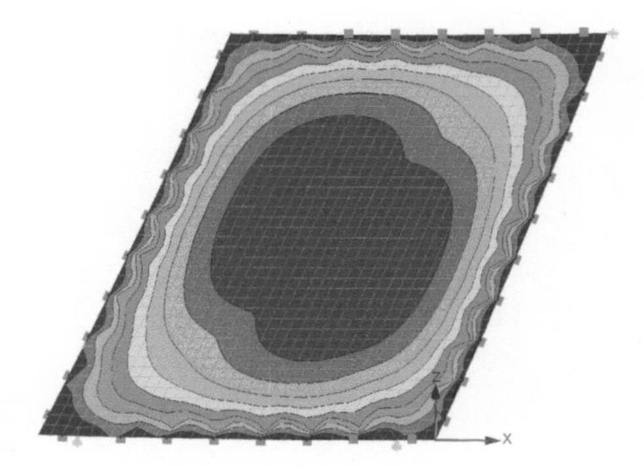

Figure 5-6 Tensions inside the bottom layer caused by wind suction, σ_1 = 7.78 N/mm², RVmax mV.

5.3 Outer edge sealing

Structural glazing requires, in addition to static proof of the IGUs, static proof of the outer edge sealing.

The geometry of the glazing was again the challenging part in dimensioning of the outer edge sealing. The dimensioning concept according to ETAG 002 [4] is valid for rectangular, flat glazing and adds a safety factor on the material resistance side of $\gamma = 6$ and at the same time uses simplified calculations of the load distribution on the sealed joints. This calculative approach does not consider those loads determined by exact calculation methods as would be expected with spherical bent glazing. Here, ETAG 002 was not applicable. Consequently, more exact calculation and proof methods, considering peak stresses in the corners caused by both short-term and long-term effects (dead load), have to be applied.

The relevant stress in the IGU's corner results from wind pressure as well as climatic load: Winter 'not in use' (Figure 5-7).

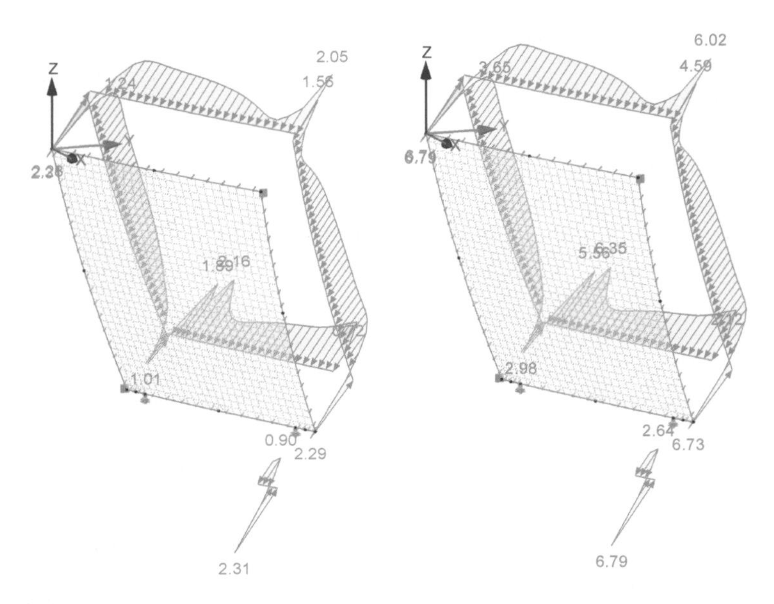

Figure 5-7 Stress (kN/m) in the outer edge sealing, left: caused by wind pressure, right: caused by climatic load: Winter ‚not in use' RVmax mV.

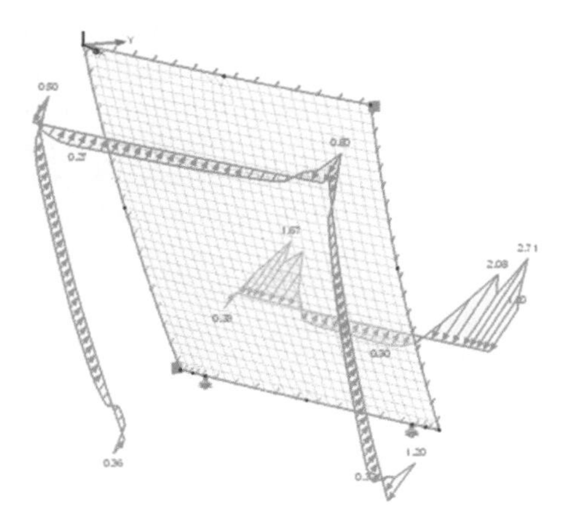

Figure 5-8 Stress (kN/m) in the outer edge sealing caused by dead load RVmax mV.

6 Remaining load-bearing-capacity

The entire analysis of the glass construction was done following European or German standards. According to these, sloping glazing within 10° from vertical are considered overhead glazing. This situation applies to the northern hemisphere for individuals inside;

the southern hemisphere for individuals outside. Since, according to DIN 18008, we do not have a regulated overhead glazing, structural component tests – for remaining load-bearing-capacity – were necessary.

Three of the, in terms of surface area, largest samples in horizontal positions were tested (Figure 5-7). Each sample, after destruction of all four layers, had to withstand dead weight (0.8 kN/m²) for 24 h. Subsequently, additional loads of 0.5 kN/m² were applied for another 24 h. Remaining load-bearing-capacity was easily proven.

Figure 6-1 Fracture pattern of the tested glazing position (© Lindner Glass & Steel).

7 Summary

Dimensioning of spherically bent insulated glazing units (IGU) is based on current normative regulations not realizable.

Such an analysis requires the engineer to have a lot of professional competence because realistic, thorough and statically useful boundary conditions must be formulated. Only with a high level of expertise is dimensioning of spherically bent glazing realizable. In this case, an iterative calculation of the system is unavoidable.

Now it is time to thank everyone involved in this project for his or her terrific cooperation and teamwork without which the realization of the spherical bent IGUs, for the construction project at hand, would not have been possible.

8 References

[1] Neugebauer, J.: The Influence of the Edge Sealing in Curved Insulated Glass, Challenging Glass 2, TU Delft, May 2010.

[2] Minasyan, M.: Randverbundbeanspruchung gebogener Zweischeiben – Isoliergläser mit Silikonschaum-Abstandhalter, S. 1-108, HCU Hamburg, Master-Thesis.

[3] DIN 18008 Teil 1 - 2, Glas im Bauwesen – Bemessungs- und Konstruktionsregeln.

[4] ETAG 002 – Teil 1: Gestützte und ungestützte Systeme, Leitlinie für die Europäische Technische Zulassung für Geklebte Glaskonstruktionen (Structural Sealant Glazing Systems – SSGS).

Adhesive Joints in Photobiogenerators – Preliminary Studies on Adhesive Materials

Bernhard Weller[1], Elisabeth Aßmus[1], Martin Kerner[2]

1 Technische Universität Dresden, Institute of Building Construction, George-Bähr-Straße 1, 01069 Dresden, Germany, elisabeth.assmus@tu-dresden.de

2 SSC Strategic Science Consult GmbH, Beim Alten Gaswerk 5, 22761 Hamburg, Germany, m.kerner@ssc-hamburg.de

Following on from the BIQ house in Hamburg [15], worldwide the first building featuring photobioreactors as external facade elements for the cultivation of micro-algae, the research project FABIG investigates the further development of integrating photobioreatcors into the primary skin of the building. Next to Arup Deutschland GmbH and SSC that have co-developed the SolarLeaf facade system with Colt International, the project team entails the Institute of Building Construction of the TU Dresden, Frener & Reifer and ADCO Technik Teams with various scientific and technical backgrounds contribute their expertise in glass, adhesives, façades, biotechnology and technical building services to design a building-integrated energy supply. A glass module containing liquid algae medium represents the core of the photobioreactor. The module consists of 3.0 m high glass panels joined by structural adhesives. In comparison with conventional glass constructions, the adhesives in photobioreactors are subject to higher mechanical, physical and chemical loads. Taking the additional loads into consideration, the authors characterize environmental conditions and present preliminary investigations on the aging behaviour of different types of adhesives. The paper describes the strategy, realization and results of preliminary investigations and draws conclusions on the selection of adhesives in photobioreactors.

Keywords: adhesives, photobioreactor, material investigation

1 Introduction

The research project FABIG develops flat panel photobiogenerators with load-bearing adhesives. The Technische Universität (TU) Dresden is part of the joint team and in charge of the glazing construction. The work package includes the design of the bondline of the glass modules as well as the investigation of suitable adhesives.

The starting point for the joint research team is to adapt the construction of an insulated glass unit for the requirements of a photobiogenerator unit. Insulated glass units are commonly known: Two glass panes confine a hermetically sealed glazing cavity that contains noble gas or dried air. The system is multiplicable up to two, three or even more air spaces. Regarding photobiogenerators the gas is replaced by a liquid photoactive microalgae medium. As a result an inner cavity containing self-reproducing biomass is created. The liquid loading has to be transferred by the inner and edge adhesive bondline. Since no

Engineered Transparency 2016. Glass in Architecture and Structural Engineering. First Edition.
Edited by Jens Schneider, Bernhard Weller.

information about structural adhesives inside photobioreactors were available, preliminary studies were executed. This article briefly introduces microalgae and photobiogenerator facilities. The main section presents the concept and results of the preliminary studies comprising tensile tests on type 1A specimens DIN EN ISO 527-2 [1] as well as the toxicity testing on raw material. Finally, the paper concludes with an evaluation of the suitability of the investigated adhesives for photobioreactors.

2 Microalgae and Photobioreactor Facilities

Microalgae are highly efficient producers of biomass. The microalgae biomass is a valuable good utilized and processed by both the pharmaceutical and cosmetics industries ([2] to [6]). The basic principle are photoactive organisms performing photosynthesis. To ensure optimal growth, microalgae require high quantities of solar energy, sufficient carbon dioxide and specific nutrients – mainly nitrogen and phosphorous. [7]

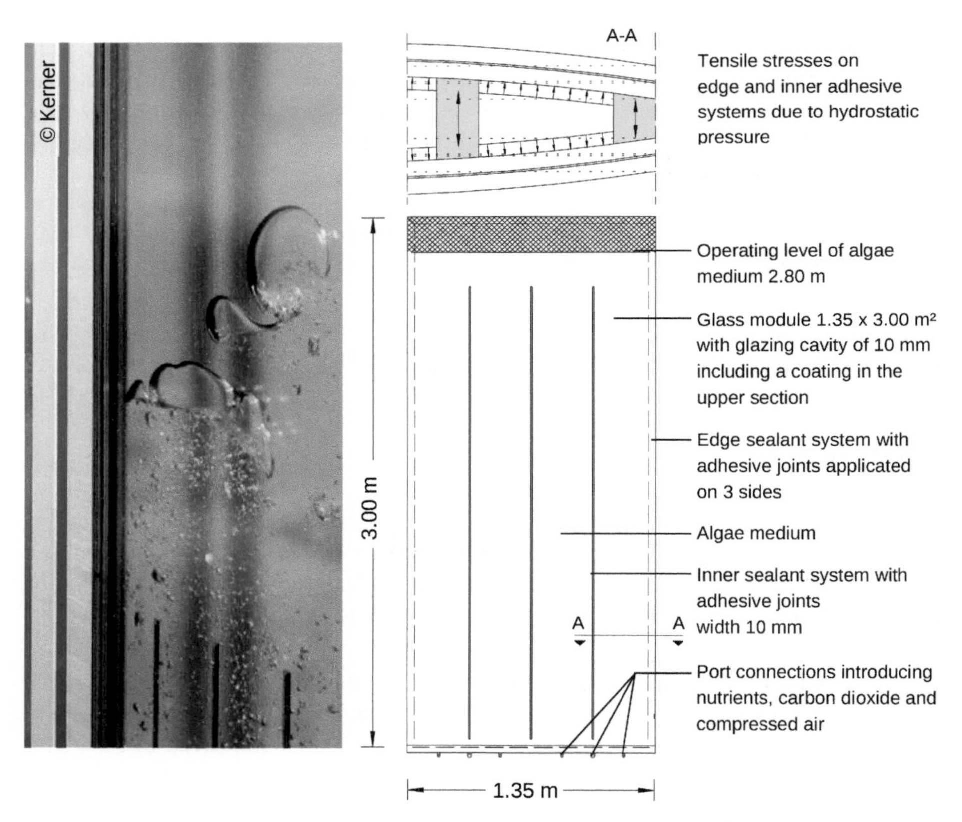

Figure 2-1 Clamped photobiogenerator panel in Hamburg.

Figure 2-2 Prototype of a photobiogenerator with inner and edge structural adhesive joints.

There are a couple of existing photobiogenerators raising and processing microalgae for commercial use. The facilities vary in composition, size and operating mode. In general, two types of facilities can be distinguished: open and closed systems. An example for open systems are "raceway ponds" stirred by paddle wheels. Open systems interact with the environment, whereas closed systems prevent the medium from any mass transfer. Examples for closed systems are pipe reactors and flat panel reactors [8]. The team of FABIG decided to further investigate closed systems. The team's new approach is the design of a flat panel reactor as a facade-integrated system. The resulting biomass can either be sold to industrial sectors, used for heating or warm water supply.

The reactor consists of two glass panels defining a glazing cavity of 10 mm (figure 2-2). On the one hand the transparent skin protects the algae medium from external impact and on the other hand it enables the transmission of sunlight. The glass modules are bonded along the edges and the inner bridge via load-bearing adhesives. The bondline will be exposed to UV-radiation. Furthermore, the reactor features various port connections for nutrients, carbon dioxide and compressed air.

A pilot project called "BIQ – Das Algenhaus" featuring flat panel biogenerators was built in Hamburg in 2013 [9]. This residential building has a secondary external shading system made of photoactive panels (figure 2-1). The modules are clamped by metal edge profiles. Currently, FABIG seeks to substitute the existing clam profiles with an attractive and appropriate adhesive bonding system. [10]

3 Test Methods

The preliminary study examines the effects of algae on the adhesive's properties. This chapter presents appropriate adhesives, describes the conditions inside the full-size reactor and derives ageing scenario for small-scale specimens.

Table 3-1 Adhesive systems selected for preliminary studies (left) and specimens for testing (right).

Code	Name	Chemical basis	Manufacturer
EP01	Araldite AW 4858	Epoxy resin 2-C	Huntsman
EP02	Araldite 2015	Epoxy resin 2-C	Huntsman
EP03	DP 490	mod. Epoxy resin 2-C	3M
EP04	DP 460	mod. Epoxy resin 2-C	3M
SI01	Ködisil HAC-A	Silicone 1-C	Kömmerling
SI02	Sikasil AS-70	Silicone 1-C	Sika
SI03	Sikasil AS-785	Silicone 2-C	Sika
PU01	Araldite 2029-1	Polyurethane 2-C	Huntsman
PU02	DP 590	Polyurethane 1-C	3M
HY01	760	Hybride 1-C	3M

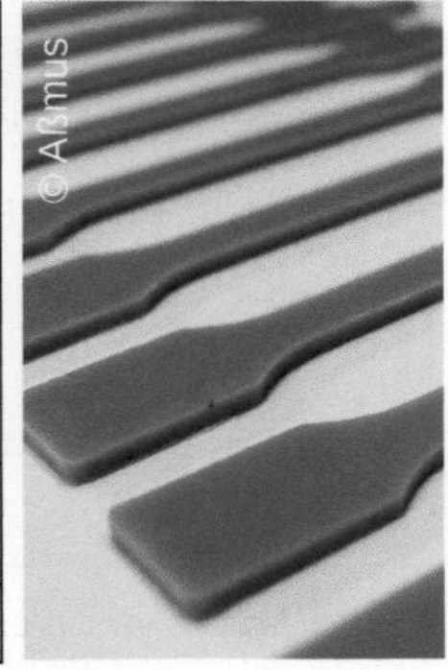

Ten different adhesive systems were preselected for investigation according to manufacturer's recommendations (table 3-1). The wide choice of products covers adhesives with epoxy, polyurethane, silicone and hybride bases. They all vary remarkably in maximum tensile strength, strain and Young's modulus. However, the chemical composition and mechanical properties have not been a selection criteria.

Specimens according to EN ISO 527-2 [1] provide the basis for uniaxial-tensile testing. The specimens are produced according to the manufacturer's instruction and cured at room temperature. All tensile tests are displacement-controlled with a cross head speed of 1 mm/min. The electromechanical testing system (Instron 5880) records applied forces, longitudinal and transverse strain via a video extensometer. Based on the results the Young's modulus and the maximum tensile strength is analysed. The Young's modulus is evaluated according to DIN EN ISO 527-1 [11], 10.3.2. All data are given as arithmetic mean values and standard deviation of the testing series. Since the algae grows within a temperature range from 25 to 35 °C, 35 °C is assumed as service temperature. Accordingly, all specimens are stored and tested at a temperature of 35 °C.

The typical conditions in a photobioreactor are assessed separately in three ageing scenarios and one reference series without ageing (figure 3-1). The specimens of the ageing series are embedded either in an acid, alkaline or hydrogen peroxide. ETAG [12] recommends a 21-day-ageing period. For preliminary studies the period was shortened to seven days. The specimens of the reference series are stored without ageing medium. For each scenario and adhesive product, five specimens are tested. No pre-testing conditioning at room temperature is performed.

The scenarios were developed according to the conditions inside the reactor: The biogenerator glazing cavity is filled up with algae medium. Moreover, the required nutrients, air and carbon dioxide are introduced constantly into reactor's inner cavity. Although the pH-value is kept at 7 the biological processes in microenvironment might create acid and alkaline conditions with a pH in the range from 5 to 12. Additionally, the generator is flooded with 3 %-hydrogen peroxide every few weeks to remove any organic coating.

Production of specimens				
Uniaxial tensile testing				Toxicity testing
Reference	**pH 05**	**pH 12**	**H_2O_2**	**Algae**
Storing in water	Storing in acid solution	Storing in alkaline solution	Storing in hydrogen peroxide solution	Storing in algae medium
7 days, 35°C	7 days, 35°C	7 days, 35°C	7 days, 35°C	2 days, 22°C
5 specimen per adhesive system	5 specimen per adhesive system	5 specimen per adhesive system	5 specimen per adhesive system	1 specimen per adhesive system

Figure 3-1 Concept of preliminary testing.

Furthermore, a toxicity test determines whether the adhesives have a harmful or even toxic effect on the algae organism. Adhesive specimens were stored in an algae medium *kirchneriella subcapitata* for 48 h at 22 °C and under light-exposure. The photosynthesis activity as a parameter for toxicity was studied by the method of Puls-Amplitude Modulation. This method detects the effective quantum yield of the microalgae photosystem II and thus gives information about the vitality of the microalgae. An effective quantum yield of below 0.2 causes irreversible cell damage [13]. The investigation was measured with a PAM-MAXI chlorophyllfluorometer.

4 Results and Discussion

The discussion focuses on the ageing behavior of the adhesives. The analysis includes the effect of different ageing conditions on the maximum tensile strength and the Young's modulus as well as the detection of toxicity towards algae. The unaged testing series serves as reference series ($E_{ref,}$ σ_{ref}). The aging behavior is rated as excellent, good, fair and poor in relation to the reference series (see caption figure 4-2 to 4-5).

Figure 4-1 illustrates the results of the tensile tests considering adhesive EP03 as an example. The diagram includes the stress-strain-behaviour of the reference specimens (ageing scenario: non) and the artificially aged specimens (ageing scenario: pH 05, pH 12 and hydrogen peroxide). The reference specimens have a Young's modulus of $E_{Ref} = (1203 \pm 73)$ N/mm² and a mean ultimate stress level of $\sigma_{ref} = (24 \pm 0.6)$ N/mm². Evidently, the average stress level decreases due to ageing conditioning. In particular the accelerated ageing with hydrogen peroxide decreases the maximum stress to (8 ± 0.51) N/mm² and the elastic properties to $0.25 E_{ref} < E_{age} < 0.75 E_{ref}$.

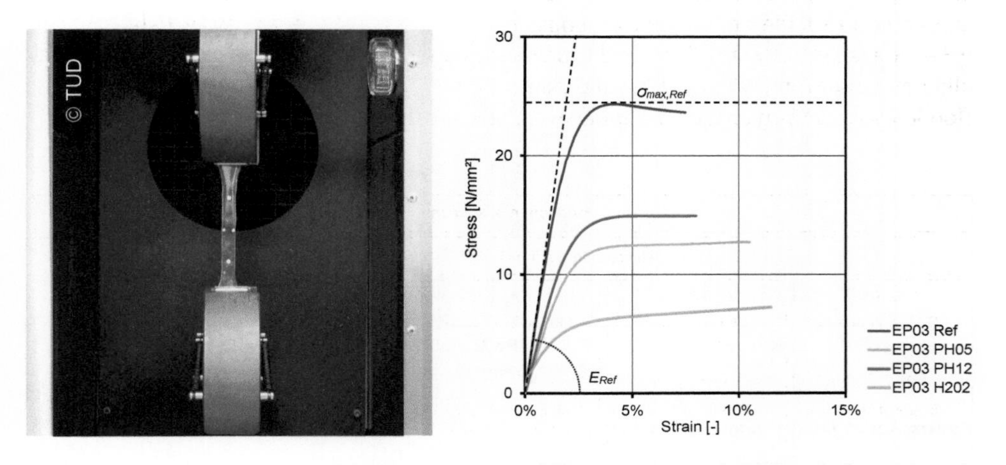

Figure 4-1 Influence of aging scenarios on the stress-strain behaviour of EP04.

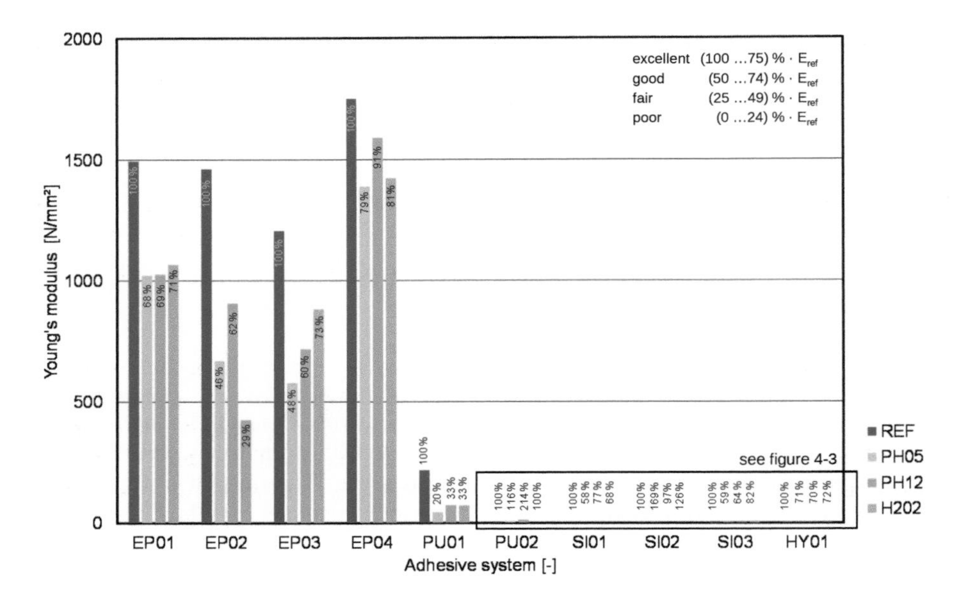

Figure 4-2 Young's modulus E ≤ 2000 N/mm².

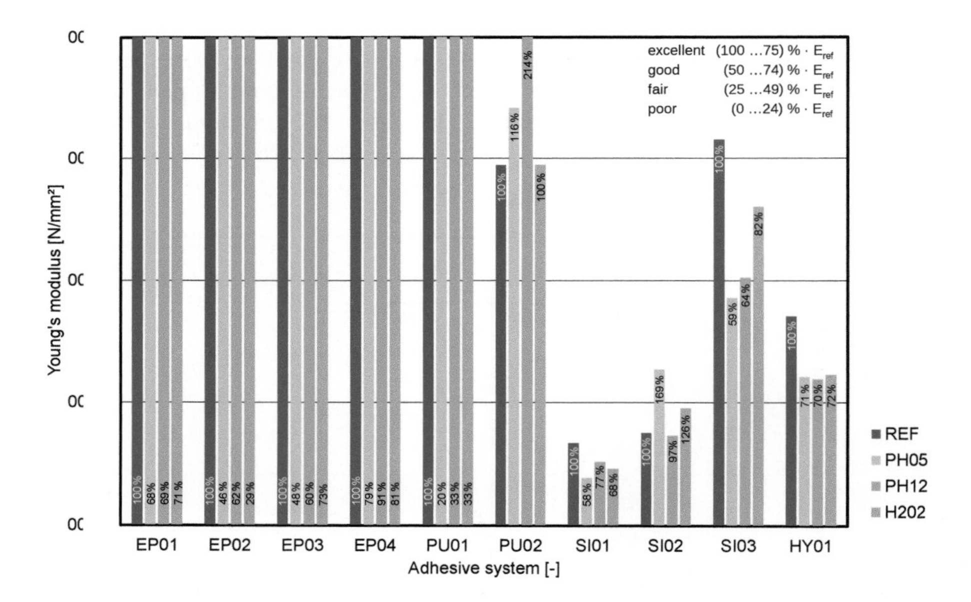

Figure 4-3 Young's modulus $E \leq 8$ N/mm².

Likewise EP03 all test sequences were evaluated. The following bar charts provide an overview of the tensile test results. Figures 4-2 and 4-3 indicate the mean Young's modulus and figure 4-4 the mean maximum stresses.

Due to different chemical basis the mechanical properties of the adhesives vary even without any ageing conditioning. The Young's modulus of unaged epoxies ranges from $E_{ref,EP} = (1200 \pm 59 \ldots 1750 \pm 104)$ N/mm². Polyurethanes remain below $E_{ref,PUR} < (250 \pm 20)$ N/mm². Silicones as well as hybride adhesives reach Young's modulus smaller than $E_{ref,EP} < (10 \pm 4.6)$ N/mm².

The analysis confirms the effects of the ageing on the elastic properties of the tested adhesives. EP04, PU02 and SI02 prove to have an excellent ageing behaviour rarely losing stiffness ($E_{age} > 0.75\,E_{ref}$). EP01, SI01, SI03 and HY01 show slight deviation in Young's modulus ($0.50\,E_{ref} < E_{age} < 0.75\,E_{ref}$). In contrast the vulnerability of the adhesives EP02, EP03 and PU01 against ageing becomes obvious. The Young's modulus drops down to $E_{age} < 0.50\,E_{ref}$.

Figure 4-4 presents the mean maximum tensile strength. Epoxies range from $\sigma_{ref,EP} = (23 \pm 0.3 \ldots 31 \pm 1.2)$ N/mm² and polyurethane specimens achieve maximum tensile strengths of $\sigma_{Ref,PUR} \leq (18 \pm 1.2)$ N/mm². The results of silicone and hybride

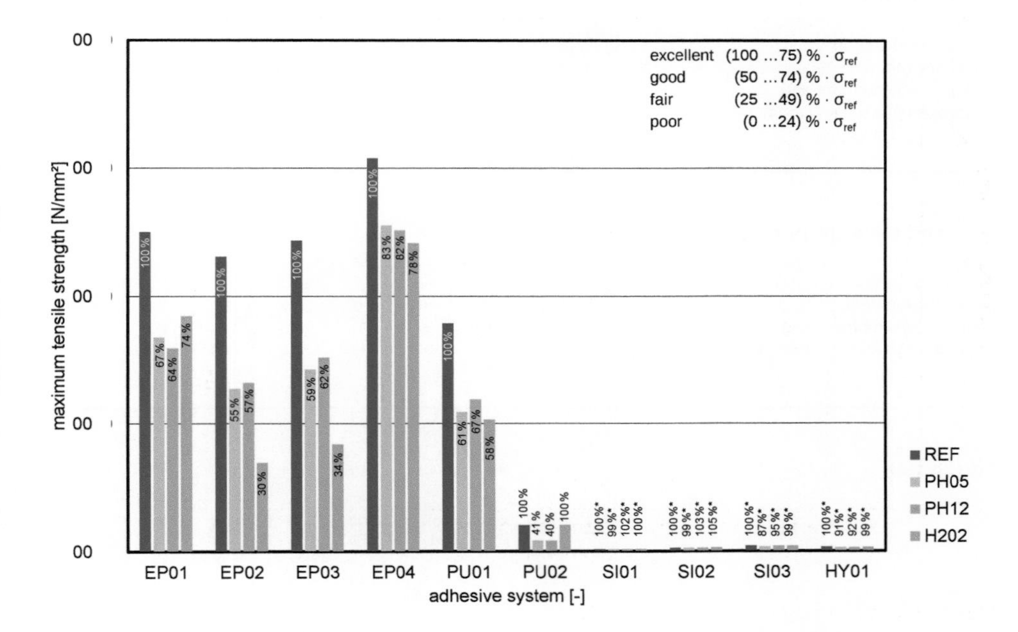

Figure 4-4 Maximum tensile strength.
*) testing execution limited to 50 % strain, no failure of specimens occurred

specimens are of limited relevance since they were not tested until failure. Instead, the testing was limited to 50 % strain due to time restrictions of the test set-up. With $\sigma_{50\%,SI,HY} < (10 \pm 0.06)$ N/mm², the tensile strength of silicone and hybride adhesives is considerably low.

The measured tensile strength was affected by the ageing, too. EP04 performed excellent after all ageing scenarios with $\sigma_{age,EP04} > 0.75\,\sigma_{ref,EP04}$. Likewise, EP01 and PU01 show good resistance to chemical attack ($0.50\,\sigma_{ref} < \sigma_{age} < 0.75\,\sigma_{ref}$). In contrast, EP02 and EP03 are vulnerable in contact with hydrogen peroxide ($0.25\,\sigma_{ref} < \sigma_{age} < 0.50\,\sigma_{ref}$). The tensile strength of PU02 decreases to $0.50\,\sigma_{ref} < \sigma_{age} < 0.75\,\sigma_{ref}$ due to chemical impact. Silicones and hybride adhesives achieve 100 % stress level at 50 % strain.

Figure 4-5 gives an overview of measured elastic and strength properties including a rating of ageing properties. The outcome proves evident effects on adhesives properties even with a shortened ageing period. The results confirm only partially the commonly assumed.

	EP01	EP02	EP03	EP04	PU01	PU02	SI01	SI02	SI03	HY01
Young's modulus [N/mm²]	**1495**	**1462**	**1203**	**1751**	**218**	**6**	**1**	**2**	**6**	**3**
Ageing behaviour in…										
- acid environment,	+	o	o	++	-	++	+	++	+	+
- alkaline environment and	+	+	+	++	o	++	++	++	+	+
- hydrogen peroxid environment	+	o	+	++	o	++	+	++	++	+
Max. tensile strength [N/mm²]	**25**	**23**	**24**	**31**	**18**	**2**	**0,2**	**0,3**	**0,4**	**0,3**
Ageing behaviour in…										
- acid environment,	+	+	+	++	+	o	n.s.	n.s.	n.s.	n.s.
- alkaline environment and	+	+	+	++	+	o	n.s.	n.s.	n.s.	n.s.
- hydrogen peroxid environment	+	o	o	++	+	++	n.s.	n.s.	n.s.	n.s.
Toxicity test	++	++	++	++	++	++	++	++	++	++

++	excellent	(100 …75) % · E_{ref}	or	(100 …75) % · σ_{ref}
+	good	(50 …74) % · E_{ref}	or	(50 …74) % · σ_{ref}
o	fair	(25 …49) % · E_{ref}	or	(25 …49) % · σ_{ref}
-	poor	(0 …24) % · E_{ref}	or	(0 …24) % · σ_{ref}
n.s.	not specified			

Figure 4-5 Rating of structural adhesives for application in a photobioreactor.

ageing resistance of silicones [14]. Epoxies and polyurethanes did not provide a unified image of properties. EP01, EP04 as well as SI01, SI02, SI03 and HY01 are considered suitable for an application in photobiogenerators.

The information in the adhesive's datasheet concerning E_{Ref} and σ_{Ref} were approved only partially. The deviation is due to the water exposure of all specimens including the reference series and the testing temperature of 35 °C.

The toxicity tests revealed that during the exposition of the microalgae species *kirchneriella subcapitata* to the different adhesives the photosynthetic activity did not change and remained high at about 0.7 similar to the control without adhesives. Hence it can be assumed that none of the investigated adhesives are toxic for microalgae.

5 Conclusions

The paper described the design and function of a photobiogenerator, characterized distinctive demands due to the algae filling and established a concept for preliminary studies of the adhesive systems. The evaluation of the preliminary studies included the elastic modulus, maximum tensile strength and toxicity towards algae.

Ten adhesive systems were investigated. Six out of of ten examined adhesive products were considered suitable for an application inside a biogenerator. The tests provided essential information on the ageing behaviour, mechanical properties and key characteristics. The results provide the basis for main investigations based on ETAG [12], experimental studies on large-scale elements and FE-modelling.

6 Acknowledgements

The authors gratefully acknowledge the financial support of the Federal Ministry for Economic Affairs and Energy (BMWi) of Germany. Adhesives were provided free of charge by the manufacturers mentioned in the paper.

7 References

[1] DIN EN ISO 527-2:2012-06: Plastics – Determination of tensile properties – Part 2: Test conditions for moulding and extrusion plastics. Deutsche Norm. Berlin: Beuth Verlag, 2012.

[2] Klein, B.; Walter, C.; Lange, H.; Buchholz, R.: Microalgae as natural sources for antioxidative compounds. In: J. Appl. Phycol. Vol. 24, 2012, pp 1133-1139.

[3] Pulz, O.; Gross, W.: Valuable products from biotechnology of microalgae. In: Appl. Microboil. Biotechnol. 65, Vol. 6, 2004, pp 635-648.

[4] Rechter, S.; König T.; Auerochs, S.; Thulke, S.; Walter, H.; Dörnenburg, H.; Walter, C.; Marschall, M.: In: Antiviral activity of Arthrospira-derived spirulan-like substances. Antiviral Res. Vol. 72, 2006, pp 197-206.

[5] Matsukawa, R.; Hotta, M.; Masuda, Y.; Chihara, M.; Karube, I.: Antioxidants from carbon dioxide fixing Chlorella sorokiniana. In: J. Appl. Phycol. Vol. 12, 2000, pp 263-267.

[6] Chu C.Y.; Liao, W.R.; Huang, R.; Lin, L.P.: Haemagglutinating and antibiotic activities of freshwater microalgae. In: World J. Microbiol. Biotechnol. Vol. 20, 2004, pp 817-825.

[7] Bley, Th.: Biotechnologische Energieumwandlung: Gegenwärtige Situation, Chancen und künftiger Forschungsbedarf. Berlin, Heidelberg: Springer-Verlag, 2009.

[8] Kaltschmitt, M.; Hartmann, H.; Hofbauer, H.: Energie aus Biomasse: Grundlagen, Techniken und Verfahren. Berlin, Heidelberg: Springer-Verlag, 2016.

[9] European Patent Nr. EP 2 228 432 A1: Bioreaktor und Verfahren zum Betrieb eines Bioreaktors. 08.03.2009.

[10] Wurm, J.: Die bio-adaptive Fassade, The bio-responsive facade. In: Detail Green. Vol. 1, 2013, pp 62-65.

[11] DIN EN ISO 527-1:2012-06: Plastics – Determination of tensile properties – Part 1: General principles. Deutsche Norm. Berlin: Beuth Verlag, 2012.

[12] ETAG 002-1: Guideline for european technical approval for structural sealant glazing systems (SSGS), Part 1: Supported and unsupported systems. Brussels: European Organisation for Technical Approvals, 1999.

[13] Glembin, P.; Kerner, M.; Smirnova, I.; Cloud point extraction of microalgae cultures. In: Sep. Purif. Technol. Vol. 103, 2013, pp 21–27.

[14] Habenicht, G.: Kleben – Grundlagen, Technologien, Anwendungen. Berlin, Heidelberg: Springer-Verlag, 2009.

[15] Hinterlüftete Fassadenkonstruktion aus Photobioreaktoren. Abschlussbericht Forschungsprojekt. Fraunhofer IRB: Stuttgart, 2013

Engineering Design of the Field of Rods using Adhesive Attachment of Glass with TSSA

Lawrence D. Carbary[1], Michael A. Ludvik[2]

1 Dow Corning Corporation, PO Box 994, Midland MI, 48686, USA, l.carbary@dowcorning.com

2 M. Ludvik Engineering PC, 55 Washington St, Ste 555, Brooklyn NY 11201, USA, mal@mludvik.com

Eric Owen Moss Architects has designed a glass cylinder roof for the 8511 Warner Drive project in Culver City, CA. Curved laminated glass tubes suspended from a structural frame provide a striking structure unequalled in structural glass design. Their pursuit of discovering methods that bring projects to public prominence is evident in this design of hanging tubes. The project uses nearly 200 curved laminated tubes (of various lengths) suspended from a structure where mechanical fastening of the glass tube by drilled holes and bolts is architecturally unappealing. The top of the tube is designed with conventional structural silicone and a "Yoke" inside the middle of the tube is attached with TSSA. The Yoke is designed to rest on a steel cable supporting the middle of the tube. The Yoke is bonded to curved steel 89 mm x 165 mm rectangular patches with TSSA. The curved fittings attached to the glass have to be machined to meet the glass that has a 305 mm radius. The engineering challenge is to use the TSSA to resist the rotational stresses between the two halves of the cylinder during wind and seismic events. This paper describes the engineering concept that allows this design to be feasible. What is also shared is the process of how such a complex design should be considered and evaluated. The project is on schedule to be finalized in 2017. The design team is in the process of executing the design through mockup and validation.

Keywords: structural bonding, glass bonding, hot bent glass

1 Introduction

Glass is used in modern architecture for building aesthetics, structural contribution, lighting, and tenant comfort. The glass panes may be fixed to the supporting structure by either linear or point bearings. The linear supports attach the glass pane to the substructure on two, three, or four sides. Point bearings can be classified into fixing clamps and point-fixed supports. The point-fixed supports are typically positioned in the vicinity of the corners, quarter, and/or midpoints of the glass pane and retain the glass pane either mechanically (metal bolts penetrating the glass) or adhesively as noted by Seibert and Hagl [1, 2]. Mechanical fixing of glass panes furnished with holes requires the designer to pay attention to the placement of the holes in order to meet the requirements of national standards. For instance, ASTM C 1048 Specification for Heat-Treated Flat Glass [3] specifies that the hole must be placed at a distance of at least 6.5 times the thickness of glass away from the corner. Furthermore, in order to deal with unavoidable stress concentrations around the fixing holes, heat strengthened or tempered glass must be used.

Engineered Transparency 2016. Glass in Architecture and Structural Engineering. First Edition.
Edited by Jens Schneider, Bernhard Weller.

Bonded point-fixed supports in contrast to mechanical point supports offer a number of advantages, such as no or less visibility from the exterior, a "smooth" transfer of the load into the glass pane (avoiding stress peaks), and the elimination of drilling holes into the glass as noted by Tasche, Weller and Hagl [4-7]. Contrary to the glass panes, the adhesive fixing used in either linear or point bearings may experience both out-of-plane and in-plane loads, depending on whether the dead load of the glazing element is carried by mechanical setting blocks into the building envelope substructure.

Structural silicone sealants have been used in linear adhesive fixing of glazing elements at a tertiary structural level since the 1960s as noted by Parise [8]. The room-temperature-vulcanizing (RTV) structural silicone sealants used in linear fixing of glazing elements (structural silicone glazing) display a low Young's modulus (generally in the range of about 1.0-2.5 MPa in tension) and a high elongation at break (generally in the range of >100 % when measured in tension on a tensile-adhesion joint with dimensions as defined in ISO 8339 [9]). The resulting joint design allows compensation of thermally induced movements and dimensional tolerances between the substrates, which is a necessity for linear structural bonded bearings. However, for adhesively bonded point-fixed bearings, Hagl reports [6] a higher Young's modulus is desirable to achieve higher stiffness with a smaller bonding area that still allows carrying significant out-of-plane loads.

TSSA is an optically clear structural silicone adhesive designed to be applied between glass and metal at 1mm thickness to provide flush smooth options compared to drilling and mechanically fastening glass. Physical properties of this material and engineering applications have been established and published by Sitte et al and Carbary et al. [10], [11]. The higher Young's modulus of 9.3 MPa of TSSA along with its 1mm thickness limits compensation of thermally induced movements. These publications providing physical properties and hyperelastic modeling properties allow the façade designer to engineer a truly unique transparent uninterrupted wall.

Flat glass facades using point support with TSSA have been fabricated using insulating glass units due to the energy efficiency, smooth facade appeal and uninterrupted glazing is as shown if Figure 1-1.

Figure 1-1 TSSA point supports on facade Feluy Belgium.

Recently interests has sparked in using TSSA on curved glass. The glass flower designed by Arup puts the TSSA in compression, Tension and shear to create this self-supporting structure that used bent glass as shown in Figure 1-2.

Figure 1-2 Self-supporting Glass Flower Sculpture using TSSA on Bent Glass.

The unique design of 8511 Warner, in Culver City California has taken this to the next step. The design uses hot bent annealed laminated glass in a 305 mm radius and machined attachments to match the radius. The architectural rendering is shown in Figure 1-3.

Figure 1-3 8511 Warner, Culver City California rendering.

2 Structural Concept

The typical trusses span 22 m (72 ft) between primary structure supports. The slumped cable with an approximate slump depth of 2.1 m (7 ft) works as a tension arch between supports. The top chord resists the cable tension, and the glass cylinders work as compression struts. Figure 2-1 highlights the schematic design concept.

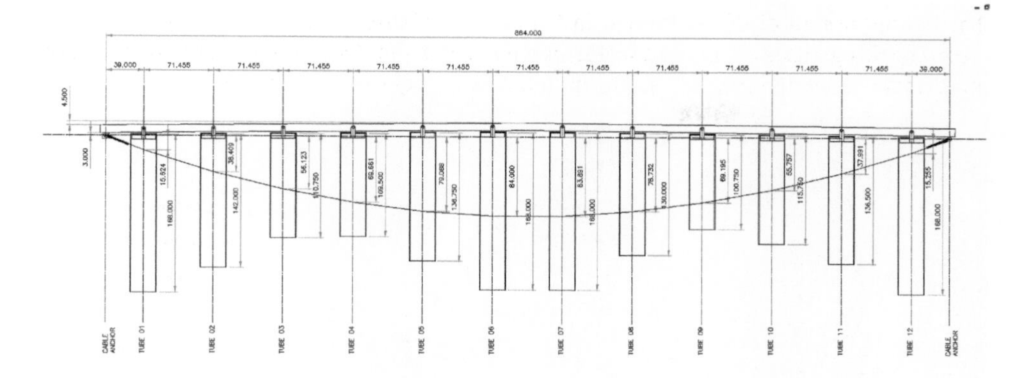

Figure 2-1 Design Concept of the Field of Rods.

The cylinders cantilever down from the top chord under wind load, and have a heavy structural silicone connection at the top to accommodate this. This is the most critical part of the structural engineering of the system. The cable is 18 mm diameter galvanized steel. A cable friction clamp will be provided at each cylinder which will be monolithic with the yoke. The yoke could be made from either galvanized steel or stainless steel. The yoke will connect to the TSSA buttons with an adjustable connection. This is shown in figure 2-2.

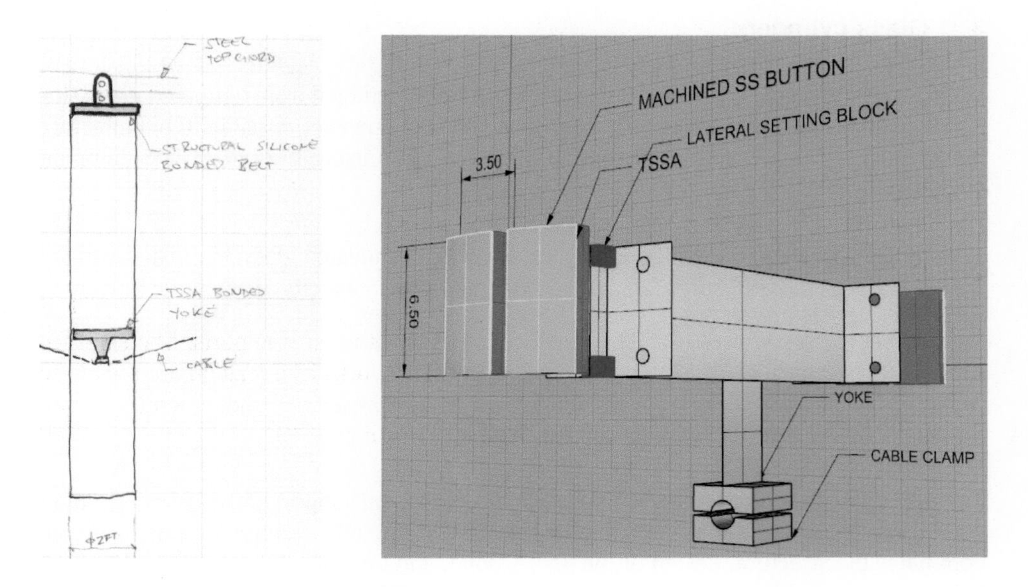

Figure 2-2 Cable and Yoke concept illustrated.

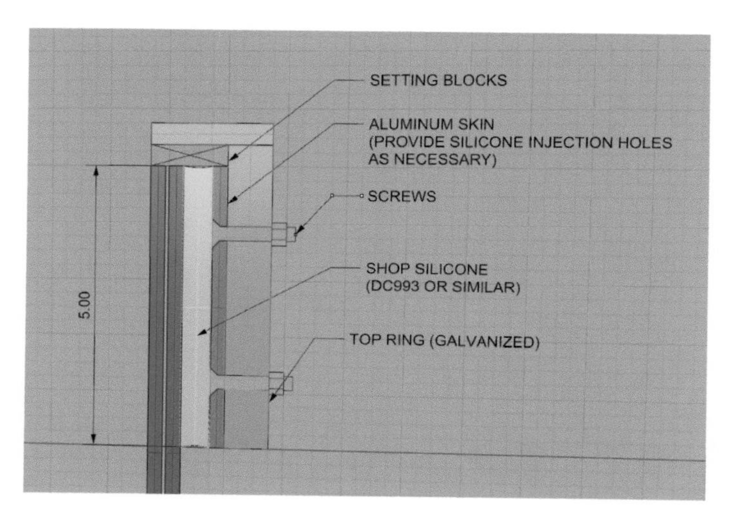

Figure 2-3 Structural Silicone concept attachment at top.

The outside diameter of the glass is set to 635 mm (25") to allow for a standard 556 mm (22") schedule 80 pipe as the top ring belt for a 12.5 mm (0.5") thick structural silicone bond. This concept is shown in figure 2-3.

3 Glass cylinders

Whilst it would be preferable for the cylinders to be constructed from tempered glass, this is not available within present industrial capacity. It is however possible to manufacture the 180 degree "clamshell" segments in annealed glass using traditional hot slumping methods.

The design concept glass will be 2 plies of 6 mm (¼") annealed glass, laminated with an SGP interlayer. Glass processors have confirmed their capacity to manufacture these items up to 4.2 m (14 ft) long.

3.1 Wind and Seismic Load Resistance

3.1.1 Load Perpendicular to the Truss

The primary structural challenge in this design is the lateral support of the cylinders under wind loads perpendicular to the line of the truss. As presently configured, this load case dominates the structural design of the TSSA bond, and the connection of the cylinder to the top chord. A 4.2 m (14 ft) cantilever must be developed under wind loads, which is achieved through a 125 mm (5") deep structural silicone belt to the top.

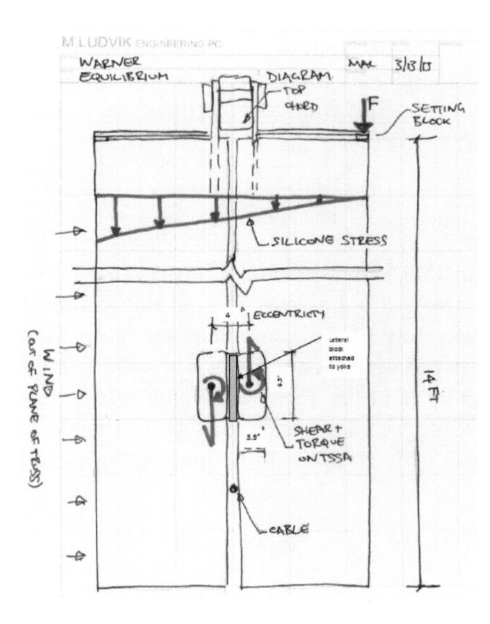

Figure 3-1 Composite action concept.

In order for the silicone belt to act as a monolithic 360 degree circle, forces must be transmitted through the TSSA bond. These are vertical (oriented to gravity) and because there is a distance between the two pads an in plane twist develops in the pads. This twist results in the majority of the stress. The weight of the cylinders is relatively secondary. The TSSA bond between the two cylinder halves is also important for the glass strength. Some composite action between the two halves is necessary in order to achieve the cantilever of 4.2 m (14 ft). This concept is shown in Figure 3-1. As a result of this heavy wind load resistance connectivity, the structure is much safer under dead loads than it would be otherwise.

3.2 TSSA

TSSA is a high strength and optically clear structural silicone product is intended for forming the connection between these glass and steel. The material has relatively high strength and low stiffness, which makes it at good application for glass because it will not transfer stress into the glass under relative thermal movements. Presently the TSSA "buttons" are configured with two 165 mm x 89 mm (6.5" x 3.5") attachments per half cylinder. Other configurations are possible, though will have a larger number of attachments. It was considered that this approach was preferable both aesthetically – less hardware and visual complexity – and from a fabrication perspective. Each attachment will require quality effort to place upon the glass and maintain alignment in the autoclave, so fewer attachments will be less work and less opportunity for mistakes. The attachments will be made from 316 stainless steel, and machined to the same radius as the glass. The TSSA will absorb tolerances between machined attachments and glass. The attachment will be mounted to the yoke with an adjustable connection. The connection will be rigid after being locked down. This was shown in figure 2-2.

3.2.1 Factor of Safety

The allowable shear and tension stresses on the TSSA of 1.3 MPa (190 psi) [10]. These correspond with a factor of safety of approximately 4.0, however the current technical manual states a minimum guaranteed factor of safety of 2.5 guaranteed by fabrication time destructive testing. It corresponds to a better than 95 % confidence level. The connections are generally designed to use less than 60 % of the TSSA allowable values, implying factor of safety against the guaranteed strength of 4.17, and 6.67 against the likely strength. This stress level was chosen so that any code compliant required testing would reach a factor of safety of 4.0.

4 Redundancy

The design is engineered with multiple layers of redundancy and they are addressed as follows:

- Failure of single TSSA bond. Each half cylinder will have two TSSA bonded plates. The second plate will be able to assume the dead load and offer a reduced lateral wind load resistance, allowing for repair in a normal maintenance mode.
- Failure of all TSSA bonds on a single cylinder. The glass will be able to hang from the structural silicone at the top chord. The cable will not kink at that cylinder so will straighten. The cable straightening will be relatively small (~0.1")
- Glass Failure One Ply or two Plies. An SGP interlayer is specified to provide for better post breakage performance. If cracked, the annealed glass will not shatter like tempered glass, and will carry compression across the crack. The SGP will continue to provide a tension connection and bending resistance can develop, allowing for repair in a normal maintenance mode

5 Summary

This design concept is proceeding to the mock up phase to validate the application of the TSSA placement and quality assurance process. Flat fittings attached with TSSA have been applied to flat glass in the typical designs as shown in figure 1-1. Flat fittings were applied to curved glass in the sculpture in figure 1-2 and also used on the Airborne Project described at a recent GlassCon Global conference by Carbary et al [11]. The Field of Rods project takes the next step of engineering design using TSSA with curved fittings onto tight radius curved glass. The engineering, mockup testing and quality control requirements are in place in all of these cases. The next mock up phase will include factory fabrication, adhesive confirmation through quality assurance, aesthetic evaluation and structural validation. As these steps are completed and validated, the project can be realized into production.

6 References

[1] Siebert, B., “Safety Aspects of Point-Fixed Glass Constructions,” Glass Performance Days, J. Vitkala, Ed., Glaston Finland Oy, Tampere, Finland, 2007, pp. 432-436.

[2] Hagl, A., “Punktuelles Kleben mit Silikonen,” Stahlbau, Vol. 77, No. 11, 2008, pp. 791-801.

[3] ASTM Standard C1048, 2004, “Standard Specification for Heat-Treated Flat Glass – Kind HS, Kind FT Coated and Uncoated Glass,” Annual Book of ASTM Standards, ASTM International, West Conshohocken, PA.

[4] Tasche, S., 2007, “Strahlungshärtende Acrylate im Konstruktiven Glasbau, Ph.D. Thesis, Technische Universität,” Dresden, Germany.

[5] Weller, B. and Tasche, S., “Experimental Evaluation of Ultraviolet and Visible Light Curing Acrylates for Use in Glass Structures,” Durability of Building and Construction Sealants and Adhesives, 3rd Volume, A.T. Wolf, Ed., ASTM International, West Conshohocken, PA, 2010, pp. 135–156.

[6] Hagl, A., “Bonded Point-Supports: Understanding Today – Optimizing for the Future,” Challenging Glass 2, Conference on Architectural and Structural Applications of Glass, F. Bos and C. Louter, Eds., Univ. of Technology, Faculty of Architecture, Delft, The Netherlands, 2010.

[7] Hagl, A., “Silicone Bonded Point Supports – Behaviour Under Cyclic Loading,” Engineered Transparency – International Conference at Glasstec, Düsseldorf, Germany, J. Schneider and B. Weller, Eds., Technical Univ. of Dresden, Dresden, Germany, 2010, pp. 139-148.

[8] Parise, C. J., Science and Technology of Glazing Systems, STP1054, ASTM International, West Conshohocken, PA, 1989.

[9] ISO Standard 8339, 2005, Building Construction – Sealants – Determination of Tensile Properties (Extension to Break), International Standardization Organization (ISO), Geneva.

[10] Sitte, S., Brasseur, M., Carbary, L., Wolf, A., Preliminary Evaluation of the Mechanical Properties and Durability of Transparent Structural Silicone Adhesive (TSSA) for Point Fixing in Glazing, Journal of ASTM International, Vol. 8, No. 10 Paper ID JAI104084.

[11] Carbary, L., Clift, C., Jeske, B., Zhong, F., “Airborne America San Diego, Transparency through Innovation and Engineering”, GlassCon Global Conference, Boston, USA, July 2016.

Connections in Glass

Bruno Kassnel-Henneberg[1]

1 Glas Trösch Euroholding AG & Co. KGaA, Pilsener Straße 9, D-86199 Augsburg, Germany

Enclosed envelopes without any visible supporting structure or stairs which seem to float on air are the dreams of many architects and designers. Glass is one of the few materials that can allow this dream to become a reality. For this reason full glass structures can appear to be practically invisible and the observer experiences the wonderful sensation of seeing a floating, weightless and totally transparent structure, as if it's almost not there. To fulfill these demands but also to enable the integrity of the whole glass structure, the connections between the elements represent one of the most critical aspects for this type of glass design. These connecting elements represent the "remaining visible" parts and therefore these details become the central focus of interest when looking at an all glass structure. Simple borehole connections often do not satisfy the architect's demands these days. In this article we will show a few of our recent all glass structures with a keen focus on the all important "connecting parts". There will be shown a solution with special glass treatment for a customized structure as well as applications for transparent glued connection elements which can be used by the industry in general product applications. The generalization of connecting details in the use of load carrying full glass structures could open new fields of application for full glass structures apart from the typical and well known façade industry.

Keywords: glass, metal to glass connections, laminated inserts, bonding

1 High loaded customized mechanical glass to metal connections

The glass canopy of the main entrance is located on the north side of the building 20 Fenchchurch Street in London. The free cantilever length of the glass roof measures 5.2 m. The cantilever glass fins have a distance of 3.0 m between each other and they are built up by 5 layers of fully tempered glass laminated with the popular SentryGlas, the glass thicknesses are as follows: 10-10-15-10-10. The roof glass has a maximum width of 3.2 m. The length of the glass is approx. 5.2 m. Each panel is freeform shaped. The laminated panels consist of a quadruple glass made of heat strengthened glass with the thicknesses of 8-12-12-12 (top to bottom) all laminated with SentryGlas. The ceramic frit is located at position 2 from top side. The front edge of the canopy is generated by a so called "fascia panel" with an identical build up to the roof glass.

Glas Trösch was commissioned by Josef Gartner GmbH to supply the glass for the canopy. In addition to this Glas Trösch was incorporated intensely into the design phase of the detailed connections. Beside the high architectonical demands the details had to be designed in a way so that they were able to fulfill the requested long term requirements in regards to load transfer and sliding ability.

Engineered Transparency 2016. Glass in Architecture and Structural Engineering. First Edition.
Edited by Jens Schneider, Bernhard Weller.

Figure 1-1 Total view of the glass canopy, Design by: Rafael Viñoly.

Part of the glass supply was the precise positioning and grouting of the detail for the connection points. Due to the early collaboration of all parties which were involved in the realization of the project it was possible to coordinate the planning in an optimized way, so that the production of the glass in combination with positioning and grouting of the details worked supremely well.

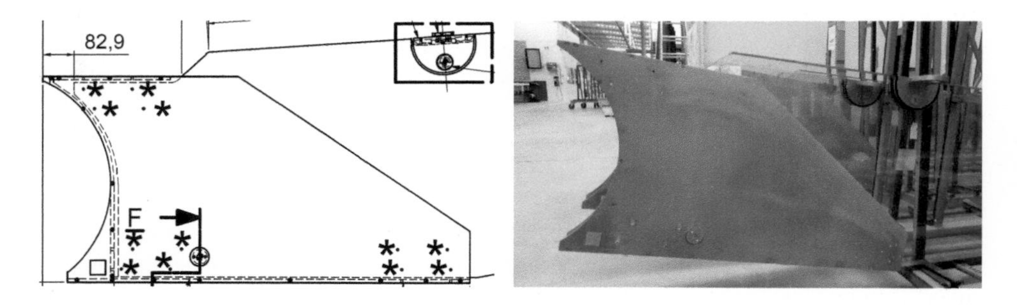

Figure 1-2 Connection of the cantilevering glass fin.

An essential point of the structure is the constraint connection of the glass fins to the main steel structure. Here the decision was taken to fix an "inner" steel shoe to the glass already in the glass production facility. Beside the protection of the glass this solution provides the execution of a tolerance free connection, which ensures that the enormous loads due to the constrained connection will be transferred exactly at the defined locations from the glass fin into the steel structure.

Figure 1-3 Front connection point between the glass fin – roof glass – fascia panel.

The image on the left shows the glass fin at the location of the front edge connection between the glass fin, the roof glass and the front fascia panel. The outer glass layers of the glass fin are milled according to the shape of the metal shoe. The offset between the single layers in the laminated fin is < 1 mm. The wall thickness of the steel shoe of 10 mm illustrates the intensity of the loads which are transferred through the connection details into the glass and this explains why the tolerances in the connection had to be reduced to a minimum.

Figure 1-4 Laminated countersunk point fixing in the roof glass.

The connection of the roof glass to the glass fin is realized by laminated countersunk point fixings. As shown in the image above, the top glass layer is not penetrated by the point fixing, and the ceramic frit on Pos. 2 hides the connection element partially so that the optical appearance is reduced. The defined space for displacement of the point fixings allows for the compensation of tolerances and movement due to temperature changes. The connection however also provides the stabilization of the glass fin by the roof glass.

2 TSSA connections for glass stairs and bridge applications

An alternative connection technique which deliberately resigns to use mechanical connection elements is the transparent gluing technique with TSSA. The abbreviation “TSSA” stands for “Transparent Structural Silicone Adhesive”. The material is developed by Dow Corning. Glas Trösch is one of the leading processors of TSSA worldwide. The special optical attraction which can be realized by this type of connection is the precise reason why there exists increasingly high demand by architects and designers. Besides the high structural capacity, this type of gluing technique is characterized by excellent strength and durability. In addition to the classical façade applications we consider this gluing technique in particular for the use in the Interior Design and Exhibition Stand Construction where structural requirements need to be fulfilled. Due to the close distance of the user to the detailed connections (for instance in stair applications), the outstanding optical appearance is fully acknowledgement and appreciated. Glas Trösch has realized several stairs with this technique in Switzerland as well as a complex exhibition project.

2.1 TSSA application in full glass stairs

First we will introduce a glass bridge which is a “Pilot Project” in our fabrication facility in the Swisslamex in Bützberg from 2013.

Figure 2-1 "Pilot Project" of a full glass bridge with structurally bonded connection elements.

The glass bridge spans across the fabrication line of the Glas Trösch AG Swisslamex in Bützberg. The width of the bridge is approx. 4.5 m. The side stringers are manufactured from triple laminated glass. The connection of the treads and the glass floor between the stainless steel angles and the glass is realized utilizing the aforementioned TSSA gluing technique. The image below on the left shows the used connection type in this application. The stainless steel angle seems to float in front of the glass stringer. The gluing is crystal clear and even the finest surface texture of the metal is distinct and plainly visible. The clarity of such a connection cannot be achieved by any other type of connection.

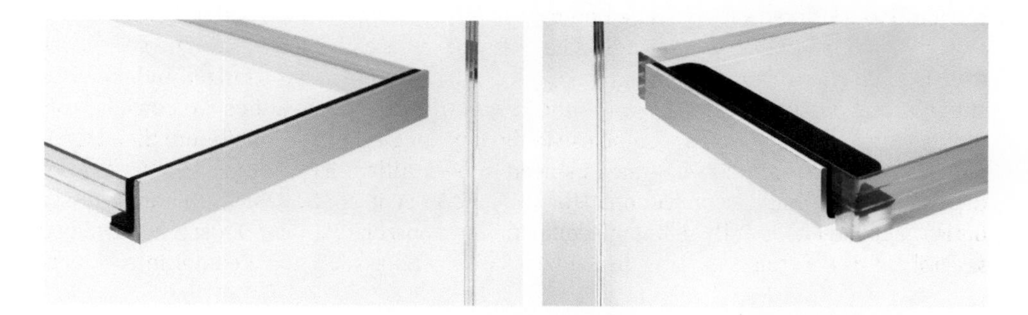

Figure 2-2 Examples for connections between glass treads and glass stringers.

The detailing in the connection between the glass tread and the stainless steel angle, and between the angle and the glass stringer can be shaped in many ways. As shown in the pictures the connection can be modified geometrically. Glass Trösch can assist the design phase to coordinate the architectural demands with the structural and manufacturing requirements. Apart from the examples detailed in this article it is also possible to create completely different structural types of connection.

3 Bonding techniques and laminated inserts in full glass structures

To laminate metal parts into the glass is an excellent method to generate connection details with the highest architectonical demands. These details give an impression of a perfectly smooth connection element. On the basis of an extraordinary full glass structure there are shown some examples how TSSA bonded elements and laminated inserts can be used.

The full glass structure (model shown below) is part of an exhibition stand which is assembled every year for a one week usage during the exhibition "Basel World" in Basel / Switzerland. On the basis of the architects design idea's Glas Trösch developed the overall structural system and the concept for all details. Glas Trösch performed the full design of the glass structure including the assembly and replacement strategy. The red numbers indicates the locations of connection elements which will be explained on the following pages. The details are TSSA bonded elements and laminated inserts or combinations of both techniques in one connection element.

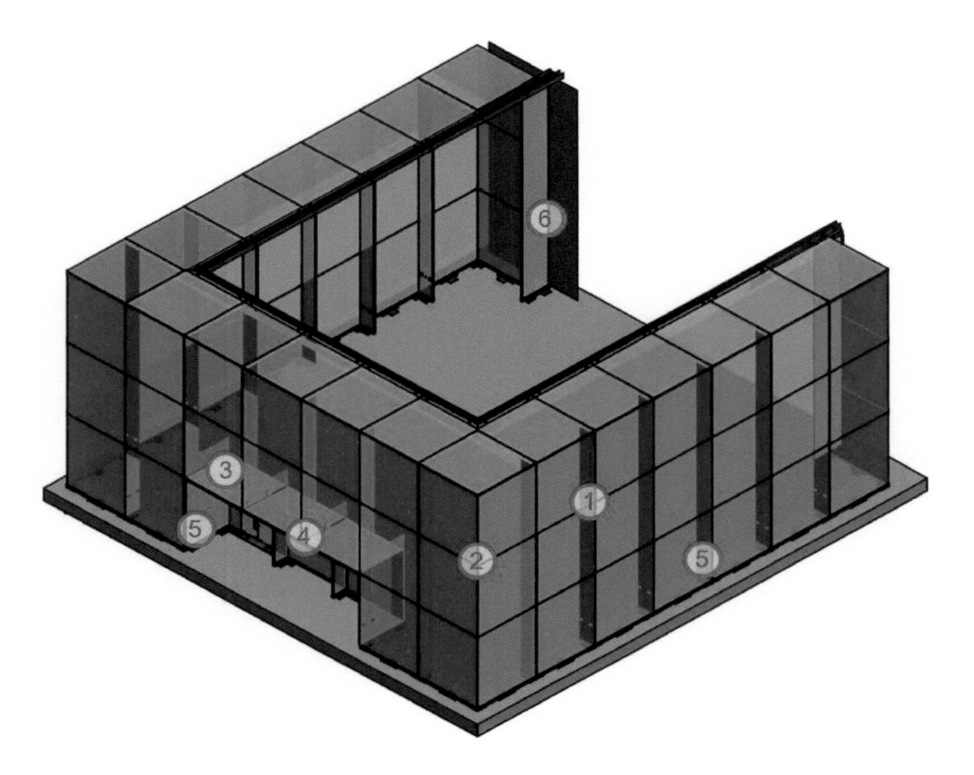

Figure 3-1 Full glass structure for an exhibition stand, architectonical idea by Ottavio di Blasi.

3.1 Point 1, laminated inserts in façade panels and glass fin

The very high precision in positioning the details enables to reduce the tolerance dimensions to a minimum. Although the connecting elements look extremely simple the reality may be different. The internal "live" of the details sometimes is made of little machines to allow for tolerances and defined sliding movements or rotational degree of freedom. Therefore very often complex milling parts are necessary to achieve the desired functionality. Also the internal stress stage in the glass and the interlayer needs to be analyzed especially when annealed glass shall be used. The shape, geometry and size of the laminated inserts are the parameters regarding the internal stress stage of the laminated inserts.

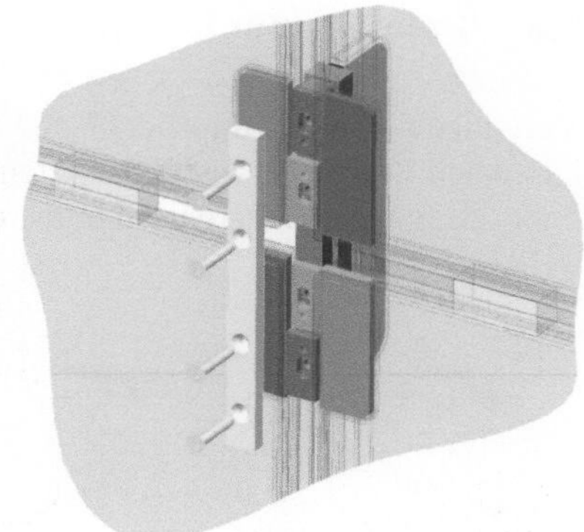

Figure 3-2 Connection between four façade panels and a glass fin with laminated inserts.

3.2 Point 2, full glass edge generated with laminated inserts

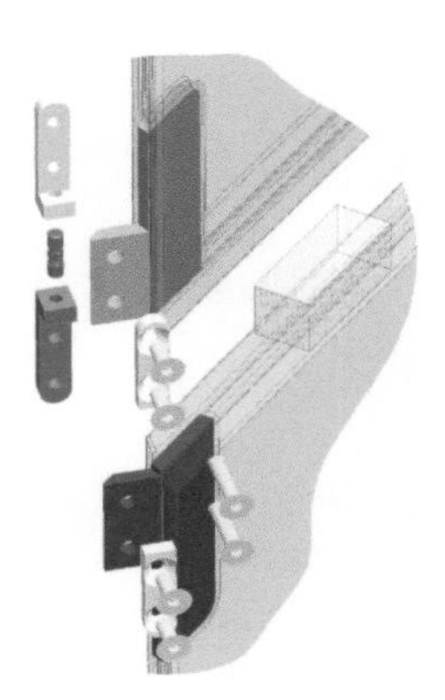

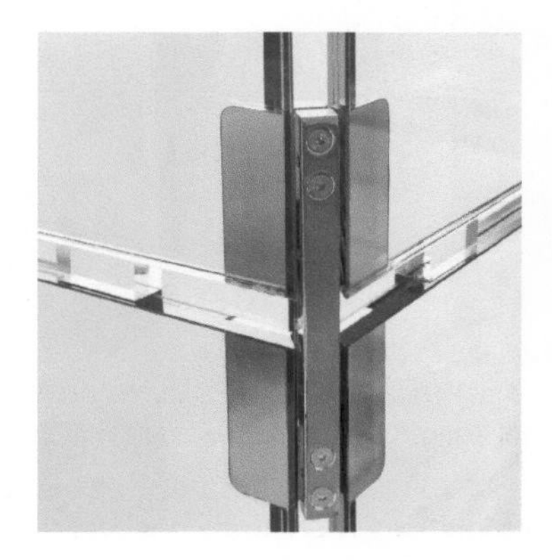

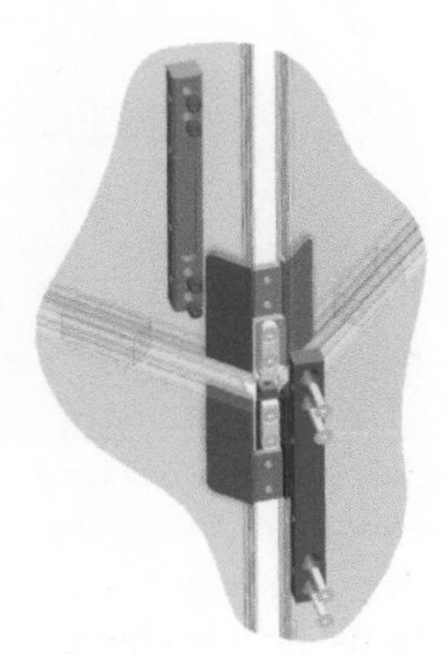

Figure 3-3 Edge connection between four façade panels self stabilizing.

To enable the structural integrity of the complete full glass structure without a glass fin behind the façade panels it is required to connect the façade panels in a certain manner. Each glass panel column (three façade panels on top of each other) at the corner is connected together so that in the façade plane these three panels act as one big element. This requires that the connecting detail is able to transfer compression and tension forces between the panels in one column but may slide free towards the adjacent edge panels due to slab movements. Shear forces between the corner panels needs to be transferred so that the panels are able to stiffen each other.

3.3 Point 3, door fixings with TSSA bonding and laminated inserts

Very nice detailing can be achieved also if the TSSA bonding method is combined with laminated inserts for instance to generate the very simple and elegant façade panel and door fixings which were used in the entrance area of the exhibition stand.

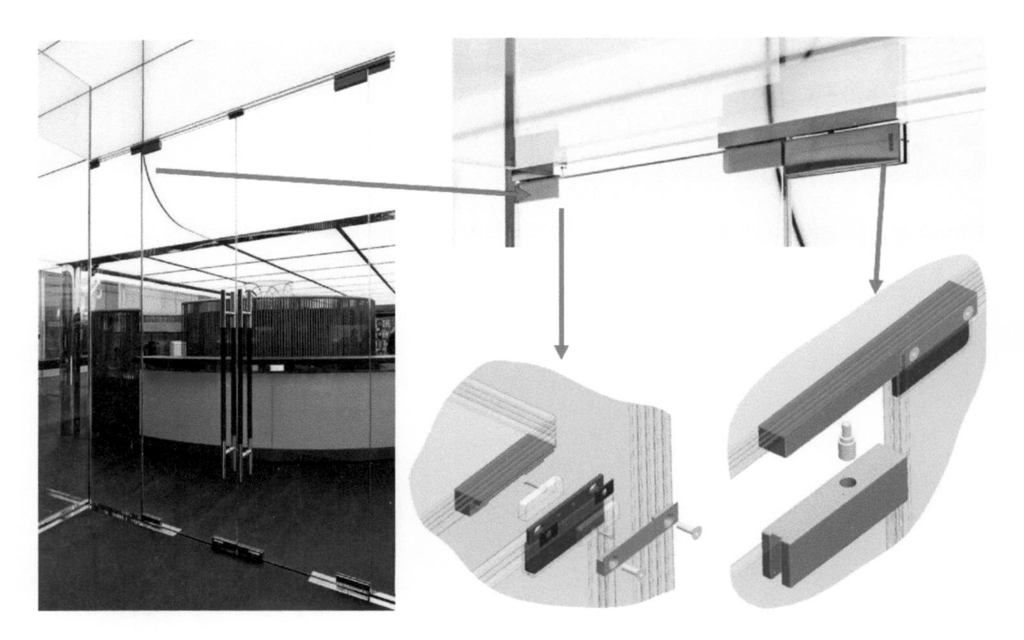

Figure 3-4 Door and glass panel fixing, TSSA details in combination with laminated inserts.

In the picture above there is shown a glass entrance where the roof glazing provides the lateral support for the façade and door panels. The fixing point is bonded with TSSA to the bottom surface of the roof glass. The façade panels are connected by laminated metal inserts to the bonded metal part at the roof. The door is connected by using a standard door fitting assembly. This shows that these highly customized design elements can easily be combined with standard connection components if required.

3.4 Point 4, laminated insert for a heavily loaded element

This detail shows a highly loaded laminated insert which is additionally secured with a pin in the borehole. The detail (bottom left) is vertically loaded with 2.1 tons permanently. The load comes from the façade glazing above the entrance area with the dimension of approx. 6 m x 3 m and the horizontal overhead glazing which spans between the special shaped (Z) glass fins. Therefore this point needs to be considered as an overhead condition. This is why there was decided to use an additional borehole fixation for residual strength criteria's. The structural design considered a total loss of adhesion so that the complete loading also could be transferred through the borehole in case of an extraordinary load case.

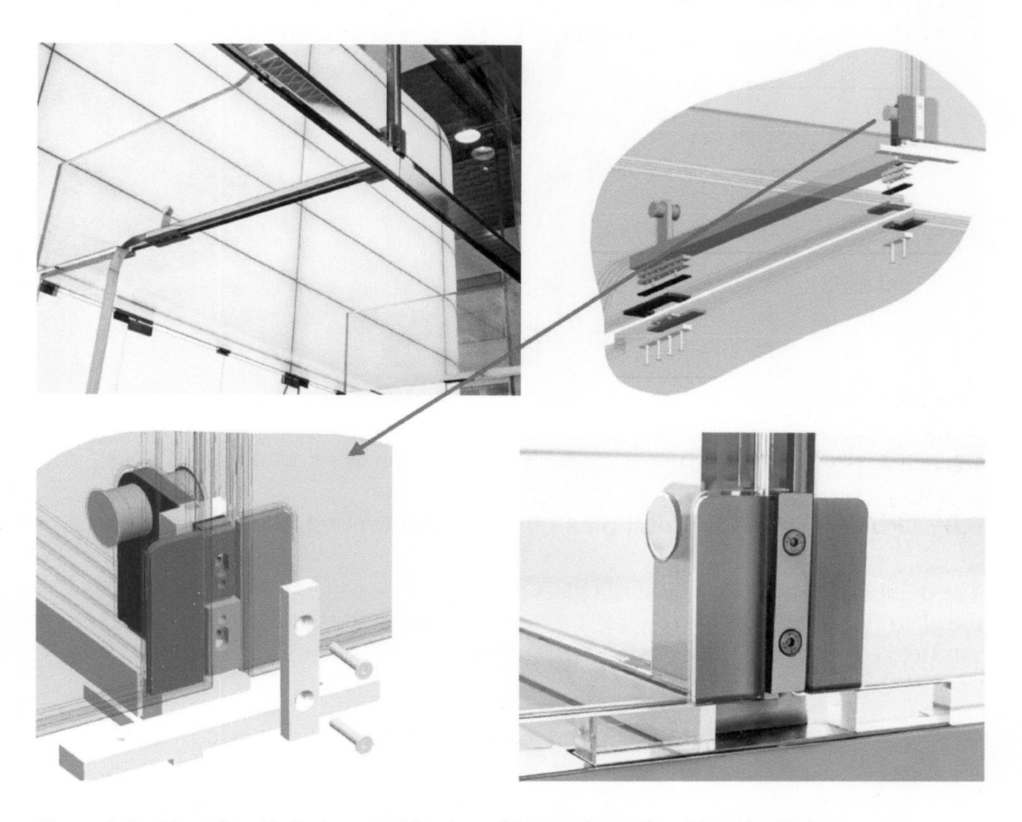

Figure 3-5 Glass fin with Z-shape and laminated inserts for roof and façade glazing.

3.5 Point 5, TSSA application in high loaded exhibition shelves

The image below shows the connection of a showcase. The maximum load of one connection point is designed to withstand more than 500 kg. The numerous showcases also are installed within the glass structure. This underlines in an impressive way a globally versatile solution but also the robustness of the connection as this is required for the use of the detailing in international trade fairs. The use of TSSA was not only decided to achieve the highest level of optical appearance at the detail connection but also to enable the use of annealed glass for the glass fins. A borehole connection would have required to use heat treated glass which was no option for the client's architectonical demands.

Figure 3-6 High transparent TSSA glued connection of a show case to a full glass structure.

3.6 Point 6, TSSA bonded glass fin connection with laminated inserts

The point 6 in figure 3-1 indicates a glass fin connection which is bonded with TSSA to the glass fin. The double sized glass fin forms the end of the glass structure and the passage to the larger part of the adjacent part of the exhibition stand.

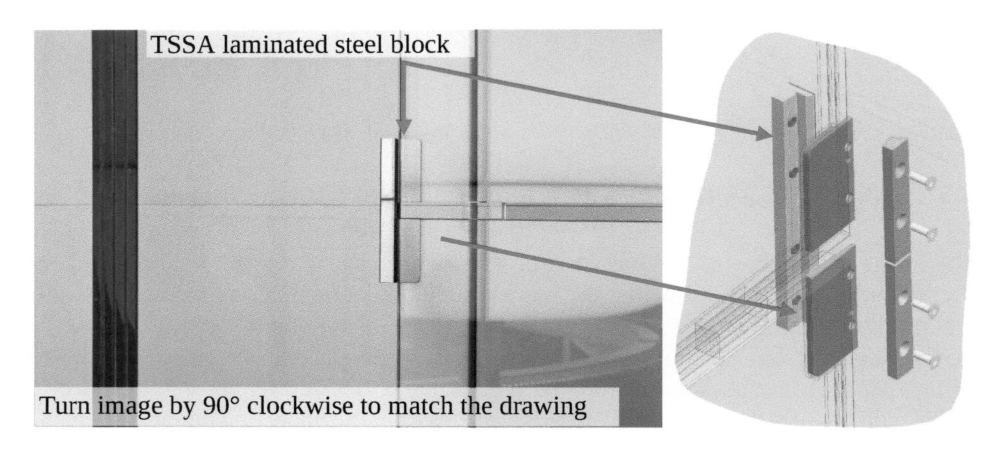

Figure 3-7 TSSA bonded elements in substitution with laminated inserts.

Similar to the point 3 shown above this detail shows again that the laminated inserts and the TSSA bonded elements complements perfectly in many different application. Very often end field situations or corner situations cannot be properly resolved only strictly using one technique. The picture above tells us that the use of both bonding techniques perfectly fits together in one detail.

4 Conclusion

The connection methods shown in above applications give an overview of possible solutions for structural connections between glass elements using metal connectors. The decision which of the connection methods may be used is mainly based on the architectonical demands and the structural requirements. However, besides these elementary criteria further criteria such as structure movements, tolerances resulting from fabrication and onsite assembly, fabrication methods, internal or external application of the structure and finally also costs should be considered.

To achieve an optimized solution regarding all these parameters it is essential that the glass fabricator is incorporated in the design process at an early stage of the project. Glas Trösch can provide substantial design support respectively if requested also a full structural design for these highly specialized connection methods.

Fully glazed high-rise – an architectural vision stays reality!

Paul-Rouven Denz[1], Andreas Beccard[2], Lars Anders[1,2,3]

1 Facade-Lab GmbH, Am Wall 17, 14979 Großbeeren/Berlin Germany, paul.denz@facade-lab.com

2 priedemann building envelope consultants, PO Box 472, Teddington/London, TW 11 1DE, United Kingdom, andreas.beccard@priedemann.net

3 priedemann fassadenberatung GmbH, Am Wall 17, 14979 Großbeeren/Berlin, Germany, berlin@priedemann.net

In classical modernism the dream of a fully glazed high-rise was within reach thanks to the emerging curtain walls. Mies van der Rohe already designed his vision of a glazed high-rise in the early 20th century. Nowadays realizing such a single-layered, fully glazed façade at the Festo Automation-Center, considering current sustainability criteria and energy regulations, illustrates the enormous technological accomplishment and the combination of top performances by different trades and project partners.

Keywords: fully glazed façade, single layer façade, air-exhaust facade, internal sun-shading, active blind, façade planning, user comfort, energy efficiency, electrochromatic glazing

1 Festo – philosophy and project goals

Festo is the leading company in automation technology and world-leader in technical training. Its new iconic building in Esslingen, close to Stuttgart in Germany, reflects Festo's high standards for innovation and high quality engineering. The 67 m tall high-rise office building is the new Landmark of the Festo campus, visible from almost everywhere within the Stuttgart metropolitan region (see Fig. 1-1).

On 16 rhombus shaped floors the AutomationCenter covers 12,000 m² of usable area, 400 working stations, a cafeteria, a direct underground connection to the TechnologyCenter of Festo, its own entrance lobby and a conference centre on the top floor.

The modern character of the building and its combination of aesthetics, technology and energy efficiency show the demand for quality and progress in every single detail. From the start of the project the AutomationCenter was supposed to set a milestone – in architecture and sustainability. Under the premise to develop a façade that combines a high user satisfaction with technical innovation.

Engineered Transparency 2016. Glass in Architecture and Structural Engineering. First Edition.
Edited by Jens Schneider, Bernhard Weller.

Figure 1-1 Festo Campus Esslingen.

Together with the architectural practice Jaschek, the structural engineer Schlaich Bergermann und Partner and the building service engineers Pfeil & Koch the priedemann building envelope consultants received the mandate to transform the guidelines for design, construction and innovation into a new façade.

2 Architectural goal

According to Festo's request the architect planned a transparent building in a crystalline optic. This important design decision obviously excluded opaque panels or external sun shading from the façade options. However considering the EnEV (German energy saving regulations) a fully glazed façade for an energy-efficient high-rise building seemed to be impossible.

Only with an intelligent new overall system the balancing act between EnEV conformity and a user friendly all-glass façade was achievable.

Figure 2-1 AutomationCenter Festo.

3 Concept – principal and function

In contemporary office buildings cooling accounts for the majority of the energy consumption because of internal heat sources. High solar income during summer or at low winter sun heat up the inside of these buildings additionally.

As is known solar control glazing alone is not sufficient to guarantee summer heat protection and meet the guidelines of workplace regulations concerning sun protection and anti-glare protection.

Based on the idea of the air-exhaust façade, which was yet successfully realized by priedemann at the „Treptowers" high-rise buildings in Berlin in the early nineties, a new concept of an internal blind was developed to act as adequate sun shading. The mandatory inner blind for anti-glare is being activated and becomes a separation layer in the façade system to generate an air-exhaust corridor between blind and external glazing. The exhaust air from the office space passes through this interspace being sucked in at the lintel (see Fig. 3-1).

In this corridor on the surface of the blind the solar radiation is being absorbed and long-wave heat radiation led away by the air flow. Thus preventing unnecessary heating-up of the indoor office space. In addition the surface of the blind facing the interior almost has the same temperature as the room air. The same temperature for all room surfaces enable a new quality in comfort because it prevents radiation asymmetry. As a result you need less cooling energy and gain a higher user comfort. This façade solution can be operated individually and thereby creates a dynamically g-value of the system regardless of weather conditions.

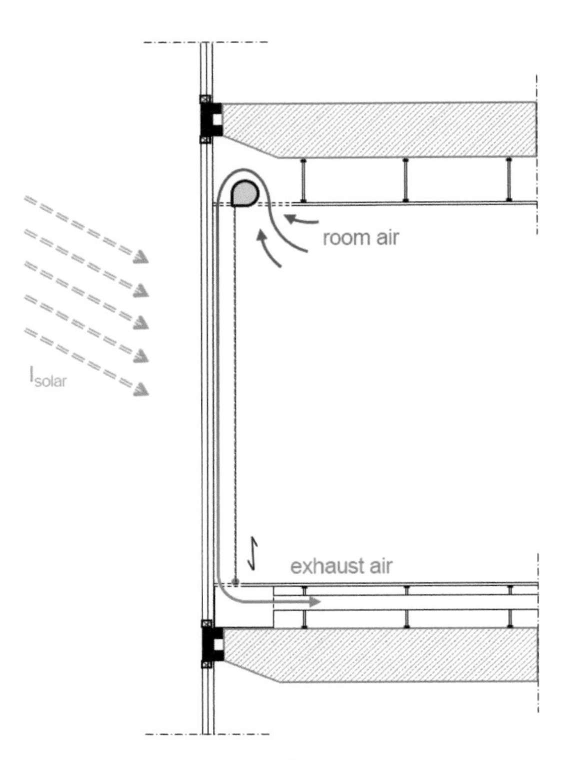

Figure 3-1 Concept blind based air-exhaust façade.

4 Tests and Mock-Up

For this new type of air-exhaust façade there has not been any reference so far. With the support of Pfeil & Koch several scenarios have been simulated to narrow down the options and to ensure the integration of this new system into the overall building services concept. To finally develop a reliable façade system and to define future operation scenarios a testing series had to be carried out. During the planning process the construction, dimension, operation and functionality of the façade system was researched with the help of in-situ tests. This testing was done by the „Facade-Lab", priedemann's R&D department and façade competence centre, and the Fraunhofer Institute on Building Physics in Holzkirchen, Germany. Together with Warema as industrial partner for the internal blind a big variety of construction options and operation scenarios were measured at the test stand of the Fraunhofer IBP (see Fig. 4-1).

Figure 4-1 Festo façade at test stand of the Fraunhofer IBP.

Different textiles for the blind, the impact of dimensions of the air-flow channel and other parts on the whole system as well as different air-flow speeds and volumes have been investigated. Functionality, efficiency, design and user comfort were the main evaluation criteria.

Testing results showed a reduction of energy consumption for cooling by 10 to 30 per cent in comparison to a standard office building depending on plan layout and air exchange rate.

5 Construction and innovations

The normally rather passive building envelope is transformed into an active building component reacting together with building services dynamically to changing environmental conditions.

This high performance façade is based on a conventional single layer unitized façade system. The façade consists of two alternating fields adding up to 8,500 m² in total: One parallel-opening window and a fixed glazing element (see Fig. 5-1 and 5-2).

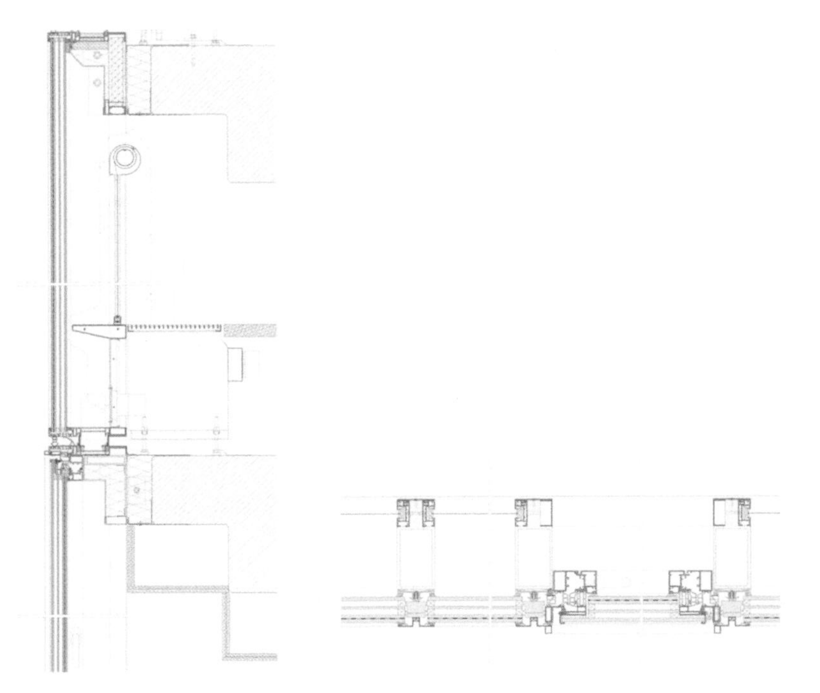

Figure 5-1 Vertical section. **Figure 5-2** Horizontal section.

The smaller element measuring 0.70 x 4.00 m sets a new standard in manually openable windows and glazing technology. Its parallel opening window can be operated by the user increasing its comfort also from a psychological point of view. The floor to ceiling double insulation glazing consists of a two-piece panel of electrochromatic glazing. This new glass technology allows changing dynamically the g-value between 12 and 40 per cent by various steps and therefore lowering the solar radiation and heat income into the building. This technology also prevents from glare and at the same time ensures a high visibility trough the glazing at any setting. Festo also sets a new milestone in this field realizing the largest façade area of electrochromatic glazing in Europe.

The bigger façade element is 2.70 x 4.00 m big and is carried out as one fixed triple insulation glazing panel in combination with the newly developed blind based air exhaust façade system as described. To secure the performance of the air exhaust corridor a new blind holding was developed tightening the blind at the site to the profile so the air can only be sucked in on the top or through the textile structure of the blind. When not using the blind a high quality sun control glass ensures necessary façade properties and at the same time a great view from the inside. The main difference to a comparable office building façade is the relocation of exhaust air ducts into the façade area and small adjustments in detailing and controlling the system.

Figure 5-3 Erection of unitized façade.

This façade solution ensures that already in the first summer since putting the AutomationCenter into operation in July 2015 the inside temperature never exceeded 26 °C even at outside temperatures of up to 40 °C. Besides the high energy efficiency, the increase in user comfort, ensuring high job performances, the developed single-skin unitized façade was realized with a depth of only 75 mm of glass construction increasing the usable floor space in the building. This sets the newly developed façade typology clearly apart from double-skin or Closed Cavity Façades.

Figure 5-4 View from inside.

To keep this newly generated absolute view clear at all times a new cleaning concept was incorporated. A cleaning robot was adapted to the special needs of the Festo Automation-Center. It can move along the whole façade like a gecko thanks to vacuum technology only held by a roof trolley at the top of the building for security reasons, energy and water supply (see Fig. 5-5). This 2 m^2 small robot cleans the whole 8.500 m^2 of fully-glazed façade in only one day. All mechanical parts and interior surfaces can be reached and therefore maintained and cleaned easily from the inside.

Figure 5-5 Closed blinds and cleaning robot.

Finally the façade was designed to incorporate the building into the Festo campus and the companies CI: technological innovation, pneumatic air systems, intelligent control systems and last but not least the companies colours blue and grey – using these colours for the façade profiles. This commitment to the company is highlighted throughout the whole day: a LED-system was integrated into the façade profiles letting the façade sparkle in Festo blue by night as well.

6 Conclusion

The combination of smart solutions add up to a building which is cost and energy efficient, low-maintenance, user-friendly and creates identity thanks to a lot of highly designed and developed details and a smart controlling system of the building services.

This result derives from the combination of façade engineering, building service detailing and new fluidic and physical findings. All these design and technological innovations were only feasible thanks to the openness of the architect, client, industry and research institutes towards new approaches by the façade planner and its leading position in the planning and construction of the façade.

This close collaboration made it possible to fulfil all the project targets, enqueuing in previous ground-breaking construction projects by Festo and realizing innovations as a reference for buildings to come.

Figure 6-1 Façade Festo AutomationCenter.

Comparative energy and comfort analysis of façade passive strategies in the Mediterranean region

Lara Mifsud[1], Francesco Pomponi[2], Alice M. Moncaster[3]

1 Department of Engineering, University of Cambridge, Trumpington Street, CambridgeCB2 1PZ1, ldm37@cam.ac.uk, fp327@cam.ac.uk, amm24@cam.ac.uk

Façades are a key element to manage effectively heat flow and transmission between indoors and outdoors environments. Passive strategies can play a critical role in façade design to reduce the operational energy consumption and carbon dioxide emissions of buildings. In hot countries, architects and engineers incorporate passive strategies into the building façades to mitigate solar gains and reduce internal overheating, with the final aim of reducing cooling loads and operational costs and diminish or nullify the need for mechanical ventilation or air conditioning. This paper reports on findings from a case study in the Mediterranean region (Malta) of a newly designed University building which incorporates horizontal louvers combined with a curtain-wall façade. Operational energy figures are obtained through dynamic thermal modelling in IES VE. Additionally, thermal comfort assessment has been undertaken in compliance with the adaptive comfort approach and the TM52 assessment methodologies. Results show the effects such passive strategies have on energy consumption, natural ventilation, and indoor comfort.

Keywords: horizontal louvers, passive shading systems

1 Introduction

In the Mediterranean region, which is characterised by a warm humid weather, most buildings require ways for cooling the indoor environment. The Maltese commercial sector consumes around 30 % of the electricity of the island. Energy loads in office buildings, especially cooling loads, are regarded as the main component of the national electricity peak demand in the warmer seasons (MRA, 2006). Malta has energy-saving targets that must be reached by 2020, one of which is the nearly zero energy building. All new builds have to reach "an energy performance not exceeding 290 kWh/m^{2}yr for [commercial] buildings" (Ministry for Energy and Conservation of Water 2014). In southern Europe, it has been reported that mechanical cooling alone consumes around 130 kWh/m^2 annually (Calleja 2012). Furthermore, sources indicate that that the mean annual temperatures in Malta have been gradually increasing by 0.230 °C per decade due to climate change. Thus Malta's climate is gradually becoming slightly drier and warmer (The Environment Report 2008, Climate Change Comittee for Adaptation 2010).

Ralegaonkar and Gupta (2010) suggest that there are four main design factors which influence the solar gain inside a building. These are general orientation of the building mass;

Engineered Transparency 2016. Glass in Architecture and Structural Engineering. First Edition.
Edited by Jens Schneider, Bernhard Weller.

orientation, size and materials of openings; the ratio of window to wall area, and finally, the incorporation of shading devices. The latter is the focus of this study.

Passive shading devices are used to reduce the solar gains, and hence cooling loads, in buildings. However, building envelopes are often people's first impression of a building and so there is a large pressure to ensure that such systems are designed to contribute to the architect's aesthetical vision. Apart from that, such shading devices greatly influence the interior environment and the user's perception and interaction of the space. Freewan (2014) recommends five main parameters which one should consider whilst designing such systems:

1. Block direct sunlight and associated glare
2. Improve internal daylight quality
3. Control internal daylight level
4. Enhance user's interaction
5. Maintain a good view to the outside

Nonetheless, for louvers and other shading systems alike, there are other factors which have to be regarded if the respective shading devices are to perform successfully and reduce internal temperatures.

Effective louver systems are dependent on correct orientation, the inclination angle of the louver and finally the louver size in relation to the glazed area (Palmero-Marrero and Oliveira, 2010).

Datta (2001) examined the performance of external fixed horizontal louvers in four Italian cities with different climates. All shading systems were applied on the south openings and the largest reductions in cooling loads were noted for the warmer climates, such as Palermo. Minor variances in the energy load were noted between louvers at an inclination of 30°, 45° and 60° for Napoli and Palermo. Furthermore, no major difference was noted as the depth of the louver decreased from 0.2 to 0.1 m. However, though the optimum inclination angle and size were investigated, the distance between the louvers was kept constant, as was the distance between the louver and the opening. By investigating the distance between these two elements, one could establish if the system would benefit from an improved airflow, and thus possibly reducing further the internal heat gains.

Alzoubi and Al-Zoubi (2010) attempted to compare vertical and horizontal shading devices installed on south facades in Jordan. Their simulation found that for vertical louvers, higher illuminance levels were achieved with lower heat gains. However, associated glare was not investigated and being on the South façade, horizontal louvers might perform better in eliminating direct sunlight, hence reducing glare. This could influence greatly the user's interaction with the designed space.

Palmero-Marrero and Oliveira (2010) also investigated two layouts of louvers, horizontal louvers for the east and west façade and horizontal louvers laid as a canopy for the south façade. This configuration was simulated in five different cities with increasing latitudes, ranging from Mexico to London. For the south façade the angle of inclination was the same as the respective latitude whilst for the East façade, the angle changed from 20° to 60°. A 20° inclination angle was found to be beneficial in all climates, since, with a higher angle, the cooling loads increased. However, the same thermostatic control was used for all climates, which can be disputed since other research suggests that the thermal comfort threshold is dependent on the inhabitant's climate (Nicol and Humphreys, 2002). Despite the lack of climatic-specific design, their research suggests that energy savings related to lower solar gains occur nonetheless.

Energy savings for lighting and cooling loads were also noted in Hammad and Abu-Hijleh's (2010) studies. Horizontal louvers for south facing facades and vertical louvers for east and west facades were simulated for a building in Abu Dhabi. Dynamic louvers were compared to static ones, and it was concluded that the performance difference was minimal when the angle of inclination for the horizontal static louvers was kept at 20° inclined upwards.

The aim of this study is to investigate the performance of horizontal louver systems in the Mediterranean region and how the use of such systems could possibly reduce the cooling loads required. It also investigates the different performance achieved by comparing a natural ventilation strategy to a mixed mode system.

2 Methodology

This research question initially stemmed from a real life project the author was involved in. As a result, though different research methods were used throughout, this study is primarily rooted in a case study approach. A case study is defined as a research strategy "where the emphasis is towards investigating a phenomenon within a context" (Proverbs & Gameson 2008). Campbell considers the "real world value" as a considerable factor which contributes to successful research (Robson 2002).

This study uses two case studies both of which are located in the University of Malta campus in Msida, Malta. For the base case, the recently completed Faculty of ICT, designed by TBA Periti, was used (Figure 2-1). This building is centred around a large open courtyard where a fully glazed curtain system was used for the facades. This façade has a retractable horizontal louver system installed which can be raised or lowered according to the users' demands.

Figure 2-1 The faculty of ICT.

The second case study is based on a new envisaged building called the Sustainable Living Complex (SLC) located in the University of Malta's (UoM) campus in Msida, Malta. This is a new complex that will house the faculty for the Built Environment, the faculty of Education and several other institutes. Team Two architects, an architectural design team of which the author forms part of, designed this project.

The whole complex may be split into three main parts, the lecture rooms, the laboratories/studios and the offices. The two office blocks were designed as individual buildings connected on each floor. The primary difference between the two is that one is centred around an enclosed atrium whilst the other is designed with a central open courtyard. The façade of these offices are also proposed to be fully glazed curtain wall systems. Furthermore, different shading treatments were incorporated for the different façade orientations, including external fixed horizontal louvers for the South facade, as indicated in Figure 2-2.

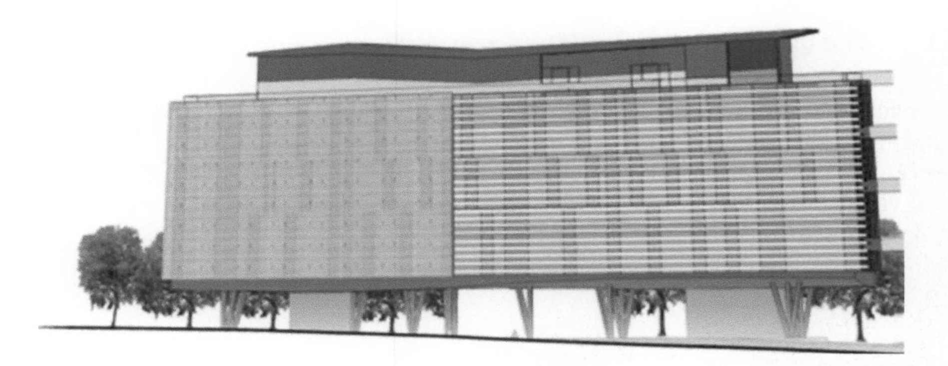

Figure 2-2 The South façade of one of the office blocks (Team Two Architects).

By using these two buildings as the basis for this study, the performance of these shading systems used in the Mediterranean context was investigated by establishing the operational energy for each respective case study.

2.1 Operational Energy

Operational energy (OE) is the energy consumed when the building is in use. Space cooling/heating, lighting, office equipment and servers are all energy uses which form part of OE. However, this study is primarily interested in evaluating the performance of shading systems in terms of space cooling/heating. Therefore, lighting and associated glare, which is also heavily influenced by the building envelope, will not be investigated in this study as it falls beyond the scope of this research.

In order to establish the OE, the Integrated Environmental Solutions Virtual Environment (IES-VE) tool was used. This software was chosen since it provides a range of analysis options and capabilities which have been tested and validated by external verifiers (IES, 2010). Furthermore, several successful studies have used this as the main simulation program (Hammad & Abu-Hijleh 2010; Kim et al. 2012; Stevanović 2013; Freewan 2014; Pomponi et al. 2015).

2.2 Assessed Scenarios and Configurations Considered

The functional unit of this study is a 1.5 m (W) x 3.75 m (H) portion of the South façade, with an openable area of 30 %. The offices oriented towards the South façade were modelled, since such facades normally experience the highest heat gains. The height of this unit is based on the floor-to-floor height envisaged for this office block. This façade is primarily made up of two main elements:

i. The glazed system
ii. The shading system

2.2.1 The glazed system

This system is based on the façade installed in the ICT building in the University of Malta campus. This is a double glazed system with argon gas with a total U-value of 1.1 W/m^2K. However, this sort of system is permanently closed and so does not allow any sort of natural ventilation to take place. Therefore, this system was modelled which allows the glazed windows to open through a sliding mechanism.

2.2.2 The shading system

The Base Case

The louvers in the base case were simulated as external louvers which were lowered when the solar radiation exceeded 600 W/m^2. This value was obtained after analyzing the building without any sort of shading devices and establishing the lowest solar radiation which caused an uncomfortable environment. The model was then simulated to include louvers when the solar radiation on the glazed facades reached that level, for the operational hours.

Figure 2-3 The façade of ICT with retractable external louvers.

Louver System 1

Two different systems were investigated in this study. The first system consists of horizontal louvers with a length of 500 mm, a pitch of 750 mm and angled at 20°.

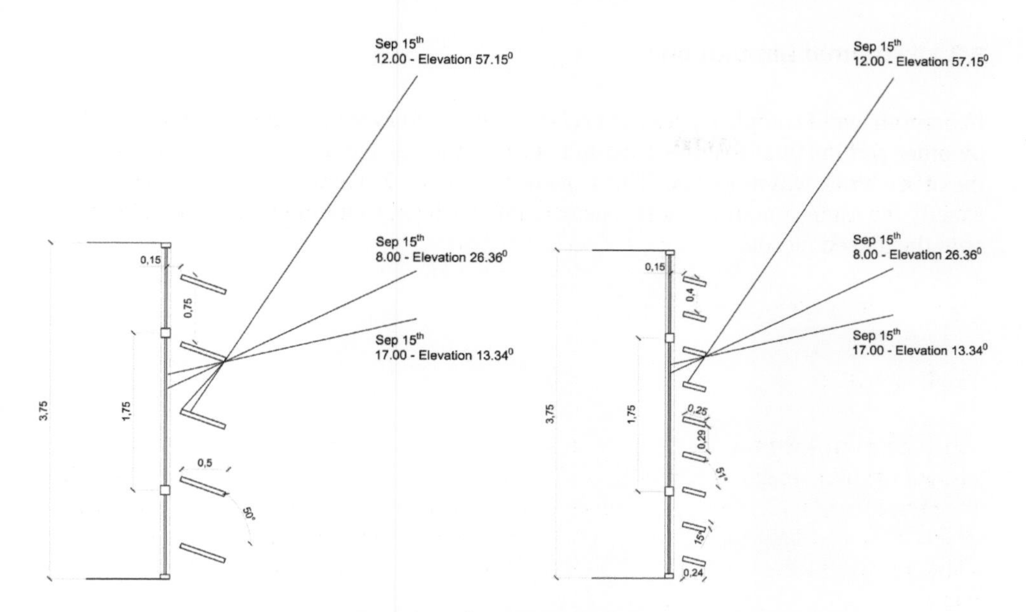

Figure 2-4 Louver system 1 (left) and louver system 2 (right) (Author's own).

Louver System 2

The second system consists of horizontal louvers with a length of 250 mm, a pitch of 400 mm and angled at 15°.

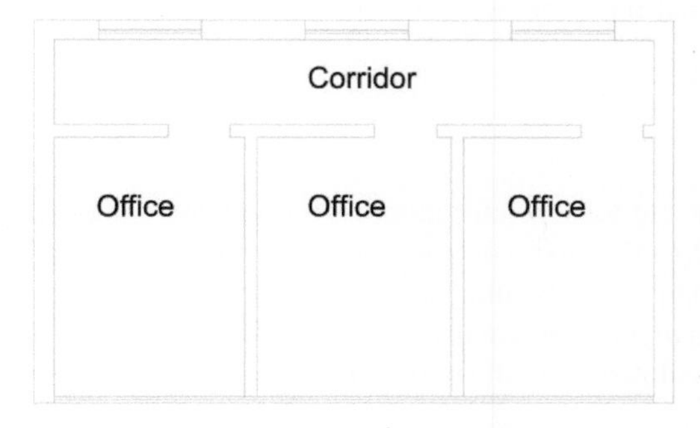

Figure 2-5 Part of the office block which was modelled (Author's own).

In both systems, when the sun's angle of elevation is greater than 50° and 51° respectively, direct sun is obstructed. Furthermore, for both systems, aluminium and steel louvers were analysed.

2.3 Thermal Simulation

In order to avoid complex modelling on IES-VE and to avoid the results being influenced by other architectural features used in the SLC, such as the atrium effect, only a part of the office block was modelled. This is shown in Figure 2-6 and 3-1. The simulation, with exactly the same conditions, was repeated with each respective shading system modelled onto the glazed façade.

Figure 2-6 Part of the office block which was modelled (Author's own).

Using the local weather file provided by the Institute for Sustainable Energy of the University of Malta, running mean external temperature according to EN15251 (BSI 2007) was calculated. The allowable comfortable temperature range was then calculated using the TM52 (CIBSE, 2013) for each month.

Using this information, two types of ventilation modes were modelled in IES. These were:

i. Natural ventilation
ii. Mixed Mode Ventilation

For the natural ventilation mode, during the local winter months, the windows opened when the inside temperature exceeded the minimum comfortable temperature, provided that the outside temperature was lower than the maximum comfortable temperature. During the summer months, as the internal temperature reached 20 °C, the windows opened provided that the outside temperature was lower than the maximum comfortable temperature.

For the mixed mode system, the windows opened when the internal temperature was within the acceptable comfortable temperature range. Once the internal temperature exceeded 24 °C, then the cooling mode was switched on. Cooling was set to the operating temperature of 24 °C as specified by TM52. A fresh air supply rate of 10 l/s per capita as recommended by CIBSE Application Manual 10 – Natural Ventilation in Non-domestic buildings (CIBSE, 2005) was included.

3 Analysis and Discussion of Results

This next section presents the results from the simulations which were carried out to obtain the OE figures for the shading systems analysed.

In order to be able to compare the performance of the different strategies, the Predicted Mean Vote (PMV) (CIBSE, 2013) comfort scale was analysed for the occupied hours. The PMV scale is based on thermal sensation and is split into seven votes, as described in table 1.

Table 3-1 The PMV Index.

PMV index	
+3	Hot
+2	Warm
+1	Slightly warm
0	Neutral
1	Slightly cold
2	Cool
3	Cold

3.1 Natural Ventilation Strategies

None of the louvers systems satisfied the criteria requested by TM52. A small difference was noted between the base case and the Louver systems. A higher percentage was noted for the cool index, while practically no difference was noted between the four louver systems.

The difference between the base case and louver strategies could be attributed to the louver systems acting as an insulating barrier, leading to the base case losing heat quicker from the glazed façade and cooling at a higher rate.

However, when analysing all the louver systems, it is clear that for more than 60 % of the time, occupants would feel cool sensations. This suggests that in the Mediterranean climate, such louver systems may significantly reduce overheating instances.

3.2 Mixed Mode Strategies

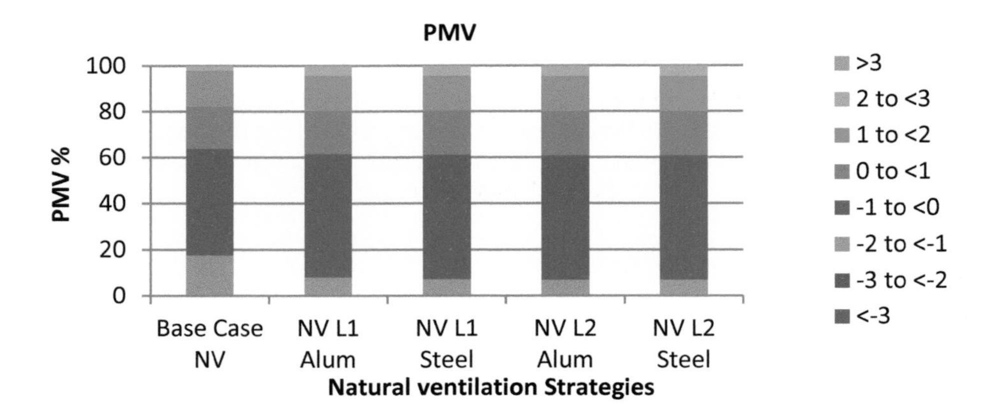

Figure 3-1 The PMV Index for Natural Ventilation Strategies.

For the mixed mode strategies, the PMV was also analysed. In this case, the louver systems compared slightly better than then base case since a higher percentage was noted for the neutral index. For the cool vote, the base case had more than double the votes when compared to the other louver systems. This further reiterates the possibility that the gap between the external louvers and the glazed façade is acting as an insulating barrier slowing down the rate of heat loss.

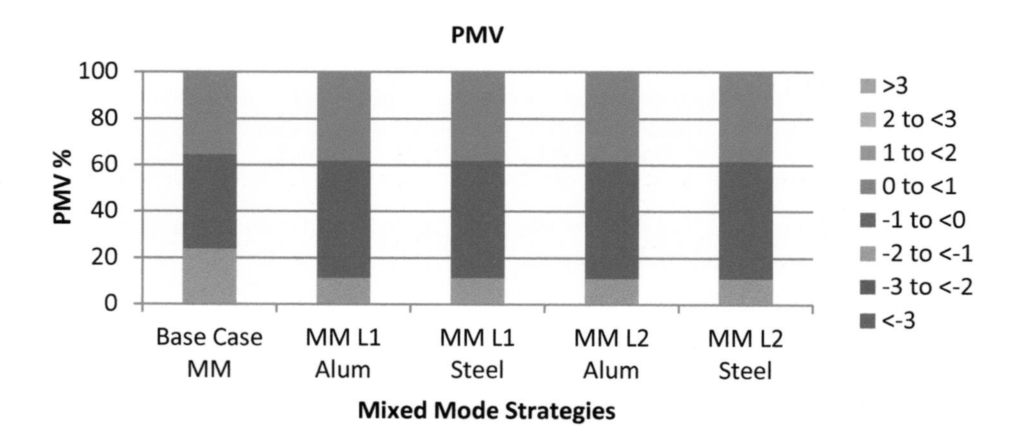

Figure 3-2 The PMV Index for Mixed Mode Strategies.

These PMVs correspond to the results given for the space cooling loads. A significant lower value was attributed to the base case, when compared to the other shading strategies. However, it is important to note that these are cooling loads, and so no heating was included in the analysis. Since the base case had an additional 10 % in the cool sensation, when compared to the louver systems, then this would also mean that higher heating loads would be required to heat the space.

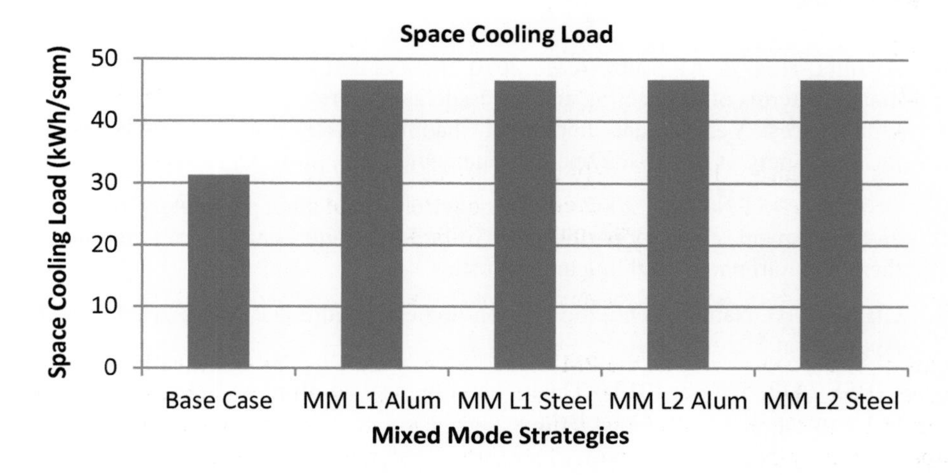

Figure 3-3 The Space Cooling Load for Mixed Mode Strategies.

4 Conclusions and recommendations for further research

This study attempted to analyse different shading strategies used in office buildings for the Mediterranean climate. Though the louver systems varied in dimensions, these varied in a proportionate manner possibly explaining why no significant difference was noted between the systems. The choice of material makes no difference in the operational energy consumed. A significant difference was noted between the base case and the horizontal louver systems. This difference could be attributed to the fact that in the horizontal louver systems, a 150 mm gap was included between the glazed façade and the louvers. This gap could have acted as an insulating barrier preventing the façade to lose heat at a faster rate. However, it is also important to note that through the simulation software used, the only way to model retractable louvers is to use the software pre-sets and indicate the solar incidence at which these will be activated. Therefore, no geometrical details are included by the user and the IES User guide clearly states that for the external shading devices modelled through the software pre-sets, there is no distinction in terms of the performance between shutters and louvers.

Further research could possibly investigate the difference in associated glare and lighting requirements between the different shading systems. This study also suggests that different space heating loads are required for these systems, therefore this could be analysed in greater detail. Furthermore, additional shading systems such as overhangs, vertical louvers and meshes could be modelled and analysed accordingly.

5 References

[1] Alzoubi, H.H. & Al-Zoubi, A.H., 2010. Assessment of building facade performance in terms of daylighting and the associated energy consumption in architectural spaces: Vertical and horizontal shading devices for southern exposure facades. Energy Conversion and Management, 51(8), pp.1592-1599.

[2] BSI 2007. BS EN 15251:2007 – Indoor environmental input parameters for design and assessment of energy performance of buildings addressing indoor air quality, thermal environment, lighting and acoustics.

[3] CIBSE 2005. Natural Ventilation in non-domestic buildings. AM10:2005. CIBSE Application Manual.

[4] CIBSE 2013. TM 52: 2013 – The limits of thermal comfort: avoiding overheating in European buildings. Great Britain.

[5] Calleja, H., 2012. Passive Cooling Strategies for a Digital Creative Industry Hub in Malta. In Proceedings of 2nd Conference: People and Buildings.

[6] Climate Change Committee for Adaptation, M., 2010. National Climate Change Adaptation Strategy.

[7] Datta, G., 2001. Effect of fixed horizontal louver shading devices on thermal perfomance of building by TRNSYS simulation. Renewable Energy, 23 (3-4), pp. 497-507.

[8] Freewan, A.A.Y., 2014. Impact of external shading devices on thermal and daylighting performance of offices in hot climate regions. Solar Energy, 102, pp.14-30.

[9] Hammad, F. & Abu-Hijleh, B., 2010. The energy savings potential of using dynamic external louvers in an office building. Energy and Buildings, 42 (10), pp.1888-1895.

[10] Kim, G. et al., 2012. Comparative advantage of an exterior shading device in thermal performance for residential buildings. Energy and Buildings, 46, pp.105-111.

[11] MRA, Malta Resources Authority, (2006). Malta's Annual report to the European Commission regarding Directive 2003/54/EC and 2003/55/EC. Valletta: MRA.

[12] Ministry for Energy and Conservation of Water, Government of Malta (2014). Malta's National Energy Efficiency Action Plan (NEEAP).

[13] Nicol, J.F. & Humphreys, M.A., 2002. Adaptive thermal comfort and sustainable thermal standards for buildings. Energy and Buildings, 34(6), pp.563-572.

[14] Palmero-Marrero, A.I. & Oliveira, A.C., 2010. Effect of louver shading devices on building energy requirements. Applied Energy, 87(6), pp.2040-2049.

[15] Pomponi, F., 2015. Life Cycle Energy and Carbon Assessment of Double Skin Façades for Office Refurbishments. University of Brighton.

[16] Proverbs, D. & Gameson, R., 2008. Case Study Research. In A. Knight & L. Ruddock, eds. Advanced Research Methods in the Built Environment. Oxford, UK: Wiley-blackwell, pp. 15-110.

[17] Ralegaonkar, R. V. & Gupta, R., 2010. Review of intelligent building construction: A passive solar architecture approach. Renewable and Sustainable Energy Reviews, 14(8), pp.2238-2242.

[18] Robson, C., 2002. Real world research: a resource for social scientists and practitioner-researchers 2nd editio. Blackwell Publishing, ed., United Kingdom.

[19] Stevanović, S., 2013. Optimization of passive solar design strategies: A review. Renewable and Sustainable Energy Reviews, 25, pp.177-196.

Experiments with Kiln-formed and Hot glass

Georg Rafailidis[1]

1 The State University of New York at Buffalo, grafaili@buffalo.edu

Working first-hand, in an experimental manner, with glass presents several challenges. Unlike materials like concrete, plaster, and even ceramics, glass can only be formed with exposure to extremely high temperatures, and requires special facilities, tools and specialized skills. This paper documents the outcome of an architectural design studio dedicated to glass in architecture, taught for two semesters; first-hand experimentation with glass resulted in prototypes at the architectural component scale. Research was conducted with the cooperation of the Corning Museum of Glass (Cog) and their state-of-the-art Glass Lab facilities for hot glass work. Simultaneously, work with kiln-formed glass was conducted at the university, using refractory molds and both ceramic and glass kilns.

Keywords: warm glass, hot glass, architectural glass, design studio, architectural pedagogy, Corning Museum of Glass

1 Introduction

This paper documents experimentations with glass to prototype architectural components as an offshoot of a research project that began in 2009 into the development of a glass block facade component filled with PCM (phase change material). For most of these tests, we used the specific wax PCM, Rubitherm RT21 (http://www.rubitherm.de/english/download/techdata_RT21_en.pdf). The initial design of the "thermometric façade" glass block and research into PCM was presented in the facades session at Engineered Transparency in 2012.

While the first phase of the research, done in collaboration with the chemical company BASF, focused on empirical testing of wax PCM, the current phase focuses on glass itself, and reveals how challenging it is to work with the material in an experimental way. Working first-hand, in an experimental manner, with glass presents several challenges. Unlike materials like concrete, plaster, and even ceramics, glass can only be formed with exposure to extremely high temperatures, and requires special facilities, tools and specialized skills. This paper documents the outcome of an architectural design studio dedicated to glass in architecture, taught for two semesters; first-hand experimentation with glass resulted in prototypes at the architectural component scale. Research was conducted with the cooperation of the Corning Museum of Glass (Cog) and their state-of-the-art Glass Lab facilities for hot glass work. Simultaneously, work with kiln-formed glass was conducted at the university, using refractory molds and both ceramic and glass kilns.

Engineered Transparency 2016. Glass in Architecture and Structural Engineering. First Edition.
Edited by Jens Schneider, Bernhard Weller.

The results of the student experiments revealed the limitations of one-off artisanal glass production techniques – both warm and hot – in making prototypes that require extremely high degrees of precision.

With the Thermometric Façade we investigated the intriguing material properties of wax phase change material and developed an architectural proposal from the material specificity of wax phase change materials. Glass turned out to be an intriguing phase change material in its own right and allowed us to speculate on its' architectural potential in a material specific manner. In the following I will describe four student projects selected from twelve projects that emerged over the course of two design studios.

2 Graduate Design Studio Projects

Variation instead of repetition in kiln formed glass (student team: Kim Sass, Steve Smigielski)

The first project began by tackling the challenge of creating forms that are typologically related but geometrically different. In kiln formed glass casting, molds typically produce identical forms and formal differences or aberrations between the cast pieces are seen as undesirable. The same holds true for the fabrication of aggregated building elements. Bricks, for example, should have the same dimensions to be laid as a brick wall in an efficient manner.

This project, however, looked how glass casting could lead to a family of forms which are clearly identifiable as belonging to a single formal typology, with individual, geometric variations within that typology. Trees and icicles served as precedents for this project. Although each tree is geometrically distinct, we can clearly read them all formally as trees. All trees also follow the same structural principle, although they are geometrically distinct. Icicles are also all formally different, although we identify them in a generalized category as icicles. Icicles also share some distinct qualities with glass: they are inanimate, they are translucent and they are brittle. Combined with the cold weather in the spring term in Buffalo, NY, ice was an affordable and accessible substitute material for the students to use to investigate the research question further. Similar to the process of ice formation, the project utilizes material-specific processes to yield variety and differentiation, in this case by looking at the age-old technique of kiln-formed glass in a mold. Instead of designing a precise form, this project develops a glass-specific fabrication process that generates forms that are geometrically different but typologically identical. In a second step, students investigated how these non-identical elements could be clustered through an aggregation logic that allows for high tolerance similar to Velcro surfaces that stick together without the need for precise placement. First tests with ice sintering produced agglomerations of icicles that were branchy and triggered thinking about modules that could start interlocking in a similar manner to the anti-tank barrier known as Czech Hedgehog.

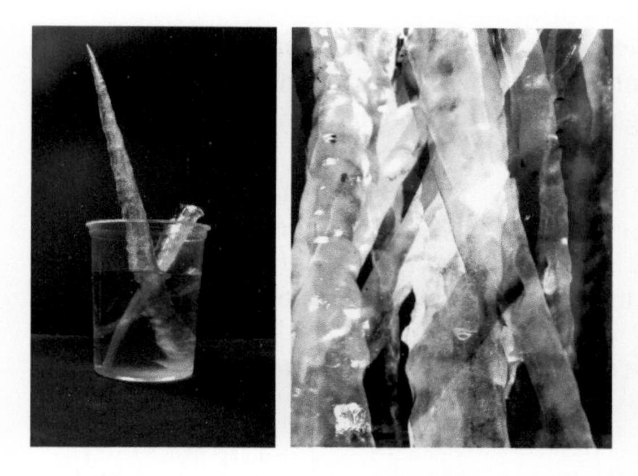

Figure 2-1 Sintering of Icicles. The icicles were between 19 cm and 24 cm long. (Image: Steve Smigielski)

Sintering is the process of creating a singular solid from several parts through heat or pressure without reaching the melting point of the material (Hobbs and Mason [1]). The students learned to produce branchy modules that where all geometrically different while at the same time belonging to the same family of forms. The modules also demonstrated the ability to interlock in unexpected and seemingly random ways. The idea of aggregating non-identical modules into larger, stable assemblies was then tested with branch knots. Tests demonstrated that as the respective branches pointed in opposite directions, the aggregated mount became higher.

Figure 2-2 Scaling of identical modules creates spatial pockets. Each module measures 4 cm x 4 cm x 4 cm. (Image: Steve Smigielski)

In a next test, students investigated the potential of scaling. Similar to the investigations by Eiichi Matusda under the direction of Michael Hensel and Achim Menges, the students dropped modules on top of each other as an aggregation method (Matusda et al. [2]). But in contrast to Matusda different scaled modules were poured on top of each other. This test produced interesting spatial pockets, without the need for formwork to create enclosed space. Scaling addressed not only the structural aggregation of modules but also the creation of space. The knowledge gained from these preliminary tests was then combined and introduced into glass-making process which involved:

- devising a fabrication process that creates a family of related forms instead of a fixed form;
- using an assistive mold similar to the icicle cup instead of a mold that fully enforces its shape onto the cast;
- using a hierarchy of radically different scales to create larger aggregations and inherent spatial pockets;

The students used glass rods and started with simple bisque cups for the assistive mold. Because the mold had a circular section, it was difficult to place it into the kiln without moving the glass rods into an all-parallel bundle. As the glass rods tended to cluster in an all-parallel bundle in the round cup mold, different mold geometries were then tested. The students proposed a hexagonal mold into which the glass rods were tossed. This mold geometry produced a wider variety of fused glass clusters and could be moved without changing the arrangement.

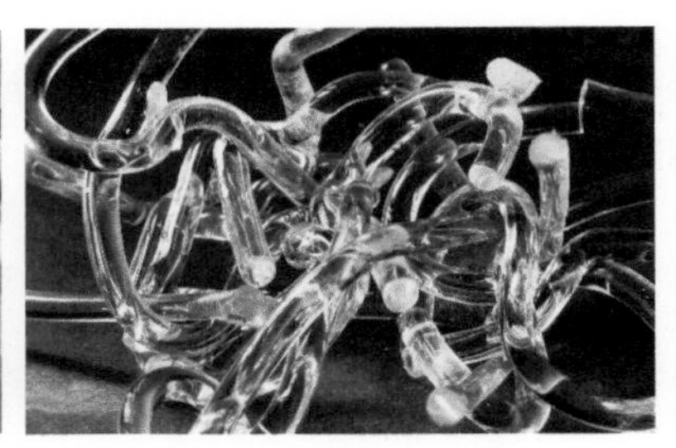

Figure 2-3 Assistive mold before and after firing. Aggregation of irregular modules shown at the right image. Each mold cup measures 6 cm in diameter. (Images left: Steve Smigielski, Image to the right: Georg Rafailidis)

As the glass tacking temperature is higher than the slumping temperature, the sintering outcomes were very different from the earlier ice studies, producing droopy forms with bent branches that aided the aggregation process, similar to Velcro surfaces. To scale-up this approach to an architectural scale, two additional materials were introduced. Modules over 2 ft. (61 cm) were made of wood and modules over 8 ft. (2.5 m) were made of steel. The construction sequence followed the hierarchy of scales and materials. In the documented model, first, three steel modules with a diameter of 30 ft. (9.14 m) are dropped on top of one another. Next, wood modules with an 8 ft. (2.5 m) diameter are dropped onto the assembly. The third pour consists of glass modules with diameters ranging from

2 ft. (61 cm) down to 4 in. (10 cm), creating an outer crust. In contrast to standard constructions, this assembly does not require labor skills, is rapid and needs no fasteners. It is nearly self-assembling. In winter, snow and ice contribute to making a sealed envelope. In summer, growing vines provide shading. Each assembly/pour creates a geometrically different but typologically related structure. The assembly has no required tolerances.

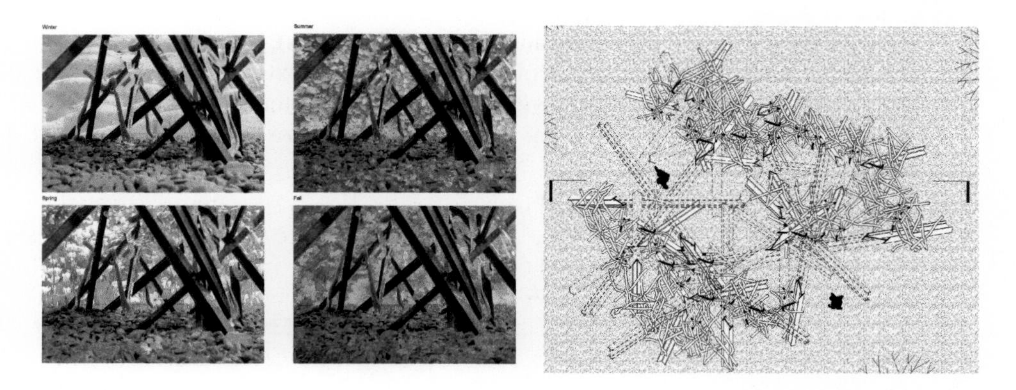

Figure 2-4 Self assembling spatial proposals with varying seasonal envelopes. (Images: Kim Sass, Steve Smigielski)

Viscous glass as structural form-finding (student team: Crystal Schmoger, Kamillah Ramos)

The following project pushed the idea of harnessing the physical attributes of warm glass, together with an assistive mold, to define form, rather than casting warm glass into a predefined form in a conventional mold even further than the previous project. Most materials change their phase from solid to liquid rapidly at their melting temperature. Glass, on the other hand, changes gradually from brittle to elastic, viscous and liquid over a rather large temperature range. This glass-specific behavior is used in hot glass techniques to create complex forms without the need for any molds. This project uses heat similarly as a form-definition tool, but in controlled "black box" of the kiln.

Figure 2-5 Flat glass sheet transforms into surprisingly long glass droplet in the kiln. The artifact measures 30 cm in width. (Image: Crystal Schmoger)

Heat, together with a simple assistive mold, and a particular type of glass, offered formal transformations in glass that led to a language of melted-glass droplets. The students made initial tests with flat sheet glass placed on top of an expanded metal grille to create three dimensional glass artifacts in the glass kiln. With several tests, and carefully calibrating the working temperature and working time, surprisingly long glass droplet columns emerged. The students identified the relevant variables of this system, which included:

- glass kiln program (working temperature and working times);
- glass composition: CoE (coefficient of expansion), glass mixture, thickness of glass sheets and number of glass sheets layered on top of one another;
- geometry: metal grille opening form, thickness of steel, size of opening, height of droop.

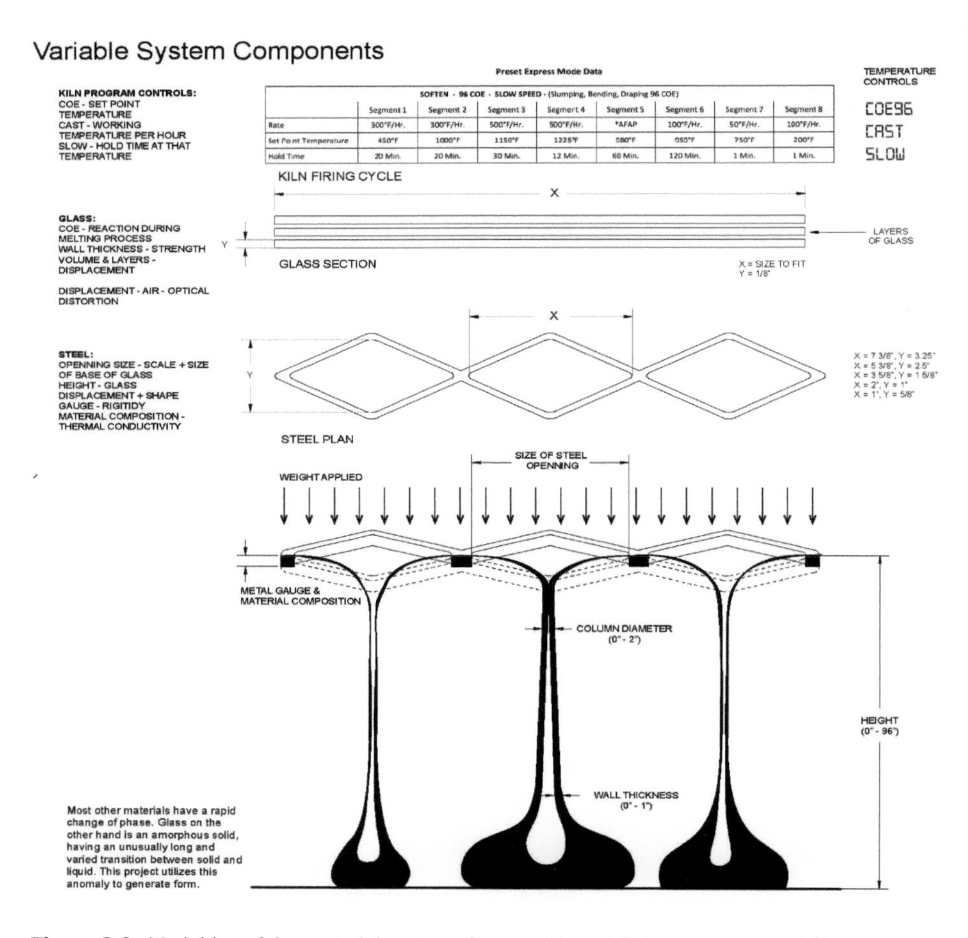

SOFTEN - 96 COE - SLOW SPEED - (Slumping, Bending, Draping 96 COE)								
	Segment 1	Segment 2	Segment 3	Segment 4	Segment 5	Segment 6	Segment 7	Segment 8
Rate	300°F/Hr.	300°F/Hr.	500°F/Hr.	500°F/Hr.	*AFAP	100°F/Hr.	50°F/Hr.	100°F/Hr.
Set Point Temperature	450°F	1000°F	1150°F	1225°F	980°F	950°F	750°F	200°F
Hold Time	20 Min.	20 Min.	30 Min.	12 Min.	60 Min.	120 Min.	1 Min.	1 Min.

Figure 2-6 Variables of the material system. (Image: Crystal Schmoger, Kamillah Ramos)

The students tested the structural capacity of the resulting droplet or mushroom column geometries by loading the glass droplets with bricks. With a diameter of just 1/8 in. (3.3 mm), a glass droplet column carried a brick load of 750 g. This surprising capacity (glass weight to brick weight ratio) led to speculations about whether this process was suitable for structural form-finding. It draws formal similarities to Frei Otto's studies in liquid threads by mixing glue and syrup (Otto [3]). In its viscous state, the glass is pushed down by gravity, putting it into purely tensile stress. By placing weight on top of the glass artifact, it seems that the resulting shape is mainly in compression, providing an advantageous force flow from the top plate down to the thin column shaft and distributing it to the footing. Further variables the students tested is the inclusion of copper fibers and strand to increase heat resistance as in wire mesh glass or to potentially act as reinforcement. They also tested the luminous qualities of these glass geometries which showed that light entering from the top is dispersed horizontally due to the refractory properties of the mushroom geometry.

Figure 2-7 Stunning structural performance of the glass droplet. Each brick measures 10 cm long, 3 cm high and 5 cm wide. (Image: Crystal Schmoger, Kamillah Ramos)

The students used the preliminary testing results to speculate on the potential maximum size of this process to create building-scale columns and speculate on a potential onsite fabrication method. Sheet glass is easy to transport, and moved commonly to building sites. The idea of shipping flat glass and casting it into a three dimensional form onsite is promising, and challenges common notions about the size of architectural glass elements, their structural potential and the idea of onsite fabrication. Ultimately, the students proposed a de- and re-mountable onsite kiln that could produce glass umbrella columns up to 10 ft. (3 m) high, cast directly on refractory masonry footings. The overall spatial proposal is a field of glass columns. As precedents, students looked at the Johnson Wax Headquarters by Frank Lloyd Wright and the High Life Textile Factory by Felix Candela (Faber [4]).

Figure 2-8 Spatial proposal showing the drooping glass forms frozen in time and the subtle differences between each cast. Model scale is 1:25. (Image: Georg Rafailidis)

Monolithic glass structure through electric arc welding (student team: Taras Kes, Andrew Kim)

While most projects took glass in various forms, and investigated techniques and formal outcomes of glass forming, the third project documented here investigates the architectural potential of glass making, turning sand or silicon dioxide into a range of vitreous or glassy substances using both the kiln and electric arc welding. This type of electric arc furnace was documented in depth by the French chemist Henri Moissan in 1904 (Moissan [5]). Students worked with silicon dioxide or silica, rather than glass, and experimented with various additives including soda ash and lime stone to lower the melting temperature of the sandy mixture, and to save energy. Initial experiments were conducted in a plywood "sandbox" using graphite rods in an electric arc welding process. The electric arc welder can produce temperatures up to 6500 degrees F (approximately 3600 degrees C), which is well above the melting temperature of sand. First tests resulted in fulgurites, small pods characterized by a vitreous interior and a chimney much like those created by lightning found in nature.

Figure 2-9 Typical electric arc fulgurite. Diameter 5 cm. (Image: Taras Kes, Andrew Kim)

In an architectural context, this process opens-up a series of glass-specific potentials and questions:

- Glass making is slow due in-part to long annealing times. Glass making with an electric arc is instant. Are there techniques that make stable glass artifacts through this process that could eliminate the process of annealing?
- Glass making happens in large industrial settings. Could the electric arc technique allow for localized, low-tech, onsite production?
- Glass elements are typically connected through cold forming techniques like gluing. Could the electric arc allow for glass to be used as an adhesive, creating truly monolithic glass structures?
- Glass could be produced where sand is, avoiding the transport of the raw material and allowing the monolithic structures to disintegrate back into the natural setting after their use.

The initial experiments investigated different material mixes and used both the glass and ceramic kilns as well as the electric arc for heat sources. Results from the electric arc process consistently produced dark, smooth vitreous bodies encased in sandy shells (fulgurites), whereas kiln-based processes managed to fuse the sand into solid, crumbly swatches, but didn't transform the material visibly into a glassy, vitreous body.

The students continued by conducting experiments into how to modify the shape of the initial, typical fulgurite pod or egg. Tests demonstrated that when the electrodes were close to the sand surface, the hot gasses form a chimney to the sand surface where the gases escape. By placing the electrodes further down in the sandbox, egg shapes or pods are formed with two openings which connect to the graphite rods conducting the electrodes. Tests also demonstrated that the electrodes can be moved after fulgurites are made, to create interconnected, longer, vitreous artifacts. The graphite rods were used together with steel tools to pry-open the fulgurites when their glassy interiors were still molten; the results were patches of dark, multicolored glass with a sandy, crusty underside. The electric arc was also used to weld these glass patches or pieces together without any added material, creating monolithic glass artifacts. Further tests investigated how pre-formed glass pieces could be attached, like glass marbles or glass rods.

Figure 2-10 Fusing of glass fulgurites through the proposed electric arc method. (Image: Taras Kes, Andrew Kim)

The electric arc proved to be a suitable tool to form glass instantly in small batches from raw materials. How could one think of creating architectural form from this specific glass fabrication process? If used for fabrication in areas where sand is already naturally occurring (coastal areas, deserts), then it might be helpful to think of sand also as a formwork material to stay true to a pure, monolithic glass concept of construction. Students conducted test to see what forms sand adopts when it accumulates through pouring. When centrically poured, cones form. The angle of the cones vary depending on the humidity of the sand. Dry sand form cones with 30-degree angles. Wet send can generate sand with steeper angles, up to 50-degrees. Electric arc fulgurites can be formed underneath the existing sand surface, withdrawn, pried open, and placed upon the formed sand cones that act as formwork. The glass artifacts would then be welded together through electric welding, forming a monolithic structure that consists of the same material as the surrounding natural environment. By digging out the interior sand, the glass shell remains. Damages to the structure could be repaired using the electric arc technique with the available material onsite. The structure could be demolished onsite by simply breaking it into smaller particles and mixing it back into the surrounding sand. Glass has an exceptionally long life span. When not mixed with other material, glass can retain this long material lifespan of thousands of years.

Figure 2-11 Spatial proposal for a monolithic glass structure. The right model is made with the proposed electric arc welding process. Model scale 1:25. (Images: Georg Rafailidis)

Wood mold as glass joint (student team: Kyle McMindes, Matt Meyers)

The final project documented in this paper is one of the many blow-glass prototypes fabricated for us by the skilled glassblowers and gaffers at the Corning Museum of Glass in Corning, NY. For two years, staff of the Hot Glass Programs at CMoG have worked with us, reviewing drawings throughout the semester, and then fabricating select pieces using molds made with fruit wood, constructed by students. Unlike projects that used the glass kiln, to which students had daily access, projects that involved blown glass prototypes weren't subject to a process of experimentation and trial-and-error through the semester. The experience nevertheless yielded unexpected results that students could extrapolate on through drawings. The project shown here takes both the remains of fruitwood molds and the blown glass components both as part of a structural assembly. Molds – not just in glass fabrication, but more widely in many fabrication processes – are often left as the invisible template that defines a form. They might be kept in an archive for further form-reproduction, or, if damaged or "exhausted" after several castings, are often tossed away. This project takes advantage of the fact that a fruitwood mold, in glass blowing, exhausts after, on average, 6-5 glass units are blown. The simple design of the mold allows it to be disassembled into a number of sticks that can be used to form wood joints in a ridged, bulbous glass block assembly. Because the glass is formed directly against the wood, the pieces of wood fit perfectly in the notches of the ridged glass units. Joints, in an all-glass assembly, are an issue because of the hardness and fragility of glass; wood, in contrast, is softer, is more flexible and able to absorb structural stresses. Fruitwood molds, in glassblowing, if they are made out of several pieces, have to be connected mechanically. The fact that the mold is constructed using mechanical fasteners (the sticks are screwed on a baseplate) also makes the disassembly of the mold quick and easy, without causing damage or change to the wood members. The wood joints would offer a number of opportunities in how to use and configure the glass assembly – the assembly could be tied-into a wood primary structure, for example. The combination of the "scrap" pieces of mold wood become an asset in an otherwise very formal and rigid type of construction (glass), enabling a glass construction to be created more flexibly and casually as a wood construction. It also questions the necessity for a material hierarchy in fabrication techniques, in which many fine, re-usable materials are needlessly tossed to produce a certain "finished" object or product.

Figure 2-12 Hot glass prototyping of the glass block with respective wood mold/joint at the GlassLab of the Corning Museum of Glass. The wood strips of the mold are used as joints between the six blown glass artifacts. (Images: Georg Rafailidis)

3 Conclusion

An fascination with the seeming inaccessibility of glass as a material for architectural experimentation lead to two years of working with students and glass through a range of techniques, including kiln-formed warm glass, the elemental process of glass making with sand and high heat, and glass blowing with wood molds, at the prestigious Corning Museum of Glass. The original motivation for this experimentation and collaboration was the design of our "Thermometric Façade" unit – a temperature-responsive glass-block unit filled with wax PCM, whose performance relies on an extremely precise interior cavity. Through handling glass and witnessing its behaviors and potentials first-hand through two architectural design studios, we can begin to imagine ways of generating a glass block prototype, a long-awaited proof-of-concept with architecturally true materials. The experiments and design proposals that came out of the glass studios nevertheless stand on their own as design research into a new paradigm for architectural glass as a highly plastic, tactile, elemental and three dimensional material in architecture. In architecture glass is typically used for being invisible and flat. Architectural glass either disappears through transparency or by reflection. It is typically not considered suitable as a structural element and is regarded as an energetic "problem" due to low insulation values. The documented projects question these architectural preconceptions. The rich history of glass fabrication and glass components, as well as contemporary material developments suggest alternative readings of glass, as a material with a much more malleable and variable materiality than generally thought.

4 References

[1] Hobbs, P.V; Mason, B.J.: Sintering+Adhesion of Ice. In: Philosophical Magazine. Vol. 9, 1964, pp 181 - 197.

[2] Matusda, E.; Hensel, M.; Menges A.: Aggregat Gefertigter Partikel 01. In: ARCH+. Vol. 188, 2008, pp 80-81.

[3] Otto, F.: IL 28. Stuttgart: Institute for Lightweight Structures, Stuttgart University, 1994.

[4] Faber, C.: Candela, the Shell Builder. New York: Reinhold Pub. Corp, 1963, pp 88ff.

[5] Moissan, H.: The Electric Furnace. Easton, PA: The Chemical Publishing Company, 1904.

Transparency „saving" ruins

Styliani Lefaki[1]

1 Architect, Aristotle University of Thessaloniki, Greece, slefaki@arch.auth.gr

Ruins are threatened by decay and oblivion, forces to which they must be exposed in order to keep their meanings, and from which they must be protected in order to survive as ruins. This makes their preservation, a paradoxical attempt, which often uses transparency in order to confront this contradiction. Transparency, as a simultaneous presence and absence, offers hypothetically an invisible protection and its applications, mostly glass constructions, try to eliminate their presence in order to be integrated into the genius loci of an evocative decay. The gained experience proves that this simultaneous fulfilment of theoretical and technical requests is sometimes an over demanding design goal.

Keywords: invisible protection, authenticity, „as found" condition, in situ, interpretation context

1 Introduction

Ruins are defined as the disjunctive product of the intrusion of nature upon an edifice without loss of the unity produced by the human builders (Hetzler [1]). When ruins are imbued with messages from the past, they are the valuable traces of former worlds. The common responsibility to safeguard them for future generations is internationally recognized, as well as our duty to hand them on in the full richness of their authenticity.

1.1 The frame of ruins' protection / the state of questions

Authenticity of ruins is difficult to describe not only because the term has various interpretations but also because of the ruins' multiple natures that range between the visible and the invisible. Ruins, defined by the predominance of decay, designate the void as their indispensable core. Their visible presence points to durability even if *„that which is, is no longer what it was*" (M. Roth [2]) Thus their authenticity refers not only to an original state but also to material and immaterial values, to procedures between human intervention in nature and nature' s intervention in the human. Therefore, they are inevitably connected with their *„as found*" condition, namely the condition in which they exist at the moment of their evaluation as worth protecting. Any intervention to prolong this existence carries the risk to import a disturbance in authenticity of ruins and their genius loci.

Hence, the preservation and protection of ruins is one of the most difficult problems that experts confront (Athanasiou et al. [3]). The various disciplines that are involved into the

Engineered Transparency 2016. Glass in Architecture and Structural Engineering. First Edition.
Edited by Jens Schneider, Bernhard Weller.

safeguard procedures, including art historians, archaeologists, architects, engineers, museologists, communication experts and so on, are envisaging invisible interventions, without material presence if possible, simultaneously present and absent. Transparency that serves such visions has been applied in plenty of examples in the recent past, juggling always between theoretical and technical demands. The equilibrium cannot be easily achieved despite the wide spectrum of solutions that are offered.

2 Dealing with ruins – contemporary directions and aspects

Ruins have been the central issue of two conflicting ideologies in the 19th century that caused a big debate between the supporters of stylistic restoration and those of its „romantic" approach. The first wanted to rebuild the ruined structures and recreate a formal unity of an ideal pure style, even if this did not exist before, while the second considered impossible to restore anything without abolishing its authenticity. The latter appraised the immaterial values of architecture, such as beauty and memory that were created through the dialectic response of the structures to time. This contradiction needed a unified theory to be resolved, which appeared later with the theories of historical and scientific restoration, and in the fifties, with the critical approach to it.

The „critical restoration" – „*restauro critico*" in Italian – was based apart from the historical and scientific documentation, on the holistic interpretation of the historical legacy and the interdisciplinary collaboration for any intervention on it. This theory was expressed in a period of a starting criticism to the modern movement through New Brutalism, an ism that indicates a path to new insights and forms of the existing architecture and refers to the clear exhibition of structure and the valuation of materials for their inherent „*as found*" qualities (Smithson A. and P. [4]). It supported the reevaluation of the esthetic and historical qualities of ruins and their transparency in a metaphorical sence: the revealing of their inner structure and the interesting patina of their materiality.

Transparency, literal or phenomenal, (Rowe and Slutzky [5]) was also a main issue for architecture during the post-war period using glass or the new translucent materials provided by the expanding industries. No wonder why the contemporary architects used it for responding to the increasing need for ruin protection due to the war and the expanding archaeological excavations. As a result transparent protecting constructions using glass, or translucent materials, showed a big variety, which is difficult to categorize. Though one can distinguish two main groups referring to criteria connected with the general architectural form, and its scale.

The first includes shelters or enclosures while the second refers to other structures that, apart from protection, support the functionality of the historical site and offer ways of communication with the public.

2.1 Shelters and enclosures

The first solution that comes to mind for protecting ruins in the "*as found*" condition is a cover or an enclosure, that prevents rainwater, snow, wind, temperature extremes, air pollution, ensures a steady microclimate for them and arranges an infrastructure allowing their „reading", id est visitors to experience them. Shelters and enclosures can offer ways for keeping precious findings in situ, inseparable from their context, an important condition for their authenticity. But inevitably they intervene into this context even if they are minimal or highly transparent and thus any of their forms opens a big discussion about its architectural language, its scale, its openness, its response to light and to the historical environment.

The technical requirements are complicated, as any new structure is practically not permitted to touch the ground. Constructions' spans should be big to keep up with the continuity of the historical remains whose unique order doesn't usually allow the regular arrangements of columns or other bearing elements. Foundations can be arranged only with great difficulty and after thorough investigation, as the ground, part of the interpretation context, is equally sensitive and most of the time under continuous archaeological research.

Apart from finding where to „stand" new constructions have to take into account static and dynamic loads, resistance to winds and earthquakes, as well as safety standards for the working personnel and the visitors. They must consider not only the current state of preservation of the ruined structure but also its estimated restorative prospective, (Birtachas [6]) the formation of a necessary vital space for further works and restoration activities. There are also problems of transporting materials and erecting a structure because most of the time the possibility of movement is limited through excavation grids, or through particular environmental arrangements. These limitations lead engineers to propose static solutions bridging wide spans, or using extreme cantilevers, as the proposed open-air museum of the Roman Villa of Herodes Atticus in Kynouria in Greece. (Karydakis et al. [7]) The proposed roof that covers an area of 4796 m^2, with possible expansion of another 811 m^2, is based only on 12 columns. Likewise the shelter for enhancement of the archaeological site of Aristotle's Lyceum in Athens, that is designed for covering the "Palestra" (wrestling school) area, a surface of 61 x 61 m. at a height of 9 m. with an arc shaped roof, like a huge umbrella, is supposed to be suspended by 6 arches and founded practically on one side (Gantes [8]). Both examples, tried to be "light" though having the transparency issue as a secondary design goal. Although they declared that they had taken restrictions of perception into account they were invasions, which scale arose many doubts. Transparency could eliminate the new intervention and keep better the continuity and character of the historical site. Besides a historical landscape that gained qualities through centuries, cannot be easily put under shadow, while ruins can easily loose their meaning and historical value under an opaque ceiling.

2.1.1 Transparent shelters with contemporary morphology

This requirement guided many solutions to choose transparency as the main issue in their concept like the case of the roman baths in Badenweiler (Accardi [9]). Although these ruins, short after their discovery at the end of the 18th century, were protected with shelters, they were inscribed in public memory, as part of an open archeological space. This openness, the respect for the environmental context and for the continuity of the historical site, still under research, lead to a glass shell, almost as an enclosure, that spaned approximately 68 x 40 m in plan, roofing a very important part of the museological narration of the archeological site.

For the same reasons, various glass shelters were erected in many archeological sites in Greece as well as on the acropolis hill, trying to install an invisible protection to valuable findings that would not interfere into an environment of high importance. Though the mediterranean climate and strong sun light carried unpleasant surprises with reflections, rise of the temperature underneath and an altering behaviour of glass, that needed a level of maintenance not easily reached.

Opaque structures that have been used instead were undoubtfully more invasive even if they aimed to lightness through sophisticated structural systems. Mostly resulting from emergency and temporary measures established a permanent devaluation of the esthetical and historical „load" of a place and alienated the ruins of their context. A characteristic example is the textile canopy that protects a unique world heritage monument, the temple of Apollo Epicurius at Bassae in Greece, which downgrades the qualities of the temple that depend on its environmental references. The case study at the Delft University (Veer et al. [10]), took the challenge to design an „invisible" glass shell with a 24 m. span, which meets the strict requirements of space and safety for the visitors and the on-going archeological works. The study offers an interesting pioneering solution, that has to respond to complicated environmental factors, answer questions of perception and avoid the danger of eliminating the importance and character of the ruined fabric through the magnitude and impressiveness of its „shiny" technology. A compromising solution balancing transparency and demands for physical light with sun protection and a moderate exposure of its technology, shows the shelter for the terrace house 2 in Efessos, with vertical surfaces of polycarbonate louvers and a roof covering 4000 m^2 with a textile membrane strengthened with fiberglass and Teflon coating. The result is light and its semi-transparency helps it to fit better into the surrounding archeological landscape. (Ladstätter and Zabrana [11]).

2.1.2 Shelters and enclosures reinstating original forms

A challenging use of transparency happens when shelters or closures establish a dialogue with the historical fabric as a reinstatement of its original form, presumed to exist before ruination. In this way they try to enrich the frame of reference and meaning of the *„as found"* condition. The reinstatement of the form, thanks to transparency, stays on an abstract level that can bear many readings, far from a univocal reconstruction.

This approach was firstly introduced in the fifties by Franco Minissi (1919-1986), the architect whose work is nowadays still connected with the visions of transparency and its use as a hermeneutical tool for the interpretation and musealisation of historical evidences. His projects remain an important reference and theoretical support for a plethora of subsequent related interventions. One guiding example is the protection of the ruined structure and the precious mosaics at Villa del Casale at Piazza Armerina in Sicily, (1958, 1963-67, 1980). Minissi evoke the contours of the original volume of the villa with a transparent protection, thus setting up a museum in situ. A supporting metal structure fixed at the basement of the ancient walls was covered not with glass but with perspex, a poly-methyl- methacrylate material that ensured a certain degree of transparency and lightness. Glass because of its weight was used only for vertical wall coverings, filling the space between the remaining columns. The project stood up to the challenges of a period highly concerned with the theoretical and interdisciplinary research in the field of architectural restoration, but the technology of the time didnt prove mature enough to support its profoundness. The same happened with the transparent protective structure of the Greek Theater at Heraclea Minoa (1960-1963), the most emblematic example of the „virtual" design of Minissi (Vivio [12]). This coating, like an abstract redesign of the monument, was shaped over the historical fabric on the basis of pre-existing mouldings of acrylic resin re-stating the form of the theater. The material used betrayed the architect as with time its transparency started to fail because of incoherent deposits and its deterioration.

Technological improvements of structural glass were tempted to continue Minissi's ideas. The protective glass and steel building, over the ruins of the roman baths of Colonia Ulpia Traiana, 17 m high and 108 m long, together with the Roman Museum, (Accardi [9]), built on the excavated foundations of the entrance hall of the public baths that reflect the design and dimensions of the original roman structure, can be considered as a revival of the Villa Casale's principles. The construction in this case doesn't show failures but instead of evoking the past tries to rebuild it in an alienated context, insisting on the volume against other qualities of the architectural remains and raising thus questions of authenticity.

The synergy of theoretical aspects with the challenges of the tecnology of structural glass vindicated Minissi in other cases like the glass roof over Juval Palast in South Tirol (Wurm [13]). A transparent roof protects the ruined part of the castle reinterpreting its original form. Other projects followed the principles of the fifties with improvements in the technological options and compromises. The transparent cover for the remains of the odeum of the great thermae of Dion, in Northern Greece, for example, covered with perspex, (Karadedos [14]), reinstates the original form of the koilon, in an abstract way and without touching it. It is elevated from the ancient fabric and this gained height creates a luminous ventilated space underneath that eliminates problems of microclimate and allows further scientific research. At the same time the cover can be used as a theater and as a „reversible" structure doesn´t put the authenticity of the historical remains into question.

2.2 Musealisation and auxiliary glass constructions

The management of a historical site seeks ways of communication between ruins and public. Arrangements like catwalks over or next to precious fragments, the framing of fragile pieces, structures in direct interaction with the historical fabric, partial enclosures and in general the arrangement of a feedback system for interpretation are elements that influence the experience of visitors. These elements cannot avoid transparency as it allows the protection of the historical fabric and its unimpeded view at the same time. Their relatively small scale can be controled more easily for avoiding failures.

Figure 2-1 Ancient Theater at Side, Attalya (photo archive G. Zoidis, April 2014).

In this field of applications Minissi again opened the debate with a unique approach that also failed technically, but allowed a research prospective. In 1952 at Capo Soprano at Gela in Sicily he introduced an innovative solution for saving the vulnerable Greek wall, brought up to light between the years 1948 and 1954 and threatened by its sudden exposure to the atmosphere. He armoured the delicate mud-brick upper zone with a glass structure substituting the pressure exercised for centuries by the earth layers. This pressure belt consisted of tempered glass plates laid in successive rows on both sides of the sensitive area and tightened together bilaterally with rods of stainless aluminium alloy through screw hubcaps with a plastic interface (Vivio [12]). A metal construction covered by perspex offered an extra protection to the system, but was removed during the eighties by the „Soprintendenza dei Beni Culturali" of Caltanissetta for esthetical reasons. That increased the already problematic penetration of water and together with the high temperatures induced by the transparent surface and the complete lack of maintenance created a greenhouse effect between the glass and the ancient structure. The constructions are today removed and substituted by a textile canopy.

Without this function of reinforcement, glass plates are today widely used as showcases to protect valuable surfaces on floors or walls. A common use is glass bridges, like the walkways in the Basilica of Aquileia in Italy that allow an unimpeded view of the mosaics, while at the same time protecting them against wear and tear. A slender stainless steel structure is suspended from the new roof construction bearing a walking surface of sheets of laminated glass (three 12 mm layers, with a 6 mm top layer that can be replaced at regular intervals), (Baldassini [15]).

A transparent floor of laminated glass seems a common solution when remains of another historical phase have to be saved and revealed for enriching and completing the meanings of a place. The theater in Agora in Thessaloniki has a glass orchestra that exposes its former structural phasis; in the Cathedral of Atri in Italy a glass floor consisting of four oversized showcases offers a strictly controlled environment to the excavated remains that are permanently displayed to the public (Aslan [16]). The New Acropolis Museum is also using a similar concept. Enriching its museological program through the findings of the ancient town, adopted glass floors for their protection and exhibition. In any case glass and transparency formed the main concept of the museum that justified its place and its existence. The archeological treasures that could not be kept in situ are exhibited in the closest possible connection to their original context, a relation made possible through the transparency of glass. The transparent glass floor was also the only solution proposed for saving the remains of the main road axis of the byzantine era in Thessaloniki that came to light during the excavation for the town's metro and are still firing various debates.

Figure 2-2 Ancient Patara, Turkey, Odeum (photo archive G. Zoidis, April 2014).

3 Conclusions

Today' s struggle to save the valuable remains of the past in their situ and in their *„as found*" condition has to confront multiple demands at theoretical and technical levels. Transparency as a design principle plays an important role and its theoretical possibility of simultaneous absence and presence offers a way to protect the historical remains *„in situ*' and in the *„as found*" condition. Solutions are case-specific but offer an undenied experience, which shows that transparency has to be further explored. Its applications prove that the results can be easily ambivalent but can be helpful for the development of a specific methodology based on a profound analysis, documentation and evaluation of the existing situation without giving a matrix of answers. Though the requirements of a dominant theoretical approach with visionary shades create always a challenging background for pioneering technical solutions.

4 References

[1] Hetzler, M. Florence: Casuality: Ruin Time and Ruins.; LEONARDO, Vol.21, No.1, 1988, pp.51-55.

[2] Roth S., Michael, Irresistible Decay: Ruins Reclaimed; Michael S. Roth with Claire Lyons and Charles Merewether, Getty Research Institute, 1997, p.p. vii.

[3] Athanasiou, A. Fani; Drossou, P. Constantina; Theoharidou, L. Kalliopi: Shelters protecting archeological sites: Problems in their unity, principles and parameters in their design. Proceedings of ΕΤΕΠΑΜ conference meeting: Shelters and Enclosures for the Protection of Archeological Sites, 1st June 2007, Thessaloniki 2008, pp. 12.

[4] Alison & Peter Smithson: The "as Found" and the "Found". The independent Group: Postwar Britain and the Aesthetics of Plenty. Edited by David Robbins. The MIT Press 1990, pp. 201.

[5] Rowe, Colin; Slutzky, Robert: Transparency. Birkhäuser Verlag 1997.

[6] Birtachas, A. Panagiotes: Investigating the potential multi-functional use of a shelter over ruins with application to the Temple of Apollo Epicurius at Bassae. Proceedings of ΕΤΕΠΑΜ conference meeting: Shelters and Enclosures for the Protection of Archeological Sites, 1st June 2007, Thessaloniki 2008, pp. 65.

[7] Karydakis P., Anagnostaki C., Tsokanis K.: Steel structure for the open air museum of the "Roman Villa of Herodes Atticus", Proceedings of 6th national conference meeting on Metall Constructions, Ioannina, 2-4 October 2008, pp. 254-262.

[8] Gantes, Charis: Nonlinear Finite Element Analysis, Encyclopedia of Earthquake Engineering, DOI 10.1007/978-3-642-36197-5_138-1, Springler-Verlag Berlin Heidelberg 2014, pp.13, 14 of 35.

[9] Accardi, Aldo Renato Daniele: Architectures on "ruins" and ambiguous transparency: The glass in preservation and communication of archeology, Journal of Cultural Heritage 9, 2008, pp. e 107-e 112.

[10] Veer, Fred: Oikonomopoulou Phaedra, Bokel Regina: Designing a Glass Pavilion to Protect an Ancient Greek Temple. Challenging Glass 3, Delft University of Technology, Bos, Louter, Nijse, Veer (Eds.), IOS Press, 2012, pp139-150.

[11] Ladstätter, Sabine; Zabrana, Lilli: The shelter Construction for Terrace House 2 in Ephesos: A Unique Museum and Scientific Workshop. The Protection of Archeological Heritage in Times of Economic Crises. Elena Korka (Ed.), Cambridge Scholar Publishing, 2014, pp. 234-242.

[12] Vivio, Beatrice: The "narrative sincerity" in museums, architectural and archeological restoration of Franco Minissi, Science Direct, Frontiers of Architectural Research (2015) 4, pp. 202-211.

[13] Wurm, Jan: Glas als Tragwerk. Entwurf und Konstruktion selbsttragender Hüllen. Birkhäuser Verlag AG 2007. 4.3 Scheiben und Stabförmige Verwendung in Tragwerken. pp.104-105.

[14] Karadedos, Giorgos: The Odeum of the Great Thermae of Dion. Study for its protection, conservation, restoration and enhancement (2007-2008). G. Karadedos Architectural Projects. Aristotle University of Thessaloniki 2009 pp. 211-212.

[15] Baldassini, Nicolo: Bridges of Glass. DETAIL 8 Dezember 1999.pp. 1428-1429.

[16] Aslan, Zaki: Protective Structures for the Conservation and Presentation of Archeological Sites. Journal of Conservation and Museum Studies, November 1997, 3(0), DOI: http://doi.org/10.5334/jcms.3974 pp.16–20.

[17] Schober, Hans: Transparente Schalen. Form Topologie Tragwerk, Wilhelm Ernst&Sohn, 2015.

Towards 50 % saving of embedded energy in glass envelopes

Mick Eekhout [1]

1 Professor at TU Delft and director of Octatube, Delft, m.eekhout@octatube.nl

Glass facades became quite fashionable amongst architects in the last decades. They symbolize transparency in organizations and display also the structural composition of buildings. However, transparent glass facades require thick insulated glass units with a high amount of embedded energy per m². How can this ecological investment be radically reduced, for example to 50 % of the current amount over 5 year's time? The ambition of the use of only half the embedded energy in the total chain of the building envelope would include a quest involving all related parties in the building process: the glass recyclers, the glass melters and float glass producers, the glass production processors, the coaters, the glass panel manufacturers, the architects, the structural engineers, the façade engineers and façade producers, the norm committees, the governmental approving bodies, the investors, owners and users. This challenge or call would lead to an industry-wide collaboration on international scale.

Keywords: minimal embedded energy, extra thin glass, accepted flexibility under loading

1 Introduction

In the 1980-ies all-glass facades were realized, after the glass houses of la Villette (1986) by Peter Rice, Ian Ritchie and Martin Francis (RFR). Two clear examples of all-glass envelopes are the Glass Music Hall in Amsterdam (1990) architect Pieter Zaanen, sized 9 x 22 x 9 m³ see fig. 1-1 (left) and the Santander Cube in Madrid (2008, architect Alphonso Millanes, sized 30 x 30 x 21 m³, see fig. 1-1 (right).

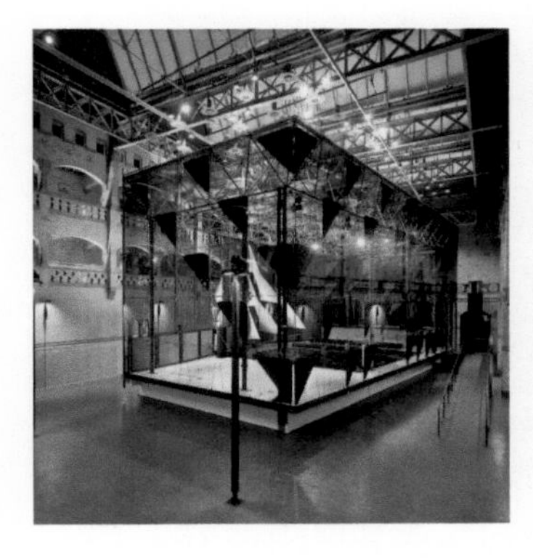

Figure 1-1 Glass Music Hall, Amsterdam (1990, left), Santander Cube 30 x 30 x 21 m, Madrid (2008, right).

Engineered Transparency 2016. Glass in Architecture and Structural Engineering. First Edition.
Edited by Jens Schneider, Bernhard Weller.

wo last projects are prototypical for the increased amount of glass used in these applications. The increased amount of glass used in almost 20 years time is alarming in ecological sense. The 8 mm thin fully tempered glass panels, suspended from the top to the bottom had a size of 1.8 x 1.8 m^2. The IGU glass panels of the Madrid Glass cube had a size up to 2.5 x 3.5 m^2, and a total glass thickness of 32 mm. Architects dream of using ever larger panels in their designs. This initiates a wake-up call for the façade industry.

2 The current ecological problem of glass facades

It was only 20 years ago when the author visited the glass factory of Vegla in Aachen and its then director Dr. Schmidt said proudly about the production capacity of the factory *"That Vegla consumed more energy than the entire city of Aachen"*. Understandable in its time. More than half the production went into car glass, less than half was architectural glass. About spending energy in the production process we think differently these days. European policy enforces large savings in energy and CO_2. The development of structural use of glass since the 1980-ies and the further development of all-glass structures has enabled technical designers to make building envelopes completely out of glass with only minor metal fittings for connections, but at a price. The director of Octatube has been engaged for almost 30 years and the professor of Product Development at the faculty of Architecture TU Delft for 25 years in the high-tech quest for more intensive use of glass in architecture, purely from the intellectual perspective: how far can we go with this brittle material to use it for constructions and even for structures? See several Octatube projects, fig. 2-1 to fig.2-2 [www.octatube.nl]. In 1992 the quest for 'Zappi' was started: the yet unbreakable structurally reliable transparent material. It took almost the entire appointment time as a full professor to chase this ambition, before the successful end of the quest could be proclaimed. And to start a new challenge: the 50 % reduction of embedded energy in all-glass facades and roofs.

Figure 2-1 Atrium glass roof Municipal Museum, The Hague (left); Cable net facades Market Hall, Rotterdam (right).

Figure 2-2 Exterior Van Gogh Museum, Amsterdam (left); Interior Victoria & Albert Museum, London (right).

The quest for better understanding how to realize glass structures still continues and has many colleagues making interesting proposals and realizations. But these days a morale cloud is hovering over this technical research & development. Times have changed, slowly but inevitably. This and the coming generation of engineers experience the limitations of materials and of fossil energy. We are surrounded by the world wide problem of sustainability and growing scarceness of materials. These limitations conflict with the high tech developments in the architectural structural glass technology of the last decades. Will high-tech end? Basic materials will become more scarce as the consumption continues at the current scale. We should restrain ourselves in our designs. A new future would need another mind set. As Albert Einstein said: *"You cannot solve a problem with the same mind that created it"*. The entire world, also the underdeveloped countries will copy our technology, so the problem will enlarge inevitably if we do not develop other, more sustainable solutions.

All the processes of float glass, pre-stressing and laminating are separated from each other. High temperatures and body-heated materials are central characteristics of the glass production. The total investment in fossil energy is a problem that regards the building industry, but it especially effects the so-called 'all-glass facades' or frameless glazing facades and roofs and all stakeholders engaged with it. It could very well be that all-glass facades will be banned in future out of ecological reasoning on European level. Certainly a fitting topic for this Engineered Transparency Conference in 2016. We cannot look away from this problem. We have this wake-up call as a challenge. With reasoning on the limited material resources in the world and investing less energy in the production of building components, a further quest for a far more minimal ecological footprint has to start. The phase of denial of the problem will be over in some more years. The use of scarcer materials is one element of concern, the embedded energy in each high-tech glass panel is a second and even more important issue.

3 Increasing thickness of glass panels

In the last 25 years of glass developments at the design & build Octatube company in Delft the glass thickness usage went from a modest 8mm (in the Glass Music Hall, Amsterdam, realized in 1990), to a currently more regular thickness of 32 mm (in the Santander Bank head quarters Glass Cube, Madrid, realized in 2008). The m^2 use of fuklly tempered glass has quadrupled in 20 years time! Honest to mention that 8 mm was for inside and 32 mm for outside purposes, but the reality is challenging us!

4 Growing sizes of glass panels

Also the sizes of the glass panels have become bigger: in the Glass Music Hall in Amsterdam 1.8 x 1.8 m^2 was used. In Madrid the size of the panels became 2.5 x 2.5 m^2 with top panels of 2.5 x 3.5 m^2. For wind loading of facades the size of the glass panels when supported at the edges matters in the equation $M = 1/8\ QL^2$. Frameless glazing have higher bending moments that line supported glass panels in framed glazing facades. So when size increases, the bending moments increase and normally also the thickness of the glass panes increase. Smaller sized panels would need smaller sized constructions to hold it up, which also have to be studied. Architects want bigger and larger glass panes, just to show that they were the first to be so smart. Since Apple instructed the construction of large autoclaves for their Apple headquarters in California, insulated glass panels of 3.2 x 15 m^2 are possible. It is rumored that regular production of 3.2 x 18 m^2 will be possible. But they go with the production and transport price that Apple only can afford. One could state that Steve Jobs not only popularized I-phones, but has indeed indirectly been extremely important for the architectural glass production industry.

5 Stiffness of glass panels

But designing buildings with larger sizes and obligatorily following the bending moment formula $M = 1/8\ QL^2$ constantly as was already done more than one century, means that bending moments from perpendicular loadings on the glass will increase and that, depending of the required stiffness the thickness of the glass panels will increase. Stiffness of glass panels, or the lack of stiffness in many cases, is a problem during design and engineering. It is also prone to national or European acceptability, laid down in (inter)national norms. These norms have once been established as a certain safety factor on the possible breakage of annealed glass and an average of personal opinions of representative of the glass industry. The (bending) strength is different for various types of glass like annealed glass, laminated annealed glass, heat-strengthened and fully tempered glass, single or laminated. Thermal pre-stressing of glass has strengthened float glass 3 to 4 times, but unfortunately its stiffness has not been increased in the same manner. The higher the pre-stress, the higher the glass panels can be loaded perpendicularly, as the strength is ok, but at the same time more deflections will show. An elephant could stand with one leg on a glass panel, but this panel would deform 200 mm under the elephant's foot. No breakage but much more deflection. We would enter from the technical domain into the psychological domain. Pre-stressed glass panels can deform largely under perpendicular loading without breaking. Is that actually a problem? Why don't we take the flexibility instead of the stiffness of glass as a point of departure for design engineering? The surrounding details have to continue their performance as well, which affects the construction of the details. The question is how large deflections in all glass panels with flexible silicone sealant seams are allowed. In the 1990-ies early point fixed applications rotating nodes were designed, so as to show the possible deformation of the glass, which never happened. What would happen in glass facades when these deformations, allowed in these 'rotating joints' would indeed occur? If the engineering would ensure that not a single element, component or material will break or fail, could we extend the normal inflexibility of structural deformation? Departing from enjoying the speed that is generated when wind deforms the sails of a sailing boat a deformation could be enjoyable. As sailors we will leave this thought for later in this contemplation. The maximum allowed deformation from bending has a technical base, a psychological base and a societal base. The technical aspects require study, proposals, mock-ups and tests, not only of the behaviour of glass planes but also of the surrounding details. The psychological aspects point at the visual acceptability for users of buildings, visitors and pedestrians. The societal base is the outcome of an open debate between all stake holders of glass in buildings and this communal opinion as a result.

Figure 5-1 How strong and how stiff is pre-stressed glass. 'Can an elephant stand on a pane of glass?'

This plea would be to take the low rigidity of glass panels as a starting point of design engineering and to start thinking in deformable or deformed glass structures. A complete other starting point as the normal attitude to allow maximum deflections similar to the panel thickness used, which are indeed personal. The Octatube engineers are used to engineer in this mode. This line of thinking is pursued in the proposals of paragraph 14.

6 Increasing insulation value

On top of the pure thicknesses of the employed glass constructions the required insulation values of glass constructions are requested to be increased. The use of glass went from single to insulated glass and from clear glass to coated glass and to low-E coated glass. In future glass panels could include triple or more layers of glass to increase the insulation value. This matter is, although quite important for the building and the energy performance over time, has to be solved with thin intermediate layers of glass.

7 Production of float glass

Since Pilkington invented and initiated the 'float glass' production process in 1959, glass is melted in an oven with recycled glass added and spread out on a tin bath for equal thickness and cooling off slowly. This process has been optimized in different float glass factories. The product essentially are 'jumbo'-panels sized 6.00 x 3.14 m, a size which is derived from regular transport, cutting and handling machines in other factories who cut these jumbo panels in smaller sizes and upgrade and finish them by pre-stressing, coating and assembling as insulated glass units. Available glass panel thickness in soda lime glass is 3 - 19 mm with special exceptions from 0.4 to 25 mm. At this moment extra thin glass (0.5 to 2 mm) is industrially produced for Smartphone's and tablets. It could initiate the solution in the quest for a dramatic energy reduction in facades and roofs of buildings.

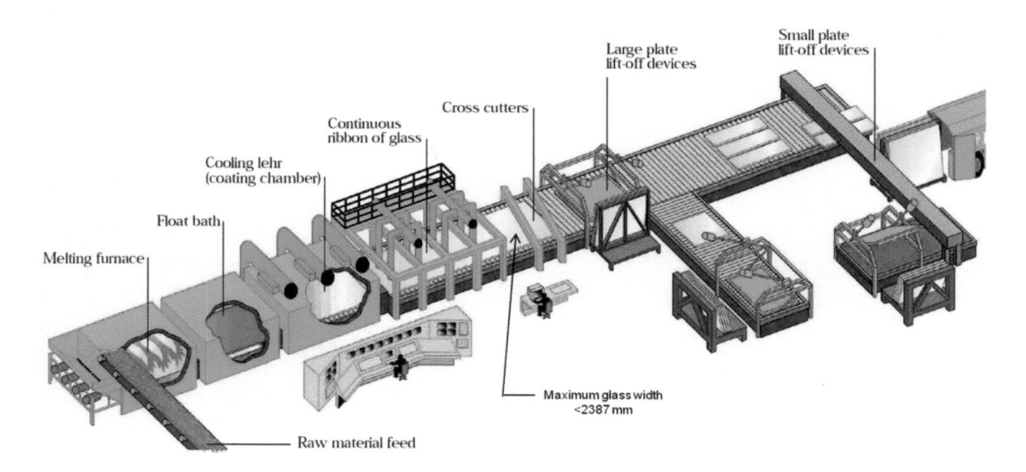

Figure 7-1 Float glass processing (photo: zbindendesign.wordpress.com).

8 Safety of glass panes

Two heat-strengthened panels laminated on top of each other will never have the same tearing pattern, so the fully broken laminated panel will always have a certain residual strength. Extra thin glass has a breaking pattern much alike heat-strengthened glass. Alternatively to thermal pre-stressing chemical pre-stressing is also possible, but mainly restricted to industrial products. The pre-stress layer in this glass is thinner; the pre-stress is higher, so chemical pre-stressing is very well suited for thin glass. No thermal treatment is involved, but chemical treatment: calcium ions are substituted by natrium ions. Using thin glass in a large scale would involve new machinery to pre-stress thin glass panels. In Europe only a few chemical pre-stressing plants are in use. The ecological effects of the production process are high and quite negative. Used temperature is 500 °C; this process is not well suited for soda lime glass and better for aluminosilicate glass (such as Leoflex). The industry is quite secretive on the ecological impact of chemical treatment of glass. The process is longer and has nasty chemical byproducts. The sizes of glass panels are limited at this moment, but this can change when the market potential will have been revealed. In the recent years extra thin glass has been developed for smart phones. This is chemically a different type of glass compared with the usual 'soda lime' float glass. The thickness of the front glass of smart phones is less than 1 mm. These glass panes are produced as float glass in different thicknesses from 0.5 m to 2.0 mm and larger sizes. AGC brings this on the market as 'Leoflex 'glass in maximum sizes of 1.200 x 750 mm^2, see fig 8-1. The stiffness is similar to soda lime glass, the pre-stress is up to 20 times stronger. Meaning that using this type of glass in architecture, smaller glass panel dimensions have to be used, and another, more flexible approach of maximum deflections. It is rumored that tempering ovens are developed which are able to produce very thin glass in an air-floated oven. Air is used for cooling off and may also be used for driving the glass panel airborne.

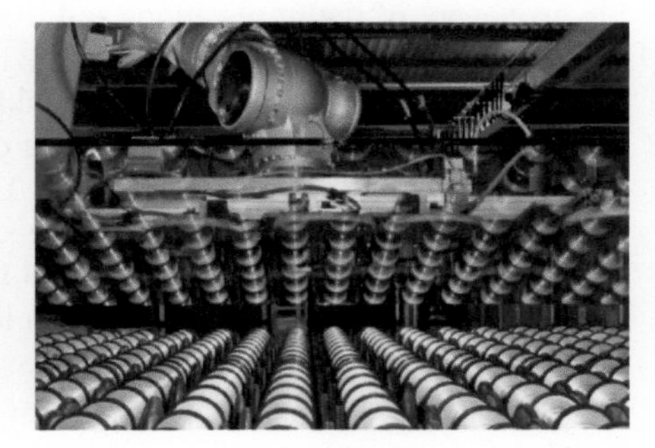

Figure 8-1 1 mm thin glass 'Leoflex' by AGC (photo: http://ceramics.org/) (left); breaking pattern of thermally pre-stressed glass (with nickel sulphite inclusion) (right).

9 Lamination of glass panes

For reasons of safety – for users of the building and the built environment – after breakage lamination of two or more panes will be a solution to keep the broken glass panel in position, whether the glass panels are laminated as fully tempered or heat-strengthened glass panels. The laminated package will have to stay in place, never falling out, never falling down. The lamination of the extra thin glass panes has to be processed without disturbing visual distortions of the thin laminated package. Thin glass can be laminated with thin PVB films or with the thicker recent lamination films. Perhaps new films have to be developed and applied for extra thin glass.

10 Façade systems

The outside pane of insulated glass panels has to take up the external loadings, has to be safe, not falling out or falling down, most probably laminated and fixed with certified chemical or mechanical fixings. The internal panes are less intensely loaded, and could be thinner. The internal panel again has to be reinforced against falling out or falling down and hence be laminated. Thin panels have to be connected with glued connections as the thin glass material is not suited for bolted connections, be it only in shear and friction. Line supported glass seems better. The better approach would be to change shear forces into membrane forces. In which case the glass will be stressed in tension instead of in bending, four-sided support of glass panels could be very efficient.

11 Architects wishes

One trend in architecture is to fight gravity and to wish the glass panels as large as possible, irrelevant of the implications and the costs. This is the route Apple is taking. This goes with exceptional costs and high environmental effects see the much published and debated Apple Cube in New York. Frankly speaking Steve Jobs has had, unknowingly, an enormous influence on the potentials of the glass industry by stimulating and partly financing larger scale glass machinery. Time for a laminated glass sculpture for Steve Jobs paid by the glass industry? We need a product champion like Steve Jobs for this quest also. The other route would be to have a more modest ecological impact and to develop glass constructions where much lower glass deadweight is employed and hence much lower embedded energy in the glass. The vocabulary of low eco glass facades will have to be developed before architects accept it and step in.

12 Engineering requirements

The engineer has to analyze the propositions of the architect in a structural way, combine them with the production possibilities of glass producers to reach a certified and safe use of glass. Engineers are used to take the dimensions of glass on the safe side. Alternatively a series of tests may have to be executed to show the load bearing capacities and the post-breakage behavior of glass constructions. Large considerations will be which norms the user has to answer to, which he can ask for exceptions, for example for the deflections. Often the deflections in the norms are taken as individually targets, rather thicker than thinner as to the producers. Consumers may be frightened by seeing bulging out or hollowing glass panels, due to the very visible reflections. Although we are accustomed by seeing the cushion formed inflated EPDF coverings nowadays, which do not explode more than their current shape. Could this lead the way to 'waving glass facades'?

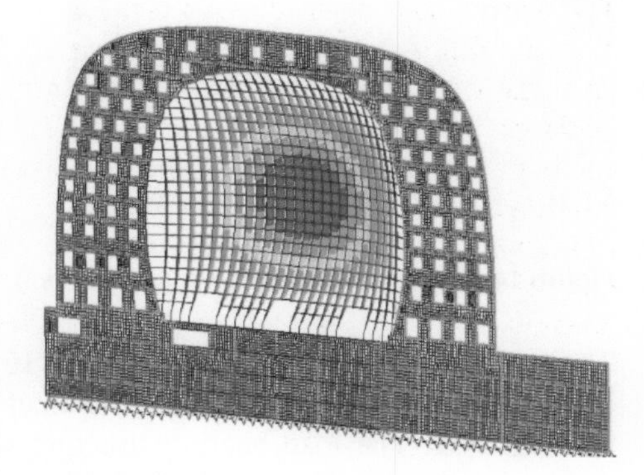

Figure 12-1 The cable net glass facades of the Market Hall in Rotterdam, 42 x 34 m (left). Deformation diagram of a fully loaded 12 Bft wind loading and a deflection on 700 mm inward and outward (right). In case thin glass would have been used, the individual pales would also bulge in or out in the cable net structure.

13 International norms

Revise the glass norms for extra thins glass after ample tests have been done in more flexible rigidity. At Octatube we usually employ a deflection maximum of the thickness of the outer glass pane, normally 10 to 12 mm. But the allowable deflection could be much more. Two mm thin glass panels could deflect easily 20 to 100 mm without breaking. The permanent functioning of the spacers and the sealant has to be warranted. A lot of development work. Safety factor before breaking of 4 would be sensible.

14 Global proposal

The proposal of these considerations is to stop the increase in the glass consumption, to evaluate the alternative possibilities and to go back to a much lower, but for all concerned parties acceptable level of embedded energy consumption in all glass facades. In these considerations on different spots the possible use of extra thin glass is proposed. The author has taken 3 different realized projects in which alternatives are sketched for all three how extra thin glass could have been applied.

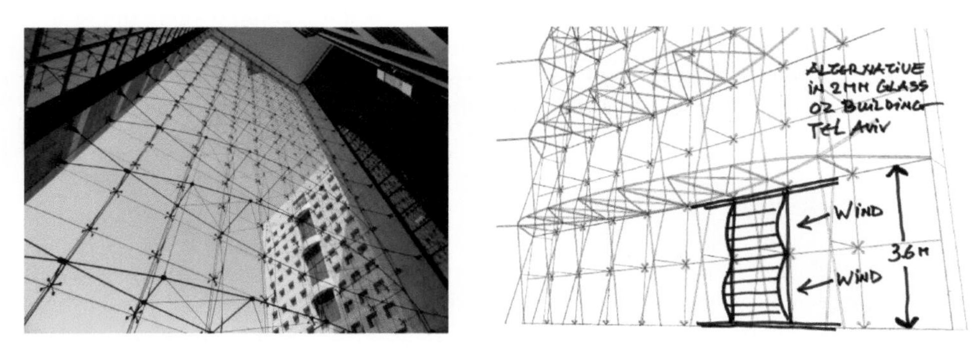

Figure 14-1 OZ building façade in Tel Aviv (left), current and alternative situation (right).

The first example is the 52 m high and 16 m wide frameless glass façade for the OZ building designed by Avram Yaski in Ramat Gan, Tel Aviv, realized by Octatube in 1995, see fig. 14-1. The load bearing structure was composed of horizontal tensile trusses; deadweight is brought upwards via deadweight suspenders to the roof structure. The glass panels 1.8 x 1.8 m^2 were 8.8.2 laminated fully tempered glass panels with 4 corner holes for M16 bolts. The tensile trusses are located at the floor levels @3.6 m. The intermediate node is stabilized by a zigzagging pair of tensile rods. The alternative in extra thin glass would be composed of panels 1.8 x 3.6 m^2, fixed at the 4 corners and at the two longer middles by glued connection saucers, glued on the inside; At the upper and lower rim the glass panels have glued stainless steel profiles which can be connected to the upper and lower panel. The glass panels are post-stressed vertically. This post-stressing will reduce the deformation due to wind loading, inward or outward or turbulence and wind rollers vertical and horizontal.

The MAS museum in Antwerp, 2011, designed by Neutelings Riedijk architects and Rob Nijsse as the glass façade designer. The current glass panels are made as laminated annealed glass. See fig. 14-2. Octatube had no involvement. This project asks for an attractive comparison. Extra thin laminated glass panels could be cold formed from the same horizontal curved steel strips that hold the glass at this moment. The cold bending towards a quarter cylinder (one inside, the next one outside directed) would require connections in the longer sides to neutralize the inward and outward bending stresses.

The third project, the Fletcher hotel near Amsterdam, architects Benthem Crouwel, 2009, has a cylindrical shaft as a second skin (no ecological purpose, only esthetics) around hotel floors. The laminated and screen- printed glass panels were produced in China.

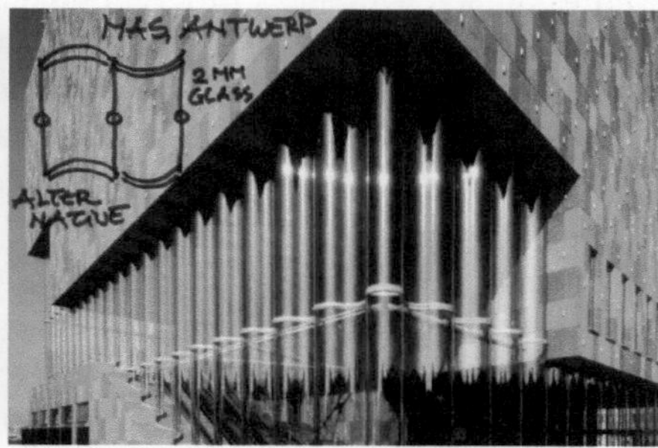

Figure 14-2 MAS museum, Antwerp, current (left) and proposed (right) situation.

The third project, the Fletcher hotel near Amsterdam, architects Benthem Crouwel, 2009, has a cylindrical shaft as a second skin (no ecological purpose, only esthetics) around hotel floors. The laminated and screen-printed glass panels were produced in China. Cold bending or laminated extra thin glass would have been quite easy in the horizontal direction in sizes of 1.5 x 3.0 m^2. The middle support would push the flat panels into the curved shape and the metal clamp lines glued on the glass would ensure the horizontal post-stressing of the façade. The short turnbuckles between the vertical strips would ensure required post-stress situation of the glass panels.

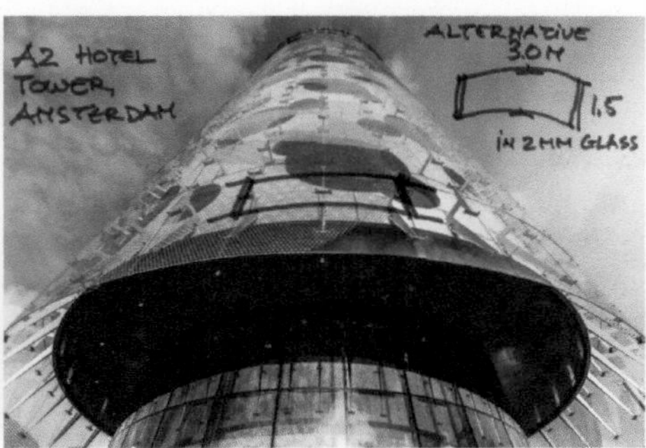

Figure 14-3 A2 Fletcher hotel in Amsterdam current and proposed situation.

15 Strategy

This requires much co-operation between different players in a platform-approach for this sort of fundamental developments. Logically it will have to become a building branch wide approach with many different stakeholders involved, both on the producing, the designing, the engineering, the approving and the consuming side of facades, see fig 19. All parties have to be awakened and activated. This is a first attempt to initiate an ecological evolution in the usage of glass in all-glass facades based on extra thin glass usage and it will have its by-effects in other glass facades.

This master plan proposal for a platform R&D project leads to an integrated chain project. Ideally, it involves many different players. Beginning with the many different glass companies with external collaborations of architects, engineers, local authorities, norm institutions, building owners and users. Setting out the master strategy with the potential participants, the willingness of potential participants to really participate will show a subdivision of the total chain (master project) into separate links (separate projects) or pieces of links (connected projects). Individual targets will be set, identifying the most suited players and inviting the external players. The emeritus professor of Product Development TU Delft Mick Eekhout is only involved as the initiator of the challenge.

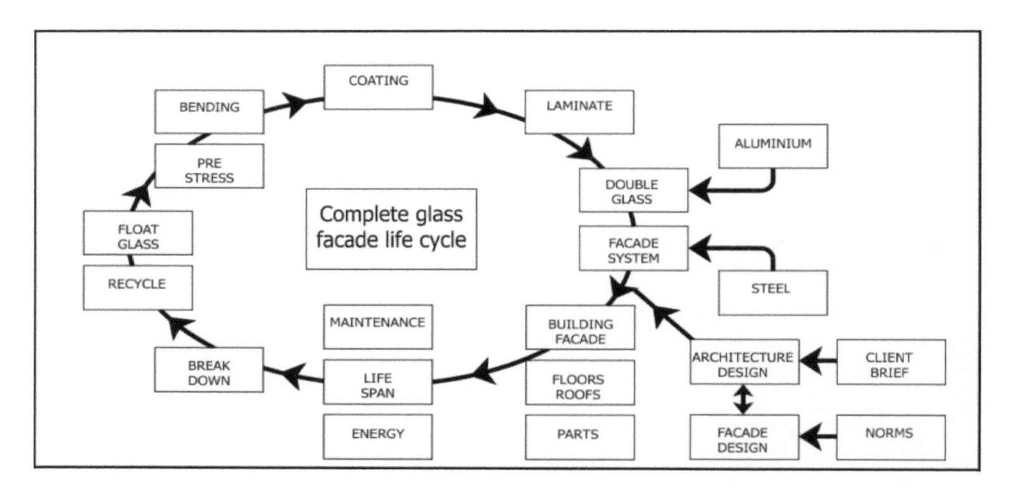

Figure 15-1 Industrial collaboration towards 50 % embedded energy reduction for all-glass facades.

16 Conclusion

This ecological transition of all-glass structures prevents a possible future societal dislike of glass facades when material supplies are becoming scarcer and saving fossil energy becomes more prominent than it is now. Proposals like this have the tendency to take a long time. This proposal of Mick Eekhout as the emeritus full professor of Product Development in Architecture at the TU Delft has the target to use only 50 % of the embedded energy in glass facades, and it takes an example in extra thin glass. So that glass facades also can be employed in the architecture of the future without objections. It requires an open collaboration of all concerned stakeholders, from the glass producers, the façade makers, the designers up to the financers, the owners and users of glass buildings. I am prepared to come back in 4 years time in 2020 and scout how much progress has been made by different stakeholder parties towards the goal of 50 % reduction of embedded energy in the all-glass facades and to show that indeed, as the collective glass façade industry, we could realize the dream of many architects to continue designing transparent buildings.

Prototype of an adaptive glass façade with vertically prestressed cables

Christine Flaig[1], Dr. Walter Haase[1], Michael Heidingsfeld[2], Prof. Werner Sobek[1]

1 Institute for Lightweight Structures and Conceptual Design (ILEK), University of Stuttgart, Pfaffenwaldring 14, 70569 Stuttgart, Germany, christine.flaig@ilek.uni-stuttgart.de

2 Institute for System Dynamics (ISYS), University of Stuttgart, Waldburgstraße 17-19, 70569 Stuttgart, Germany, michael.heidingsfeld@isys.uni-stuttgart.de

This paper discusses a prototype of an adaptive glass façade. The load-bearing structure of the 4.9 m high façade consists of double cables. Horizontal deflections are reduced by placing active components in between two vertical cable lines. These components are acting in the direction perpendicular to the cable axis resulting in a decrease of required prestressing forces and thus reducing the amount of material in the structure. The paper focuses on the structure and the functions of the adaptive façade.

Keywords: façade, vertical cables, adaptive system, glass, static loads, dynamic loads

1 Introduction

1.1 Glass façades with vertically prestressed cables

The development of structural glazing and highly efficient load-bearing systems opened up new fields for glass façades. Engineers like Peter Rice, Jörg Schlaich, and Werner Sobek have set new standards for the design of efficient load-bearing systems such as flat cable-nets or vertically prestressed cables. These new standards enable engineers to realize highly transparent façades with a minimal amount of material used for the load-bearing system. [1]

To increase the efficiency of glazed cable facades, Werner Sobek developed and realized a type of glass façade with vertically prestressed cables only. The first façade of this type was realized at the centra1 foyer of Bremen University (Figure 1-1, left). In comparison to earlier approaches to cable supported glass facades, the prestressing forces limiting the horizontal deflections of the glazing surface, are reduced. All cables were supported on the lower side with stay spring bearings to guarantee a constant prestressing force in the cables. Additionally, the construction of the springs includes a stopper to limit the horizontal deflections of the glazing surface under high wind loads, accepting high forces in the cables. [1]

Engineered Transparency 2016. Glass in Architecture and Structural Engineering. First Edition.
Edited by Jens Schneider, Bernhard Weller.

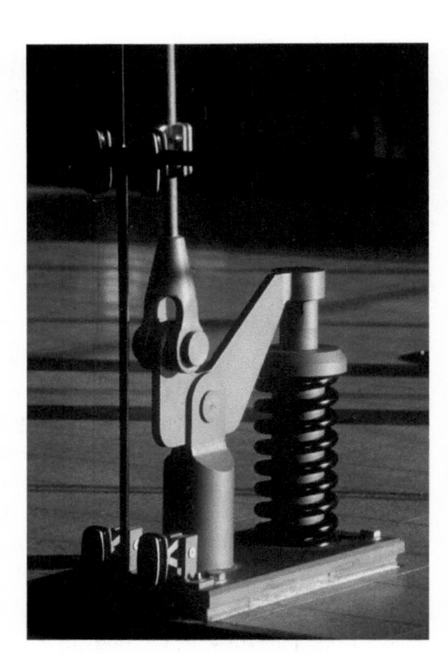

Figure 1-1 left: central foyer of Bremen University, right: stay spring bearings © Andreas Keller.

In a following project, the Lufthansa AG Headquarter in Frankfurt/Main, the façade system was refined by adding a second layer of vertically prestressed cables ensuring a combined load transfer of high loads. This was necessary due to the heavy weight of the glazing, which consisted of 1.45 m by 3.45 m panes of insulated glass connected eccentrically to the cables. [1]

Even though both examples of vertically prestressed cable facades exhibit a highly efficient load-bearing behavior, the high prestressing forces required for the reduction of horizontal deflections and warping of the glass panes result in a significant increase of the cables' cross sections. Thereby, the restriction of the horizontal cable deflections due to wind loads is determining for the design. According to European standards, a horizontal deflection of the cable length divided by 50 should not be exceeded to ensure the serviceability of the structure. The required high prestressing forces subsequently result in high reaction forces at the cable supports and the necessity for additional material to anchor these forces. In addition to the increase in material at the supports, time-dependent effects, such as creeping of concrete, have to be anticipated in the design. In order to meet this effect, the magnitude of the cable prestressing force has to be increased even further.

According to European standards structural components are designed to meet the demands of serviceability limit state (SLS) and ultimate limit state (ULS) for maximum life-loads, that statistically occur at least only once in a 50-year period and are additionally increased by a design factor. Therefore, a significant amount of material is used only in a small fracture of the structure's lifetime.

1.2 Adaptive Structures

Considering the fact that the construction industry has contributed significantly to the shortage of resources, it is essential that new approaches for design and construction are developed and implemented to save resources and reduce embodied energy. Adaptive structures stand for such an approach.

The load bearing system of adaptive structures is designed to withstand such load cases that occur during the whole lifetime of a structure or with a comparatively high frequency. Extraordinary load cases are not dealt with by adding more material, but by actively manipulating the load-bearing behavior of the structure by means of additional components called actuators. Actuators introduce a defined state of deformation or stress into the structure to reduce deformations and to homogenize stress fields. The actuation is always executed temporarily and in accordance to a present load case and, therefore, replaces the need for additional material for scarce load cases with the use of energy. To trigger and control the actuators and to ensure that their reaction corresponds to the present load case, two other components – sensors and a control software – are necessary in an adaptive structure. Sensors will continuously measure the state of the structure whereas the control software evaluates the sensor information and signals the actuators to give the requisite reaction to a present load case. [2]

In previous investigations concerning an adaptive shell structures (see Stuttgart SmartShell in [3]), actively movable supports were used to reduce and homogenize stresses and, thus, reduce the deformations in the structure. However, due to the virtually not existing bending stiffness of cables, glass façades with vertically prestressed cables will exhibit an increase in deflection when stresses are reduced. Therefore, it is necessary to develop a concept of adaptive glass façades with prestressed vertical cables.

This paper discusses the adaptivity concept for glass facades with prestressed vertical cables and the prototype of a newly developed type of adaptive façade with vertical prestressed double cables currently being researched at the Institute for Lightweight Structures and Conceptual Design (ILEK) in cooperation with the Institute for System Dynamics (ISYS) at the University of Stuttgart. The idea of active deflection-manipulation in façades was proposed by Haase in 2012. The paper concludes with the simulation and evaluation of the structure load-bearing behavior in active (adapted) as well as passive modes.

2 Adaptivity concept for glass façades with vertically prestressed cables

Since high prestressing forces in glass façades with prestressed cables are necessary to reduce horizontal deflections when subjected to high wind loads, the adaptivity concept developed in [4] aims at reducing prestressing forces and horizontal deflections. Since the highest deformations in these facades are introduced by highly dynamic wind load cases, both static and dynamic stresses and deflections have to be counteracted by active components in the system.

Two possibilities have been identified to manipulate the prestressing force and the deflections of cables: active system components acting a) parallel to the cables axis and b) perpendicular to the cables axis (compare Figure 2-1, right). Parallel to cables axis, forces can be decreased or increased, for example, with active supports. The increase of the cable stresses results in a reduction of horizontal deflections of the façade. This presents the possibility for a low initial prestressing force that is only increased when necessary. While reducing time dependent effects such as creeping in concrete supports, the maximum stress in the cables – to limit the deflections in the highest occurring load cases – remains the same as in a passive system. With the second option of adaptation, in a direction perpendicular to the cable axis, horizontal deflections in the front cable can be reduced without the need for high prestressing forces. Attention should be paid to the dynamic deflection when the prestressing force is reduced because the structure will be more flexible. Therefore, the actuators can generate an artificial stiffness in the structure.

The adaptivity concept includes active components acting both parallel and perpendicular to the cables. The system differentiates between three system modes (see Figure 2-1: The passive system is dimensioned for low gust speeds with a high probability of occurrence (mode A Figure 2-1). For higher gust speeds with a smaller probability of occurrence (mode B Figure 2-1) the system is active. Stresses in the cables will be adapted in accordance to the deflection state of the system in order to achieve a minimal horizontal deflection of the glazing surface. The active components' action is performed perpendicular to the cables to avoid high prestressing forces in this mode. The third system mode is only activated for extraordinarily high loads to assure the structures stability (mode C Figure 2-1). The prestressing forces will be reduced in order ensure that added load of prestressing force and wind forces to not exceed the strength of the cables. In this case, large deflections are accepted even when they exceed the limits of the serviceability limit state.

This paper focuses on the implementation of the system mode B into a protoype of an adaptive façade. The development of the adaptivity concept for glass facades with prestressed vertical cables will be published in [4].

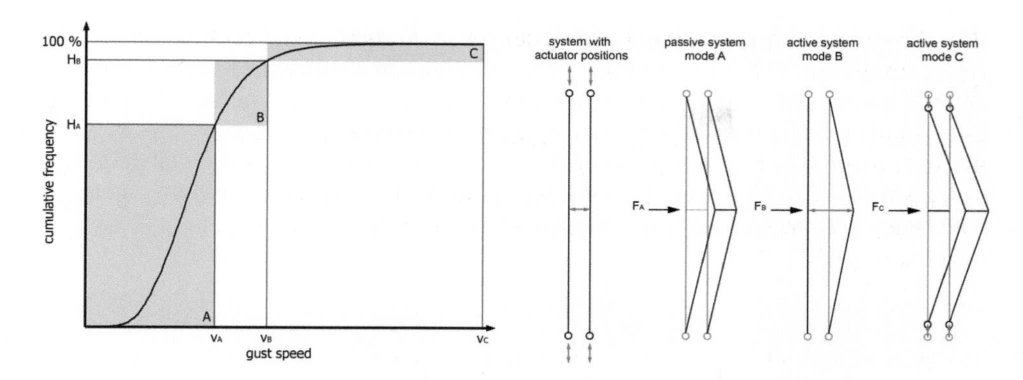

Figure 2-1 Description of modes A, B and C of the adaptivity concept, left: mode as a function of gust speed, right: schematic representation of the active components action in mode B and C.

3 Prototype of an adaptive glass façade with vertically prestressed cables

3.1 Structure

With the presented prototype, the active manipulation of deflections due to static and dynamic loads are researched. Therefore, only the active components acting perpendicular to the cables (system mode B, Figure 2-1) were integrated in the façade. The prototype is 4.9 m high and 3.4 m wide. It consists of two columns of toughened safety glass panes installed on a steel frame with three double cables supporting the panes at their corners. Each column has three glass panes with the dimension of 1.5 m x 1.5 m in size and 8 mm in thickness (Figure 3-1).

Figure 3-1 Rendering of the prototype.

The three double cables (spiral strands) have a metallic cross section of 22 mm² (Ø 6.1 mm) each. The front and back cables are connected with active struts, which can actively extend and contract between the cable clamps of the front and back cables (Figure 3-2). These components are acting in direction perpendicular to the cables axis and glazing surface. The prototype in passive mode exhibits a maximum horizontal deflection of 0.23 m (≈L/20) for a wind pressure of 925 N/m². The actuators are dimensioned to counteract these deflections with a maximum extension of 0.23 m.

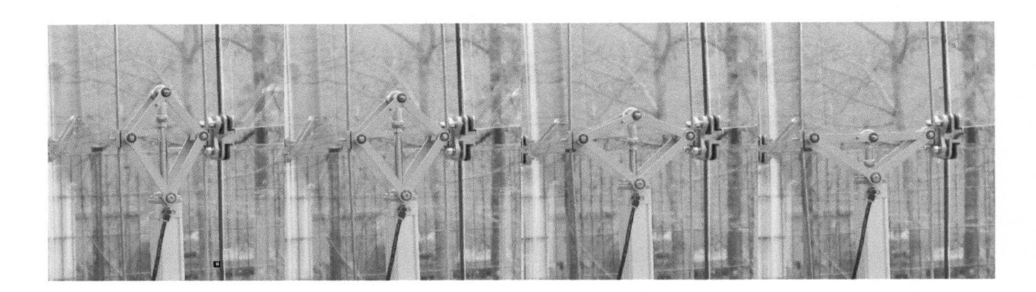

Figure 3-2 Active struts – pneumatic actuators.

The movement is performed by pneumatic cylinders with a high reaction rate to manipulate dynamic as well as static deflections (Figure 3-3, left). The measurement of the system state is provided by two different types of sensors: strain gauges at the cable fittings and angle transmitters at the actuators.

To avoid slacking of the cables due to dead loads in the passive state the front cables are prestressed with a force of 1.5 kN and the back cables of 0.2 kN. Due to the low prestressing force, the system shows large deformations when subjected to wind loads in the passive state. The pneumatic cylinders actively manipulate these deformations with an artificial stiffness. The system will therefore be temporarily stiffer as a function of the wind pressure.

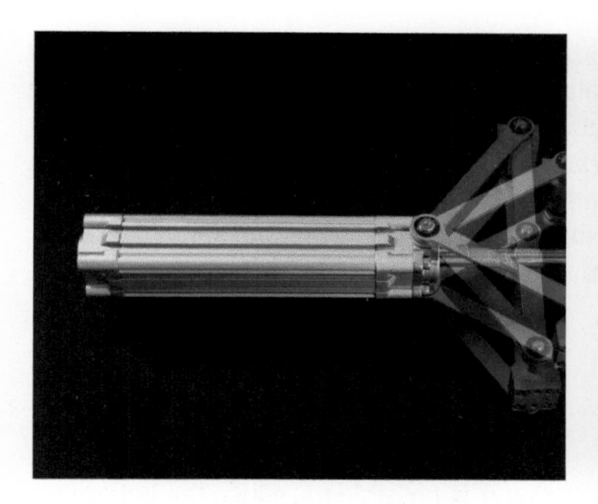

Figure 3-3 left: actuator with pneumatic cylinder and scissor in two positions (neutral and extend), right: glass holder.

The glass panes are fixed at the front cables with stainless steel glass holders with EPDM-pads (Figure 3-3, right). The glass panes are only clamped at the edges of the glass to avoid peak stresses in the corner regions of glass panes. The glass holders are pinned at the cable clamps to the front cables. Since the prototype is designed for the ULS dimensioning of wind load according to Eurocode 1 [5, 6], only the maximum stresses and not the deformations of the system were considered for the dimensioning of the cable and substructure cross sections.

3.2 Function

For a manipulation of the system it is necessary to continuously monitor the system with the strain gauges which measure the forces in the cables. To estimate the current extension of the struts, the angle transmitters measure the angle of the scissor of the actuator. If cable forces rise up due to the horizontal loading on the glass panes, the actuators will be activated and horizontal deflection in the front cables will be reduced. To actuate the pneumatic cylinders for the static and the dynamic deflections a control system is essential to initialize the translation in longitudinal direction as well as its speed with sufficiently high precision.

3.3 Analysis

For the design of the prototype and the investigation of different possibilities of structural adaptivity, numerical simulations were conducted with the FEM software ANSYS – Mechanical APDL. The prototype was designed for the ULS without considering the limits of deflection demanded by the SLS. Therefore, the wind pressure of 925 N/m² for free-standing walls according to Eurocode 1 [5, 6], the dead loads and the prestressing force of the cables were used to dimension the components of the structure. The prestressing force of the cables was chosen to avoid slacking for the dead loads (see chapter 3.1). The prestressing force was induced by a temperature load in the simulation model. The wind and dead loads were applied as single loads at the connecting nodes of the cables. The static and dynamic behavior of the prototype was analyzed with non-linear numerical methods, where the horizontal deflections and the stresses in the passive and active mode were examined. To determine the optimal extension of the scissor actors, an optimization tool was developed by ISYS. This algorithm will later be integrated into the structure to allow real-time control.

4 Results

The results of the numerical simulations show that it is possible to reduce the horizontal deflections of the glazing surface to nearly zero by means of actuation. When the system is subjected to wind loads the maximum cable force increased to 11.5 kN in the passive mode and 15.9 kN in the active mode (Figure 4-1, Table 4-1).

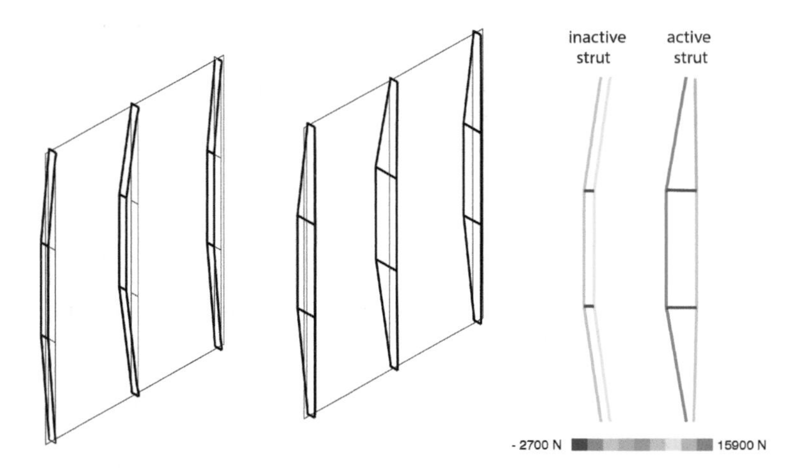

Figure 4-1 Left: deformation plot of the passive prototype, middle: deformation plot of the active prototype, right: cable forces of the passive and active double cables in the middle.

Table 4-1 Comparison of system reactions of the passive and active system mode.

System mode	Max. cable force	Max. deflection of the front cable
Passive	11.5 kN	0.23 m
Active	15.9 kN	≈ 0 m

A comparison of an adaptive façade as the prototype with a typical glass façade with vertically prestressed cables with an equal size shows that prestressing forces can be reduced as much as 87 % for a constraint of maximum horizontal deflection of L/50 for the given load case. Thereby, it was possible to reduce the cable cross section to 42 % and in the maximum reaction force to 57 % (Table 4-2).

Table 4-2 Comparison of the prototype and a passive façade-system.

Façade	Max. prestressing force	Max. cable force	Cross section of cables
Active	1.5 kN	13.0 kN	22 mm^2
Passive	11.5 kN	18.9 kN	38 mm^2

5 Discussion

The numerical simulations of the presented prototype showed that it is possible to manipulate the horizontal deflections in glass facades without high prestressing force and that the glass surface undergoes close to no horizontal deflections. A horizontal deflection of nearly zero would not be realizable with a conventional glass façade with prestressed cables. The flexible system resulting from the low prestressing force in the cables can be manipulated with an artificial stiffness generated by the actuators.

Compared to a purely passive system designed according to code, the amount of material in the cables in this newly developed system could be reduced as much as 42 per cent. Furthermore, the cross sections of the supports can be downsized because the maximum reaction forces have been greatly reduced.

The glass panes are dimensioned in this prototype for deflections and warping according to the codes. The use of the new system permits that warping of the glass panes can be neglected. With the presented results, it was proven that the glass panes could be thinner.

The results presented in this paper do not consider the amount of energy for the operating phase and the embodied energy. This will be the concern of further publications [4].

6 Conclusion

A newly developed system for adaptive glass façade with vertically prestressed cables was introduced. A new approach for the design and the construction of glass facades with a minimal amount of material and a high performance was shown. The structure is supported temporarily by energy, which replaces material in the structure. This occurs with active components which are acting perpendicular between the double cables and manipulate the deflections due to wind loads in the structure without high prestressing forces in the cables. Due to the radical reduced prestressing force and thus the small stiffness of the structure, the stiffness will artificially increase with the actuators as a function of the wind speed. Therefore, time dependent effect such as creeping of concrete supports can be reduced.

Furthermore, it was shown by means of the non-linear numerical simulations that horizontal deflections in the front cable due to static wind loads are reduced to nearly zero. Due to the minimal deflections in the front cables the warping in the glass panes is reduced and can be neglected in the design.

Further research will concern the deformability behavior of the prototype and the comparison with the numerical simulations. In addition to this, the adaptivity concept of glass facades with vertically prestressed cable will be discussed in detail in [4].

7 References

[1] Sobek, W. and Rehle, N „Beispiele fuer verglaste Vertikalseilfassaden. Herrn Prof.Dr.-Ing. Udo Peil zur Vollendung des 60. Lebensjahres gewidmet," Stahlbau, vol. 73, no. 4, pp. 224-229, 2004.

[2] Neuhaeuser, S, Weickgenannt, M, Haase, W, Sawodny, O, „Adaptive Tragwerke – Aktuelle Forschungen im Ultraleichtbau: Stahlbau," Stahlbau, vol. 82, no. 6, pp. 428-437, 2013.

[3] Neuhaeuser, S, Weickgenannt, M, Witte, C, Haase, W, Sawodny, O, Sobek, W, „Stuttgart SmartShell – a full scale prototype of an adaptive shell structure," Journal of the International Association for Shell and Spatial Structures, vol. 54, no. 4, pp. 259–270, 2013.

[4] Flaig, C, „Untersuchung verglaster, adaptiver, vorgespannter Seilfassaden," PhD thesis, Institut für Leichtbau Entwerfen und Konstruieren, Universität Stuttgart, Stuttgart, (in preparation).

[5] DIN EN 1991-1-4: 2010-12-00-Eurocode 1: Einwirkungen auf Tragwerke – Teil 1-4: Allgemeine Einwirkungen – Windlasten; Deutsche Fassung EN 1991-1-4:2005 + A1:2010 + AC:2010, 2010.

[6] DIN EN 1991-1-4/NA: 2010-12-00 – Nationaler Anhang – National festgelegte Parameter – Eurocode 1: Einwirkungen auf Tragwerke – Teil 1-4: Allgemeine Einwirkungen – Windlasten.

Silicones enabling crystal clear bonding

Valérie Hayez[1], Dominique Culot[1], Stanley Yee[2], Markus Plettau[3]

1 Dow Corning Europe S.A., Parc Industriel Zone C, B-7180 Seneffe, Belgium, valerie.hayez@dowcorning.com

2 Dow Corning Corporation, 2200 W. Salzburg Road, Midland, MI48686, USA

3 Dow Corning GmbH, Rheingaustrasse 34, D-65201 Wiesbaden, Germany

A façade is quite literally, the face of a building, the signature of its owner or the architect, with much consideration dedicated to its conception. Major advances in glazing and façade technology over the past 30 years have enabled fully glazed sustainable designs providing the demanded aesthetic, whilst respecting occupant benefits such as natural daylight and integration of energy efficiency systems. This explains the significant and growing interest for increased glass in facades in commercial buildings. The use of bulky frames limiting the transparency of a façade can be minimized through point fixation or structural glazed fixation systems. Yet even these have their limitations since the adhesive being used are not necessarily both transparent and durable. This paper discusses the design possibilities offered by established silicone technologies as well as a new generation of optically clear structural silicones, as a film or a hotmelt. Already well-established for façade exteriors, a high strength and optically clear elastic silicone film adhesive is designed for point structural bonding of glass in a variety of shapes. A similar high strength clear film laminate for interior is also available for point and area lamination connecting glass to glass. Its durability, high strength and elastic properties provide significant advantages, targeting applications like structural bonding of glass stair cases, glass beams and other interior decorative glass connections. A very recent breakthrough technology is an optically clear hot-melt silicone for linear structural bonding and sealing which can be done on site. Application examples are: total vision glass sealing, structural glazing, glass bonding of closed cavity facades, shadow boxes, partition walls and even for specific insulating glazing designs this new technology has generated interest. Strength and durability combined with its unique aesthetics and transparency open up a new dimension in architectural design freedom.

Keywords: optically clear silicone, transparent design, high strength, silicone film, point fixation, linear bonding

1 Introduction

Structural glazing is a well-known technology enabling glazed façades for over 40-years. Structural glazing has been used to enable architectural design through the reduction – and in some instance the elimination – of exposed metal, by reducing sightlines and smoothing the surface texture and tautness of the exterior glass façade. In response to the

Engineered Transparency 2016. Glass in Architecture and Structural Engineering. First Edition.
Edited by Jens Schneider, Bernhard Weller.

continued architectural quest to maximize facade transparency and break the visual barrier between the interior and exterior environments, optically clear silicone bonding technologies enhance and enable the aesthetics of a façade.

Already well-established for façade exteriors, the clear and high strength silicone film (transparent structural silicone adhesive or TSSA) provides higher permanent design strength when compared to conventional structural glazing silicones. It is designed for point structural bonding of glass in a variety of shapes (round, rectangular, triangular, etc). Similarly, a clear film for interiors (transparent structural silicone laminate or TSSL) is also available for point and small area lamination. Its durability, high strength and elastic properties show significant advantages, especially in interior applications like structural bonding of stair cases, glass beams or other interior decorative glass connections.

The latest breakthrough technology is an optically clear silicone for structural glazing of glass facades, weather sealing and other glass bonding applications, where clear aesthetics is required. Examples are: glass bonding in closed cavity façade designs with invisibly fixed exterior glass, total vision glass, weather sealing, partition walls, etc. This technology maximizes transparency and opens up a new dimension in architectural design freedom.

A review of the technical attributes and applications of these different clear silicones will highlight how they can be utilized to enable the architectural quest for filigree facade structures with high degrees of transparency.

2 Transparent structural silicone film adhesive and laminate

2.1 Performance

Transparent Structural Silicone Adhesive (TSSA) is an optically clear and high strength structural silicone adhesive film designed to be applied between glass and metal supporting components at 1mm thickness to provide flush and smooth options in comparison to those strategies that require drilling and mechanical fastening of glass. By eliminating the traditional need to drill through the glass for placement of retaining bolts and the use of gaskets to retain air tightness at the point of fixation, it ensures superior durability and longevity, as the gas-filled insulating glass cavity remains untouched. Therefore, it contributes to a more thermally sustainable insulation of the façade. Physical properties of this material and engineering applications have been established and published [1]. Based upon extensive testing of durability according to ETAG002 [2] requirements, TSSA is approved for exterior glass façade applications with dynamic design stress of 1.3 MPa and static design stress of 0.6 MPa [3]. These design values are about 50 times higher than those allowed for conventional structural silicone sealants [4]. The higher Young's modulus of 9.3 MPa along with its 1 mm thickness limit the need for compensation of thermally induced movements. The silicone film adhesive is supplied in sheets or in a pre-cut shape ready to use on the point fixation system.

Transparent Structural Silicone Laminate (TSSL) is specifically engineered for interior applications and shows strong advantages in strength and flexibility which allow accommodating much more movements and vibrations occurring on glass stairs and railing glass beams for instance. The TSSL is supplied on a roll only. The dynamic design stress is 1 MPa and the static design stress is 0.6 MPa [3].

2.2 Examples of applications

The Institute for Research and Treatment of Cancer is a private non-profit cancer research organization based in Torino, Italy. This project comprised an extension and renovation to the existing building which included the construction of a new tower and a glazed walkway. It was important to enhance building aesthetics and create a better indoor environment to improve the health and well-being of the building occupants by providing a balance between natural daylight provision and façade durability. The façade was constructed using insulating glass units that were gas-filled and dual sealed around the perimeter. The insulating glass units were attached to the curtain wall façade by means of a point fixed glazing system which utilized TSSA (Figure 2.1). With almost invisible bonding points, the use of TSSA creates a homogeneous, sleek facade appearance and improves the overall building aesthetic.

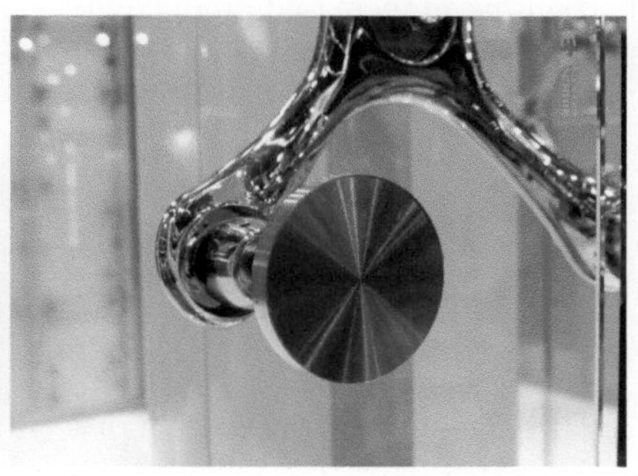

Figure 2-1 Left: Example of TSSA (Institute for Research and Treatment of Cancer) application. Right: detail showing the optical clarity of the silicone film.

The use of TSSL in a glazed staircase is illustrated in Figure 2.2. A stainless steel L profile has been laminated to the glass stair stringer using TSSL clear thin high strength silicone film. The assembly of the glass stair to the glass stringer on site was typically done with a conventional, manually applied black silicone. This can be now replaced by a clear structural bonding (see paragraph 3).

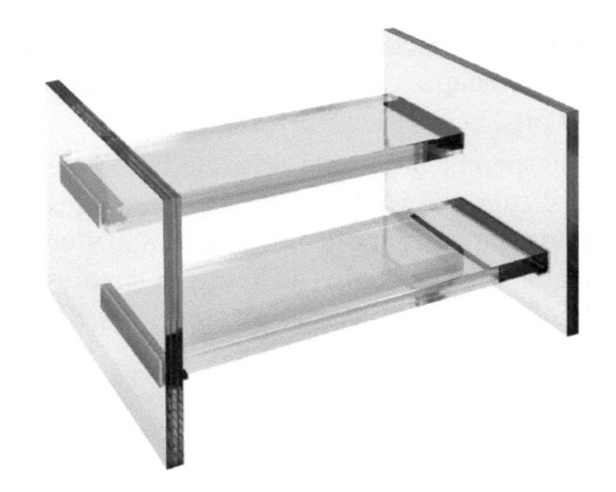

Figure 2-2 Example of Interior glass-metal bonding using TSSL (picture from Glass Troesch: Swissstep Bond).

3 Transparent hot-melt structural silicone sealant

For more than 15 years, development of an optically clear hot-melt silicone adhesive has evolved and yielded different generations of the adhesives to be commercialized, each targeting various applications within the residential fenestration and light commercial sectors (e.g. bonding of windows and doors). The primary advantage of this technology is provided by the high initial tack (also known as green strength) enabling quicker movement of the bonded units resulting in increased productivity – in comparison with other technology platforms used in this market sector [5].

The earlier generations of generic hot-melt technology are typically limited in creep resistance, making them unsuitable for structural glazing applications. Recent developments address the creep observed in the incumbent technology and give rise to a broader, and more stable, strength application.

4 Performance

4.1 Features and benefits

The clear hot-melt silicone HM-2400 is a 100 % silicone neutral-cure mono-component adhesive. The cure is obtained by reacting with moisture in the atmosphere.

The hardness reaches 60 on the A scale. The sealant is comparatively elastic, reaching an ultimate elongation of more than 100 % according to ASTM D412 [6] and has an allowed movement capability of ±50 %. Due to this high movement capability and elasticity, joint dimensioning can be different than with structural glazing sealants. Whereas structural glazing requires a minimum joint of 12 by 6 mm^2, a silicone hot-melt joint of 10 by 3 mm^2

may be recommended to resist the loads typically occurring in structural glazing applications. However a case by case evaluation might prove to be necessary. Independently of the thickness of the joint, mechanical properties will remain the same. The clear silicone hot-melt sealant develops primerless adhesion to glass and anodized aluminum. Thanks to the use of a polysilicate resin in the formulation, adhesion starts with instant initial tack which is much higher than standard structural sealants. The initial tack is determined by measuring glass-glass lap shear adhesion strength at 6 mm/min (figure 4-1). Whereas only 300 Pa initial tack is measured at the time of application for mono-component structural glazing sealant [7], more than 10.000 Pa is developed by the hot melt. With its > 30 times higher initial tack, it provides a significant advantage in handling safety. After 24 hours, both technologies reach about 0.2 - 0.4 MPa. This strength is in theory enough to handle and ship glued units. However, moving a unit glued with a mono-component structural glazing sealant before its full cure can lead to macroscopic distortions and separations between the cured and uncured phases. The polymer network would not be able to develop in an optimal way and the final properties of the cured sealant will be much less than in normal curing conditions without movement. On the other hand, the polysilicate resin significantly increases the cohesive strength of the hot melt formulation such that the sealant bead maintains its mechanical integrity during transport, which ensures optimal final strength properties. Thanks to this effect, units glued with hot melt technology can be shipped after 24 hours and productivity improvements can be obtained. Further adhesion build up in tension and in shear are shown in figure 4-1.

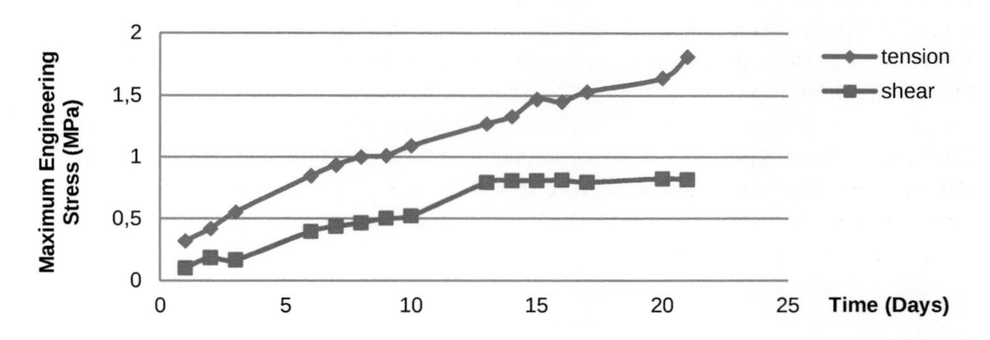

Figure 4-1 Adhesion build up (maximum engineering stress in MPa) in tension and shear of Hot-melt glass-glass H-pieces (12 x 3 x 50 mm³). Pulling speed 6 mm/min.

Using similar joint dimensions, a mono-component structural glazing sealant needs more than 15 days to reach 1 MPa tensile strength. On the other hand, the hot-melt technology reaches this strength already after 8 days of cure.

Table 4-1 shows the stress-strain values for the hot-melt as measured on 12 x 12 x 50 mm H-pieces and dynamic load (6 mm/min) during curing of the hot-melt (1 week, 2 weeks and 3 weeks at room temperature RT). Force and elongation were measured after different curing times (1 week at 60 °C and 4 days exposure at UV) and ageing conditions (4 days UV exposure, 1 week heat exposure at 60 °C).

Table 4-1 Tensile tests: measurement of stress/strain for 12 x 12 x 50 mm H-pieces and dynamic pull (6 mm/min).

	intersection with the line 0.5 MPa-50 % elongation		A		C		D
	Force (Mpa)	**% Elong**	**Force max (MPa)**	**Elong at Fmax (%)**	**Force (MPa) at C**	**Elong at C (%)**	**% Elong max at F=0**
1w 23C	0.466	3	1.05	12.72	0.67	225	280
2w RT	0.4	4	0.82	12.84	0.55	350	350
3w RT	0.444-0,555	4.7-5.6	0.98	11.95	0.56	150	220
4d UV	0.46	4	0.99	12.03	0.67	225	250
1w 60C	0.444	2	0.99	14.56	0,6-0,67	200	240

The line at 50 % - 0.5 MPa is required by the norm EN 1279-4 [8] as minimum below which a sealant shall not fail. In practice, this means that point C of the curve cannot fall inside the green triangle. The characteristic points defining the stress-strain curve of the hot-melt sealant do not deviate significantly before and after ageing. A maximum tensile around 1 MPa is obtained, while the elongation at failure is well above 200 %. These values were used to graphically portray the trend of the stress-strain curve for both the hot-melt and the mono-component SG sealant as shown in figure 4-2. This graphic is a simplification showing only the behavior of both sealants in the lower deformation region (100 %). In this region, the mono-component behaves more or less linearly but it will reach a much higher elongation at break (±500 %).

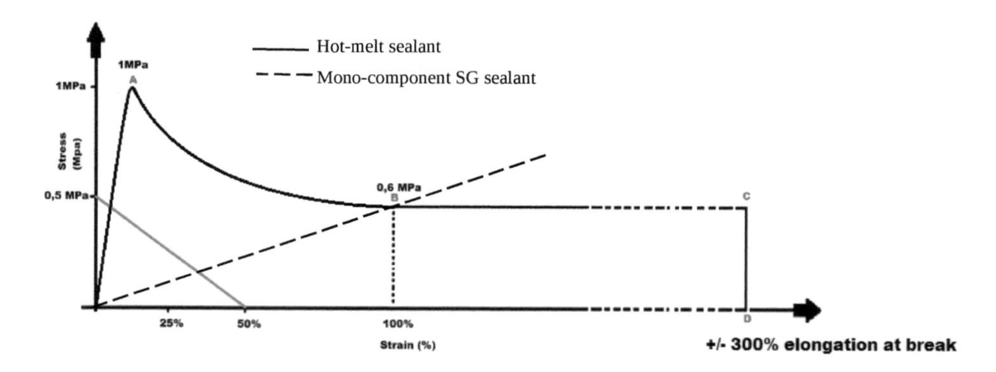

Figure 4-2 Graphical comparison of hot-melt clear sealant and mono-component structural glazing sealant under dynamic load, standard joint dimensions (3 weeks cure curve).

Both curves intersect in point B at 0.6 MPa stress. At dynamic movements representative of real life situations (12.5 - 50 % strain), the surface under the hot-melt curve is large which indicates a good ability to dissipate energy. The extreme dynamic load resistance of the hot-melt technology was evaluated by performing cyclic testing representative of hurricane exposure. Therefore monolithic laminated glass was bonded in a 1.2 by 1.8 m aluminium frame using 12.7 mm hot-melt joint. The unit was subjected to cyclic static pressure differential loading based upon ASTM E1886 [9]. The test could not be passed when using previous versions of the hot-melt technology, due to bad elastic recovery. However, units bonded with 0.51 mm bondline of the latest developed hot-melt passed 9000 cycles at ±3.8 kPa. Deflections from center of glazing of 8 mm or 6.4 mm were measured during positive respectively negative cycling. Finally, the behavior of the hot-melt sealant under static load was evaluated. Lap shear samples (15 x 30 mm x 2 mm) on anodized aluminium were subjected at RT and 50 % RH to incremental loading (weight and duration) as illustrated in figure 4-3.

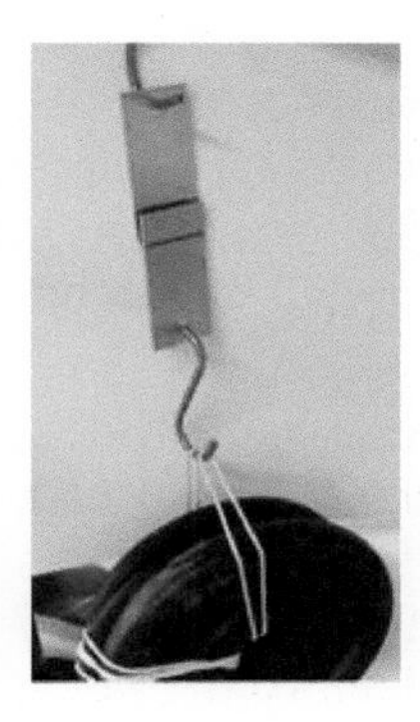

Figure 4-3 Dead load resistance test set-up: lap shear anodized aluminum samples subjected to incremental loading.

The resulting dead load resistance is illustrated on figure 4-4.

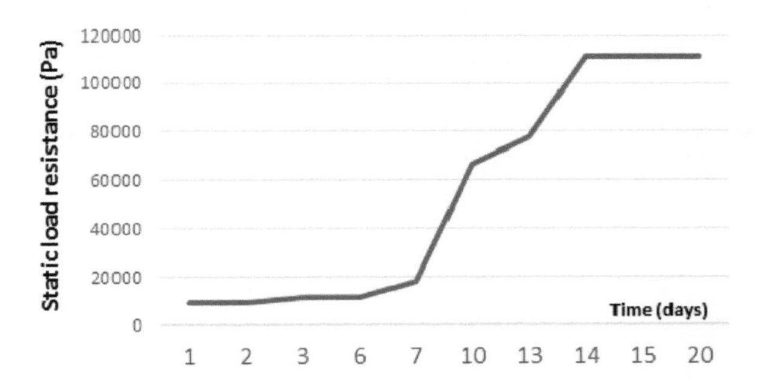

Figure 4-4 Static load resistance build up (Pa) with time (days) for the hot-melt clear sealant.

Figure 4-4 shows that after 10 days cure, the hot-melt sealant already has developed a shear stress resistance of 70.000 Pa. After 3 weeks cure, this increases to 90.000 Pa and at 4 weeks of full cure an ultimate static load resistance of 110.000 Pa is reached.

The presented data confirm that the hot-melt sealant has the potential to be used as a transparent structural silicone sealant. Additional testing according to ETAG 002 [2] will help to further confirm this assumptions. Following a procedure similar to standard structural sealants, it is possible to determine the allowable maximum design strength in dynamic and static load for the hot-melt technology by using a safety factor.

4.1.1 Design Dynamic and Static Load Resistance

In order to determine the design dynamic and static load resistance for the hot-melt technology, the approach used in structural glazing was followed. The maximum tensile strength measured on the hot-melt (1 MPa, Table 4-1) is reduced by a safety factor. In the case of dynamic short duration loading such as normal wind load stress, a safety factor of 6 is typically used in Structural Glazing [2]. This results for Hot-melt in about 0.16 MPa dynamic design load. However, since the measured forces are in the same order of magnitude as structural sealants, the same design dynamic strength of 0.14 MPa was selected. This design load corresponds with a slightly higher safety factor (7 instead of 6) than for structural glazing sealants.

Structural glazing typically applies for design static load resistance a standard safety factor of 60 to the maximum tensile strength, resulting for the hot-melt sealant in about 16000 Pa which is in the same range as mono-component SG sealants. The safety factor is arbitrarily chosen and typically used when no detailed static load measurements are available. Since it is known that the strength of the hot-melt technology varies with the speed of deformation, an additional safety factor of 2 was applied, bringing the static load design stress to 8000 Pa. The validity of this static design stress and of the safety factor was verified by performing static load measurements. Figure 4-5 shows that after full cure, the hot-melt resists to a static load of 110.000 Pa for several weeks without movement. The defined static load design stress (8000 Pa) hence represents a proven safety factor of 14 which is well beyond what is commonly used in other industries. As an illustration [9], buildings commonly use a factor of safety of 2.0 for each structural member, automobiles use 3.0, and aircraft and spacecraft use 1.2 to 3.0 depending on the application and materials. In order to further improve dead load resistance, it will be recommended to utilize independent dead load supports similarly to a majority of SG applications.

4.1.2 Optical Clarity

The clarity of the hot-melt technology is illustrated in figure 4-5. This sample is build up with 2 mm hot-melt sealant sandwiched between two glass plates of each 4 mm thickness. No distortion nor color change is observed between the glass and silicone.

Figure 4-5 Illustration of optical clarity of hot-melt silicone.

In order to quantify the clarity of the sealant, the transparency [10, 11] was measured after 4 weeks of cure (fresh) and no ageing using a spectrometer (figure 4-6). This was repeated after ageing of the sample at different conditions. No significant difference was recorded after exposure to dry heat (1000 hours at 100 °C), high temperature and humidity (85 °C and 85 % RH) or thermal shock testing (exposure for 30 min at -40 °C or +125 °C with 10 sec switching time between temperatures). Between 450 and 800 nm, differences of 2-7 % were observed.

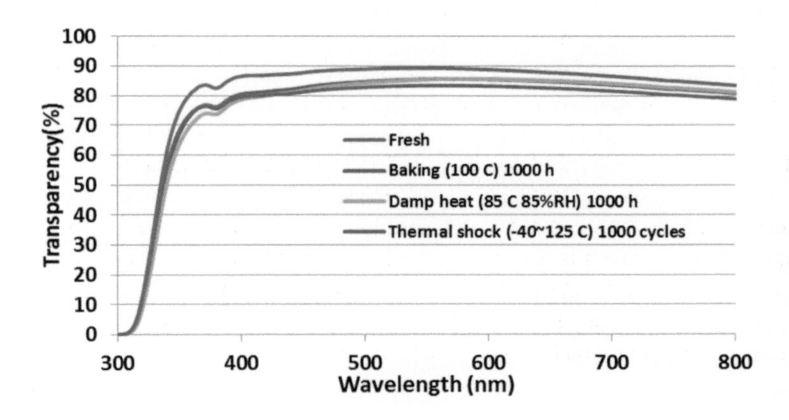

Figure 4-6 Measurement of transparency after several ageing tests.

No yellowing was observed after 10000 hours exposure to UVA-340 lamp as illustrated in the figure 4-7 [12].

Figure 4-7 Transparent Structural Sealant after 10000 hours UVA exposure.

4.1.3 Accessories

To further enable and harness the technical and architectural advantages of the hotmelt adhesive, its use on a façade can be combined with other optically clear materials to provide backing or act as a spacer, whilst still being transparent. First trials of manufacturing transparent spacers and gaskets have been made with clear, optical grade moldable silicones. These silicones have specifically been formulated for the lighting industry where they are, for example, being injection molded in 3D shapes such as lenses or led luminaires. Their specific formulation ensures high temperature, UV and weathering resistance without yellowing and guarantee perfectly clear views. Furthermore their scratch resistance is excellent, making it easy materials to handle in a factory [13].

4.2 Examples of potential applications

4.2.1 Structural bond

Although additional testing is needed, the presented results have indicated the potential capacity of the clear hot-melt technology to fulfill the role of a traditional structural glazing sealant. The combination of strength and transparency might be especially attractive for double skin façades or glassfin applications. Glassfins are typically bonded to the face glass along the height (long dimension) with a black structural sealant whilst the top and bottom of the glassfin is mechanically fixed in a U channel profile. Therefore in this kind of application the joint is only subjected to dynamic wind loading and no deadload making it an ideal application where hot-melt technology could bring a clear aesthetic advantage. Glassfin dimensions of 30 mm thickness are typical. In order to respect the ratio between width and thickness of the joint, a thickness of 10 mm would typically be applied. The full width of 30 mm will not always be filled with black silicone, but the use of a middle backerrod could be further to the detriment of the aesthetics. Switching to clear hot-melt means the thickness could be significantly reduced without using a backerrod. Aesthetic effects as illustrated in the figure 4-8 can be achieved.

Figure 4-8 Aesthetics in glassfin fixation using transparent structural silicone adhesive.

4.2.2 Secondary seal

Architectural trends tend towards larger glasspanes increasing transparency and light of the façade. However, the use of a black primary and secondary sealant in the insulating glass unit (IGU) edges prevents obtaining completely transparent units. The use of hot-melt technology could improve this situation partially.

Table 4-1 already provided the mechanical properties before and after ageing. Below we discuss the water vapor transmission rate (WVTR) which is one of the essential parameter to ensure the durability of an IGU.

The water vapor transmission rate of the hot-melt clear adhesive was measured according to EN 1279-4. An average permeability of 14.9 gr/24h.m^2 was measured on 2 mm thick membranes [14]. This represents an improvement of up to 30 % in comparison with a non clear structural mono-component sealant used in IGU applications [15] and 15 % improvement in comparison with a non-clear structural 2-component sealant [16].

Taking into account the above results of the clear silicone hot-melt, this material could be used in combination with optically clear spacers made of moldable optical silicones to further lighten the insulating glass units.

First trials (figure 4-9) have been performed whereby a clear spacer and the clear silicone hot-melt were used on the verticals of the insulating glass units where the clarity allows pure transparency. On the horizontal top and bottom a traditional insulating glass design is used, with a metallic spacer containing desiccant, butyl and a silicone secondary sealant.

In the past, projects with air-filled units and transparent spacers (e.g. PMMA) have been realized. The risk of condensation and glass corrosion has been assessed by calculating the maximum moisture loading for the desiccant allowable over a certain period. However to get to a long-term durable solution, more research and development needs to be done to get an almost gas-tight primary sealant. So far crystal clear designs in IG are limited to certain niche applications for exterior and interior clear glass designs.

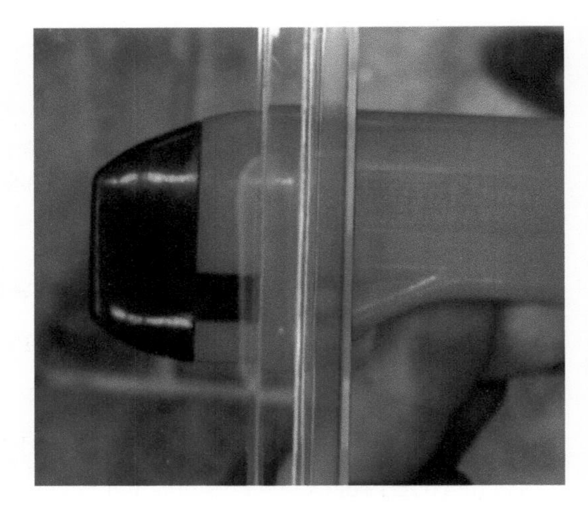

Figure 4-9 First prototype of insulating glass unit using the clear hot-melt on the verticals.

4.2.3 Weatherseal

Using the hot-melt as a weatherseal to increase the transparency of a curtain wall seems to be an obvious application. Butt joints sealing 2 monolithic glasses provide a unique transparency. As the silicone is suitable for interior and exterior applications with a good UV-, temperature and weather-resistance, it opens many opportunities to connect glass to glass or glass to metal. Good examples are double skin facades or closed cavity facades, where a floating and clear aesthetics can be achieved combining structural properties and weather sealing function.

5 Conclusion

As silicone-based structural glazing solutions have gained an ubiquitous role in the modern commercial architectural realm, we can see that the technology and its application have evolved to address the challenges brought forth by more sleek and transparent facade designs. Designs that, in many instances, can only be realized with clear silicone bonding solutions. The unique combination of solutions comprises higher design load capacity, a high movement capability to accommodate joint movement paired with the higher instant strength enhancing handling and productivity. It opens new design options and fabricator benefits all to produce a more visually understated transparency on behalf of well stated architectural design intents.

6 References

[1] Sitte, S., Brasseur, M., Carbary, L., Wolf, A., Preliminary Evaluation of the Mechanical Properties and Durability of Transparent Structural Silicone Adhesive (TSSA) for Point Fixing in Glazing, Journal of ASTM International, Vol. 8, No. 10 Paper ID JAI104084.

[2] Guideline for European Technical Approval for Structural sealant glazing kits, Part 1: supported and un-supported systems, November 2005, EOTA.

[3] IFT Rosenheim test report n 11-000515-PR01.

[4] Silicone Structural Glazing Manual, www.dowcorning.com.

[5] Hautekeer J-P. et al, New Bonding Technologies for Window Glazing and Assembly Bring Benefits Throughout the Entire Industry Value Chain, Glass Processing Days 2005.

[6] ASTM D412, Standard Test Methods for Vulcanized Rubber and Thermoplastic Elastomers-Tension.

[7] Dow Corning® 895 structural glazing sealant, www.dowcorning.com.

[8] EN 1279-4, Glass in building. Insulating glass units. Methods of test for the physical attributes of edge seals.

[9] Burr, A and Cheatham, J: Mechanical Design and Analysis, 2nd edition, section 5.2. Prentice-Hall, 1995.

[10] ASTM E1886, Standard Test Method for Performance of Exterior Windows, Curtain Walls, Doors, and Impact Protective Systems Impacted by Missile(s) and Exposed to Cyclic Pressure Differentials.

[11] ISO 15368:2001 Optics and optical instruments – Measurement of reflectance of plane surfaces and transmittance of plane parallel elements.

[12] ASTM C1648-12, Standard Guide for Choosing a Method for Determining the Index of Refraction and Dispersion of Glass.

[13] De Buyl F. et al., Dow Corning® Moldable Silicone Leading Innovations in LED Light Fixtures, for The Society of Silicon Chemistry Japan, N°31, pp23-38 (2014).

[11] INISMA TEST REPORT N° 2016B MAS 26887on Dow Corning® HM-2400.

[12] IFT test report 655-32769/1e, EN 1279-4 on Dow Corning® 3793.

[13] INISMA TEST REPORT N° 2013B VEC 17502-3a, EN 1279-4 on Dow Corning® 3363.

Dip-Energy – Glass Energy Performance Calculator for Dip-Tech Printed Glass

Andy Shipway[1], Dorit Regev[1], Niv Raz[1], Michael Dovrat[1], Shirley Segev[1], Ziv Cahani[1]

1 Dip-Tech Ltd., zivc@dip-tech.com

Due to the growing interest of architects in the functional characteristics of digitally-printed glass to support sustainable architecture, Dip-Tech has developed a software tool that predicts the energy-performance characteristics of digital in-glass printed images printed with Dip-Tech's printers and ceramic inks. This document describes this software tool, as well as the physical models and calculation models it uses.

Keywords: sustainable design, green architecture, sustainable building, architectural glass, energy calculator, dip-tech, printed glass, sustainable architecture, energy performance

1 Introduction

Dip-Tech develops, manufactures and sells leading digital printing solutions for the global flat glass industry. Dip-Tech's printers, used by glass processors worldwide, employ colorful ceramic-pigmented inks with glass frits that fuse seamlessly into the flat glass during the tempering process. This process enables the printing of colorful, vivid, high-resolution images on large glass panes for architectural applications.

In recent years, more and more architects have become interested in using digital in-glass printing not only for the aesthetic beauty of the printed glass, but also for the functional characteristics that support sustainable architecture and meet environmentally responsible architectural goals[1]. In response to this growing need, Dip-Tech developed a tool called "Dip-Energy".

The most striking advantage of digital printing on glass versus conventional methods (e.g. screen printing) is the ease with which complex and unique designs can be printed. However, this freedom of artistic expression means that models used to calculate energy-performance parameters for single-material coating of glass are insufficient.

Applying knowledge about the ink physical characteristics and printing process, Dip-Tech's energy performance tool calculates and predicts energy-performance characteristics of images printed with Dip-Tech's printers and inks. Among the parameters calculated are the visible light transmission (VT) and solar heat gain coefficient (SHGC) [2]. These parameters are predicted by analyzing the entire façade's images down to the colors used, ink layer thicknesses used in printing, ink physical properties (light reflection, absorption and transmission properties), as well as the glass type and presence of additional

Engineered Transparency 2016. Glass in Architecture and Structural Engineering. First Edition.
Edited by Jens Schneider, Bernhard Weller.

insulated glazing (IG) layers. These calculations and estimates can be used by architects, air-conditioning engineers, interior designers, lighting engineers and other interested disciplines to meet their environmental goals in terms of lighting and energy efficiency, and meet green-building standards and regulations. For example, printed glass can be used to obtain LEED [3] and BREEAM [4] certifications by contributing to a building's overall energy saving and influencing key parameters evaluated by these organizations. Additional examples might be an interior designer who wants to locate textile products at a place they would not suffer from fading phenomena caused by UV or a lighting consultant who wishes to light the room at the desired level.

2 Theory and Model

The solar light spectrum ranges from ultra-violet to infra-red light, with wavelengths ranging from approximately 280 to 4000 nm. Since photons (light energy quanta) of different energies interact differently with ink, it is beneficial to separate the light into several wavebands, for which the interactions are calculated separately. The calculation results for the different wave bands are subsequently summed up, and weighed according to their relative strength in the solar spectrum.

Light that is incident on any material may either be reflected, transmitted, or absorbed; light that is absorbed by the ink layer (or glass layers) will eventually heat the glass window itself; and light transmitted through the ink and glass, both visible and infra-red, will eventually raise the temperature behind the glass, since the energy allowed to enter will be absorbed inside the room or building. Light reflected by the ink or glass windows is essentially redirected back out and does not affect the internal temperature. In addition to the light energy dependence, the ratio between reflected, absorbed, or transmitted energy depends on the light's angle of incidence, so either a single "standard" angle must be chosen, or an average over several angles should be considered.

When trying to calculate the energy-performance parameters of an arbitrary multi-color image printed on glass, standardized energy interaction parameters available for single-color coated glass cannot be readily applied. This stems from the fact that in any given design, different areas on the glass will generally have different ink thicknesses and colors. Thus, a more complex model is required in order to estimate the light interaction characteristics of arbitrarily thick ink layers of arbitrary ink mixtures, present at arbitrary locations on the glass. Light that enters the ink layer undergoes a complex set of interactions that is not approximated well by the application of the Beer-Lambert law of attenuation or simple scattering models. The fused (tempered) ink layer is not homogeneous – pigment particles are not identical in material (color), shape or size – and the exact distribution of these properties is generally not known. In addition, pigment sizes as well as the distances between them in the fused glass ink layer are in the order of the wavelength of light – invalidating (again) the mentioned laws' assumptions about homogeneity and independence of scatterers in the medium. Moreover, the fused glassy frit matrix has a refractive index different from that of the glass substrate, effectively resulting in an additional physical interface between the fused ink layer and the glass substrate.

The digital printing process itself also adds complexity to the layer structure. For example, since the digital printing process is discrete (each pixel may receive, or not, a drop of jetted ink), the requirement for very low ink coverage essentially results in a distribution of discrete dots, since not all pixels may be covered with ink. Conversely, high-ink coverage results in thicker, more homogeneous layers of ink. Consequently, areas with low volumes of ink may possess a rougher, more scattering surface, compared to the smoother and more reflective surface produced by areas dosed with higher volumes of ink. These characteristics of the printed ink layers may lead to rather counter-intuitive results: in some cases, reflectance increases with increasing ink layer thickness, while in other cases, reflectance decreases with increased ink layer thickness, depending on the interplay between scattering at the front surface of the ink, and absorptance of the ink layer.

Complexity also arises when digitally-mixed colors are used. These mixtures are composed of individual dots of color side by side as well as on top of each other, and thus behave to some extent more like multilayers of different materials with additional interfaces rather than like mixtures that may be described by effective medium approximations.

The complexity of the system suggests that a purely theoretical model describing the interaction of light with printed ink layers on glass that will account for all the phenomena mentioned above is very difficult to obtain. Consequently, in order to predict the interactions of light energy with the ink layer, aphenomenological model based on empirical measurements and observations was developed.

First, printed layers of each of the inks were prepared at various thicknesses. By use of a spectrometer equipped with an integrating sphere, the full transmission and reflection spectra (including both specular and scattered components) were measured to be between 300 and 2500 nm. In addition, measurements were made for various digitally-printed mixtures of inks.

Light attenuation was modelled by an equation similar to the Beer-Lambert law with an additional term accounting for the observed skew from the original equation. The solar spectrum was divided into seven wavebands, and for each of the wavebands, the parameters in the equation were fitted to the measured data for each ink. In addition, other parameters were defined that relate to the relative "strength" of each ink when used in mixtures. The parameters in the resulting model (35 for each ink) were optimized until the best fit with the experimental data was achieved, finally allowing for the characteristics of any arbitrary ink mixture and thickness to be estimated.

3 Implementation

Based on the considerations described in the previous section, the obtained model describing the interaction of light with the printed ink layers, and the standard formulas for fenestration calculations of light passage [5] describing light passage in glass windows, a complete model for the passage of light through printed glass windows was obtained.

This model is available via the energy performance tool, which is a web-based software application accessible from any PC or portable device with a connection to the internet. The tool is hosted on Dip-Tech's customer support site and is readily available for use by its customers. The tool comprises the following modules that offer unique functionality.

3.1 Image File Analyzer

This module is used to read a graphic file of any format and estimates the coverage percentage of each of the ceramic inks. Since graphic files are usually represented in RGB (red, green, blue) colors to be displayed on computer screens and Dip-Tech's ink performance formulas are calculated using specific printed ink colors, there is a need to perform a transformation of the graphic files from the original RGB color space into BGWORK (blue, green, white, orange, red, black) color space used by the printers and the energy-performance model. After performing the transformation, the composition of printed inks and their dose in each area of the designed is obtained.

3.2 Ink Performance Calculator

This module calculates the energy-performance parameters based on the ink composition and dose data received from the Image File Analyzer. The model described in Section 2 is applied for every area of the image and the results are summed up for the entire image. The calculated parameters are the solar heat gain coefficient (SHGC), visible light transmittance (VT), visible light reflectance, total solar spectrum light transmittance and reflectance. The printed ink layer is usually very thin (tens of microns) compared to the glass window assembly itself (which may be composed of several glass panes with a thickness of a few mm). In addition, the ink layer (fused glass) possesses emissivity similar to that of glass (not of low-E type), and Dip-Tech's color pigments present in the ink layer do not absorb infrared radiation. Thus, the printed ink layer does not affect the window's heat transfer properties (*U*-factor) in a significant manner and the U-factor is not calculated by the tool.

The energy performance tool' development was based on Berkeley Lab WINDOW calculation procedures as known for its established results and common usage.

3.3 Multi Glass Fenestration Definer

This module allows the user to define the glass panes composing the window, the physical parameters for each of the panes (derived from the glass specifications given by the glass manufacturer), and the location of the printed layer in the assembly.

3.4 Performance Estimation Engine

this module combines all input from the sub-modules described above, and calculates and outputs the estimation of the energy-performance parameters for the entire window glass assembly including the printed layer.

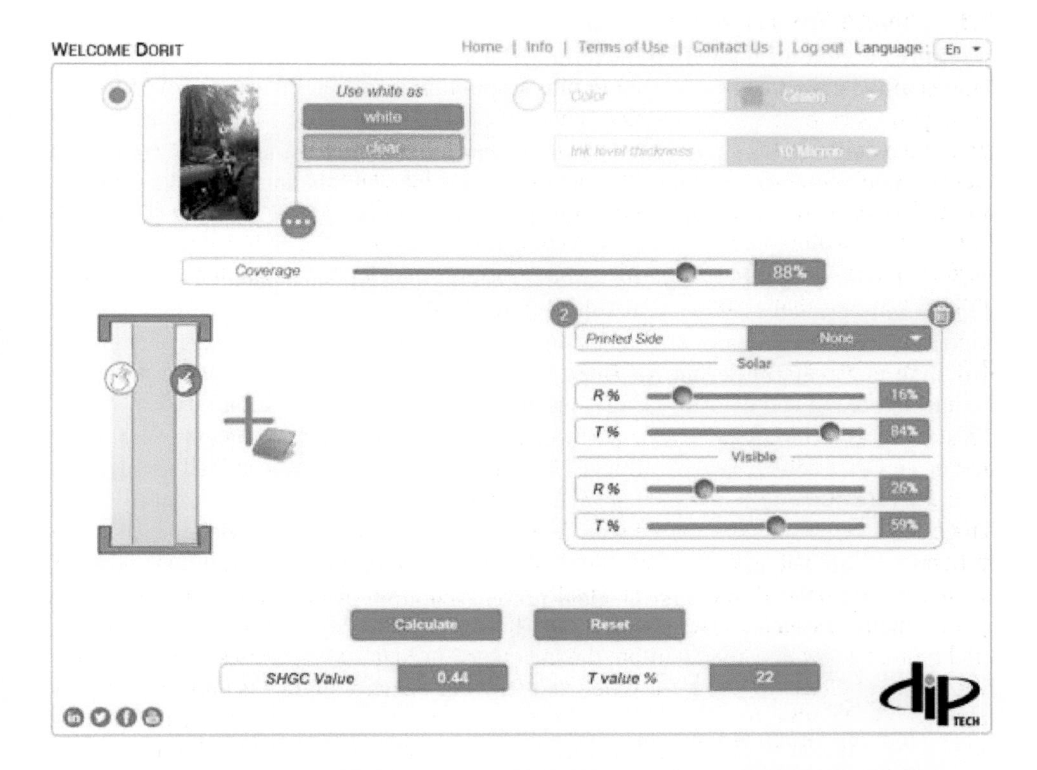

Figure 3-1 Dip-Energy Tool – Dip-Tech's Glass Energy Performance Calculator – User-Interface.

4 Conclusion

Due to increased interest in energy-efficient architecture using digitally-printed glass, Dip-Tech developed a web-based calculation tool for predicting the energy-performance parameters of digitally-printed façades.

Dip-Tech developed a phenomenological model based on empirical measurements and observations that takes into account the non-uniformity in layer thickness, color, blending of colors and surface properties that are not considered in existing calculators for smooth and uniform coatings. This model describes the light interactions with specific digitally-printed ink layers.

"Dip-Energy", the web-based software tool, combines this ink-light interaction model specially tailored for digitally printed images and inks with standard fenestration calculations of the glass layers–taking both into account in the result.

The discussed tool was successfully used where in some places, architects could have used printed glass instead of low-E coated glasses. Also, the calculator's results were successfully referenced by other glass performance calculators.

This tool, which simplifies the calculation of energy-performance parameters for the entire digitally-printed window assembly and makes them easily accessible, may serve to increase the involvement of architects, interior designers, lighting engineers and air-conditioning engineers in design for energy efficiency and sustainability while producing beautiful looking and colorful façades printed with digital printing solutions.

5 References

[1] http://glassmagazine.com/article/fabrication/dare-go-digital-1311080, Dare to Go Digital, Digital printers for glass present exciting opportunities along with new challenges, Katy Devlin May 1, 2013.

[2] http://www.commercialwindows.org/, developed by Center for Sustainable Building Research (CSBR) at the University of Minnesota and the Windows and Daylighting Group at Lawrence Berkeley National Laboratory (LBNL).

[3] http://www.usgbc.org/leed, United States Green Building Council (USGBC), LEED (Leadership in Energy and Environmental Design).

[4] http://www.breeam.com/, BREEAM (BRE Environmental Assessment Method), BRE Global Ltd.

[5] Berkeley Lab WINDOW 4.0: Documentation of Calculation Procedures., Finlayson E., Arasteh D.K., Huizenga C., Rubin M.D., Reilly M.S., July 1993.

Stiffening multi-story timber-glass composites façades in tall buildings

Alireza Fadai[1], Matthias Rinnhofer[2], Wolfgang Winter[3]

1 Department of Structural Design and Timber Engineering, Vienna University of Technology, Karlsplatz 13/259-2, A-1040 Vienna, Austria, fadai@iti.tuwien.ac.at

2 Department of Structural Design and Timber Engineering, Vienna University of Technology, Karlsplatz 13/259-2, A-1040 Vienna, Austria, m.rinnhofer@iti.tuwien.ac.at

3 Department of Structural Design and Timber Engineering, Vienna University of Technology, Karlsplatz 13/259-2, A-1040 Vienna, Austria, winter@iti.tuwien.ac.at

The objective of several research projects of the Department of Structural Design and Timber Engineering (ITI) of the Technical University of Vienna (VUT) was to develop stiffening glass fronts, which partly replace stiffening elements of tall buildings. With the purpose to meet the ecological compatibility, the possible applications of multi-story timber-glass composite (TGC) façades as supplementary stiffening element in tall buildings were investigated. Thus, the ITI combined the advantages of glass and the rapid-assembly of timber constructions with ductile metal fasteners, which improves the structural performance.

Keywords: timber, glass, composite, stiffening, façade

1 Introduction

The need for tall buildings has arisen for sound economic, social and environmental reasons. Tall buildings are now considered as a viable solution to many of the worlds developed and developing countries and community's problems such as increased population and limited availability of land. They are efficient with respect to land use; serve many people simultaneously from single set of infrastructure and services and provide more space and accommodation.

Tall buildings are exposed to horizontal forces such as wind, seismic actions or inadvertent inclinations of load bearing components. Normally these forces are transported to the stiffening elements like walls or bracing systems over stiff slabs to lead them into foundations. TGC- façades should now take the bracing function and replace stiffening walls, so investigations were made to evaluate the possibilities and limits of such systems. Up to now TGC-elements were just used in one or two story buildings, but not in tall buildings with four stories and more, for example in combination with a core, as well in timber as in reinforced concrete buildings. A decisive point for using TGC-elements is their stiffness against horizontal loads. The horizontal actions are transferred over stiff slabs to shear walls or a core and are distributed due to their stiffness relations. Therefore, the

Engineered Transparency 2016. Glass in Architecture and Structural Engineering. First Edition.
Edited by Jens Schneider, Bernhard Weller.

TGC-elements must raise a minimum stiffness to release stresses from the existing bracing elements and to support them in an economic way.

2 State of the Art

Due to the slight potential of wooden construction to brace buildings, new developments were promoted in the last years. With the developed TGC-structures the potential of timber and glass is optimally used, the glass pane has not only the aim to provide transparent façades, it is used to transfer compression stresses forced by horizontal loads while the timber elements are mainly loaded with tension stresses. Two different systems were established which follow the same main principles. Both systems are based on a timber post and beam system for the vertical load transfer. The glass pane itself is glued to an adapter frame out of birch plywood, which is screwed to the substructure. In a research project from Holzforschung Austria (HFA) [1] a system with a toothed adapter frame was published to guarantee a narrow visible width of the post and beam substructure (patent no. 502 470 [2]). This system only uses shear stresses in the 2-components-silicone bond line to introduce forces in the glass pane. The main principle of this system can be seen in 2-1.

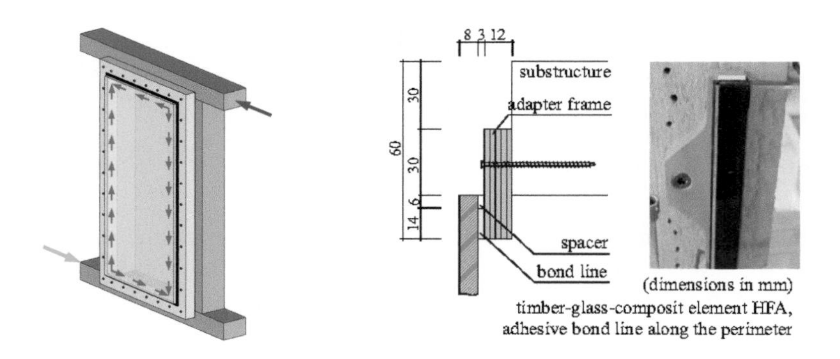

Figure 2-1 Shear bonding with adapter frame (patent no. 502 470 [2]): System (left) and detail (right).

The second system, published from ITI [3] uses an L-shaped adapter frame and, additional to the silicone bond line, blockings in the corners of the glass pane to transfer compression forces. So this systems avails the shear area as well as a compression diagonal to transmit higher forces and to provide higher stiffness (patent no. 511 373 [4]). Figure 2-2 shows a schematic display of the system.

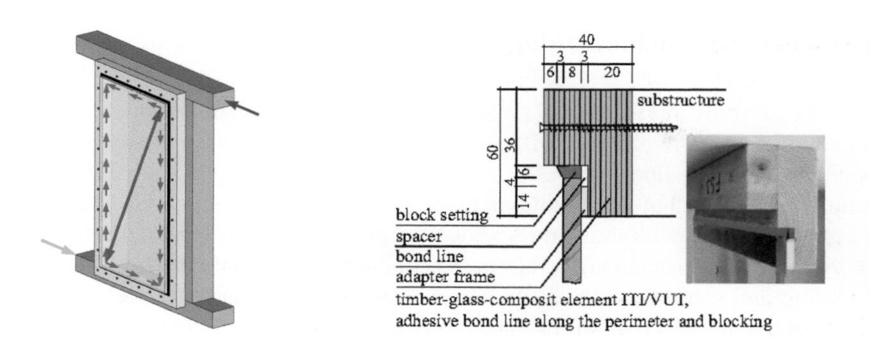

Figure 2-2 Shear bonding and blocking (patent no. 511 373 [4]): System (left), detail (right).

Based on the spring model of Kreuzinger and Niedermaier [5] regarding the shear area ITI [3] developed a linear spring model also concerning the additional blocking. Figure 2-3 shows the spring model (ITI/VUT [3]). With this model, it was possible to calculate the resistance of a TGC-element against static influences. Subsequently Weissensteiner [6] adapted the model for coupled frames, calculated that the middle elements reach a stiffness 5.6 times higher than a single element and confirmed this calculation with finite element analysis. This information is very important for the evolution of tall buildings because in higher façades it is also possible to couple the elements between the different stories and not only in horizontal direction to reach a higher stiffness. In Figure 2-3 the upper part of the model shows the springs responsible for the stiffness regarding the shear area, the lower part shows all the components of the compression diagonal (blocking etc.). These two parts are connected in parallel while the individual components are connected in series. A detailed description of the calculation of all constituents is given in the report from ITI/VUT [3].

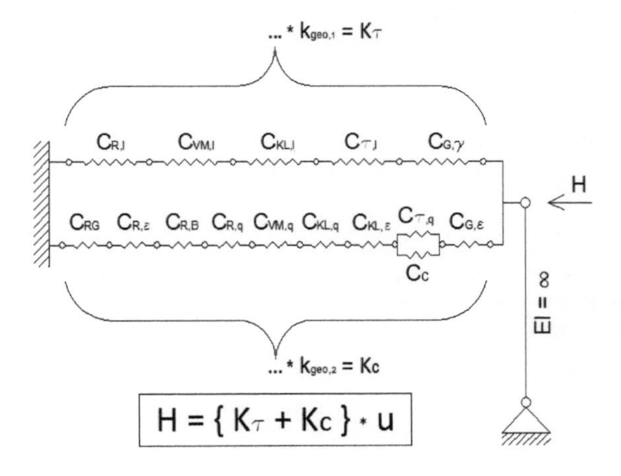

Figure 2-3 Spring model (ITI/VUT [3]).

3 Case study: Office building in Vienna

Main goal of the case study of ITI/VUT [7] was to evaluate the bracing potential of TGC-elements in tall buildings. Therefore, an office building with very simple geometric specifications was designed; the floor plan of a typical story can be seen in Figure 3-1 left. The chosen building was designed for building class 4 (escape level < 11 m; cf. Figure 3-1 right) regarding OIB 2 [8] with four stories above top ground surface. The load bearing structure for vertical forces consists of a reinforced concrete rectangular core in the middle of the building and some columns in the façades as well as in the inner part of the story. The TGC-elements, based on the system of Figure 2-2, are placed in the longitudinal as well as in the transversal direction of the building. The number of elements per side depends on the relation of the width *B* and the length *L* of the building. The bond line of the TGC-element is made of a soft silicone adhesive (OTTOCOLL® S 660 [9]), for the blocking in the corners a rigid liquid epoxy material (HILTI HIT-RE 500-SD [10]) is used. Each element has a stiffness against horizontal forces of 11 kN/cm. This value results out of experimental investigations during a former research project [3] and calculations based on the spring model of Figure 2-3. In addition, the highest load for buckling failure was calculated; therefore, a value of 34 kN horizontal force per TGC-element was proven using the Dunkerley straight line for calculating the buckling coefficients [11].

During first calculations, it got clear quickly, that the TGC-elements are far too soft to take loads from the reinforced concrete core. The next step was to connect the TGC-elements to tall stiffening shear walls, as described in Chapter 2. The stiffness was increased to 33 kN/cm for horizontal coupling and to more than 40 kN/cm for horizontal and vertical coupling. These values are only valid for middle elements, so a minimum of three elements is necessary to achieve this stiffness at one element. If there are four elements in a row, these values count for both middle elements and so on. Nevertheless, this was still not enough, so the next step was to decrease the load bearing elements in the core to a minimum number of shear walls, see Figure 3-1 middle. The reinforced concrete was still too stiff for the TGC-elements to gain a relevant minimization of the residual forces in the core.

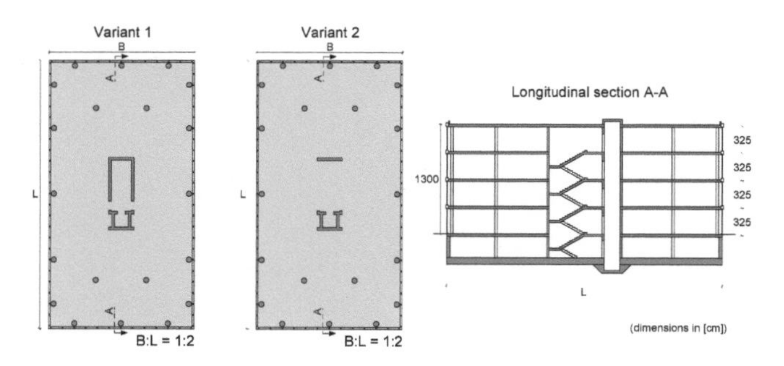

Figure 3-1 Floor plan of a typical story; Variation 1 (left), Variation 2 (middle), section (right).

The decision was made to analyze a post and beam construction without any bracing elements except for TGC-walls in the façades. This study was made with a reinforced concrete structure as well as with a timber structure.

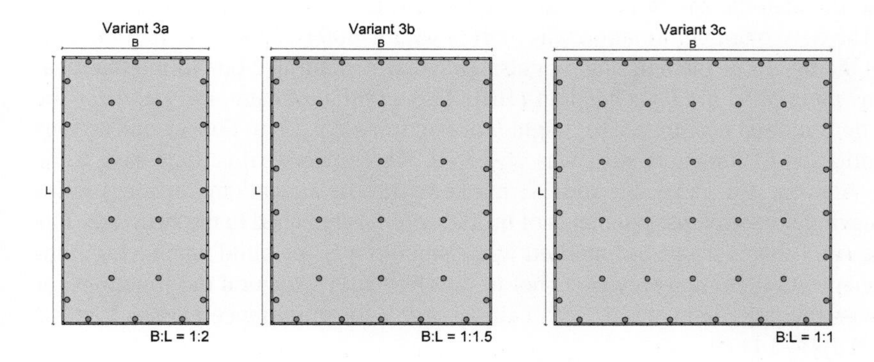

Figure 3-2 Floor plan of a typical story; Variation 3a, 3b, 3c.

The post and beam structure was used for a parametric study for the application of TGC-systems as bracing elements. The variables were the relation of width and length of the building and the number of stories. It was varied between $B: L = 1: 1, 1: 1.5$ and $1: 2$ and between one and four stories. The floor plans of a typical story for each type are shown in Figure 3-2. All options were considered in timber (Table 3-2) as well as with a reinforced concrete structure (Table 3-3). The TGC-elements were not coupled in this study, so a stiffness of 11 kN/cm was estimated. These types of buildings were analyzed regarding their load bearing behavior against wind and seismic actions. For the vertical load bearing no exact calculation was made, but the differences in structural mass were incorporated in the calculation for seismic actions. As building site, Vienna's 22nd district was chosen, because there the highest wind loads in Vienna occur (see Table 1).

For seismic actions, a ground acceleration of $a_{gr} = 0.80\ m/s^2$ according to ÖNORM B 1998-1 [12] was estimated.

Table 3-1 Load case Wind in Vienna's 22nd district according to ÖNORM B 1991-1-4 [13].

Basic wind velocity $v_{b,o}$	27 m/s
Basic velocity pressure $q_{b,o}$	0.46 kN/m²
Peak velocity pressure q_p	1.07 kN/m²

Due to a loss of normative specifications a safety factor of $\gamma_M = 1.5$ according to ÖNORM B 3716-1 [14] was chosen, because buckling of the glass pane was determined as the decisive failure of the structure. Experimental tests in [3] showed, that after the failure in the glass pane due to buckling some reserves in the usability of the system were still existing. So the material safety factor γ_M for glass was used global for the whole TGC-structure. The horizontal deformation was limited with *H/300* (ÖNORM B 1990-1 [15]) where *H* is the height of the building, but also *H/500* was examined due to international standards as mentioned in (Jayachandran [16]). These limitations are also valid for the horizontal deformation per story. The height *h* of one story is 3.25 m. For seismic actions also the limitation of interstory drift was observed, where two verifications have to be made. The first one (in Table 3-2 and 3-3 marked with the addendum "brittle") is for buildings having non-structural elements of brittle materials attached to the structure. The second one (in Table 3-2 and 3-3 marked with "smooth") is for buildings having non-structural elements fixed in a way so as not to interfere with structural deformations, or without non-structural elements. In both cases $n = 0.5$ for importance classes I and II (ÖNORM B 1998-1 [12]).

In Table 2 and 3 the values marked in blue fulfill all the criterions with high safety (level of utilization < 80 %), the green values are still on the safe side (between 80 % and 100 %) while the red values exceed the limits.

Table 3-2 Results of parametric studies based on a timber construction.

assumption	stories		1			2			3			4		
	L:B		1:1	1:1,5	1:2	1:1	1:1,5	1:2	1:1	1:1,5	1:2	1:1	1:1,5	1:2
	H/300	[mm]	10.8			21.7			32.5			43.3		
	H/500	[mm]	6.5			13.0			19.5			26.0		
wind	horizontal displacement	[mm]	2.2	3.0	4.4	8.1	10.7	15.7	16.8	22.1	31.8	28.5	37.4	53.0
	utilization grade H/300	[%]	20%	28%	41%	37%	49%	72%	52%	68%	98%	66%	86%	122%
	utilization grade H/500	[%]	34%	46%	68%	62%	82%	121%	86%	113%	163%	110%	144%	204%
	F per element	[kN]	3.8	5.2	7.7	11.9	15.4	23.0	19.1	25.6	38.0	26.6	35.8	53.6
	utilization grade	[%]	17%	23%	34%	52%	68%	102%	84%	113%	168%	117%	158%	236%
seismic action	interstory drift	[mm]	17.5	17.6	17.8	24.1	24.2	24.8	26.8	26.9	28.7	28.0	28.6	30.2
	utilization grade "brittle" $d_r \leq 0.005*h/n$ = 33 mm	[%]	54%	54%	55%	74%	74%	76%	82%	83%	88%	86%	88%	93%
	utilization grade "smooth" $d_r \leq 0.01*h/n$ = 65 mm	[%]	27%	27%	27%	37%	37%	38%	41%	41%	44%	43%	44%	46%
	F per element	[kN]	20.1	20.2	20.4	27.8	27.6	28.6	31.0	31.1	33.0	32.4	33.0	33.2
	utilization grade	[%]	59%	59%	60%	82%	81%	84%	91%	91%	97%	95%	97%	98%

Table 3-2, based on the previously described principles, shows, that a timber post and beam construction can be executed with all observed geometric specifications up to two stories. Three stories are also possible if the length and the width of the building are more or less the same. The values for three stories and more complex geometries as well as the values for four stories with $B:L = 1:1.5$ and $B:L = 1:2$ exceed the limits with 10 to 70 % especially in the calculation of the buckling load per element. This should be part of further researches, because an increase of the estimated value of 34 kN definitely seems to be possible for example by using a stiffer type of adhesive such as a modified epoxy used by Nicklisch [17].

For seismic actions, all the criterions are fulfilled. This is a sign for the good and elastic behavior of TGC-structures under cycling loads without exceeding the limit for interstory drift.

Table 3-3 Results of parametric studies based on a reinforced concrete structure

			1			2			3			4		
assumption	stories		1			2			3			4		
assumption	L:B		1:1	1:1,5	1:2	1:1	1:1,5	1:2	1:1	1:1,5	1:2	1:1	1:1,5	1:2
assumption	H/300	[mm]	10.8			21.7			32.5			43.3		
assumption	H/500	[mm]	6.5			13.0			19.5			26.0		
wind	horizontal displacement	[mm]	2.2	3.0	4.4	8.1	10.5	15.4	16.4	21.4	30.8	27.5	35.9	50.4
wind	utilization grade H/300	[%]	20%	28%	41%	37%	48%	71%	50%	66%	95%	63%	83%	116%
wind	utilization grade H/500	[%]	34%	46%	68%	62%	81%	118%	84%	110%	158%	106%	138%	194%
wind	F per element	[kN]	3.8	5.1	7.7	11.8	15.4	23.0	19.1	25.6	39.1	26.7	34.4	53.5
wind	utilization grade	[%]	17%	23%	34%	52%	68%	101%	84%	113%	173%	118%	152%	236%
seismic action	interstory drift	[mm]	29.1	29.3	29.4	38.5	39.2	39.7	33.9	34.7	35.7	29.2	29.8	31.7
seismic action	utilization grade "brittle" $d_r \leq 0.005*h/n =$ 33 mm	[%]	90%	90%	90%	118%	121%	122%	104%	107%	110%	90%	92%	98%
seismic action	utilization grade "smooth" $dr \leq 0.01*h/n =$ 65 mm	[%]	45%	45%	45%	59%	60%	61%	52%	53%	55%	45%	46%	49%
seismic action	F per element	[kN]	33.6	33.7	33.9	44.5	44.3	45.8	39.1	40.0	41.1	33.7	35.8	36.5
seismic action	utilization grade	[%]	99%	99%	100%	131%	130%	135%	115%	118%	121%	99%	105%	107%

With a reinforced concrete structure some checks are exceeded already with two stories since this structure is very dominant compared to the soft TGC-elements. The use of such elements together with a stiff core is not reasonable in consideration of economic and ecologic aspects.

In an additional study, the case of fire was investigated. Therefore, the building types of 3-2 were split into two fire compartments in the middle of the building where the assumption was made, that in one part a total collapse of all stiffening elements against horizontal forces occurs. So the other half of the building must carry all loads itself. This was possible in all geometries. The highest recorded load per element during the simulation up to three stories was 9.8 kN, which is far beneath the limit for buckling of 34 kN. So the case of fire isn't the decisive criterion in this type of building anyway, the focus of further developments has to be on the increase of the potential of carrying horizontal loads in short and long term periods and on a higher stiffness of the single element as well as of coupled elements in the façades.

4 Conclusion

Façades are very complex systems, which serve many different claims; it is increasingly in the interest of the community to use them also for static functions considering resource efficient constructions. The objective of several research projects of the ITI at the VUT was to develop stiffening timber-glass fronts, which replace wind bracings and/or expensive frameworks in tall buildings. The solutions for multi-story bracing façades were developed. The TGC-facades enable a more efficient functionality of structural glass by allowing the use of approved timber joining techniques. The results provide a marketable component system for buildings, which could optimally use timber and glass. Currently thanks to these research projects and implementations, the requirements are well known.

The newly developed system were evaluated by using linear and non-linear FEM calculation. Different types of TGC facades were compared with each other. The investigations showed that all structural systems have sufficient stiffness against horizontal displacement. The executed calculations prove the efficiency of TGC-structures in timber buildings up to four stories.

Nevertheless, the unique combination of wood products with glass components will also request deeper investigations regarding the interfaces, the compatibility and the collaboration between the different components and materials. The final issue consists in optimizing a composite system by using the best characteristics of each material with the aim to develop and advance a high competitive composite building system. The newly developed system shall be promoted through the initiation of pilot applications. Initiation of pilot projects could provide the possibility to experience the practicality and applicability of the developed composite systems, and create different design and application solutions.

Timber-glass composites façades have a lot of potential to be discovered. The development of such elements and the increase of stiffness and load bearing capacity must be the goal of further research projects to facilitate tall buildings a resource efficient façade and no further bracing systems in the inner part of the building. Furthermore, the discussion of the safety factor should proceed to enable the engineers to calculate TGC-structures in a reasonable way, based on a normative background.

5 Acknowledgement

The authors gratefully acknowledge the financial support of the Vienna Business Agency (ID: 1122338) and brandRat ZT GesmbH for funding the research project “Timber-Glass-Compound façades – Behavior in case of fire – Fire protection concepts” [7].

6 References

[1] Edl, T.; Schober, K.: Statisch wirksame Holz-Glas-Verbundkonstruktionen zur Aussteifung von Holzbauten. Holzforschung Austria, Vienna: 2005.

[2] Patent AT 502 470 B1: Verbundelement aus Glas. Registration on: 06.07.2005. Published on: 15.08.207. Patent Applicant: Österreichische Gesellschaft für Holzforschung, A-1030 Vienna.

[3] Hochhauser, W.; Winter, W.; Kreher, K.: Holz-Glas-Verbundkonstruktionen: Berechnungs- und Bemessungskonzepte. Technical University of Vienna, Vienna: 2011.

[4] Patent AT 511 373 B1: Verbundkonstruktion aus einer Glasscheibe und einer Rahmenkonstruktion. Registration on: 27.04.2011. Published on: 15.05.213. Patent Applicant: Technische Universität Wien, A-1040 Vienna.

[5] Kreuzinger, H.; Niedermaier, P.: Glas als Schubfeld, Tagungsband Ingenieurholzbau. In: Karlsruher Tage, 2005.

[6] Weissensteiner, F.: Holz-Glas-Verbundkonstruktionen im Einsatz an thermischen Pufferzonen. Protoytischer Einsatz am Beispiel eines Anbaus. Master Thesis, Technical University of Vienna, Vienna: 2013.

[7] Institut for Structural Design and Timber Engineering/Technical University of Vienna: Holz-Glas-Verbundfassaden: Verhalten im Brandfall – Brandschutzkonzepte, Technical University of Vienna: in progress.

[8] OIB Richtlinie 2: Brandschutz. Richtlinie des Österreichischen Instituts für Bautechnik, Vienna: 2015.

[9] Hermann Otto GmbH: www.otto-facade.de; Accessed: 11.07.2016.

[10] Hilti Deutschland AG: www.hilti.de; Accessed: 11.07.2016.

[11] Neumann, L.; Arnold, A.; Hochhauser, W.: Zur Stabilität von geklebten und geklotzten Glasscheiben: Beurteilung der Dunkerley’schen Geraden zur Beulwertbestimmung. Technical University of Vienna, Vienna: 2011.

[12] ÖNORM B 1998-1: Auslegung von Bauwerken gegen Erdbeben, Grundlagen, Erdbebeneinwirkungen und Regeln für Hochbauten. Austrian Standards Institute, Vienna: 2011.

[13] ÖNORM B 1991-1-4: Einwirkungen auf Tragwerke, Allgemeine Einwirkungen – Windlasten. Austrian Standards Institute, Vienna: 2013.

[14] ÖNORM B 3716-1: Glas im Bauwesen – Konstruktiver Glasbau. Austrian Standards Institute, Vienna: 2015.

[15] ÖNORM B 1990-1: Grundlagen der Tragwerksplanung, Hochbau. Austrian Standards Institute, Vienna: 2013.

[16] Jayachandran,P.: Design of Tall Buildings – Preliminary Design and Optimization. In: National Workshop on High-rise and Tall Bauildings, University of Hyderabad, Hyderaba: 2009.

[17] Nicklisch, F.; Hernández S.; Schlehlein, M.; Weller, B.: Pavillon mit aussteifender Holz-Glas-Verbundfassade. In: Glasbau 2016, Weller, B., Tasche, S. (eds.), Dresden: 2016, 47-61.

Timber-glass composite shear wall

Daniel Neumer[1], *Geralt Siebert*[2]

1 Universität der Bundeswehr München, daniel.neumer@unibw.de

2 Universität der Bundeswehr München, geralt.siebert@unibw.de

Fulfilling the need to bring natural day light within a building, glass has become a fundamental material in the design of contemporary facade structures. Currently the use of glass is limited to the mere function of covering a building. It only serves the purpose to separate the inside from the outside environmental influences and withstand its own weight. No load bearing function is applied. The remaining building structure must bear all loads. This is challenging, especially for horizontal loads (e.g. wind) because no shear walls exist. Considering the high compressive strength of glass, glass as a facade element is not fulfilling its true potential. Based on varied edge finishes and support conditions, tests have shown an impact on the load bearing capacity of a glass pane. In this research project at the Universität der Bundeswehr München a timber-glass composite shear wall is currently being developed and tested. The measurements of the glass pane in the composite element are (w x h) 2.3 m x 2.1 m. The timber frame is connected to the glass through blockings (POM). These are disposed near the edges of the glass pane as in a window. In this way the system enables the load transfer of horizontal forces through compression diagonals within the glass pane, thus transforming glass into a bracing element. The experiments and numerical studies of this research are the basis for a long term monitoring project in a real life situation.

Keywords: shear wall, load bearing, buckling, timber-glass-composite

1 Introduction

These days the use of glass as a facade element is limited to the mere function of covering a building. It only serves the purpose of separating the inside from the outside environmental influences and withstanding its own weight. This means with the use of increasingly larger glazed facades, the load transfer has to be taken by the remaining building structure, leading to over dimensional support framework. In this context it would be useful to exploit the high compressive strength of glass.

Although many research projects have proven the ability of glass as a load carrying element, there exist no guidelines or standards. For this special approval through building authorities is needed. For this reason a monitoring project shall be realized in which the bracing is conducted only with timber-glass composite shear walls. The purpose is to gain long term information of the developed system in real life situation.

The key points for the application of glass as shear element in this system are load introduction and stability issues. Initial insights are presented in this paper.

Engineered Transparency 2016. Glass in Architecture and Structural Engineering. First Edition.
Edited by Jens Schneider, Bernhard Weller.

2 Load Introduction

2.1 Investigated Parameters

For the load introduction at the glass edge three parameters (glass type, edge formation, connector material – see Table 2-1) were tested. The measurements of the glass panes and the position of the connector material are shown in Figure 2-1 (left). The glass had a thickness of 6 mm. For each combination three tests were conducted. The test setup (Figure 2-1 (right)) simulates the corner of the glass pane in a frame.

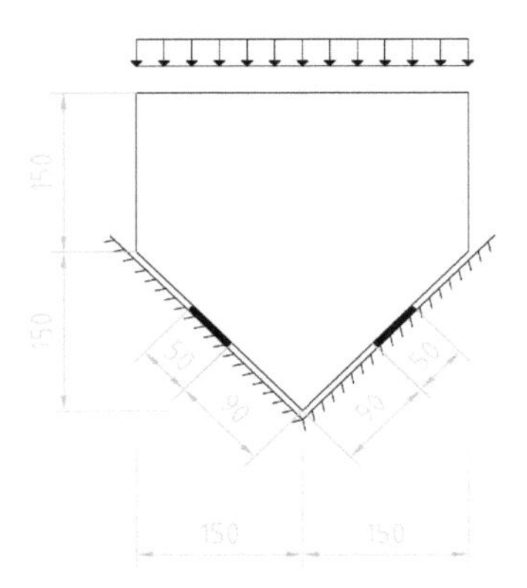

Figure 2-1 Measurements of glass panes (left) and test setup (right).

The compression load at the top of the pane was introduced along the entire length of the glass pane. As interlayer for the load introduction Teflon was used to minimize friction. The deflection horizontally and vertically was recorded. The load introduction was conducted stepwise (between 0 kN and 30 kN steps of 3 kN & between 30 kN and 90 kN steps of 30 kN) to detect deformations in the connector material. After each load step the specimen was unloaded completely. After reaching a load of 90 kN the test was aborted even if the glass pane or the connector material did not fail.

Table 2-1 Investigated parameters for load introduction.

Glass Types	Edge Formation	Connector Materials
annealed glass	raw edge	Hilti Hit 70
fully tempered glass	polished edge	POM
-	rounded edge (*C-form*)	beech
-	cut edge (*only annealed glass*)	steel

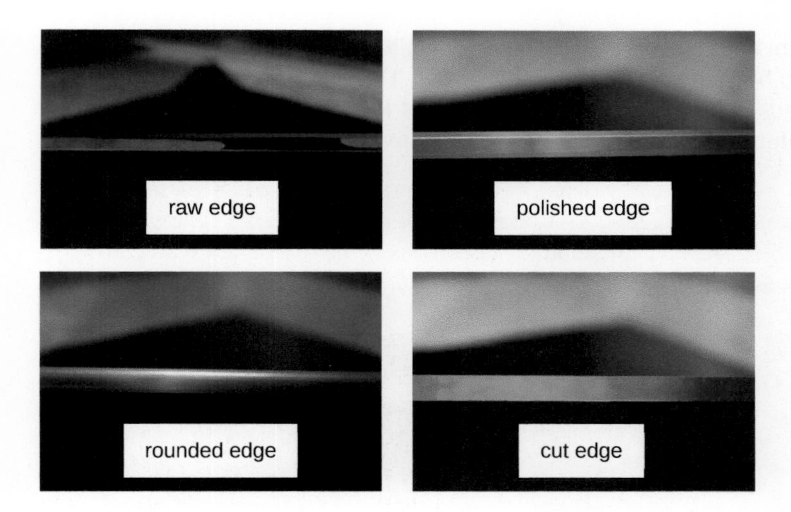

Figure 2-2 Edge formations.

2.2 Results

2.2.1 POM

In conclusion POM shows the best result for the load introduction into the glass. For all tested glass types and edge formations the connector failed; except with the "cut edge". The maximum loads show a small variance. In the used setup the connectors are able to resist about 60 kN. No difference between annealed glass and fully tempered glass can be detected. The failure did not occur immediately (plasticize of connector).

Figure 2-3 Deformed POM after testing (raw edge).

2.2.2 HILTI HIT 70

The test demonstrates deficiencies caused by stress perpendicular to the load introduction in the connector. For all tested glass types and edge formations the connector failed. The specimens with "cut edge" broke as the connectors failed. No influence of the edge formation on the load bearing capacity can be detected. The results show a high variance. The maximum loads alternate between 22 kN and 44 kN. Cracks in the connector appeared prior to failure. Here also a high variance is given. The loads alternate between 15 kN and 27 kN.

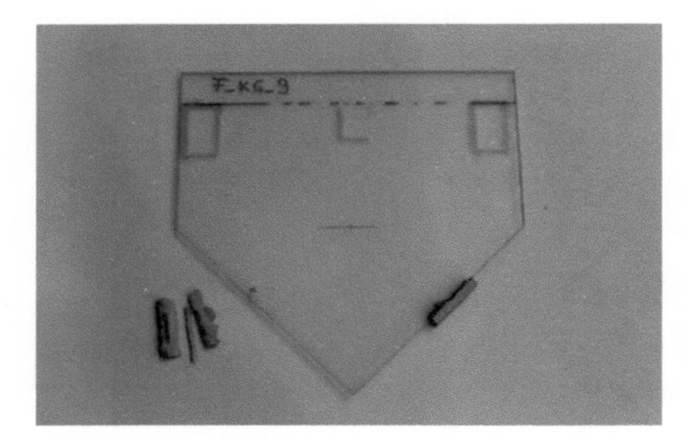

Figure 2-4 Hilti Hit after testing (raw edge).

2.2.3 Beech

In pre-tests on timber materials beech qualified for further investigation. Only tropic woods achieved better results than beech. But these are expensive and non regional. The load introduction into the timber connector occurred to the end grain.

For all tested glass types and edge formations the connector failed. No influence of the edge formation on the load bearing capacity can be detected. The maximum loads alternate between 39 kN and 47 kN. Compared to other connector materials the biggest elastic deflections exist.

Figure 2-5 Deformed beech after testing (raw edge).

2.2.4 Steel

Although contact between glass and steel should be avoided, this combination was tested to get an impression of the maximal load bearing capacity and to distinguish between the different edge formations. The edge formation “cut edge” was not tested.

It can be summarized that cracks in the glass occurred while testing annealed glass, independent of the edge formation. But it was still possible to introduce higher loads afterwards. The stepwise load increase highlighted a crack progress.

Testing fully tempered glass showed the potential to introduce load through the edges into the glass. All edge formations reached a maximum load of 90 kN. After testing plastic deformation in the connector was detected.

Figure 2-6 Plastic deformation in steel after testing (rounded edge).

2.3 FEA Location of Blocking Position

Additionally the experimental testing of a 2D finite element analysis with ANSYS Workbench (V 15.0) was conducted to investigate the influence of the location of the blocking along the corner on the stress distribution in the pane (Figure 2-7).

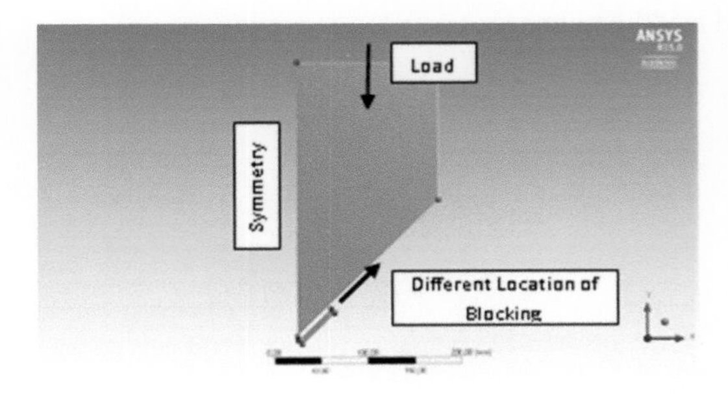

Figure 2-7 2D FEA investigation on the location of the blocking.

First results of the numerical studies are presented here. More detailed studies are currently being conducted. The here presented results were generated with the materials glass and steel (connector). The used element types for the mesh are Quad8 & Tri6. A frictionless contact between the glass pane and the connector was set. The element size at the contact area is 0.5 mm. In the following diagram the stresses at a load of 500 N are compared. The results show an almost constant compressive stress in the pane with an increasing distance between the blocking and the corner, while the tension in the beginning increases and then quickly decreases.

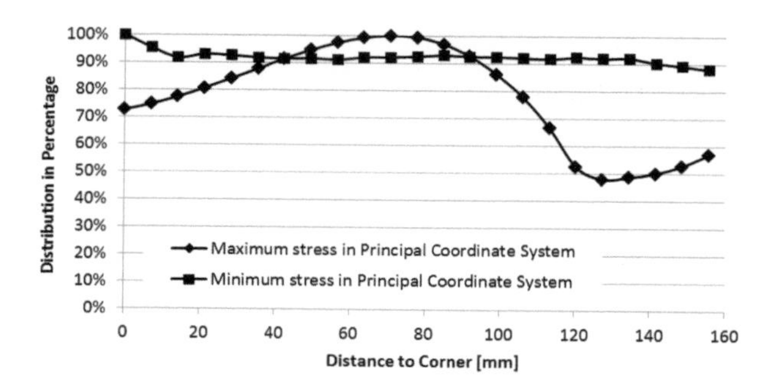

Figure 2-8 Maximum and minimum principal stress in model pane depending on distance from the corner.

3 Large Scale Specimen

3.1 System

The measurements of the glass pane in the timber-glass composite element are (w x h) 2.3 m x 2.1 m. The timber frame is connected to the glass through blockings (POM) which are disposed near the edges (Figure 3-1). For the timber frame BSH GL24h is used. The applied glass is a triple glazed insulating glass. The glass configuration is (outside to inside) 8 mm annealed glass, 15 mm cavity, 8 mm fully tempered glass (polished edge), 15 mm cavity, 6 mm annealed glass. The special characteristic is that the load introduction only occurs into the fully tempered middle pane.

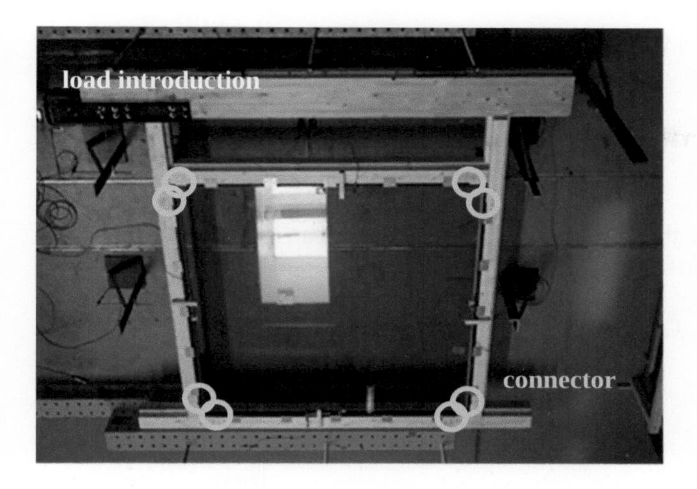

Figure 3-1 Test setup big scale specimen.

3.2 Results

Up until now only a pre-test has been conducted. The test showed that the load transfer in the timber frame needs to be optimized. Due to deformations in the frame the glass pane achieved an unfavorable stress state which caused a failure. Already loads up to 50 kN (at about 65 mm deflection at the load introduction) could be introduced into the timber-glass composite until the system failed. After an optimization the system should be able to withstand even higher loads.

The construction approval requirements of the design load (26 kN) at a max deflection of 30 mm for the monitoring project are fulfilled with the tested system. The effects of further system optimization needs to be explored.

4 Conclusion & Prospect

According to the presented results it is possible to decrease the stress in the glass and as a result increase its loadbearing capacity by modifying the support conditions; because the right choice of edge formation and blocking material allows an introduction of high loads into the glass pane. This means additional loads to the dead weight can be introduced in the glass pane through the glass edges. As well, the position of the blocking strongly influences the stress distribution in the glass pane. The here presented results show only trends. To verify these further investigations are planned. Up to now many factors have not taken into account yet. For example imperfections due to misaligned support conditions of the glass edge. The large scale specimen confirmed the introduction of additional loads to the dead weight into the glass pane. In the pre-test the glass breakage occurred due to insufficiencies in the connections of the framework. It is to await if the framework can be improved in such a way that critical stability loads for the glass pane can be reached.

5 Acknowledgement

The work summarized in this paper results from the research project "Holz-Glas-Fachwerk" which is supported by the ZIM program of the German Federal Ministry of Economics and Technology (BMWi). The authors thank the industrial partners HUF HAUS and Glas Schneider, Germany.

6 References

[1] Luible, A: Lasteinleitungsversuche in Glaskanten, Rapport ICOM 463, Ecole polytechnique fédérale de Lausanne: 2004.

[2] Wellershof, F.: Nutzung der Verglasung zur Aussteifung von Gebäudehüllen, Dissertation RWTH Aachen, Shaker Verlag: 2006.

[3] Engelhardt. O.: Flächentragwerke aus Glas, Tragverhalten und Stabilität, Dissertation Universität für Bondenkultur Wien: 2007.

[4] DIN EN 1995-1-1, Eurocode 5: Design of timber structures – Part 1-1: General – Common rules and rules for buildings: December 2010.

[5] DIN 18008-1: Glass in Building; Design and construction rules; Part 1: Terms and general bases: December 2010.

[6] DIN EN ISO 14439: Glas im Bauwesen – Anforderungen für die Verglasung – Verglasungsklötze; Deutsche Fassung prEN 14439:2007: November 2007.

[7] Neumer, D.; Siebert G.: Influence of Edge Design and Location of Load Introduction on the Loadbearing Capacity of Glass, IABSE – Conference, Nara 2015.

Insulated glass units with pressure compensation

Bernhard Weller[1]*, Mirko Köhler*[1]

1 Technische Universität Dresden, Institute of Building Construction, George-Bähr-Straße 1, 01069 Dresden, Germany, Mirko.Koehler@tu-dresden.de

The saving of energy has become a very important topic over the last years. However, not only for building constructions but also in the field of naval architecture or transportation the energy revolution changed the materials used. Insulated glass units (IGUs) are used because of their increased thermal insulation value, but there is a problem about the gas tightness of the edge sealant systems. In the worst case, insulated glass units absorb so much moisture that these fog up on the inner site of the glass panels and get unusable. Most notably, in the naval architecture we can recognise a very high transfer rate of defective insulated glass units. The insulated glass units are exposed to high stress as a result of environmental impacts. One has to consider high wind loads and the dead weight of the outer panels which are transferred into the edge sealant. Furthermore, there is the impact of the UV-radiation that causes climatic loads inside the insulated glass unit. All those loads can cause a pillow effect and weaken the edge sealant to a point of a very high moisture transfer into the insulated glass unit. One way to extend the lifetime of insulated glass units is to eliminate the influence of the climatic loads that reduce the functionality of the edge sealant system. This experimental study tries to compare edge sealant systems that are state of the art with insulated glass units which are equipped with pressure equalisation. The focus is on an artificial ageing test according to the EN 1279-2 in order to compare the moisture penetration of the different specimens. After the successful tests with specimens that can equalise the climatic pressure, there were tests to examine the type of function of the pressure equalisation. The experimental investigation showed that the climatic loads have a strong influence on the gas transfer through the edge sealants of the insulated glass units. Systems that enable the insulated glass unit to connect the gas cavity with the ambient air absorb less moisture compared with reference systems. It can also be seen that the type of the connection has an influence on the lifetime of the insulated glass units, because through this connection moisture can also penetrate into the cavity between the glass panels. The results will allow us to develop more durable sealant systems. Additionally, we can optimise the artificial ageing programme for marine glazing products.

Keywords: insulated glass units, pressure equalisation, durability, air tube

1 Introduction

1.1 State of the Art

Insulated glass units consist of two or more glass panels and a spacer (most of the time stainless steel, aluminium or plastic) that keeps the panels in a certain distance from each other (Figure 1-1). To ensure the stability and impermeability of the insulated glass units (IGU), there are two sealings between the panels and around the spacer. The primary sealant is made of Polyisobutylen (PIB) and is arranged between the panels and the

Engineered Transparency 2016. Glass in Architecture and Structural Engineering. First Edition.
Edited by Jens Schneider, Bernhard Weller.

spacer, normally with a thickness of 0.25 mm on each side of the spacer. The primary sealant is responsible for the impermeability of the insulated glass unit. On the back of the spacer is the secondary sealant. Different elastomeric materials that are used, e.g. Polysulfide (PS), Polyurethane (PU) and Silicone (S). The secondary sealant is responsible for the stability of the insulated glass unit.

Because of the penetrating moisture, the spacer contains a desiccant that gathers the humidity and extends the lifetime of the insulated glass unit. The cavity between the panels is filled with inert gas (Argon, Krypton or Xenon) to improve the thermal insulating value. Most of the time, Argon is used because Krypton and Xenon are rare and more expensive. The lifetime of an insulated glass unit is defined by the relative humidity in the cavity between the panels and the gas rate (the discussion of which is beyond this paper). If the desiccant cannot absorb more water, the moisture in the cavity will condense at the glass panels. The defective insulated glass unit is not transparent at normal room temperature anymore and has to be replaced by a new one.

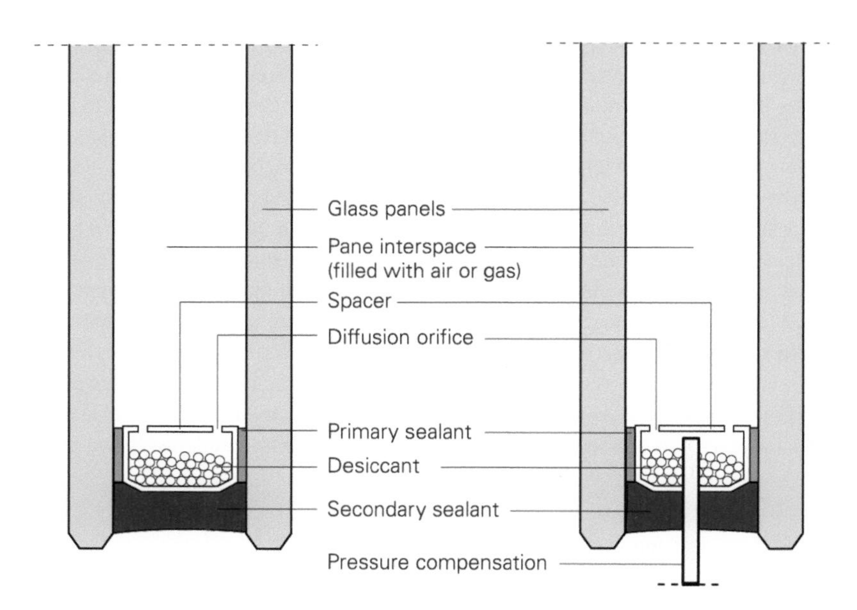

Figure 1-1 Simplified illustration of an insulated glass unit and an insulated glass unit with the possibility of pressure compensation.

In the shipbuilding industry, insulated glass manufacturers reacted to the problem of the durability of the IGUs in using more secondary sealant than the construction industry. Normally, the secondary sealant on the back of the spacer is 2-6 mm thick. The edge sealant system of marine insulated glass units has a size of up to 30 mm. Despite this solution, the insulated glass units on vessels cannot withstand the marine environmental loads and get unusable after a few years, sometimes already after a couple of months.

Other solutions for marine glazing units are not present at the moment. The aim of this research work is an improvement of these marine glazing units. As ca be seen in figure 1-1, the different research structures are equipped with pressure compensation. That should minimise the impact of the climatic loads and increase the durability.

1.2 Climatic Loads

Climatic loads have a negative influence on the durability of the edge sealant systems. Figure 1-2 shows how the temperature inside the insulated glass unit and the air pressure affect the shape of the glass panels. The concave and convex deformations of the external glass panels are the result of the climatic loads. Thus, those initiated loads damage the edge sealant system during the time of use. The movement of the glass panels stretches and compresses the sealants, thereby weakening the sealants and thus increasing the permeability rate. As a result, the insulated glass units lose their transparency and get unusable. Additionally, the insulated glass units can lose the inert gas inside the closed cavity. Thus, the thermal insulation value decreases.

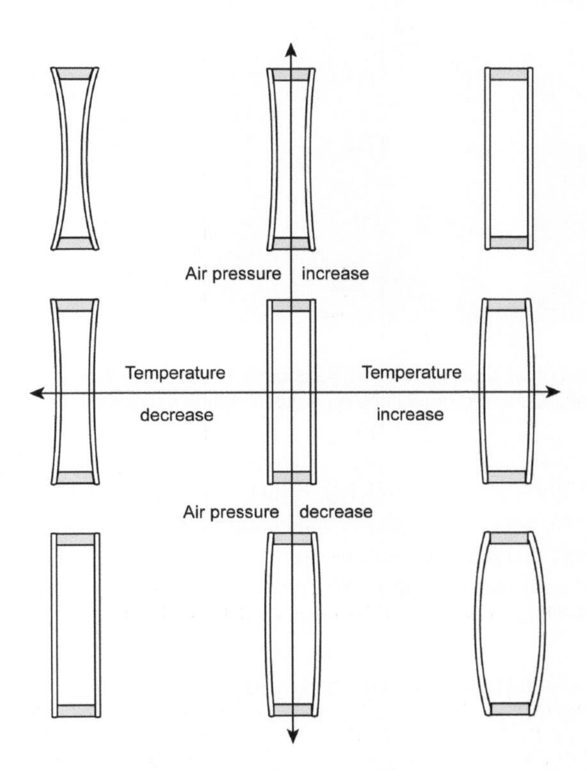

Figure 1-2 Simplified illustration of the insulating glass effect.

This paper will explain whether the possibility of pressure compensation as shown in figure 1-1 could minimise the negative effects of the climatic loads. In the present research project, different ways of pressure compensation were tested.

1.3 Requirements

The problems about the durability of insulated glass units are known. The lifetime of insulated glass units is limited by the penetrating humidity and the leaking inert gas. Marine glazings often lose their functionality after a lifetime of six to eight years because the sealants cannot withstand the impact of the maritime environment. Some insulated glass units on vessels fog up on the inner site of the glass panels (figure 1-3, left) already after a couple of months. The objective of the research project is to create an insulated glass unit with an air pressure compensation (figure 1-1) in order to avoid the negative influences of the climatic loads.

Figure 1-3 Possible damages on insulated glass units. Insulating glass unit with moisture in the closed cavity between the glass panels (left). Distorted reflections on a glass facade caused by the pillow effect (right).

The second problem is the insulating glass effect (figure 1-3, right). The glass panels of the insulating glass deform their surface, which is caused by the changing pressure inside the closed cavity of the glasses. Hence, distorted reflections appear on the glass surface of the facade. An insulated glass unit with functioning pressure compensation can avoid those distorted reflections and show a plane surface over the whole period of use.

Vessels are in use around the world, which is why the replacement of a few glazing units is connected with considerable material and personnel costs. This results in elaborate and cost-intensive repairs and regular exchanges of defective insulated glass units. The short lifetime of the insulated glass units incorporated in vessels and the expensive exchange costs are responsible for a high acceptance of novel insulated glass units with a new edge sealant system.

2 Approach

2.1 Artificial Ageing

To compare the durability of different edge sealant systems, an artificial ageing programme is useful. Thus, the conditions inside the artificial ageing chamber accelerate the damage of the edge sealant systems and increase the amount of penetrating moisture. Consequently, it is possible to compare the used specimens already after a few weeks in contrast to research projects with natural ageing programmes.

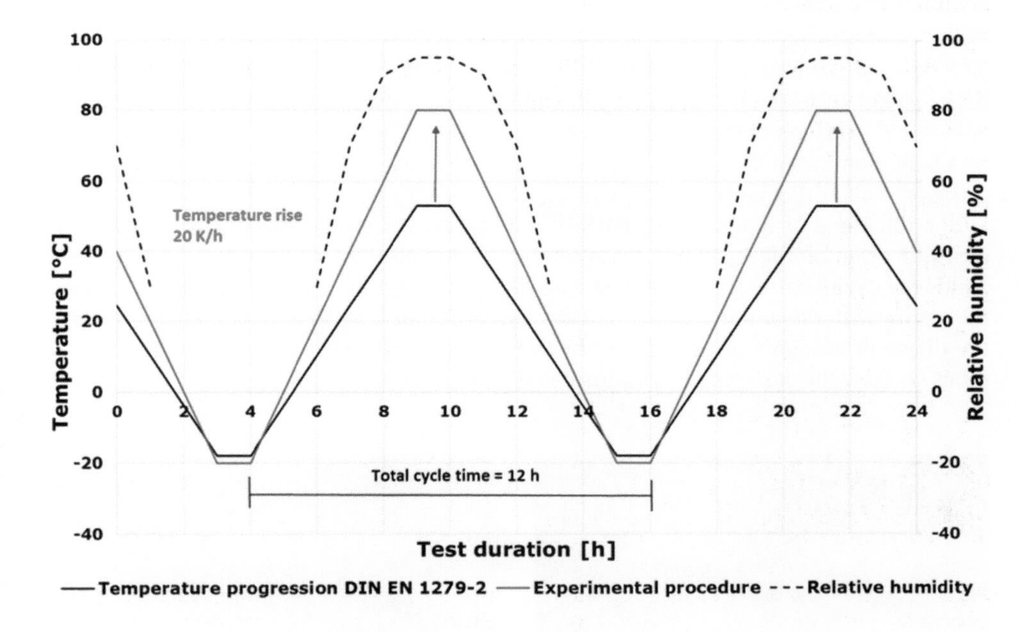

Figure 2-1 Representation of the artificial ageing according to the DIN EN 1279-2 and the experimental procedure.

Figure 2-1 shows two artificial ageing programmes. The red line shows the experimental procedure which is inspired by the mandatory test for moisture absorption in the DIN EN 1279-2 [1]. In order to save time and to accelerate the artificial ageing, the temperature range was adjusted. Additionally, it is to be expected that the influences of the maritime environment are more aggressive. That means that it is necessary to adjust the artificial ageing of the DIN EN 1279-2. The range is between a minimum of -20 °C and a maximum of 80 °C. The duration of the cycles did not have to be changed because the temperature rise was set to 20 K/h. Every cycle lasted 12 hours. The programme ran for 28 days, which rendered a total number of 112 cycles (56 days).

Before and after the artificial ageing, there was a times of conditioning which lasted two weeks. In total, the whole programme lasted 12 weeks. The artificial ageing programme of the DIN EN 1279 has a duration of 15 weeks, including the time of conditioning.

2.2 Specimens

All the specimens used showed all the same structure. The DIN EN 1279-2 defined specimen dimensions. All glass panels had a width of 352 mm and a height of 502 mm. The glazing construction consisted of two 4 mm fully tempered glass panels and a cavity with a width of 12 mm.

The topic of the research is the functionality of pressure compensation. Therefore, the edge sealant systems need to be equal in all specimens used. In this way, it is possible to compare the different kinds of pressure compensation. The edge sealant systems consisted of a bent aluminium spacer. Because of its high density, polyisobutylen was used as the primary sealant. Polysulfide (PS) was the material of choice for the secondary sealant, with a thickness of 6 mm on the back of the spacer, except the specimens of type E. In order to examine the insulated glass units with the ability to compensate the pressure inside the cavity between the glass panels, all types from B to G had an air tube to connect the cavity with the ambient air. The tubes had different lengths and diameters to control the different amounts of penetrating humidity. The next two figures show how the air tubes connect the ambient air with insulated glass units.

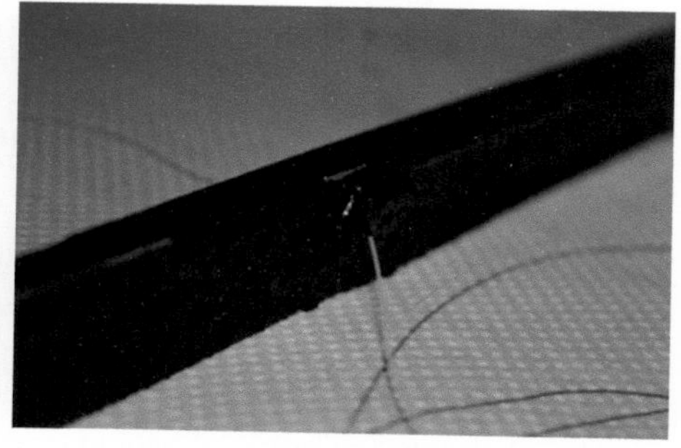

Figure 2-2 Installed air tube inside an insulated glass unit. Air tube inside the cavity between the glass panels (left). Air tube leaving the insulated glass unit trough the secondary sealant (right).

Table 2-1 shows the seven types of specimens and the parts of the structure that divided the specimens. The most important details are in the column with the dimensions of the air tubes. It was decided to choose two different lengths and two different diameters in four combinations (type B, C, F, and G). Thus, it is possible to show the effects on the climatic loads. To make sure that the climatic conditions inside the climatic chamber did not influence the pressure compensation, the openings of the air tubes ended outside the climatic chamber. In order to see whether the moisture can penetrate into the insulated glass unit through the air tube, the air tubes of type D specimens remained in the climatic chamber. The edge sealant system of type E was the only one with a reduced thickness of the secondary sealant on the back of the spacer.

The specimens of type A were the reference specimens. Type A was used in prior research programmes. Thus, it is possible to compare the results of the several artificial ageing tests. Type A are the only specimens without the possibility of pressure compensation. This is important in order to show the negative or positive effects on the other specimen structures.

Table 2-1 Tested specimens.

Type	Secondary sealant	Dimension air tube	Description	Number of specimens
A	PS 6 mm	-	Reference	11
B	PS 6 mm	Length: 2 m Diameter: 0.2 mm	Dimension of the tube	11
C	PS 6 mm	Length: 4 m Diameter: 0.2 mm	Dimension of the tube	9
D	PS 6 mm	Length: 2 m Diameter: 0.2 mm	Tube opening inside the climatic chamber	9
E	PS 3 mm	Length: 2 m Diameter: 0.2 mm	Smaller secondary sealant	9
F	PS 6 mm	Length: 2 m Diameter: 0.5 mm	Dimension of the tube	9
G	PS 6 mm	Length: 4 m Diameter: 0.5 mm	Dimension of the tube	9

2.3 Data Acquisition

For the permanent data recording, two different data loggers were placed inside the specimens to record the relative humidity, temperature and the atmospheric pressure (figure 2-3 (left image)). However, the results of the data loggers are not part of the paper.

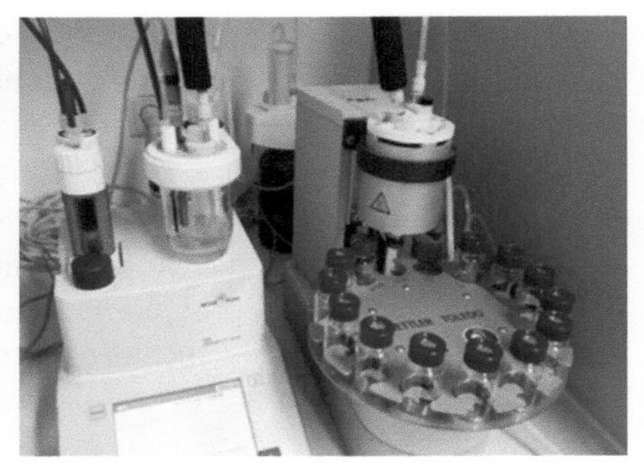

Figure 2-3 Data acquisition with two different methods. Data loggers for atmospheric pressure (left side) and relative humidity and temperature (right side) (left image). Karl-Fischer-Titrator in use with desiccant specimens (right image).

Another possibility to quantify the penetrating humidity is the Karl-Fischer titration (figure 2-3 (right image)) which is described in the DIN EN 1279-2. The titrator measures the fraction of moisture within the weight of the desiccants. To determine a start value, the moisture content of four specimens was controlled on the date of delivery. It was assumed that the desiccant took moisture while the insulated glass units were produced. The specimens that were artificially aged and the reference specimens were analysed after the artificial ageing. Table 2-2 shows the number of specimens that are used for this test.

Table 2-2 Tested specimens according to the use in the test programme.

Type	Total number of specimens per type	Thereof, specimens for (per type)		
		Start value	Artificial ageing	Reference
A - B	11	2	7	2
C - G	9	-	7	2
Total	67	4	49	14

3 Results

3.1 Test Process

The company “Polartherm Flachglas GmbH” produced the specimens for the test. “Polartherm” is close to the location of the examination and started the production of the specimens at the beginning of August in 2015. After the insulated glass units were delivered, the conditioning of the specimens at laboratory conditions (23 °C, 50 % r.h.) was started. The artificial ageing started after two weeks of conditioning and ended after eight weeks running after that. Chapter 2.1 explain the artificial ageing programme in more detail. The specimens were not disturbed while ageing inside the climatic chamber.

Two specimens of type A and B were used to determine a start value of the moisture content for this testing programme. Thus, the desiccant of these four specimens was removed and examined before the artificial ageing started for the other specimens.

After the artificial ageing, the specimens were removed from the climatic chamber and were stored in the laboratory for the next two weeks of conditioning. Except for the specimens without desiccant, which are equipped data loggers. These specimens were opened to collect the data loggers directly after the artificial ageing. The desiccant was extracted after the two weeks of conditioning. The titrator examined the moisture content of each specimen. However, during the titration some samples could not be examined exactly. Those values are not included in the results.

3.2 Absorption of Humidity

Figure 3-1 exemplifies the moisture content of the standard specimens after the artificial ageing. This graphic in particular makes it possible to compare the actual testing programme with the last test series. The specimens of type A and the artificial ageing programme were the same in both series.

The gap between the actual average value and the values of the testing series is 2.54 %. It is not clear why there is this difference between the insulated glass units despite having the same structure and being under the same conditions while ageing. All of the specimens were visually checked before the test started. Thus, it can be excluded that there was a difference in the level of quality of the specimen series.

Such differences make it difficult to compare several test series, even when similar specimens are used and the conditions are the same.

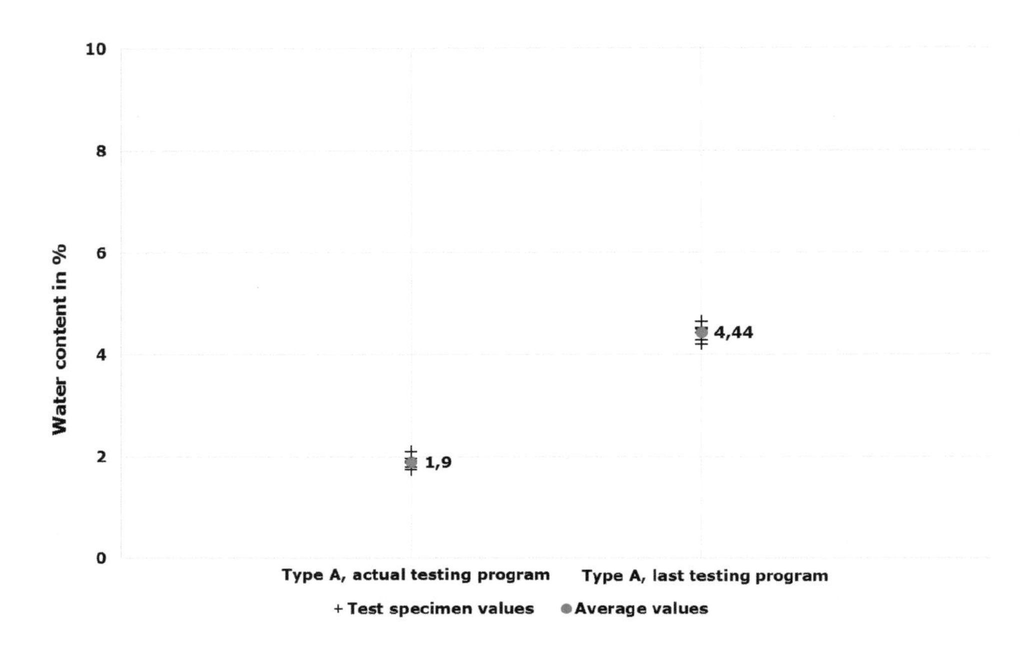

Figure 3-1 Moisture content in the desiccant of the standard insulated glass units of the actual and the last testing programme.

Figure 3-2 shows the moisture content inside the specimens before and after the artificial ageing. The standard insulated glass units can thus be contrasted with the specimens that have the possibility of pressure compensation. Additionally, it is possible to compare the different dimensions of air tubes, which are necessary for the pressure compensation between the ambient air and the cavity of the insulated glass units.

The black line in the graph shows the start value for all specimens. It is assumed that all specimens start with a moisture content of 1.38 %. The black plus symbols mark the values reached by the artificial aged specimens. The red cycles supplement these values and show the average values of the artificial aged specimens per type. Finally, the green diamonds depict the values of the reference specimens. Two specimens of each type were stored in the laboratory under normal conditions. In this way, it is possible to compare the artificial with the natural ageing of the insulated glass units.

Firstly, there is the determination that the standard specimens without an air tube take up less moisture in average than the other types of specimens. The types C, D and E exhibit a range of over 2 % of moisture content from their minimum to the maximum. That leads to the assumption that the production of insulated glass units with an air tube creates imperfections inside the edge sealant system. These imperfections can reduce the density of the used sealants.

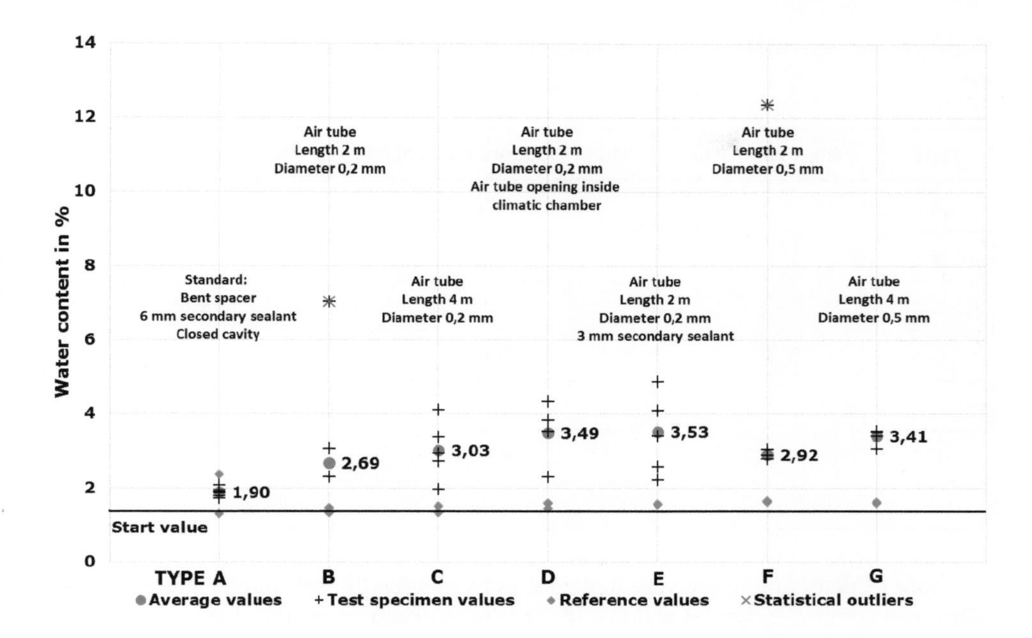

Figure 3-2 Moisture content in the desiccant of the insulated glass units after artificial ageing.

In direct comparison to the specimen types B and C, the specimens without air tubes show a higher density in this test. A problem for the evaluation is the low number of specimens of type B and F. This was due to problems with the titration and two statistical outliers that are crossed out. The good result of the standard specimens is not completely clear. In the last tests, the specimens with air tubes reached better values than the standard specimens [2]. Since it cannot be determined why the standard specimens did not react to the artificial ageing programme, other test series with the same setup need to be carried out in the future. In doing so, the reason for the quality variations can be found. The specimen types B, C, F and G are there to compare the different dimensions of the air tubes to show the influence on the durability relating to the moisture penetration. It can be seen that the length of the air tube can have a negative influence to the moisture penetration. The average moisture content of the specimens with a four-meter air tube exhibits a higher increase of the moisture content in comparison to the two-meter air tubes. Table 3-1 shows that the two linear dimensions differ in 0.61 % (maximally). This illustrates that the length of the air tubes has impact on the results. The reason for that might be the higher air resistance inside the longer tube. Thus, the air compression cannot react immediately to a change in pressure inside the insulated glass unit. It seems that the diameter of the air tubes has no major impact to the penetrating moisture. Hence, the number of specimens and the range of values were kept limited. Before the testing, the assumption was made that moisture can enter the insulated glass unit through an air tube with a bigger diameter. However, the average values demonstrated that there is no profound difference between the examined diameters and their influence on the penetrating moisture.

Table 3-1 Average moisture content in the desiccant of the insulated glass units after artificial ageing.

Type	Start value [%]	Mean moisture content [%]	Moisture increase [%]
A	1.38	1.90	0.52
B		2.31	0.93
C		3.03	1.65
D		3.49	2.11
E		3.53	2.15
F		2.92	1.54
G		3.50	2.12

One way of determining whether moisture can penetrate the specimens through the air tubes is in comparing type B and D with each other. Both types have the same structure and air tubes. However, the openings of the air tubes normally lead to the outside of the climatic chamber. The openings of the air tubes of type D lead inside of the climatic chamber. Furthermore, they were exposed to the high relative humidity generated by the artificial ageing programme. Thus, the average value of type D is the second highest of this testing series. Hence, it became clear that moisture can penetrate into the insulated glass units through the air tubes and decrease the durability. The values of type E show the highest moisture increase after the artificial ageing. Type E is the only specimen with a reduced secondary sealant. One can conclude that even if it is possible to reduce the influence of the climatic loads that the structure of the edge sealant system cannot be reduced. It seems that the secondary sealant plays an important role in the density of the insulated glass unit.

4 Conclusion

In conclusion, the actual research is not consistent with previous testing series. The specimens of type B and F with the possibility to compensate the pressure in the cavity between the glass panels reached the same amount of moisture content like similar specimens in previous testing series. It remains to be seen whether the density of the standard specimens varies in further artificial ageing tests as well. Therefore, the quality of the edge sealant systems of all specimens need to be checked carefully. The test revealed that the length but not the diameter had an influence on the functionality of the pressure compensation. On the basis of the available results, the air tubes with a length of two meters and a diameter of 0.5 mm will be used for further research. The reasons for that the values of those specimens remained constant and that the handling process is more effective.

The narrower the diameter of the air tube, the higher is the probability that the air tube will be damaged during the production of the insulated glass units. Important for the use of insulated glass units with pressure compensation is the result that moisture can enter via the air tube even it is small in diameter. For the use of these glazings on ships, it is necessary that the endings of the air tubes lead to the inner climatic room of the ships. In this way, the openings can be protected against the maritime environment.

5 Further Research

In this test realisation, the focus was on the different dimensions of air tubes in order to examine the influence on the climatic loads. Further tests should focus on the influence of the dimensions of the glass panels on the lifetime of the insulated glass unit. The next study needs to show how the pressure compensations reacts to various sizes of the glass panels. Furthermore, the time of reaction of the pressure compensation will be checked as well. Climatic loads are changing only slightly over time, but wind loads will stress the insulated glass unit in a short time. For this reason, specimens with different size of glass panels will be examined in order to find out how the surface of the glass panels will move under fast-changing temperature conditions. Thus, the measurement will show if the pressure compensation can reduce the pillow effect and the negative effect to the edge sealant system. Finally, the marine impacts on the insulated glass units have to be verified. Therefore, marine glazing units will be equipped with data loggers during a long-term monitoring study. Thus, real climatic loads will be recorded and a marine artificial ageing process will be derivate.

6 Acknowledgement

The investigations were carried out as part of a research project supported by the German "Federal Ministry for Economic Affairs and Energy". The authors would like to thank their partners "marine glazing Brombach and Gess GmbH & Co. KG" and "Polartherm Flachglas GmbH" for the close cooperation and the great support in form of technical consulting and the production of high-quality specimens.

7 References

[1] EN 1279-2: Glas im Bauwesen – Mehrscheiben-Isolierglas – Teil 2: Langzeitprüfverfahren und Anforderungen bezüglich Feuchtigkeitsaufnahme. Deutsches Institut für Bautechnik, Beuth Verlag GmbH, Berlin (2015).

[2] Köhler, M.: Experimental Study on Durability of Insulating Glass Unit Sealant Constructions. IN: Conference on Architectural and Structural Applications of Glass – Challenging Glass 5, June 2016, pp 249-260.

Cold-Bent Laminated Glass for Marine Applications

Marcin Kozłowski[1]*, Minxi Bao*[2]

1 Eckersley O'Callaghan, London, marcink@eocengineers.com

2 Eckersley O'Callaghan, London, minxi@eocengineers.com

This paper reviews the current design standards for glazed openings on large yachts then presents a concept design of a curved, glazed opening in the side hull of a large yacht by adopting laminated cold-bent glass. It also includes results of numerical studies on simulation of cold bend lamination process and the long term relaxation due to temperature variation. A case study of large opening made of cold bent laminated glass was carried out with the use of a staged analysis in Strand7® finite element software.

Keywords: cold bent, laminated glass, marine, curved panels, Strand7

1 Introduction

Curved glass has been applied in many interesting ways to create free-form transparent surfaces for facades and canopies (Neugebauer [1]). In recent years, large curved glass panels have also attracted interest from the marine industry which pursues higher aesthetic aspirations for yacht design. Clients typically desire unobstructed views of the surroundings from the inside coupled with a smooth, shiny external appearance for their vessels. Traditional glazed openings in yachts are typically restricted to fully supported framing and relatively small glass panels, see Figure 1-1 (left). In contrast, flexible, curved glass panels can form the hull line of large yachts without framing or splitting into small pieces, and therefore provide more transparency whilst offering a unique alluring appearance, see Figure 1-1 (right).

The challenges of using large glass panels in yachts are different to those in conventional glazing facade design in building industry. For example, the stability of a building is mainly determined by the foundation and the glass facade is usually deemed as a "non-structural element". In contrast, yachts rely on their glazing elements to maintain their weather and watertight integrity. If the glazing fails, the vessel will be directly open to water and hence at a high risk of sinking or capsizing. More complicated escape routes in yachts than in buildings also place a higher importance and robustness level on glazing to prevent any potentially catastrophic consequences. General marine design guidance gives design loads on yachts (due to waves) to be 100 times, or more, greater than the nominal design loads for buildings. Therefore, glass elements in marine applications require high load-bearing capacity and post-failure robustness.

Engineered Transparency 2016. Glass in Architecture and Structural Engineering. First Edition.
Edited by Jens Schneider, Bernhard Weller.

Figure 1-1 Traditional (left) and desired (right) marine glazing.

1.1 Marine code review on large yachts

According to the International Tonnage Convention [2] large yachts are yachts with length of the hull higher or equal to 24 m in use for sport or pleasure and commercial operations, with a tonnage limitation up to 3000 gross tonnage.

Two prevalent design standards are generally referred to for glazing design of large yachts: The International Convention of Load Lines (ICLL) published by the International Marine Organization in 1966 [3] and the standard ISO-11336-1:2012 Large yachts - Strength, weathertightness and watertightness of glazed openings [4] published in 2012 by the International Standardization Organization (ISO). The latter was prepared by a group of experts from yacht builders, glazing manufacturers, flag authorities and classification societies.

ICLL provides a wide range of requirements for commercial vessel, including Article 5 which excludes *pleasure yachts not engaged in trade* from its scope. However, when a yacht is offered for charter, it is considered to be in commercial use. In terms of glazing, ICLL states perfunctorily *fixed or opening skylights shall have a glass thickness appropriate to their size and position* and *sidescuttles and windows together with their glasses, deadlights and storm covers if fitted, should be of approved design and substantial construction in accordance with, or equivalent to, recognized national or international standards.*

In comparison to ICLL, ISO 11336 provides more detailed information for glazing element. It specifies technical requirements for independent glazed openings on large yachts and also takes into account the location of the opening. The standard limits glazed openings to above the highest waterline and stresses the importance of glazing for the ship weather/watertightness integrity. The code makes design pressure dependent on the location of the opening and the size of the vessel. The minimal design load for glazed opening is specified as 15 kN/m^2. However, in most of cases, the design load can even exceed 70 kN/m^2. ISO 11336 also gives very stringent design flexural strength of glass, σ_A, which is obtained by dividing the value of the characteristic failure strength σ_C by the corresponding safety factor γ. For example, thermally toughened glass $\sigma_C = 160$ MPa, $\gamma = 4.0$ thus $\sigma_A = 40$ MPa. The approach is very conservative as the calculated design strength is only related to glass type, disregarding load duration and the stress location. In other words, the design value refers to the maximum allowable stress of the weakest part of glass panels, i.e. edges, when subjected to a long term load. Despite the very conservative approach for glass strength, ISO 11336 is open to qualification of glass strength by testing. By carrying out short-term and load-specific testing, we are able to have more accurate data to justify higher design strength of glass to resist wave loads.

1.2 Use of glass in marine applications

The glazing components used in yachts should function in three major criteria, i.e. structure, aesthetics and comfort. The following chart is presented to elaborate these three roles and outline the process of selecting glass elements, see Figure 1-2. Structural glass can be categorised by four types of toughening treatment i.e. chemically toughened, fully tempered, heat strengthened and annealed glass. Only chemically toughened and fully tempered glass are of sufficient design strength in large-scale glazing marine applications. However, chemically toughened glass cannot have coatings or frits on its surface, which limits its comfort criteria. Moreover, its size is limited by chemical toughening bath. Fully tempered glass is the only remaining option but can barely meet all three criteria – it is also only available in singly-curved shapes and at great expense due to the hot-bending process.

With a strong demand for more economical and geometrically flexible strengthened glass, cold-bent laminated glass is introduced (Neugebauer [1], Belis et al. [5], Teich et al. [6], Fildhuth and Knippers [7]). The method is to force the toughened or chemically tempered glass sheets with interlayers in between against a negative mould to form the shape before lamination. The lamination process activates the interlayers which will maintain the curved shape after releasing from the mould. Cold-bent laminated panels can achieve free-formed shape in both single or double curvature. It is limited by neither the size of hot bent furnace nor chemical toughening bath. It also allows for coatings or frit and therefore is an ideal option for glazing elements in yachts.

As cold-bent laminated glass is still a relatively new technique, research work is needed to better understand its mechanical behaviour. Geometry loss is a very unique feature of cold-bent laminated glass. Initially, an elastic spring-back can be observed after release from the mould. After, the curvature gradually reduces over time, which is also temperature dependent. The long-term behaviour is mainly related to the viscoelastic properties of the polymeric interlayer. Therefore, to obtain the final desired curvature of the panel, a numerical study on the effects must be performed.

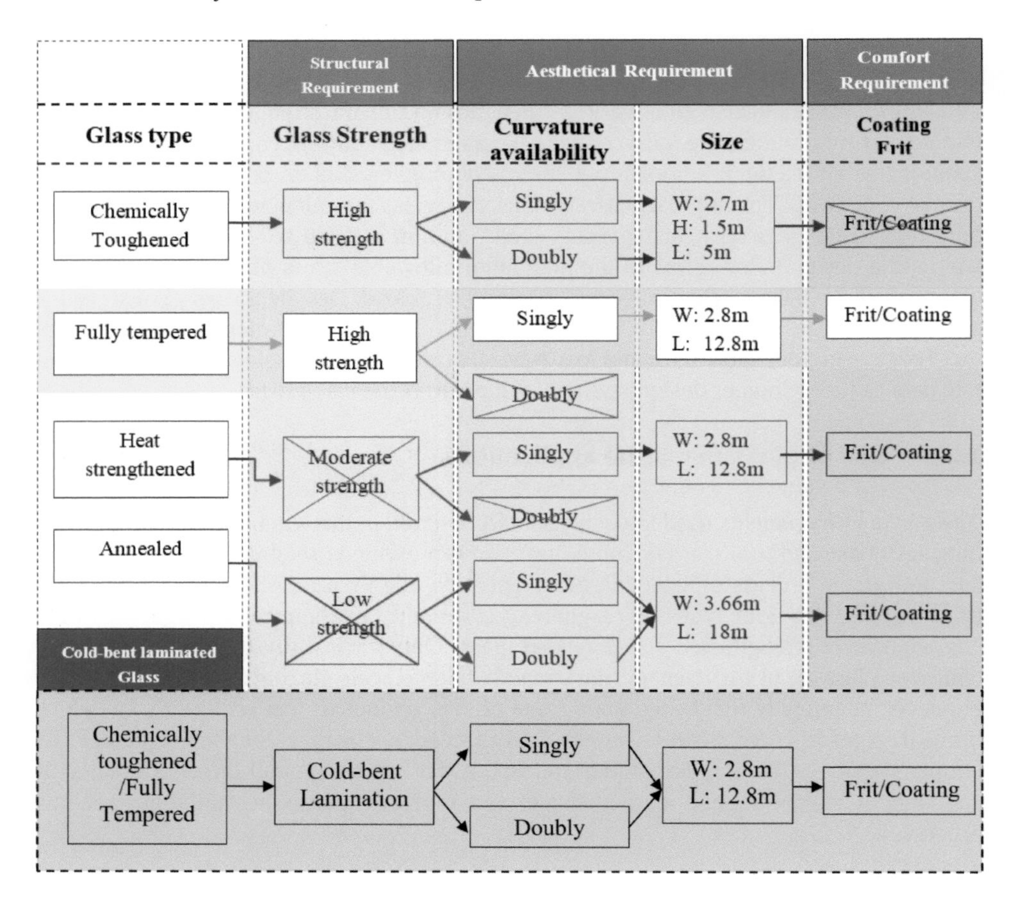

Figure 1-2 Selection of glass for marine applications.

2 Numerical studies

This section presents an initial study on cold bent laminated glazed openings in superstructure of a large yacht with single curvature geometry. The spring-back effect of the laminated cold bent glass and SentryGlas® (SGP) relaxation due to the high temperature and long term load duration are investigated. In addition, the stress in glass accumulated from long term relaxation and an external wave loading is studied.

2.1 Geometry, materials and methods

The dimensions and radius of a single glass panel are illustrated in Figure 2-1. The build-up used in the research was 5 × 12 mm fully toughened glass with 1.52 mm SGP interlayers.

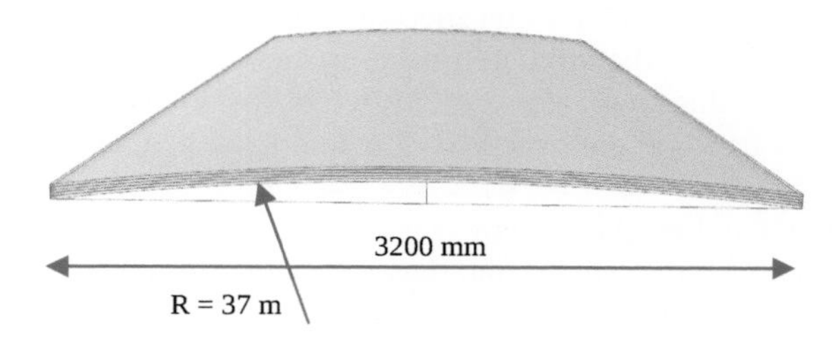

Figure 2-1 Geometry of studied panel.

The glass material used in the analysis was soda-lime, thermally toughened glass with linear-elastic physical characteristics E_g = 70 GPa and ν = 0.23 [8]. Figure 2-2 (left) provides varying values of Young's modulus of SGP E_{SGP} at different temperatures and load durations (Bennison et al. [9]). The factors in Figure 2-2 (right) are calculated as E_{SGP}/628 MPa, where E_{SGP} is the corresponding Young's modulus at the temperature at 10 years load duration basis, while 628 MPa is the initial material property of the SGP adopted in Stage 2, assuming 20 °C and 1s load duration. In cold-bent laminated glass, the glass curvature completely relies on the shear bond of the SGP interlayers; therefore, we can deem it as permanent load applied to the interlayer. The SGP was modelled with the Poisson's ratio corresponding to the temperature and load duration.

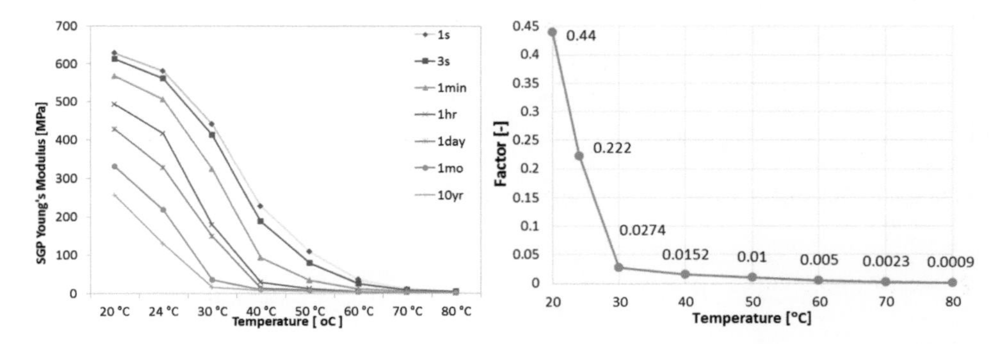

Figure 2-2 Young's modulus of SGP at various temperatures at 10 years duration time (left) and Young's modulus versus temperature of SGP (Bennison et al. [9]).

The commercial finite element programme Strand7 [10] was used in the study. To examine the laminated cold-bending process and relaxation after installation a stage analysis was carried out. To simplify the calculations, the laminated panel was considered as a symmetric plain-strain model. An initial bending moment *M'* was applied to the free end to create a consistent curvature in the panel. The model neglects manufacturing conditions, while in reality, the clamping/fixation methods may have different impact on the glass geometry. The following stages were carried out in the analysis:

- **Stage 1**: Glass and interlayer work independent of each other and bent to a predefined value with an external bending moment *M'*. By considering spring-back effect, *M'* should be approx. 20-25 % greater than the minimum bending moment *M* required to form the design curvature according to the equation $M = E \times I/R$. This initial reduction factor was obtained by a trial and error method. The interlayer had no stiffness at this stage;
- **Stage 2**: Material property of SGP at load duration of 1s and 20 °C is assigned to the interlayer. 'Morph' option was activated so the interlayer remains the deformed shape without any additional residual stress;
- **Stage 3**: Bending moment *M'* is released, glass panel springs back;
- **Stage 4**: Relaxation effect under different temperatures;
- **Stage 5**: Design (wave) load is applied.

2.2 Results and discussion

The instantaneous spring-back and subsequent relaxation effects are evaluated by glass radius and the corresponding arc rise as shown in Figure 2-3 (left). These parameters were examined in all stages. The variation of the two parameters is plotted with different temperature at 10-year load duration. It is noted that the relaxation results are calculated without any edge restraints. The arc rise loss ratio is presented in Figure 2-3 (right).

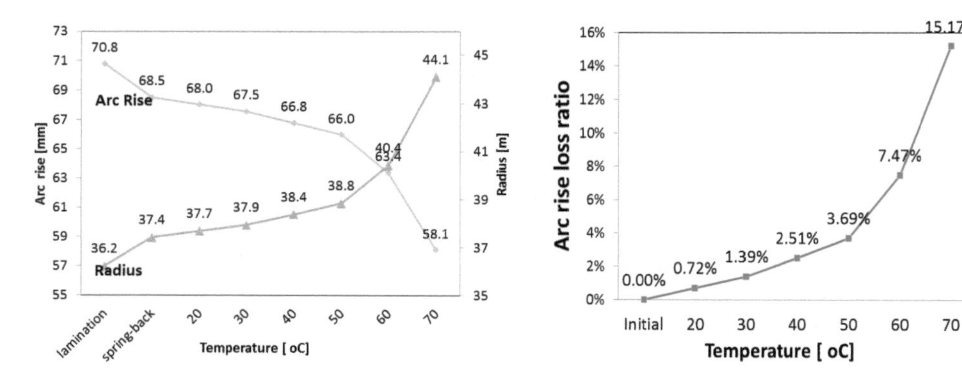

Figure 2-3 Radius and arc rise change due to spring-back and relaxation with increasing temperature (left); arc-rise loss ratio due to relaxation with temperature rise (right).

By observing the results presented in Figure 2-3, the non-accumulative arc-rise loss level of laminated cold bent glass shows a strong temperature dependency. Before temperature rises above 50 °C, the maximum loss level differential is 3.7 % which equals to 4.8 mm.

However, when the temperature exceeds 60 °C, the stiffness of interlayer drops drastically, and leads to a 15.2 % arc rise loss.

In the last stage of the analysis, an external wave load of 32 kN/m^2 [4] was applied to the curved panel. The values of E_{SGP} and ν_{SGP} correspond to 70 °C and 1s load duration time. The Figure 2-4 (left) presents a stress profile through the thickness of laminated glass panel derived by superimposition of stress resulting from relaxation and external loading. The maximal stress obtained is 59.0 MPa which is below the allowable stress limit for thermally toughened glass at short duration time of 69.0 MPa [8]. In addition, a parametric study was carried out on the value of spring-back and SGP relaxation level with varying build-up of the panel. It was found that thicker panels show lower level of spring-back and relaxation, see Figure 2-4 (right). The conclusion is in accordance with the long-term monitoring results of the recovery behaviour of cold-bent laminated glass reported by Fildhuth and Knippers [7].

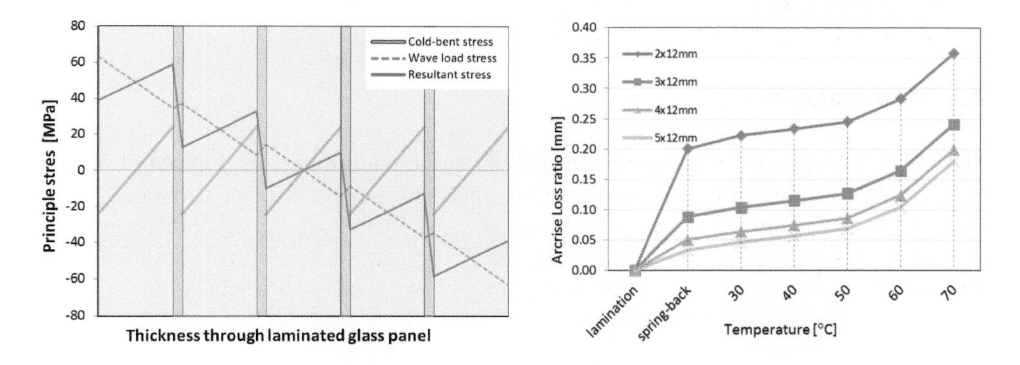

Figure 2-4 Stress profiles through the thickness of laminated glass panel (left); results of parametric study (right).

3 Conclusions and future studies

The presented study investigates the spring-back effect of a single curved laminated cold-bent glass panel, the SGP relaxation due to the high temperature and long term load duration and combined stress in glass due to an external wave loading and cold-bent lamination. Based on the numerical studies performed the following conclusions can be drawn:

- Cold-bent laminated glass is a solution offering high transparency, desirable aesthetic appearance and high load-bearing capacity of glazed openings for large yachts;
- Due to the relatively high bending stress in glass (before lamination process), laminated cold-bent glass is deemed to be a solution appropriate for subtle curved glass panels.

- Spring-back effect and further relaxation of the panel with build-up of 5 × 12 mm due to high temperature and long-term behaviour were found to be 3.2 % and 15.2 %, respectively;
- Glass design needs to take into account the superposed stresses by cold-bent lamination and external load;
- The parametric study shows that decreasing the build-up of a laminated cold-bent glass panel results in lower value of spring-back and SGP relaxation level.

The study involves a simplified plain-strain numerical model of a single curved laminated glass panel with the linear-elastic material model for SGP interlayer. To fully understand the potential and structural behaviour of double curved panels an appropriate 3D model with more advanced material model of SGP is necessary. The results presented in the paper are the output of numerical software which was based on various simplifications – thus, it should be validated by physical testing. In the near future, the authors aim to perform experimental research on the structural behaviour of laminated cold bent glass subjected to varying temperature.

4 References

[1] Neugebauer, J.: Applications for curved glass in buildings. In: Journal of Façade Design and Engineering Vol. 2, 2014, pp 67-83.

[2] The International Convention on Tonnage Measurement of Ships, International Maritime Organization, 1969.

[3] The International Convention of Load Lines. International Marine Organization, 1966.

[4] ISO-11336-1:2012 Large yachts – Strength, weathertightness and watertightness of glazed openings – Part 1: Design criteria, materials, framing and testing of independent glazed openings.

[5] Belis, J.; Inghelbrecht, B.; Van Impe, R.; Callewaert, D.: Cold bending of laminated glass panels. In: Heron Vol. 52, 2007, pp 123-146.

[6] Teich, M.; Kloker, S.; Baumann, H.: Curved glass: bending and applications. In: Engineered transparency, Düsseldorf, 2014, 75-83.

[7] Fildhuth, T.; Knippers, J.: Recovery Behaviour of Laminated Cold Bent Glass: Numerical Analysis and Testing. In: Challenging Glass & COST TU0905 Final Conference, Lausanne, 2014, 113-121.

[8] ASTM Standard E1300-12a: Standard Practice for Determining Load Resistance of Glass in Buildings. Annual Book of ASTM Standards, ASTM International, West Conshohocken, PA, 2012.

[9] Bennison, S.J; Quin, M.H.X.; Davies, P.S.: High-performance laminated glass for structurally efficient glazing. In: Innovative light-weight structures and sustainable facades, Hong Kong, 2008.

[10] Strand7 Manual – Introduction to the Strand7 Finite Element Analysis System, Edition 3, January 2010.

High-tech solutions for the smoothly refurbishment of buildings: Investigation of existing window systems with vacuum glasses

Helmut Hohenstein[1], Ernst Heiduk[2], Peter Schober[3]

1 Dr. Hohenstein Consultancy, Kinderheimstr. 3, 45770 Marl, Germany,
helmut.hohenstein@hohenstein.biz

2 TU Wien, Inst. f. Architekturwissenschaften, Abt. f. Bauphysik u. Bauökologie,
ernst-christian.heiduk@tuwien.ac.at

3 Holzforschung Austria, Wien, P.Schober@holzforschung.at

Research Partners: Professor Ardeshir Mahdavi, Ulrich Pont, Matthias Schuss, Christian Sustr, Olga Proskurnia, TU Wien, Inst. f. Architekturwissenschaften, Abt. f. Bauphysik u. Bauökologie; Hubert Pichler, Franz Dolezal, Holzforschung Austria, Wien; Rainer Vallentin, Vallentin u. Reichmann Architekten.

For the improvement of the thermal protection of historic buildings without changing their external appearance, highly insulating vacuum glass is the first possible replacement for single glazed historic windows keeping the original structure. Since a while, produced on an industrial scale in Asia, very thin, lightweight, highly insulating Vacuum (Insulation) Glass (VG) is available. In 2014/15 the Vienna University of Technology and Holzforschung Austria investigated together with Dr. Hohenstein Consultancy how this glass can be used for the thermal rehabilitation of historic but also new box windows [1]. Starting this project thermal simulations of VG in window frames were done. The influence of the edge seal, support pillars and format sizes on energy transmission and stress was examined. With the construction of two Viennese box windows with VG, thermal laboratory tests in the climate chamber and mechanical stress tests were conducted. With these results, the energy saving potential has been determined and calculated for the whole building stock in Austria. Project results: My study on VG availability shows the current dynamic market situation, specifically in Asia. The results of the evaluation of quality, the thermal simulations, the installation trials and tests of mechanical durability show very promising results. It makes sense to swap the single glass in the outer wing for VG, and conservators in Austria readily consent to this. In a follow-up project, the moisture behaviour in window frames and masonry will be studied. Conclusion: Vacuum Glass will be an important new option for the thermal rehabilitation of historic windows and not only for new windows and facades. In coordination with conservation experts, an application in the conservation area is very much demanded and possible, too. In combination with other soft and adequate renovation steps, many historical buildings could be modernized into low-energy buildings without losing architectural quality. At the same time user comfort will increase.

Keywords: vacuum glass, application case studies, processing, sustainability, environmental protection

Engineered Transparency 2016. Glass in Architecture and Structural Engineering. First Edition.
Edited by Jens Schneider, Bernhard Weller.

1 Use of vacuum glass replacing single pane glass in existing window systems

1.1 Introduction

As the development options of VG for markets and applications are in general broader than for IG and also the performance range, it has basically the intrinsic power to replace the world market of Insulating Glass (IG). VG application will change successively window and façade systems towards lighter, thinner, more transparent as well as more flexible and modular solutions. The key point is at least the tremendous contribution for energy and carbon emission saving worldwide in cold and even hot climates coming along with an increased comfort Hohenstein et.al [2,3].

Buildings are responsible for about 33 to 40 % total energy loss. In average, maximum 2 % of buildings are constructed new per year. Thus the poor performance of the building stock remains very long and is the major concern for all governmental energy policies. One can say that the energy targets will not be met as predicted without changing this.

Table 1-1 U-values of different types of insulated glass.

Type of glass	U-value [W/m²K]	Thickness [mm]
Single Glass	5,8	3 - 6
Standard IG	3,0	20 - 28
Low E Double IG, Argon filled	1,1	24 - 32
Triple IG with 2 Low E coatings	0,7	36 - 54
VG with one or two Low E	0,3	6 - 12

VG offers a unique option to alter the building stock energy losses in a wide range as single glazed units can be exchanged against thin VG. Windows count in a building for about 1/3 of energy loss and glass surface alone for 25 % loss – in old Houses the values are lower less due to a restricted glass surface per building (statistical data are given for Austria and Germany). Respective *U*-values of different glazing types are shown in Table 1-1.

Now available VG, the latest technical generation, has a very good heat transfer coefficient U_g from 0.70 to 0.34 W/m²K. With best Low E (lowest emissivity available in Europe values of 0.30 are possible). VG is available from a total thickness of 6.0 mm and bigger and has a 4vacuum gap of 0.2 mm with a solder glass edge seal of a width of 8 to 13 mm (Synergy [4]).

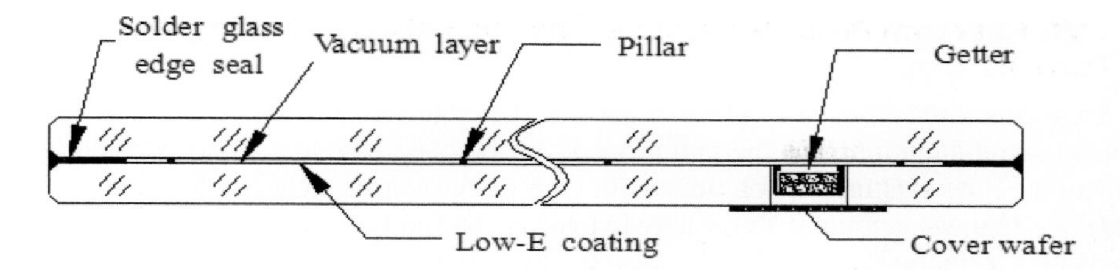

Figure 1-1 Cross section for latest VG solution from Company Synergy with glass edge seal, pillars and getter with cover plate (Tang et.al. [5,6,7]).

For saving energy and carbon emissions the contribution from VG is theoretical up to 95 % compared to single glazings. This means in cold climates up to 20 % additional energy savings in buildings. Huge savings for total energy with VG and comes together with best performance options as stated below. Just what most government bodies are searching for. In practical terms we calculated about 2-4 % of the energy bill can be achieved in a few years with best VG utilized in the old building stock as shown in a research study in Austria.

For heat retention VG solutions (solar control coatings) the savings for cooling energy can be enormous, too, even more, when soon smart glasses (electrochromic glass) can be combined with VG, a potential of more than 50 % is realistic. Investment in expensive external shading can be avoided, too. HVAC systems can be essentially reduced. Globally the energy for cooling might be even more than for heating as about 60 % of people live in hot and humid climates. Glass facades in the past worked often as collectors of heat which can be altered with VG/smart glass combinations and a very thin structure (9-12 mm). At least VG can be similar efficient like highly insulating walls. Many sample buildings are erected since 2004 in China and since 2012 in Europe [8,9,10].

Obviously the evidence for everywhere application of VG is not yet fully proved in practice yet. Four producers in Asia are existing since some years, but many more try to establish an initial production due to fast growing demands, unfortunately not one in Europe. Most are facing mass production and application problems, whereas few are close to viable solutions and one just reached a level which can serve international markets and relative wide applications. Product performance itself proved to be widely attractive as shown by many building projects over many years and tests a. o. in Europe. Acceptable equipment for an industrial mass production is not yet on offer. First because related companies lack for machine know-how and second in the "Western World" no-one built up a production line yet and still act more or less on laboratory level.

There is a comprehensive approach necessary especially from a managing and marketing level combining different forces and abilities for a quick broad and sustainable market start towards VG – away from the meanwhile insufficient and hardly improvable IG world.

1.2 VG Research Projects

Because of the direct contact of glass to glass in the edge seal, the installation situation requires special attention. The thermal property of the glass edge for VG is considering conduction below existing IG solutions with warm-edge spacer technologies. Generally this is the actual weak point of VG, where in two German and one Austrian governmental projects new techniques and material research was started recently to solve this issue and to allow at the same time much slimmer profile systems (Project e.g. [11]).

Figure 1-2 Thermography of the box window calculated by University Vienna, Austria. Thermal loss on the VG sealed edge for conventional glazing in former times – shows the task for better glazing methods and demand of different profile solutions. Clearly visible are the pillars (spacers) and more strongly visible are the thermal bridges of the edge seal.

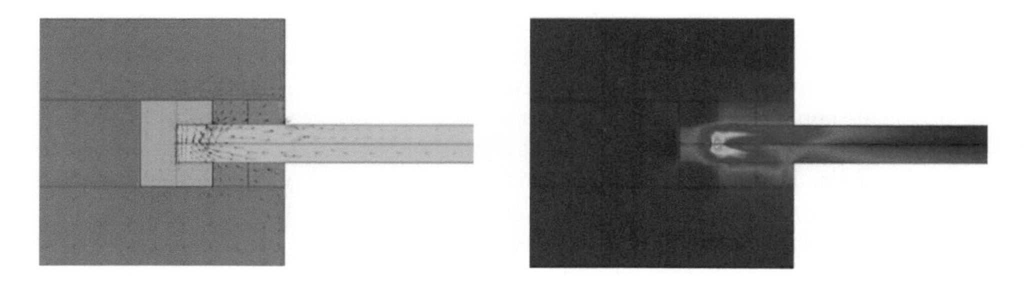

Figure 1-3 Thermal flow around the glass at the sealed edge – calculated by University Vienna.

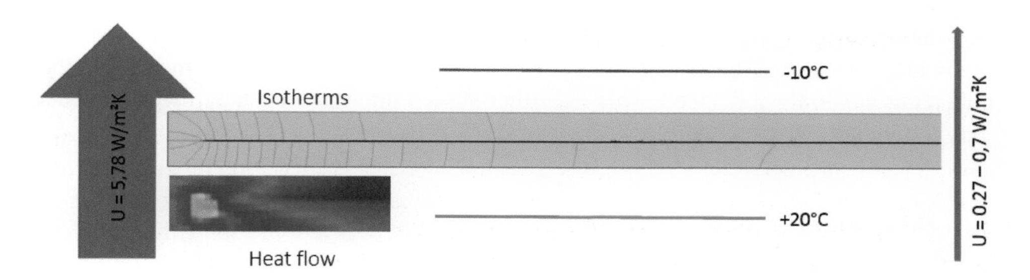

Figure 1-4 Isotherms and heat stream in VG – investigations by Prof. Ernst Heiduk, University Vienna. The direct contact of the two glass panes through the edge sealing results in a strong thermal bridge at the edge of the vacuum glazing. In the centre, on the other hand, the heat flow is strongly minimized (through the vacuum gap) and is focused only on the small spacers (pillars).

The essential criteria for significantly improved quality of VG of the 2nd generation are:

- Optimized spacers (Pillars), nearly not visible
- Tempered glass with high flatness and greater distances of pillars,
- Selected Low-E-coatings with low emissivity,
- Thin edge compound (glass frit) with 0,2 mm space
- Development of a combined flat cap, extraction port and molecule catcher (getter) for long-term preservation of high-vacuum allowing life-times of 50 years (calculated, measured without changes over 12 years)

Figure 1-5 Vacuum glazing with its components (Synergy / Beijing, PRC).

There are only inorganic sealing components in the system which are stable over very long times beyond human life times, if tightness is given during production. The predicted life span of over 50 years is due to the control of the outgazing effect of surfaces in high vacuum which lowers the vacuum without any leakage and "technical life–time" is related to the effectiveness of the used getter system. This can be calculated after measuring the outgazing materials and speed. This is further shown through long-time measurements over more than 10 years, too.

2 Research Projects of applied VG in box windows

In thermal simulations, the replacement of single glazed windows with VG for certain real window constructions was calculated. In box-windows for the inner wings and the outer wings the utilization of VG was simulated. Measurements on build windows with VG were performed, too (Mahdavi et al. [12]).

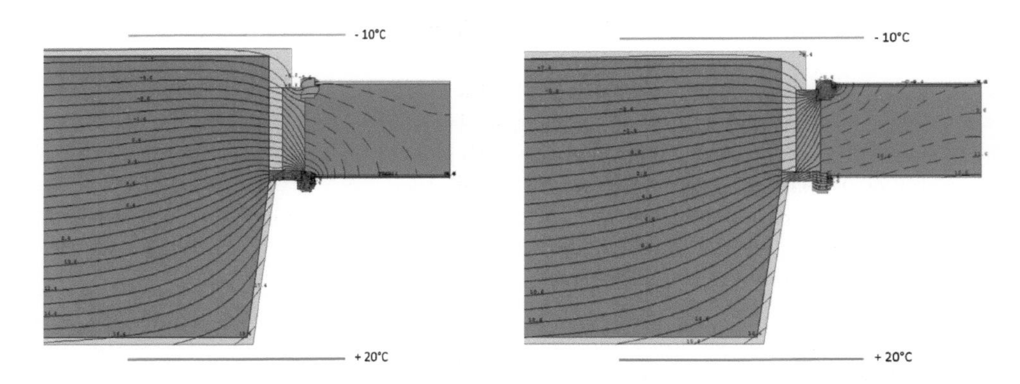

Figure 2-1 Wooden box window with VG inside (left). Wooden box window with VG outside (right).

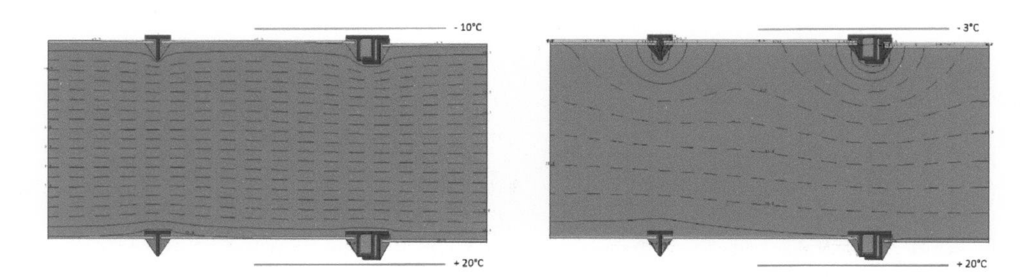

Figure 2-2 Thermal simulation of VG in box window. Metal box window with VG inside (left). Metal box window with VG outside (right).

Thermally the wooden box windows improved significantly in both variants, the occurrence of condensation, however, is very different. With VG in the outer wings, less condensation occurs and the transparency is always there.

Even with a steel/glass structure with VG outside, the thermal protection improved considerably. Depending on the outdoor temperature, ice or condensate occurs on the outer steel profiles, while there is no longer a condensation risk inside at the average Viennese winter temperature of about 3 °C.

In addition to these examples, calculations were made for:

- The installation of VG or VIG (combination of vacuum glass with insulating glass) in all types of existing windows - identifying the improvement potential.
- The heating demand reductions, carried out for VG-installation in the box windows of selected historical buildings
- The estimated potential for improvement in the building's energy balances by the use of VG and VIG for modernization in the building stock. For this, the "Typology of Austrian residential buildings" of the Austrian Energy Agency was used.

The transfer of this application potential in model calculations for the German housing Park 2010 - 2060 with the use of VG and VIG from 2020 onwards shows the reduction potential for the heating demand and, relevant to climate change, CO_2 emissions reduction, too.. This result is also applicable to Austria. It shall be noticed that for this early research Project the best *U*-value due to a lack of coatings with lowest emissivity and therefore lowest *U*-value of 0.34 W/m²K was not yet deliverable. Best version were those with a *U*-value of 0.58 W/m²K based on an Emissivity of 0.07.

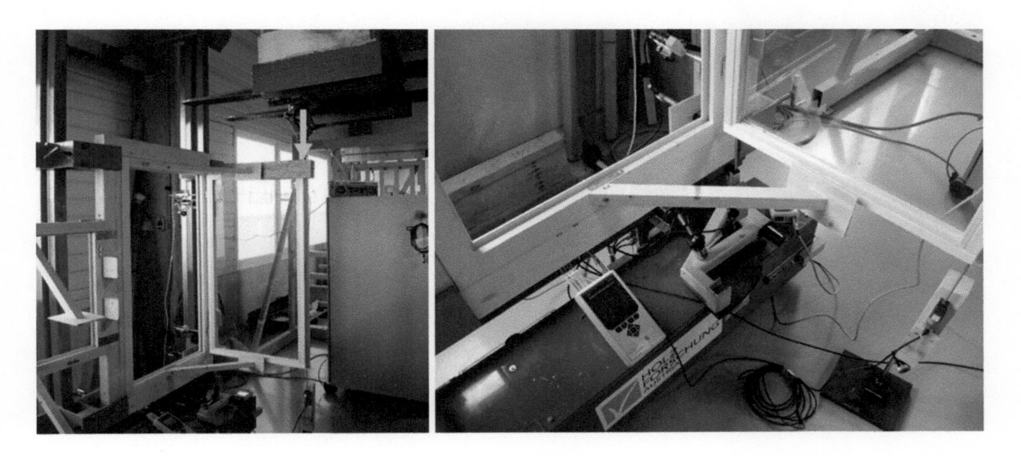

Figure 2-3 Vertical Stress in a Sash (ÖNORM EN 14608 – racking 400 and 600 N) executed at HFA Austria.

In the laboratory of Holzforschung Austria thermal/hygric and mechanical tests were carried out with mock box windows with VG in the format 50 x 110 cm and a U_g-value of

0.58 W/m²K. In the hotbox, two setups, with VG in the inner wings or in the outer wings, were tested.

For the mechanical load test conforming to DIN 14351-1 three kinds of glass fastening with glazing bead, with linseed oil putty and glass nails or glued with 3mm silicone, were tested and compared. The climate chamber tests showed a huge thermal improvement of the window. VG-use in the inside wings can lead to acceptance problems. If the joint tightness inside is not sufficiently improved, then the very cold window space in between causes the outer pane to become icy with reduced transparency. When installing VG outside, transparency is always present, only a small condensation strip at the glass edge (heat-bridge) and some condensate on the frame occur. The first mechanical test results are, especially with adhesion, very satisfactory above test requirements. The pendulum test (EN 13048) is not fulfilling the standard and is, due to the pillars, a systemic-structural weakness of VG – further investigations are planned.

Figure 2-4 Horizontal Twist/tortuosity of a test sash (ÖNORM EN 14609 -250 N) exec. at HFA Austria.

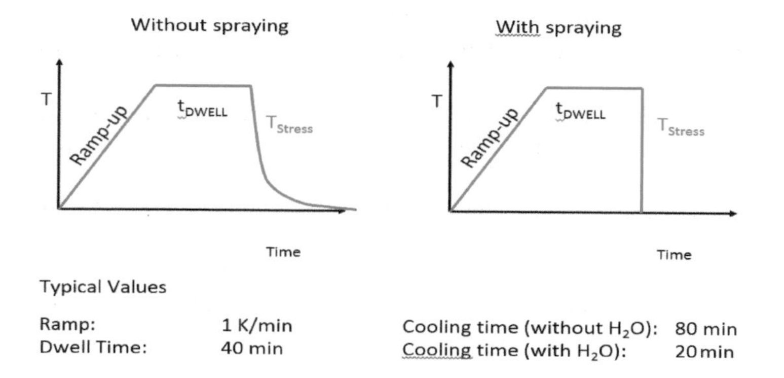

Figure 2-5 Thermal Stress Testing Method – Temperature Profile with and without spraying of water – tested by FhG ZAE Bayern (Center for Applied Energy Technology).

ZAE-Bayern in Würzburg did execute at the same time thermal stress and life time tests, on a test equipment especially developed for VG, which showed very good resistance to climatic loads. Breakage occurred much beyond typical temperatures of IG at ~117° C (Büttner et al. [13]).

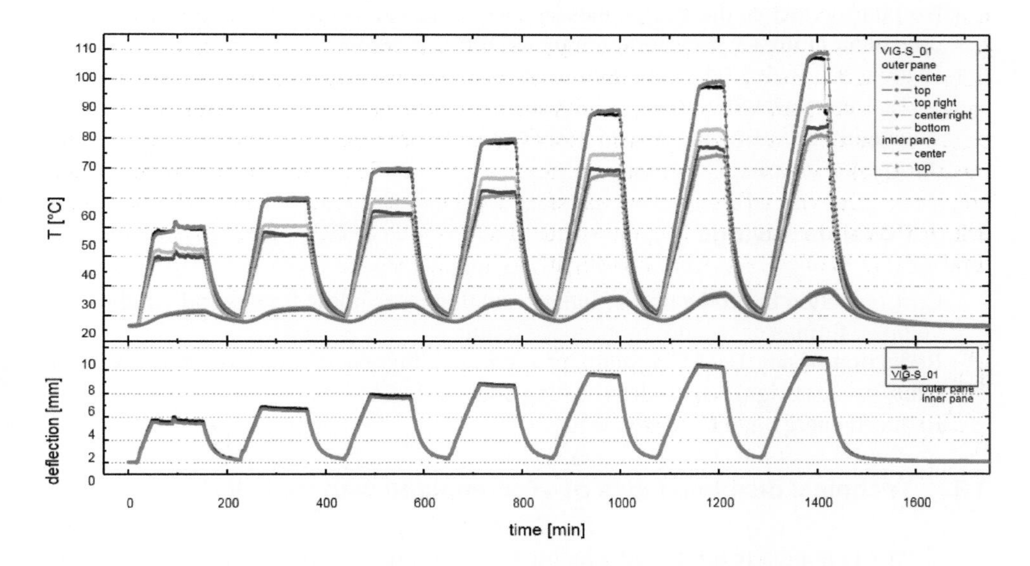

Figure 2-6 Results for thermal Stress Tests by ZAE Bayern - VG breaks only at 117 °C. Breakdown temperature = 117 °C. Breakdown deflection = 9.8 mm. Temperature shock unproblematic for stability

3 Findings and Results

- The thermal simulation results, installation trials and tests are promising. The thermal savings and climate protection potential of VG-usable in the building stock are significant.
- The simulations and tests show the strengths and weaknesses of window systems with VG. Further deeper studies of the usage of VG in stock windows is desirable and necessary.
- Intensive investigations of using VG in new windows are also desirable and necessary.
- The market potential for VG in existing windows is relevant in Austria, very relevant internationally. The biggest potential lies wherever heating or cooling is required, and even more where currently only single glazing is common.

3.1 Technical advantages of VG compared to triple IG

- Ug-value for inclined windows is not changing, whereas IG lose about 30 % due to convection in IG
- VG is much thinner, minimal thickness 6.0 mm and much lighter in weight
- No pump effect of the glass panes by temperature or pressure changes. VG can be used at all altitudes; transportation by aircraft is possible
- Regular VG with 37 dB has much better sound control than IG with 29 dB
- VG has an expected lifetime of 50 years
- The highly debated window/wall ratio can be moved towards transparent parts without energy losses.

3.2 Possible savings

- Cost reduction in production of lighter and thinner window frames and building envelopes, fittings or facades more space usable
- Reduction of any HVAC systems or even avoiding electrical powered HVAC systems, because significantly less heating and cooling energy required
- Reduced energy and CO2 emissions

3.3 Technical disadvantages of VG compared with triple IG

- Currently not many optimized window frames for utilization of VG are developed
- Only some types of Low E and sun protection coatings are proofed, yet.
- Thermal bridging behavior of the glass edge seal is insufficient and needs adapted glazing technics, not only with a deeper recess in the window frame.
- High process temperatures did not allow till recently to produce tempered VG.
- No curved shapes are mass produced yet, but this is only a machinery problem.
- A more severe restriction is the lack of lead-free solders, which are on research.

4 Conclusion and Outlook

- VG related window and façade frames and systems require precise planning, calculations and tests for replacement in new windows. In existing windows the glazing techniques need to be improved. From the first research results one can recognize that solutions can be soon found and transferred to the production level.
- VG will be an important new option for the thermal rehabilitation of historic windows and an improvement for energy and carbon emission savings for all valuable existing windows.
- In combination with other sustainable and appropriate renovation steps, many historical buildings could be transformed into low-energy buildings.

- In cooperation with conservation experts, utilization in the conservation area is very promising.
- VG allows and requires the construction of new window and façade profiles with a VG-adequate approach. Less weight and volumes can be achieved.
- The achieved Ug-values of up to 0.27 W/m²K in the laboratory are realistic for the next product generation of VG, also in mass production.
- Any reduction of product costs and any shortening of delivery times and routes enhances the application potential. Production sites should be established in EU.

VG will be the long desired choice to replace insulating glass together with heavy window and façade solutions and thus contribute in a high extend to a better performance for buildings. The discussions about window/wall-ratios can be stopped. VG with excellent thermal insulation performance, whether for new buildings or for the existing buildings, can bring great economic and social benefit. Even more this is valid for hot and humid climates saving cooling energy. Refrigerating industry has high demand of VG. Buses, trains and automotive- even not yet tested, most likely too. Side windows in busses as well as windows for wagons are options, even more OEM-parts in automotive industry.

Clear is, there is enough demand for VG, but no sufficient application oriented research activated, yet. At this stage it is very important to provide soon the required production capacities for VG in Europe and America. Chinese and Korean markets are utilizing VG with high speed.

5 Submission of contributions

This research project was sponsored by the Austrian Research Promotion Agency in the program "City of the Future".

6 References

[1] Project VIG-SYS-RENO, TU Wien, HFA Austria, Dr. Hohenstein Consultancy Study of vacuum glazing for window modernization, Ernst Heiduk, Ardeshir Mahdavi, Ulrich Pont, Matthias Schuss, Christian Sustr, Olga Proskurnina – TU Vienna, Inst. f. Architectural Sciences, Dep. f. Building physics and ecology (AT).

Peter Schober, Hubert Pichler, Franz Dolezal – Holzforschung Austria, Wien (AT) Helmut Hohenstein, Dr. Helmut Hohenstein Consulting, Marl (DE).

Austrian Research Promotion Agency – National funding institution for business-related research and development in Austria.

[2] Helmut Hohenstein, Jianzheng Tang: A new dawn rising – Magical options for windows, facades & walls with vacuum glass integrating other building innovations, engineered transparency. International Conference at glasstec, Düsseldorf, Germany, 21 and 22 October 2014.

[3] Dr. Helmut Hohenstein, Dr. Hohenstein Consultancy, A new dawn is rising. New options for windows, facades & walls with vacuum glass and other integrated building innovations, Intelligent Glass Solutions 2.2013. Publisher nick@intelligentpublications.com.

[4] BEIJING SYNERGY VACUUM GLAZING TECHNOLOGY Co., Ltd., Add: No.7, Xinghai 3rd Street, Beijing Economic and Technological Development Area, Beijing, P. R. C, www.bjsng.com, (VG general provided by Beijing Synergy).

[5] Tang Jianzheng, Dong Yong, Li Yang: "New Development of Vacuum Glass Technology", "Architectural Glass and Industrial Glass", November 2008, P5.

[6] Tang Jianzheng: "Vacuum Glass Industrialization Status and Development Prospect", "China Glazing", June, 2008, P3.

[7] Tang Jianzheng, Li Yang, Li Nan: Heat-Strengthened Vacuum Glazing – A Breakthrough in High Energy Efficiency Glass.

[8] B10-Active Plus House with VG, in Stuttgart, Weissenhof-Settlement, by Professor Sobek, www.aktivhaus-b10.de/home, [http://derstandard.at/2000004153602/Ein-Haus-als-Gast-in-der-Siedlung].

[9] Project VISIONEUM ENERGIE+ Professor Sahner Univ. Augsburg, Company Raico, Test House VISIONEUM ENERGIE+ in Königsbrunn at Augsburg, with experimental facade construction and VG from Synergy.

[10] Passive low-energy consumption residential demonstration project "Water Front". Qinhuangdao·Hebei. [J]. Journal of Green Building, 2013.

[11] Research Project – Actual Proof of VG for windows, Prof. Rogall, FH Dortmund, Dow Corning, Co. Josko and Dr. Hohenstein Consultancy, funded through „Zukunft BAU" of BMUB.

[12] Ernst Heiduk, Ulrich Pont, Matthias Schuss, Elisabeth Finz et al.: Abteilung Bauphysik und Bauökologie – TU Wien, A new dawn rising – Magical options for windows, facades & walls with vacuum glass integrating other building innovations, Fassadenbautagung 2015 – TU Wien, Zukunftsperspektiven im Fassadenbau, Einführung Univ. Prof. DI Dr. A. Mahdavi, Ordinarius für Bauphysik TU Wien.

[13] Flexibler Randverbund für Vakuumisolierglas-Systeme (VIG-S), Abschlussbericht Febr. 2016, gefördert vom Bundesministerium für Wirtschaft und Energie (BMWi), Dr. B. Büttner, S. Hippeli, Dr. H. Weinläder, Bayerisches Zentrum für Angewandte Energieforschung e.V. (ZAE Bayern).

Off the perpendicular – transparency beyond the box

Hans Frey[1]

1 Waagner-Biro Stahlbau AG, Leonard-Bernstein-Strasse 10, 1220 Vienna, Austria

As the title implies, this text is not about building the plain field. Waagner-Biro Stahlbau AG is known as specialist for complex building envelopes and pioneer in mastering challenging geometry. Early on the company has used glass as structural element and lately delivered with the Glazed Link in Manchester an exemplarily all glass structure, where the classic relationship of glass = infill and steel = structure is turned upside down. But in even less pure applications glass can truly bear an important role, physically and metaphysically. Two case studies take a closer look at projects using glass fins for a commercial and a cultural development and put them in a larger context.

Keywords: buildings and projects, steel and glass structures, structural glass, glass fins

1 Why

1.1 Striving for transparency

Figure 1-1 Dan Graham, Installation at MAMO, Marseille, FR (Image: Author).

Glass is a key design feature in building if not the most important in contemporary architecture. It bears a central role in the design. Today it serves as much more than a sole

Engineered Transparency 2016. Glass in Architecture and Structural Engineering. First Edition.
Edited by Jens Schneider, Bernhard Weller.

enclosure material, its structural properties are well known and proven by multiple applications. Despite today's challenging market environment with many uncertainties, environmental restrictions and high pressure on budgets glass remains an attractive option for creating an outstanding sculpture or for highlighting special areas of a building. Aside of the structural aspects there are also many options for coping with the additional challenges that realizing a fully functional building means vs. creating a sculpture like Dan Graham's glass works, that have inspired architects and engineers since a long time.

1.2 State of the Art

Waagner-Biro's purest glass structure to date is Ian Simpson's Glazed Link in Manchester, a sculptural link in between two historic buildings. And also in between being a sculpture and a building itself. Instead of steel bearing glass, glass bears the sculptural steel roof.

Figure 1-2 Glazed Link, Manchester, UK (Image: Trevor Palin).

Light is the theme, with a barely visible all glass structure that was engineered jointly with Eckersley O'Callaghan. But also light in terms of illumination, a spectacle day and night that challenges perception.

A recent predecessor to the floating stainless steel monocoque in Manchester was the roof of the Louvre's Islamic Arts Department by Mario Bellini and Rudy Riciotti with Hugh Dutton in Paris where maximum transparency was also achieved by avoiding the wider joints of insulated glass units and the application of fairly transparent gaskets for the vertical low-iron panels (within the roof, where not visible, all glass is insulated to achieve the proper performance). Also here light is the theme from the overall concept up to the smallest detail.

Figure 1-3 Louvre – Départment des Arts de l'Islam, Paris, FR (Image: Author).

The extra effort was well worth it in both cases as no structural element is right behind the panels, thus the joints are fully exposed to vision as can be seen in above image.

1.3 Lateral Force

Figure 1-4 Tower Place, London, UK.

The company's earliest application of glass as structural element was for Foster + Partners' Tower Place project in London back in 2002. The scope included the atrium enclosure consisting of a glass roof and a cable wall that is suspended from the roof structure as it does not reach to the ground. But that not being special enough, the façade's lateral support is provided by 4 m long glass tubes shown in above image. The glass tube field concept is by Carpenter | Lowings, another long-term source of inspiration for the sensual use of glass in architecture.

Like the glass tubes in above project the glass fins of the following case studies have primarily lateral structural duties, but are beyond that an important aesthetic factor of the overall concept, transparency with a twist. Both projects are located in Dubai. One is the refurbishment of a large shopping complex, Festival City. The other is the new Etihad museum building commemorating the Emirate's foundation. As different as their use and structural systems are, both applications are equally challenging.

2 What

2.1 Festival City, Project Salsa, Dubai, AE

Back in 2008 Waagner-Biro completed one of their first projects in the Middle East, the award winning Festival City. The project is a shopping center not far from Dubai Airport. The company's scope included the mall's roof and the so called Festival Square, the focal point within the complex. The Festival Square's toroidal roof structure was executed as classic steel frame with quadrangular panels. Likewise the main façade facing Dubai Creek was executed as a steel frame with a rather dense grid. Not long after completion ambitious expansion plans materialized that were stalled due to the financial crisis.

Figure 2-1 Festival City, Festival Square, Dubai, AE (Image: Trevor Palin).

However, in this highly competitive and short lived environment the owner needed to take action for not falling behind all the other ambitious retail developments in the Emirate. Thus project Salsa was born and Waagner-Biro secured 2014 the contract for refurbishing Festival Square and expanding the project closer towards the waterfront – proof of the customer's satisfaction with the company's previous performance for him. Following rendering shows the same view as above photo. The roof remains in place while the façade was demolished. HOK's expansion design with a rather opaque roof is clearly in reaction to the stricter environmental requirements (the earlier project never had the external shading structure it was to receive). And it also shows a very contemporary approach for how to design the main façade, a highly transparent curved feature wall.

Figure 2-2 Festival City, Project Salsa, Dubai, AE (Image: Al Futtaim).

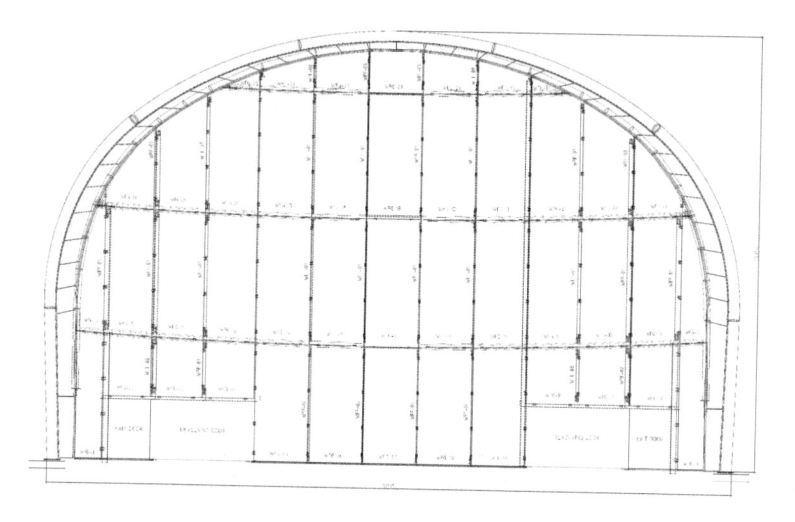

Figure 2-3 Festival City, Project Salsa, Dubai, AE – West Façade layout.

This tilted feature wall, the West façade, is at the arch's pitch nearly 21 m high and 31 m wide. The primary structure bearing the curved roof as well as providing support for the curved façade is made of steel. Two circular vertical columns divide the facade in three main sections which are subdivided by three horizontal box section transoms (and split by the entrances). How these fields are structured is clearly a sign of how technology has advanced since the completion eight years ago as well as what is available in the market, and the design has adjusted accordingly: contrary to the small rectangular steel framed panels of the original design now 6.0 m high and 2.5 m wide glass panels are laterally supported by glass fins that span between the transoms. Each low-iron fin with a depth of 400 mm is comprised of 5 layers of 10 mm FT with PVB interlayer, to which the glass panels are only locally connected. Horizontally the glass is connected back to the steel structure. The IG units have laminated inner and outer panels each being comprised of 2 x 10 mm HS. A high performance low-e coating is on level 4.

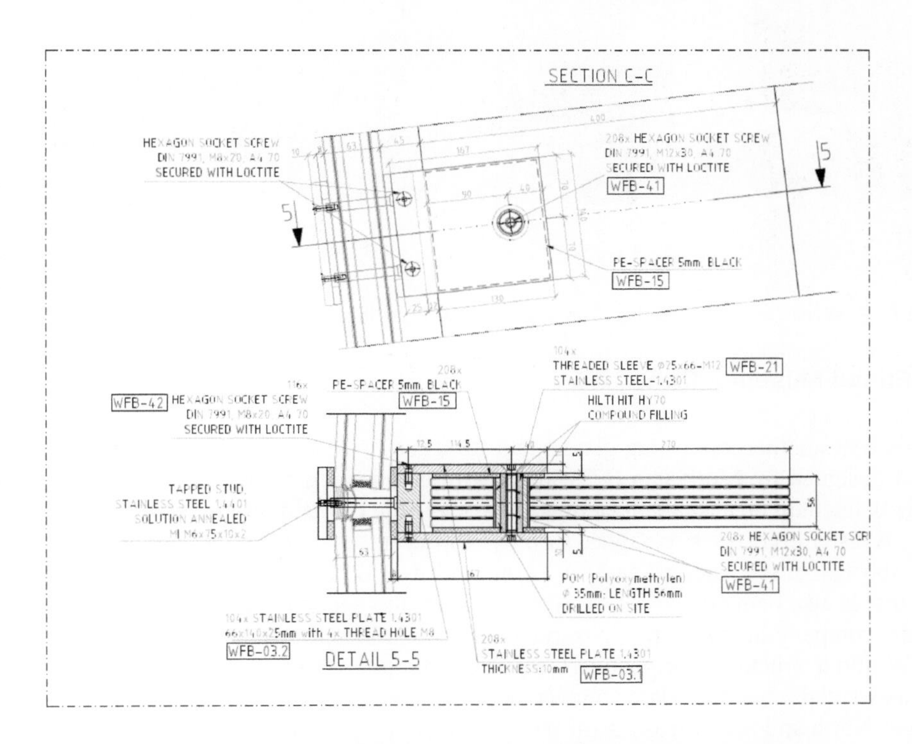

Figure 2-4 Festival City, Project Salsa, Dubai, AE – West Façade glazing details.

The result of this design effort is clearly a new and much more open relationship between inner and outer marketplace that will definitely boost to the shopping center's competitiveness. The lens on the waterfront will become the central shop-window of the whole development.

Figure 2-5 Festival City, Project Salsa, Dubai, AE – West Façade nearing completion.

2.2 Etihad Museum, Dubai, AE

Later this year another interesting glass application can be found at the Etihad Museum in Dubai designed by Moriyama & Teshima with engineering by Werner Sobek. It is currently being built on a National Heritage site where the United Emirates were founded in 1971. A sister division of Waagner-Biro has earlier installed the giant flagpole there. Whilst Abu Dhabi is creating Saadiyat Island with world class museum's such as the Louvre that is also being delivered by Waagner-Biro, this second ongoing museum project of the company in the country has much higher significance for the Emirates themselves. Within a structure representing the parchment upon which the unification agreement was written a band of glass follows the undulated surface of the top and bottom. The lower North and South facades tilt 69°; the layout of the vertical main West façade follows this angular scheme providing for a dynamic appearance. Likewise the main structure is tilted. But here, different from the hybrid lens presented earlier, the glass wall spans from top to bottom without any intermediate steel structure, it is a sole vertical/sloped system independent of the main structure.

Figure 2-6 Etihad Museum, Dubai, AE (Image: Moriyama & Teshima).

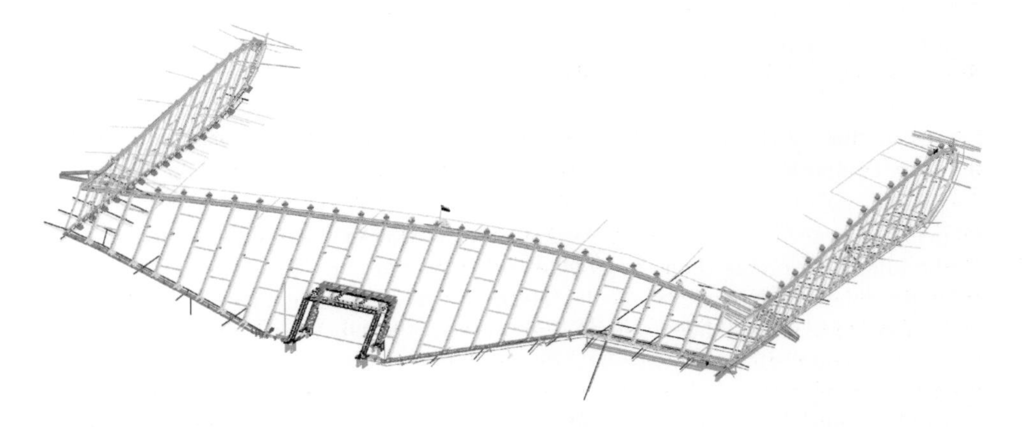

Figure 2-7 Etihad Museum, Dubai, AE – Façade model.

The West façade reaching up nearly 13 m has fins with a depth of 600 mm whereas the lower side facades have fins with 400 mm depth. Their general build up is 4 x 15 mm FT low-iron glass with 1.52 mm sentry interlayer. The facade glass panels rest on each other and are toggle fixed to the fins only for wind loads as shown in following images.

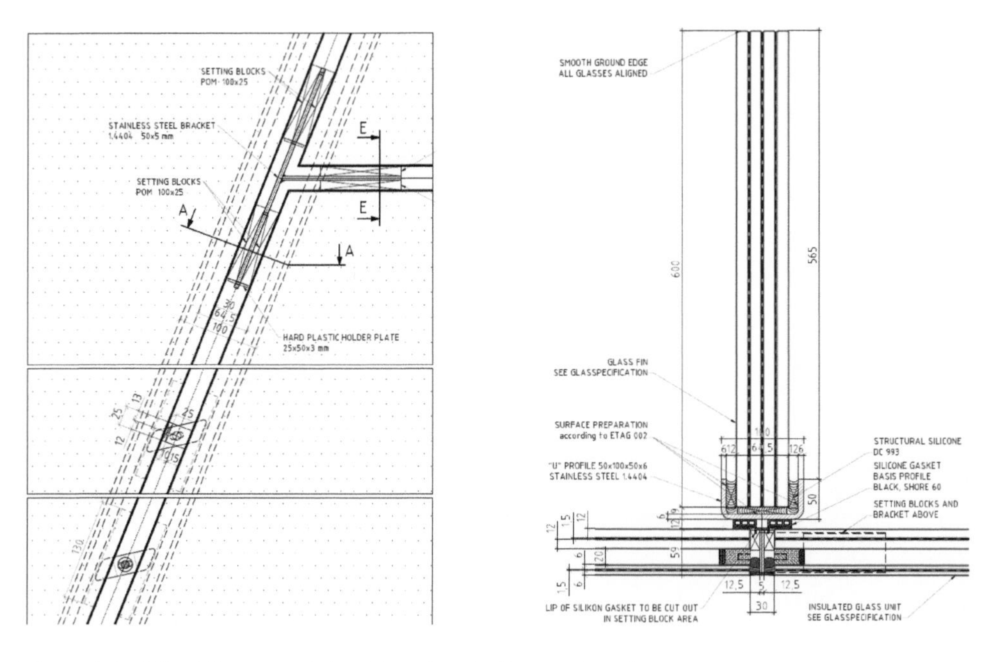

Figure 2-8 Etihad Museum, Dubai, AE – Façade Support Detail.

The parallelogram shaped glass panels of the main West façade with a side length of 5.03 m x 2.25 m are typically comprised of PVB laminated panels 2 x 6 mm on the outside and 2 x 12 mm Sentry laminated on the inside. All HS and low-iron, with the exception of the clear panel bearing the solar coating on level 4. The last bay at the corner to the South, on which the whole tilted façade 'leans on', has a very special build-up reflecting its structural performance. The outer panel is again 2 x 6 mm PVB laminated HS, but the inner composition is comprised of 4 x 12 mm Sentry laminated low-iron HS.

High transparency in such a climate naturally means dealing with the solar impact despite the building's favorable shape with wide overhangs that provide for natural shading in the upper zones. Thus in addition to the selective coating the 20 mm cavity is Argon filled and warm edge spacers have been used. A gradient decorative frit originally considered, in its basic nature similar to what has been used to create the multilayered solar control screen on Waagner-Biro's Louvre dome that creates the fascinating rain of light, was not realized. In summary the all glass structure provides for the highest level of transparency possible under the boundary conditions.

The complexity of the museum's skin is not limited to the glazing. Also the parchment's geometrically challenging GFRP cladding is part of Waagner-Biro's scope. The progress photos show already the impressive contrast between the solid shell and the large opening.

Figure 2-9 Etihad Museum, Dubai, AE – West Façade in progress.

Figure 2-10 Etihad Museum, Dubai, AE – South Façade in progress.

3 What's next

The two case studies from Dubai are not the only projects where Waagner-Biro is currently working with glass fins. On the boards is also the Lakhta Center's massive Multi-functional Building in St. Petersburg, Russia, for which the company has – amongst other items – the atrium facades under contract. Some of the tilted glass fin supported facades reach higher than 17 m.

The popularity of glass fins is evidence for transparency still being a driving force and the advances in engineering and fabrication make it easier to achieve, technically as well as economically. Certain retail projects show that where enough funding is in place also almost no limit in size exists. The further development of connection design will be the key to achieving an even higher level of transparency.

4 References

[1] Frey, H.: From bid to building. In: Glasbau 2016, Weller B., Tasche S. (eds.), Berlin: Ernst & Sohn, 2016, 33-45.

All images, unless otherwise noted, are by Waagner-Biro.

Building Doha – Qatars Iconic Facade Projects

Dr.-Ing. Dipl. Wirt.-Ing. Thomas A. Winterstetter[1], Dr.-Ing. M.Arch. Lucio Blandini[2], Prof. Dr.-Ing. Dr.-Ing. e.h. Dr. h.c. Werner Sobek[3]

1 Werner Sobek Stuttgart, Albstrasse 14, 70597 Stuttgart, Deutschland, thomas.winterstetter@wernersobek.com

2 Werner Sobek Stuttgart, Albstrasse 14, 70597 Stuttgart, Deutschland, lucio.blandini@wernersobek.com

3 Werner Sobek Stuttgart, Albstrasse 14, 70597 Stuttgart, Deutschland; Universität Stuttgart, ILEK, Pfaffenwaldring 7+14, 70569 Stuttgart, Deutschland

Qatars policy for modernizing the country is to combine construction innovation with iconic architecture. Werner Sobek Stuttgart (WSS) worked on some of the most challenging envelope projects in Doha. The present paper includes the super-transparent facades of the Qatar National Convention Centre (Yamasaki Architects, Troy/Michigan, USA) with 20m tall glass fins, the cable facades of the Doha Exhibition and Convention Centre (Jahn Architects, Chicago, USA) with their innovative support on horizontally stressed cables, and the National Museum of Qatar (Atelier Jean Nouvel, Paris, France) which is a real milestone in 3D engineering and BIM design.

Keywords: cable façades, glass fins, complex geometry, BIM modelling, FRC cladding

1 Introduction

Many of the largest, most innovative and challenging recent architectural construction projects of the world are located in the Gulf states, and Qatar is home to some of the most iconic ones. The present paper describes the stunning architectural and engineering features of the envelopes of these projects, and the Gulf region façade market experience and the local design and construction conditions are explained.

2 Super-Transparent Glass Facades

2.1 Qatar National Convention Centre

The QNCC is located in Education City, a part of the city dedicated to universities, education and the like. The architecture is dominated by a very recognizable and iconic tree-shaped megastructure across the full width of the front elevation (fig. 2-1), reminding of the legendary Sidra tree in whose shadows the scholars used to teach and read in old times.

Engineered Transparency 2016. Glass in Architecture and Structural Engineering. First Edition.
Edited by Jens Schneider, Bernhard Weller.

WSS was approached in 2009 by the architects (Yamasaki Architects, Troy/Michigan, USA) to enhance the façade design by providing engineering, design and tender documents for the super-transparent 20 m tall insulated solar-control vision glass façade in between the metallic tree megastructure.

At these times, it was clear that the options to procure 18 m tall glass fins in one piece were very limited, if any. For economic and competitiveness reasons, a design had to be engineered where the glass fins would be produced in several pieces and put together on site.

The usual option in that case is to connect the glass fin parts of the full-height assembly by means of metallic connectors, using drilled holes in the glass, to arrive at a bending-rigid continuous vertical glass fin beam element. The disadvantage of that approach is that the forces that can be transferred through such drillings are limited, and several large drillings in one row would be needed, and that the limited tension stress capacity of glass would require a certain distance between two individual drillings; both effects would have made the connection pieces very large and dominant.

To enhance the transparency and the clean look of the connections and to meet the highest architectural requirements, WSS took a different structural approach: The individual 800 mm deep laminated glass fin pieces are only as tall as the insulated glazing bays (5 m), but are connected to tiny steel profiles at the front and back of the fins, to form a vertical truss system against horizontal (wind) loads. In that truss system, the glass fins are always acting as compression diagonals, regardless of direction of wind. In case of wind pressure, there is a slender tension bar at the inward side of the glass fin assembly, to take up the tension chord loads of the truss system; in case of wind suction, the glass fins do transfer the "compression web" forces amongst themselves by compression contact force action.

By using that innovative structural system and thus exploiting the structural properties of the glass in an optimized way, it became possible to maximize architectural transparency and the aesthetic quality of the details. At the same time, the individual glass fin pieces were not excessively long and it was possible to include façade bidders from the region into the tendering process. Consequently, the façade was awarded to Metal Yapi from Istanbul, and shop design and erection were successfully made under the guidance, checking and supervision of WSS.

The opening of the building in 2012 became a huge success. In the meantime, the second stage of the National Convention Centre project is already completed, too, and does provide Qatar with one of the largest and most recognizable conference venues in the world.

Figure 2-1 Qatar National Convention Centre view from outside.

Figure 2-2 Qatar National Convention Centre – 20 m tall glass fins connections engineering.

2.2 Doha Exhibition and Entertainment Centre

The DEEC is located in the highrise business district of Downtown Doha, next to the Corniche. The building features a wide-spanning roof and two super-transparent elevations, an inclined south façade and a vertical west façade.

The architectural design intent by Jahn Architects, Chicago, was to create an appearance as transparent and generous as possible, and it was clear from the beginning that a cable façade would be the method of choice to achieve that goal.

The critical condition was that the large cantilevers of the main roof trusses were designed in a way that they were able to carry the dead loads of the 18 m tall facades, but not any large additional loads such as from pretensioning of vertical cables.

Therefore, WSS invented a design where the wind loads are carried only by horizontal cables. These cables are tensioned against bow string columns at the corners of the building, and they are only laterally supported in 18 m spacings at box section roof truss tiedown columns of the main roof steel structure.

The project features the first purely horizontally stressed cable façade in the world. The horizontal façade cables are up to 400m long, taking full advantage of the specifics of steel cable construction, and making the facades a stunning experience for visitors (fig. 2-3).

Special attention was given to all details within the cable façade scope of work of WSS. The glass clips and the cable clamps and all other connections are speaking a common design language, with curved-tapered steel plates and minimized connectors (fig. 2-4).

The design was continued by Jahn & WSS to detailed design and tender documents, and the installation was awarded to Josef Gartner GmbH from Germany. The building was successfully completed and ceremoniously opened in November, 2015.

Figure 2-3 Doha Exhibition and Entertainment Centre – 18 m tall horizontal cable facade.

Figure 2-4 Doha Exhibition and Entertainment Centre – a common detailing language everywhere.

3 National Museum of Qatar - Complex Geometry and BIM Modelling

The National Museum project is based on the design idea by Atelier Jean Nouvel (AJN), Paris, of a sculptural building compound taking the form of a desert rose, a crystalline gypsum object that is created by certain mineral and water evaporation conditions in the deserts of the Arabian peninsula. Consequently, the museum does have a very iconic, complex three-dimensional geometry made of several hundreds of disc elements of various diameters and thicknesses, up to 87 m large, which are intersecting each other in a completely irregular way (fig. 3-1).

Figure 3-1 National Museum of Qatar – bird's eye view of full project extent.

The pre-contract tender design was made by the AJN team. The design-and-build contract was awarded to Hyundai Engineering & Construction from Korea (HDEC), which included all envelope design-and-construction works. The sculptural disks which are the envelope of the project do have a cladding made of fibre-reinforced concrete (FRC). These solid panels are irregularly shaped and all have a double-curvature coming from the curvature of the individual lens-shaped disks. They are attached in an invisible concealed way to a secondary steel substructure which is on top of a waterproofing-foamglass-insulation-metal-deck roof buildup. All that entire buildup is supported by the primary steel beams and trusses designed by ARUP and produced and installed by Eversendai in the UAE.

WSS was subcontracted as façade specialist for consulting and site supervision on the entire façade design and installation, and in particular for the optimization, engineering and 3D modelling of the entire FRC cladding substructure and related steel elements. The key to designing that super-complex three-dimensional building compound was to combine all design files of all the individual companies involved into a giant 3D BIM model. That 3D BIM model is a kind of living organism, with constant data feed and updating from all parties involved, in full Level of Detail LOD400, with a permanent clash detection and design coordination process running in the background. The model was administered by Gehry Technologies in California using Digital Project software.

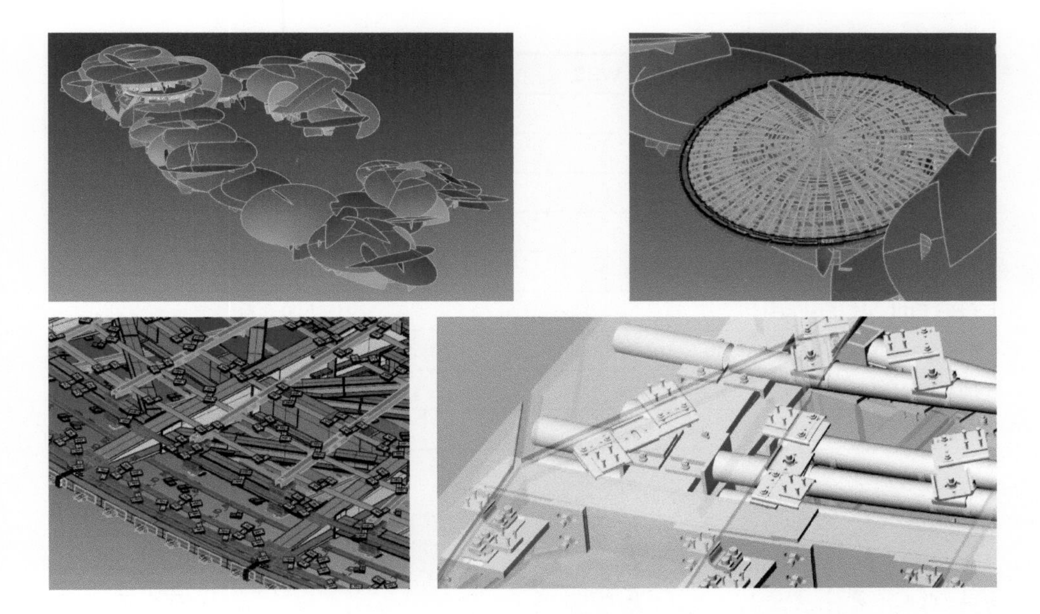

Figure 3-2 National Museum of Qatar – various stages of detailing of cladding substructure.

All details and layouts and profiles were optimized by WSS to originate from a set of common design principles, to minimize the extent of individual components and to use as many equal parts as possible. All kinds of tolerances had to be accommodated, as well as limited availability of hardware products on the local market and ease of fabrication, construction, transportation and installation.

The entire secondary steel scope was divided among no less than five producers (Gulf Steel, Hanlim, Boston Steel, PSCQ, and Eversendai). These were working in parallel on different areas of the project at the same time, for the sake of construction schedule and stability of performance. The detailed WSS 3D BIM files went to these producers who derived their cutting and drilling machines input and all other related information directly from them.

The size and complexity of the project was an outstanding challenge to all parties involved. Design teams were working on three continents, the producers and site people were from everywhere across Asia, with up to 3,000 workers on site (fig. 3-3). WSS deployed a permanent site team of up to four engineers and 3D specialists for site supervision and design coordination. Only with the combined best efforts of Hyundai and all their subcontractors, it was made possible that this bold architectural dream became reality.

Table 3-1 Selected Quantities of Secondary Steel Structure (SSS).

Item	Quantity
Amount of Individual 3D BIM Files by WSS	12,000
2D Drawings of SSS	4,000
BIM Data Volume generated by WSS	700 GB
SSS Beams, Weight in Metric Tons	1,800 to
SSS Beams, Total Length, estim.	50,000 m
FRC Embeds (Fixings)	500,000
Support Stubs Connectors to Primary Steel	23,500
Nose Fixings	7,700
Amount of FRC Panels	75,000

Figure 3-3 National Museum of Qatar – installation of panels.

The Berkeley glass pavillion

Carles Teixidor[1]

1 Bellapart SAU, Ctra. Parcelaria 32 Les Preses 17178, Spain, cteixidor@bellapart.com

The new entrance glass pavillion for The Berkeley hotel in Knightsbridge, London, is expected to be completed by May 2016. This paper describes the design, testing and construction of the pavillion, focusing on its carbon fibre structural components and its innovative glass-honeycomb sandwich panels which were developed specially for this project.

Keywords: carbon fibre, composite, glass, honeycomb, adhesive

1 Introduction

The ground and first floors of The Berkeley hotel in Knightsbridge, London, are being refurbished at the time of writing this paper. The main part of the refurbishment works is the demolition of the current entrance area and its adjacent rooms, and the construction of a new glass pavillion designed by Rogers Stirk Harbour and Partners, in collaboration with Expedition and Arup.

As shown in figure 1-1, the pavillion consists of a 29.4 m x 10.2 m glazed roof and canopy covering the entrance area and the extensions of the Blue and Collins Rooms, both cladded with glass walls.

Figure 1-1 Sketch of the pavillion (courtesy of RSH+P).

All roof and lateral wall cladding is composed of glass-honeycomb sandwich panels specially developed for this project, whereas front walls are cladded with conventional insulating glass units.

Engineered Transparency 2016. Glass in Architecture and Structural Engineering. First Edition.
Edited by Jens Schneider, Bernhard Weller.

The glass-honeycomb sandwich panels consist of two thin glass skins structurally bonded to an aluminium honeycomb core, obtaining a stiff and relatively light panel with a peculiar translucent look. Triple glazing is used in both wall and roof cladding enclosing internal spaces in order to neutralise the thermal bridge caused by the honeycomb.

All glass panels are supported by a simple isostatic structure made up of beams and columns. However, it must be pointed out that some of the main structural members were designed in carbon fibre reinforced polymer (CFRP) with a high quality surface finish.

2 Structure

The pavillion structure consists of sixteen free-form CFRP beams, measuring 10 m in length, that are connected to the building at one end and to a stainless steel (SS) column at their mid point, creating a 4.5 m cantilever over the car stop. In the entrance area, the SS columns are replaced by a 9.5 m long truss formed by two stainless steel chords and a set of CFRP V-shaped brackets acting as diagonal members (see figure 2-1).

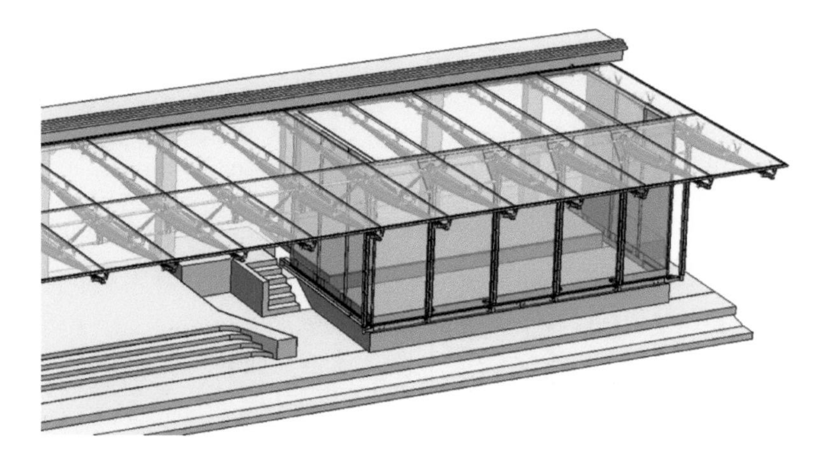

Figure 2-1 Structure of the pavilion.

The CFRP beams are spaced 1.9 m centres and cover both the canopy area and the internal spaces. Therefore, all front wall glass panels are cut in a special shape that allows the CFRP beams to cross the front walls through glass joints. This shape also permits the front wall panels to tilt inwards without any collision to any structural member, which is essential for future glass replacements.

The bottom of the columns is supported by a number of stainless steel fins welded to a continuous stainless steel channel which extends along the perimeter of the two rooms.

This channel is in turn supported by an internal mild steel structure which also supports the floor of the room extensions at an approximate height of 1 m above the street level.

Bracing is entirely provided by the front wall and roof glazing. In fact, four front wall glass panels are structurally connected to the roof glazing and to the bottom channel and columns, acting as shear walls which impede the pavillion to swing in longitudinal direction. Swing in transversal direction is not possible as the CFRP beams are directly connected to the concrete structure of the building.

The accidental situation in which one corner column and its adjacent shear wall panel are lost (i.e. due to a car accident) was analysed in order to determine its impact in the lateral stability of the structure and in the behaviour of the central door truss.

3 Carbon fibre beams and brackets

The CFRP beams were designed to carry significant permanent and imposed loads with relatively small deflections. Therefore, the top and bottom areas of the beams were designed as 17.6 mm thick chords composed of unidirectional carbon fibres oriented in longitudinal direction (0°) whereas lateral areas are 7.6 mm thick laminates with fibres oriented in four different directions (0°, ±45° and 90°). An epoxy matrix was used in all areas. In order to provide the required high quality surface finish, the complete beam was covered with a carbon fibre weave.

The connection to the stainless steel structure was solved by using threaded stainless steel inserts embedded into the beam and bonded to it with an epoxy adhesive. The total weight of the 10 m long beam is 162 kg (128 kg without SS inserts).

The beams were fabricated by manual lay-up of the carbon fibre prepregs on two 10 m long composite moulds, each corresponding to one half of the beam plus overlap flaps. Then, the two moulds were joined together and a plastic bag was inserted in the empty space inside the beam. Vacuum was created between the bag and the moulds and the whole assembly was cured in an oven for 18 h at 70 °C and an additional 6 h at 95 °C. Temperature in the thicker areas of the laminate was monitored and used as an input signal by the computerised control of the oven.

After that, the beam was removed from the mould, the SS inserts were bonded in it and a final post-curing thermal cycle in the oven was carried out in order to reach the desired mechanical properties and glass transition temperature of the laminate. Finally, the beams were polished and coated with a transparent varnish.

The fabrication of the CFRP V-brackets was similar, although these elements were cured in an autoclave at 95 °C and 2.5 atm for 10 h, approximately.

The design of the CFRP elements was performed in accordance to Cripps [1]. The resistance of the beams in ultimate limit state was verified using a combination of the Tsai-Wu [2][3], maximum stress and maximum strain criteria, whereas maximum deflections in service conditions were agreed with the client's consultants using steelwork standards as a reference. In addition, 10 % of the layers in the beam unidirectional top and bottom chords were oriented in ±45° and 90° directions in order to guarantee that no significant creep would occur according to MIL-HDBK 17 [4].

Figure 3-1 Full size prototype beam during testing.

The structural performance of the CFRP beam was assessed by carrying out three different loading tests on the full-size prototype beam. This prototype was instrumented with strain gauges and beam deflections were measured using depth gauges, as shown in figure 3-1. Test results were compared to predicted values obtained from the finite element models used for design and showed a reasonably good agreement [5]. In addition, loading tests were also carried out for the CFRP V-brackets.

4 Glass-honeycomb composite panels

The required thickness of façade and roof glass panels is defined by their maximum stresses and deflections which in turn are function of their stiffness. This is specially important in roofs, where self-weight deflections might be determinant for the thickness.

Therefore, it is always interesting to find a way to increase the stiffness of a glass panel without increasing its weight.

On the other hand, for privacy or design reasons it is sometimes interesting to use translucent glass panels which allow light to enter in a building while not showing what is happening in it.

The glass-honeycomb composite panels are intended to solve these two problems. They are composed of two sheets of glass structurally bonded to a microperforated aluminium honeycomb core by means of a 1.0 mm thick continuous layer of UV-curing transparent acrylic adhesive, creating a true structural composite panel with a peculiar translucent look.

In fact, during fabrication the liquid adhesive climbs on the honeycomb by capillarity creating a concave meniscus in each honeycomb cell that is solidified during curing. Therefore, the cured panel is composed of an array of acrylate lenses, two per cell, that distort the images transmitted through the panel (see figure 4-1).

Figure 4-1 Glass-honeycomb composite panel.

The structural advantages of the glass-honeycomb composite panels are remarkable. The bending stiffness of the panel increases significantly due to the fact that the two glass skins are 25 mm apart, with an intermediate aluminium honeycomb allowing an effective shear transfer. Therefore, both deflections in the centre of the panel and stresses around point fixings decrease compared to a conventional insulating glass unit.

These lower stresses permit heat strengthened glass (for point-fixed panels) or annealed glass (for perimetrally supported units) to be used. After breakage, the relatively large glass fragments remain attached to the honeycomb thanks to the acrylate adhesive, which allows designers to use monolithic glass components in many situations which would require the use of laminated glass if a conventional IGU was to be installed. Despite this good post-breakage behaviour, laminated glass was used in all lower glass plies of the Berkeley pavillion in order to skip the tests required by the Authorities if monolithic glass panes were used.

The light and solar transmission of the glass-honeycomb panels is high as the crossed reflections between the honeycomb, adhesive and glass cause that most incident radiation is transmitted through the panel as diffuse light. Therefore, coatings are required for an effective solar protection.

On the other hand, the thermal performance of these units is lower than a similar conventional IGU due to the thermal bridge caused by the aluminium honeycomb. Therefore, in the Berkeley pavillion an argon-filled air chamber and an additional laminated glass component with a Low/E coating was added to all panels enclosing internal spaces (see figure 4-2).

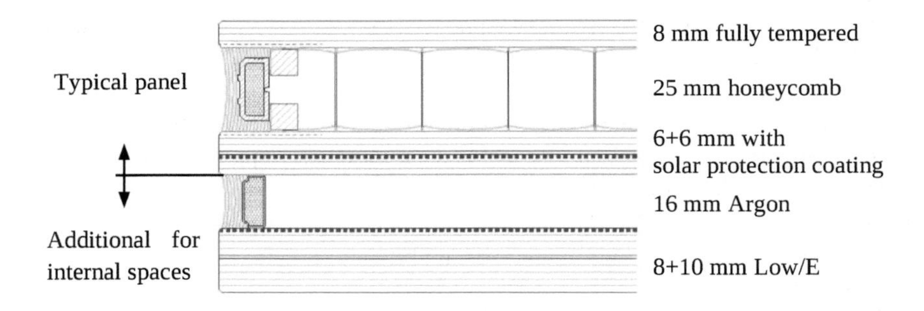

Figure 4-2 Cross section of a typical glass-honeycomb panel.

The fabrication of the glass-honeycomb panels is a delicate process that requires special equipment, skilled personnel and a clean environment (pressurised room) to avoid dust being trapped in the adhesive layer (see figure 4-3). At the time of writing this paper, the production of all panels for the pavillion has been completed. The typical size of the panels in the project is 4.60 x 1.90 m whereas the size of the largest panel produced is 4.85 x 2.30 m

Another important point for fabrication is to use the right adhesive. Apart from the adhesive structural and visual properties, viscosity and vapour emissions during curing are important factors that must be taken into account when choosing an adhesive. The more viscous an adhesive is, the more difficult it is to remove any air bubbles trapped during production. However, low viscosity adhesives tend to emit higher quantities of acrylic vapour during curing which might create condensations that cause undesired permanent textures and tears on the adhesive layer.

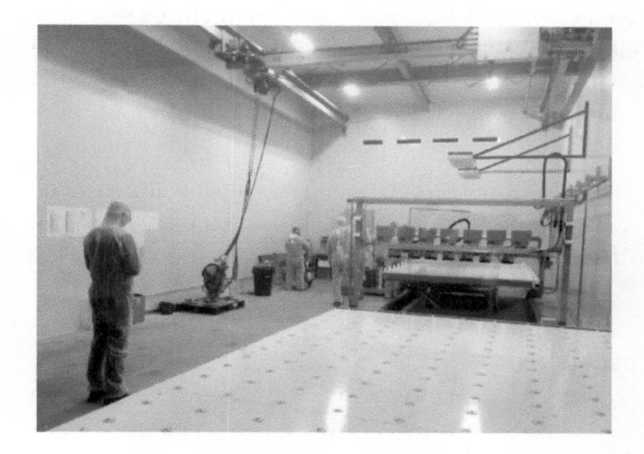

Figure 4-3 Large glass-honeycomb panel during fabrication.

In addition to all calculations, the structural properties and durability of the panels for the Berkeley pavillion was assessed by testing. A number of tests were performed which were intended to check the following points:

- Yellowing of adhesive specimens after a 2000 h irradiation in a sunlight simulator.
- Chemical compatibility of the adhesive with all perimetral sealing materials.
- Fogging of units under sudden temperature changes.
- Moisture ingress through the perimetral seal and/or pneumatic conduits.
- Effect of cyclic temperature variations on the glass-honeycomb bond.
- Tensile resistance of aged glass-honeycomb specimens.
- Compression resistance of glass-honeycomb specimens.
- Bending resistance of aged glass-honeycomb specimens.
- Cyclic bending resistance of glass-honeycomb specimens.
- Long-time bending behaviour of the glass-honeycomb sandwich (creep).
- Impact and post-breakage behaviour according to CWCT Technical Note 42 [6].

With the exception of the impact test sequence, all other tests were specifically developed for the project [7] based on:

- Internal test procedures of the adhesive manufacturer.
- Special procedures developed in conjunction with the adhesive and dessicant manufacturers.
- relevant EN standards modified to accomodate the specific needs of the project.

Although tests indicated that the glass-honeycomb bond is strong enough to bear the internal pressure loads caused by the expansion of air in the air chamber under temperature variations, as a precautionary measure it was decided to equalise the pressure of the air chamber with that of the surrounding atmosphere. Therefore, a system of flexible pneumatic conduits connecting all glass panels to a battery of six breathers was designed in order to allow the release of any significant pressure variation within the honeycomb cells. These breathers are equipped with adequate filters to avoid dust entering in the panels and sufficient quantity of self-indicating silica gel dessicant to allow a minimum replacement period of 1.5 years.

Figure 4-4 Installation on site.

5 Conclusions

The paper has shown how the use of carbon fibre composites permitted to design and fabricate 10 m long beams with the free-form shape desired by the Architect and a high structural efficiency. In fact, the weight of these beams is approximately half the weight of an equivalent variable-depth carbon steel I-beam with the same bending stiffness.

On the other hand, glass-honeycomb sandwich panels are able to meet the structural and privacy requirements of the project and to provide a unique visual appearance to the cladding of the pavillion.

The use of innovative materials in construction can provide creative solutions to specific problems, although they must be carefully analysed and tested to guarantee their durability and suitability for the intended application.

6 References

[1] Cripps A.: Fibre-reinforced polymer composites in construction. CIRIA, 2002.

[2] Tsai S.W. and Wu E.M.: A general theory of strength for anisotropic materials. Journal of Composite Materials, vol. 5 (1971), pp. 58-80.

[3] Barbero E.J.: Finite element analysis of composite materials. CRC Press, Boca Raton (FL), 2008.

[4] MIL-HDBK-17-3F Composite materials handbook. Volume 3: Polymer matrix composites. Materials usage, design and analysis. US Department of Defense, 2002.

[5] Bellapart: Full-size prototype. Description and testing. Revision 5. Olot (Spain), 2009.

[6] CWCT Technical Note No 42 Safety and fragility of glazed roofing. Guidance on specification and testing. Centre for Window and Cladding Technology, 2004.

[7] Bellapart: Glass-honeycomb composite. Description and testing. Revision 8. Olot (Spain), 2010.

Design of the solar shading screen at Tarek Bin Ziad School

Marco Vetter[1], Dirk Osterkamp[2]

1 Marco Vetter, Engineering Consultant Colt International GmbH,
marco.vetter@de.coltgroup.com

2 Dirk Osterkamp, Marketing Colt International GmbH, dirk.osterkamp@de.coltgroup.com

The building envelope is undergoing a revolution. It can now be designed to do more than just to keep the weather out and to provide an impressive appearance. It can make a positive contribution in reducing the energy consumption of the building. Modern design approaches can turn the envelope into an active component through the use of solar shading systems. This paper describes how Colt solar shading devices were integrated in the reconstruction of the Tarek Bin Ziad School in Doha, Qatar.

Keywords: solar shading systems, protection against solar heat, reducing cooling loads

1 About Colt International and Solar Shading

1.1 Shading systems saves energy

Solar shading systems have a great potential to impact on energy use and thereby to reduce the use of fossil fuels. Buildings use more than 40 % of total energy resources, of which around half is used for heating and cooling. Modern glazed commercial buildings have become very well insulated, so that less energy is required for heating them in winter. However, if no attention is paid to their shading requirements, they can use a lot more energy for cooling in summer than for heating in winter. The ESCORP/EU25 study recently demonstrated that if all buildings in the EU were properly solar shaded, 80 million tonnes of CO_2 could be displaced for cooling and 31 million tonnes of CO_2 for heating every year. Accordingly local building regulations increasingly require designers to reduce solar heat gain, with solar shading recommended as a preventative measure unless areas of glass are minimised.

Engineered Transparency 2016. Glass in Architecture and Structural Engineering. First Edition.
Edited by Jens Schneider, Bernhard Weller.

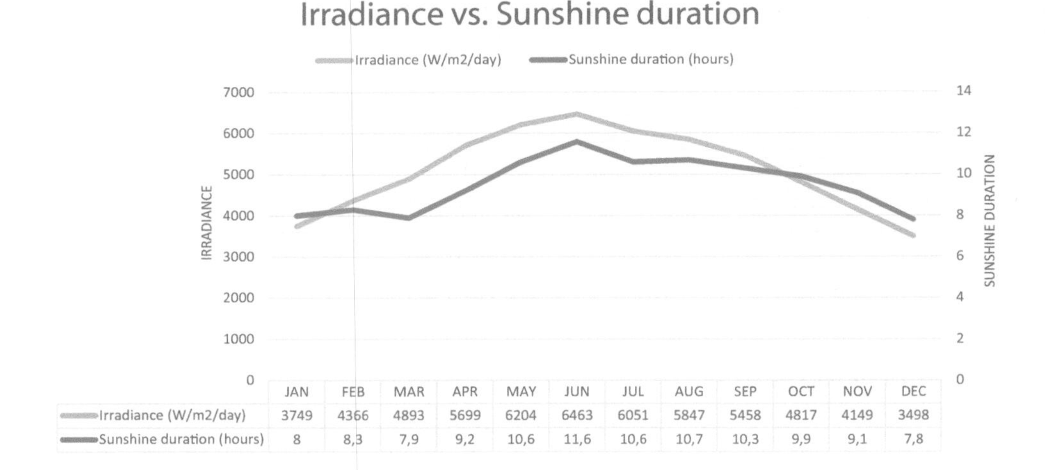

	JAN	FEB	MAR	APR	MAY	JUN	JUL	AUG	SEP	OCT	NOV	DEC
Irradiance (W/m2/day)	3749	4366	4893	5699	6204	6463	6051	5847	5458	4817	4149	3498
Sunshine duration (hours)	8	8,3	7,9	9,2	10,6	11,6	10,6	10,7	10,3	9,9	9,1	7,8

Figure 1-1 Irradiance vs. Sunshine duration.

1.2 Increase comfort and productivity

We can all appreciate the benefits of working in an environment where the temperature is comfortable. For office buildings in the summer, the optimum temperature is around 24 °C, with a range of +/- 4 °C. If the building is fully glazed but does not have an effective solar shading system, the internal temperature can shoot up as high as 35 °C during summer months, due to the effects of solar radiation. Such an uncomfortably warm environment adversely affects the productivity and concentration levels of the occupants inside. In air conditioned buildings uncontrolled solar heat gain can increase cooling loads, plant size and overall running costs. The amount and quality of natural daylight also has a positive effect on the productivity of the building's occupants. Independent studies have shown that when people sit near a window, enjoying the benefits of natural daylight, their productivity significantly improves. A lack of daylight can also result in an over-use of artificial lighting, which in turn contributes to internal heat gain. A well-designed shading system can boost the occupants' comfort and productivity by regulating the amount of heat and light entering the internal space and by reducing glare.

1.3 Provide architectural impact

Solar shading systems can be designed to provide great architectural impact as well as being highly functional. Colt solar shading systems come in a great variety of materials. Glass, metal, wood, acrylic and fabric louvres are all available to architects to create an impressive facade. The louvres can be either fixed or moveable and also can integrate photovoltaic electricity generating cells.

1.4 Why is external solar shading one of the most effective ways to control the internal conditions of a building?

Radiation from the sun is transmitted, absorbed and reflected by the louvres. As a result solar heat gain is prevented from passing into the building, minimising ventilation requirements and reducing cooling loads. If a controllable system is installed, adjustable louvres track the position of the sun, thereby reducing the numbers of days when the building overheats. Equally, in winter the louvres may be adjusted in such a way that the building benefits from the heat from the sun, and they can be closed at night reducing heat loss. At the same time daylight levels are enhanced and levels of glare are reduced.

2 Colt Shadovoltaic

2.1 A glass louvre with photovoltaic cells

Colt Shadovoltaic describes a fixed or controllable external solar shading system that incorporates glass louvres with photovoltaic cells integrated into the glass so as to generate electricity at the same time as providing shading. The louvres are available in various colours, surface finishes, patterns and coatings to meet specific design requirements. Both monocrystalline and polycrystalline cells, as well as organic or thin film photovoltaic cells may be used. The photovoltaic cells may be integrated into the glass, either by attaching them onto the reverse side of the glass panels or by laminating them between two sheets of safety glass. This system combines the functions of solar shading with the generation of electrical power. The construction possibilities are also wide. The design can be chosen by 5 different systems, Colt LS-1 – LS-5.

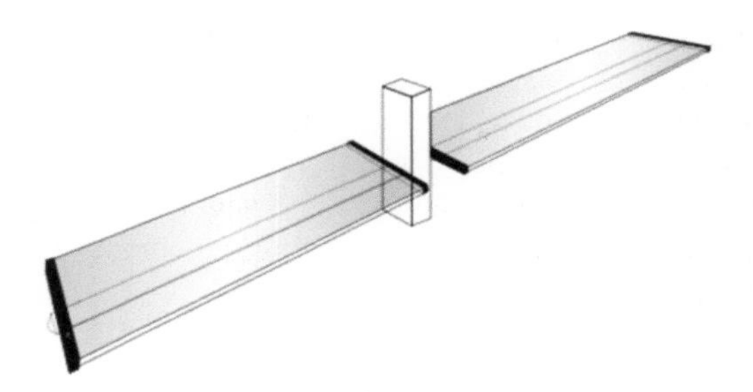

Figure 2-1 Colt LS 3.

2.1.1 Control Systems

Although fixed Solar Shading works fine for a certain angle, performance is dramatically reduced on facades which receive a large amount of sunshine during the day. A controllable shading system can best overcome this problem. Sun tracking louvres follow the path of the sun, making sure the solar shading system always optimizes the protection against solar heat gain. On dull or overcast days the louvres are controlled in such a way that if clouds pass over the building, the louvres will automatically open to maximize daylight entry and then later revert back to their original position. Colt Solar Shading systems may be controlled in two different ways: Hand control via lever or crank handle, electrically operated via actuators, which require a controller such as ICS 4-Link, SolTronic or a client BMS.

2.1.2 Colt ICS 4-LINK

ICS 4-Link is ideally suited to larger projects with more complex control requirements. It is a generic control system that can operate solar shading and natural ventilation systems. It has a wide variety of operating modes, including sun tracking, daylighting optimization and PV illumination. It responds to timers and sensors to ensure that the building 'reacts' appropriately to the sun's position and to the weather. Remote operation is available via an internal modem interface and a manual override is also possible.

2.1.3 Colt SOLTRONIC

SolTronic is ideally suited for small to medium sized projects. It is a simplified version of ICS 4-Link and can control up to 15 groups of actuators, each group containing 15 actuators, with individual motor control. Commissioning is extremely straightforward. SolTronic responds to external weather conditions automatically calculating the position of the sun, and adjusts the position of the louvres accordingly.

Figure 2-2 RWE Arena, Oldernburg, Germany – Moveable Shadovoltaic shield.

3 Design of the Colt Shadovoltaic Screen at Tarek Bin Ziad School

3.1 The Building and the requirements

The presented project is located in Al Sadd, Doha, Qatar and was developed by EHAF Consulting engineers as the design consultant, Dorsch Qatar, as the project management and supervision consultant, Bojamhoor Trading and Contracting as the main contractor. Colt International as a specialist of large façade project was appointed to build a movable solar shading screen.

Figure 3-1 Visualization of Tarek Bin Ziad School.

The idea of this project was to have a movable screen that can move autonomously around the building. Colt had experience with a movable Photovoltaic Screen that was built in Oldenburg, Germany for the EWE arena. However this shield was smaller and connected to the grid so that all energy generated could be fed into the electrical power system of the building.

3.2 The Colt Solution – a rotating Screen powered by batteries.

In order to develop a design that is capable of rotating alongside a curved building and at the same time generate the maximum possible energy out of the sunlight was a tough step for Colt's engineers. The path of the sun in Qatar is quite different to the movement of the sun in northern Europe.

The project is located at: 25.29° (North) and 51.53° (East), 10.5 m above the sea level.

The rotating screen will be positioned in front of the façade and will be mounted on bended rails that run on the roof, behind the onsite wall construction. The rails support the rotating construction segments. The segments are carried by a roller system. The weight and wind load of the rotating system are transferred through the rails into the building.

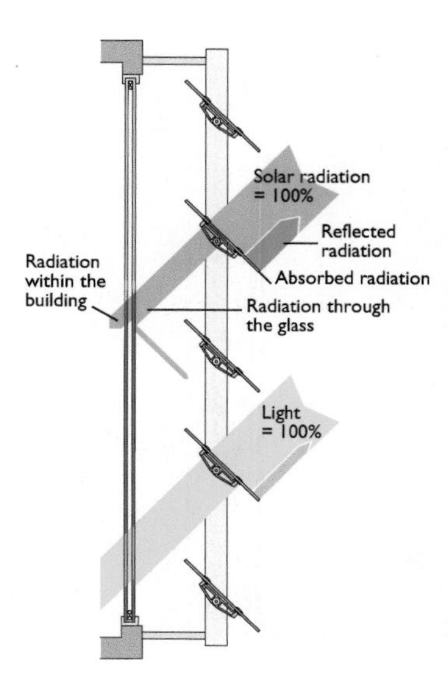

Figure 3-2 Solar Radiation and solar shading.

The segments are prepared to take up the Shadovoltaic modules as described above. Every segment consists of a 7.1 m high steel frame that carries six Shadovoltaic louvres while the top part of the system is cladded with 2mm thick aluminum sheet.

Table 3-1 Technical details of the system.

Nr. of fields	30
Field length [mm]	2400
Field height [mm]	7300
Construction surface [m²]	525.6
Nr. of louvres per field	6
Total nr. of louvres [mm]	180
Louvre length [mm]	2320
Louvre width	400
Louvre surface [m²]	167.04

The shield itself is running on a stainless steel track and is driven by one worm gear engine of ZAE with a driving torque of 1,680 Nm and an electrical capacity of 3.000 W. The shield is gliding on top of two rails, one on the roof and one on the façade of the building.

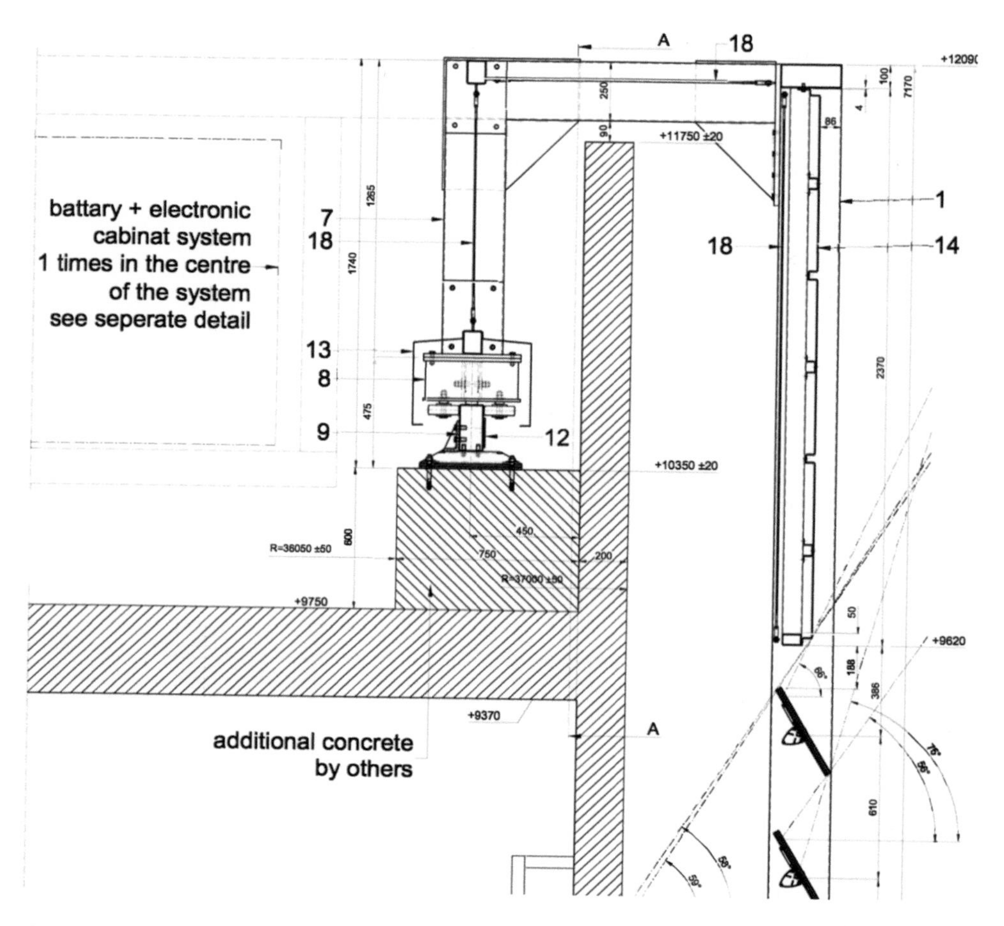

Figure 3-3 Vertical section of the screen.

The shield construction rolls on high performance wheels, similar to the track of a roller-coaster. In order to meet the high heat conditions in the desert of Doha, the wheels were tested in advance by Colt to provide maximum safety. The wheels were heated up to 93 °C and then loaded with 1,200 Kg to form an equivalent condition as it would be present on top of the roof in Doha to test possible deformation and rolling friction

Colt did a Solar Shading study to determine the amount that could actually be generated by the Colt Shadovoltaic louvres. For example on June 21, the worst case scenario would allow us to generate a daily energy generation of ca. 37.6 kWh. This energy is produces in 5 of the 30 elements that are always in the focus of the sun. The other PV elements also generate energy, but this is accounted as an extra back up. The energy is needed to move the shield 120 m along the façade.

4 Photovoltaic Energy Generation and Storage

4.1 PV Cells

- Outside blade: 10 mm “Low iron” + heat soak tested hardened glass
- Interlayer: 2 mm PVB, Photovoltaic Cells,
- Inside blade: 10 mm half hardened float glass
- Louvre adjustment (fixed): to meet best solar radiation
- Torsion tube: aluminium
- Surface treatment: polyester powder coated according to a RAL standard color 27 cells in one louvre.
- PV-modules
- Cell type: Monocrystalline cells, black
- Power output: the layout of the cells is engineered to generate enough energy for the batteries to be able to move the shield autonomously on the façade
- Rotating Screen and Solar movement

4.2 Battery Box

The batteries for storage of the power will be placed inside a container with climate control. This container is integrated in the rotating construction. There will be enough backup power to rotate the system approximately 2 times over the full traveling distance in case there is no sunlight available.

5 What solar shading can offer you

In the past 40 years solar shading design solutions have been developed in an outstanding pace as energy consumption and awareness have been on the rise. A holistic approach harnesses the natural elements to create energy efficient buildings and contribute to a sustainable built environment. Its systems work together to reduce the building’s cooling requirements with solar shading, harness the cooling properties of air and water, and use the building envelope to generate electricity. With operating companies located worldwide, Colt tries to adjust to different environmental situations and requirements with a broad product portfolio. Colt was not only the first to incorporate electricity generating photovoltaic cells into solar shading louvres. Colt tries to make people understand that low energy building fails on its weakest link, so it can provide integrated solutions that cover many aspects of a design, including solutions to enhance the use of natural daylight and natural ventilation.

Figure 5-1 Colt Shadovoltaic modules.

Energy Forms: Balancing Energy & Appearance

Ben van Berkel[1], Tom Minderhoud[1], Astrid Piber[1], Ger Gijzen[1]

1 UNStudio (Member of the Construct-PV research consortium), Stadhouderskade 113, 1070AX, Amsterdam, The Netherlands, t.minderhoud@unstudio.com

In cooperation with the partners from the Construct-PV consortium:
Züblin, D'Appolonia, Fraunhofer-ISE, NTUA, AMS, SUPSI, ENEA, TU Dresden, Meyer Burger and Tegola Canadese under grant agreement no 295981 [www.constructpv.eu]

The perception of Photo-Voltaics (PV) modules has long been dependent on production grade panels mounted on houses and in large solar plants. This, however, is going to change, as the sun and other renewables become more commonly used and start to replace conventional energy sources such as hydrocarbons and other fossil fuels. As this PV technology becomes more widely recognized, it will be one of the main new materials being introduced into the public realm. Not only as a method for producing electricity, but as a building integrated material that can be adapted to the design, local context and characteristics by architects and designers. This objective has continued to be elusive because until recently the focus has been solely on energy production. This is changing and UNStudio is now involved in a EU-sponsored research project under the FP-7 program. This research aims to come to a cost-effective design interpretation of the latest PV-technology, keeping in mind both the actual modules and the manufacturing process in a standardized production line. Dissecting the physical make-up of the existing product, we developed a layered design approach together with the project partners that would allow design freedom and flexibility in the expression of the module. This flexible adaptation in the layering was developed keeping effective energy production in mind. The result of this layered approach is a range of colors, patterns and glass finishes in different and endlessly varying constellations. The latest improvements in this setup are focused on optimizing the energy output in relation to the visibility and effect of the patterns on the exterior of the glazed PV module. This paper provides the latest overview on how these designs can be manipulated and controlled in CAD software and how this is currently being tested to find a balanced and optimized output in electricity yield and visual appearance.

Keywords: building integrated PV (BIPV), PV module, solar architecture, energy forms

1 Introduction

1.1 UNStudio Design

UNStudio is active in a wide variety of design fields undertaking work ranging from infrastructure, masterplans and architecture to the detailed scale of innovative products. The name, UNStudio, stands for United Network Studio, and refers to the collaborative nature of the practice. The main idea behind this approach is derived from the network based organisation within UNStudio where teams of architects and designers are paired with

Engineered Transparency 2016. Glass in Architecture and Structural Engineering. First Edition.
Edited by Jens Schneider, Bernhard Weller.

leading specialists from industry partners. The objective of this approach is to reach integrated solutions with a strong design driven background, while optimally interpreting and utilizing available state of the art technology and materials.

This approach has been developed within UNStudio's architectural practice through a highly flexible methodological approach which incorporates parametric designing and advanced 3D-modeling that allows continuous iterations and improvements on the evolving design. UNStudio first became involved with the topic of Photo-Voltaic Energy when approached for the fundamental research into Building Integrated PV (BIPV) for an FP7 call of the European Commission. The research project - Construct-PV - aims to integrate aesthetic and performance driven designs in a cost sensitive way.

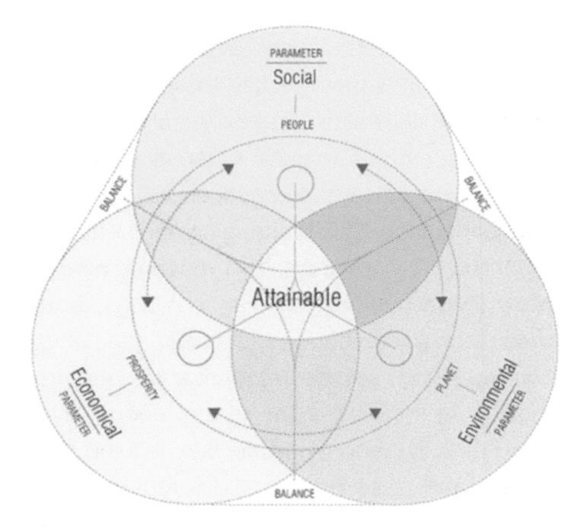

Figure 1-1 UNStudio's approach to sustainability.

1.2 Economic Rational

The strict requirements set by clients, contractors and end-users as far as costs are concerned, were used as a baseline to develop an innovative and cost effective BIPV design concept. Given the competitive nature of the PV-sector in the current economically turbulent times, UNStudio formulated a catalogue approach that allows flexibility in pricing. This strategy makes different levels of complexity possible, thereby providing a link from straightforward mass-production modules to a more bespoke approach with specifically designed and adapted modules. The idea is that, based on the available budget, the product can either be chosen from a pre-defined catalogue or can be more custom designed to various extents, changing a fixed number of variables including glass finishes, colours and cell designs.

The goal is not only to come to a realistic, improved and more aesthetically pleasing appearance of PV in the built environment, but also to provide local manufacturers within the European Market with a competitive edge. Requirements set by the building industry in terms of logistics and quality make it more likely for BIPV to be sourced locally. In addition both the Building Integrated PV functionality and the improved design provide a market premium for manufacturers in an industry where it is a challenge to rise above the standard mass production industry grade PV product.

1.3 Design Catalogue

Since the PV's aesthetic appearance represents the visual materialisation of the building's envelope, it makes sense to choose a product with a high quality design finish. A professional client or developer will insist on a materialisation that reflects the high standards of the entity intending to utilize the building. For this purpose typical industrial modules used for mass electricity generation are inadequate. UNStudio has used state of the art PV technology provided by industry partners. This technology was dissected and reinterpreted into separate layers that could serve as building blocks for an architectural catalogue providing design flexibility for future designs. With the catalogue-based approach, we aim to maintain our ambition to deliver a range of exciting, new interpretations in framing, cell placement, cell and module colours, patterns and material finishes in various combinations.

The layering approach essentially consists of three main design layers that can be manipulated and incorporated into various more elaborate layered combinations resulting in seemingly infinite variations and a look and feel which is closer to bespoke material solutions than the standard PV appearance. This limited collection of layering principles has inherent cost saving advantages that become possible with mass manufacturing and assembly of the separate and flexibly produced layers proposed in the design catalogue.

1.4 Suitability for integration in New Designs

For the Construct-PV research, UNStudio has committed itself to developing new design strategies for Photo-Voltaics. This has resulted in the described layered approach as presented in this paper and in an earlier paper (*2014 PVSEC: 6DO.7.2 Design Innovation from PV-Module to Building Envelope*). Although the aesthetic requirement to improve the appearance of the buildings envelope starts to play a more important role, the production of energy remains the core objective of Photo-Voltaic modules. Therefore, together with a focus on a high quality design outcome, we have striven to keep energy efficiency in mind while working on this concept.

For an architect with a large portfolio of work and new projects being secured on a regular basis, the value of the commitment to this design concept lies in the integration of these innovative PV principles into new designs. In a number of new designs for facades PV functionality has been proposed either directly to the client or to the developer marketing the project. From previous experience UNStudio has learnt that the best way to ensure the inclusion of additional design functionality is to plan and design this holistically and integrate these features early in the architectural process.

The desired outcome of the Construct-PV project is therefore to include both the research technical background together with captivating and inspiring design images to inspire stakeholders and other parties in the building process. This result would enable interested parties to get an initial idea of expected requirements in terms of material and installation. This paper gives an insight in the process to date, the latest results and the forecast until the completion of the project. The text is supported by technical diagrams, conceptual images, design visualizations and photos of real life mock-ups

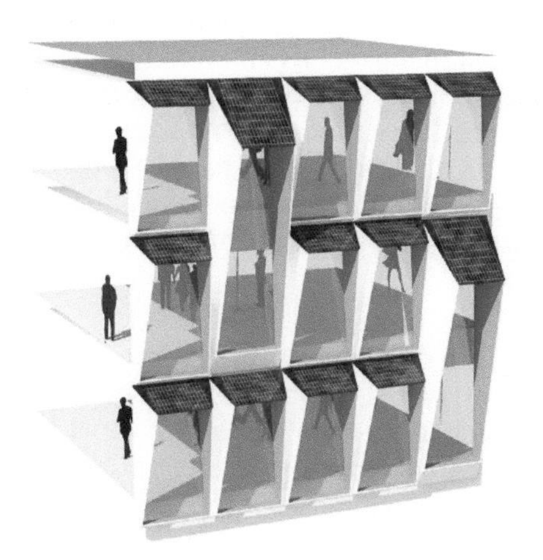

Figure 1-2 Hanwha, Integrated PV in a curtain wall façade with PV geometry and functionality.

2 Project Team

2.1 Design Background and Team

Construct-PV is a European research project in which UNStudio is part of a consortium that includes universities, contractors, consultants, manufacturers and research institutes. These partners are leaders in their respective fields and with their combined knowledge, design parameters – ranging from technical and material properties to methods of design workflow and ways of production – were investigated in order to reach a coherent approach covering the various disciplines.

The following European partners have been involved in this research since 2011:

- Züblin, Stuttgart, Germany
- D'Appolonia, Genova, Italy
- Fraunhofer-ISE, Freiburg, Germany
- NTUA, Athens, Greece
- AMS, Athens, Greece
- SUPSI, Lugano, Switzerland
- ENEA, Naples, Italy
- TU Dresden, Dresden, Germany
- Meyer Burger, Thun, Switzerland
- Tegola Canadese, Vittorio Veneto, Italy

Additionally, from an architectural point of view, the various architectural scale levels were researched to enable design solutions best suited to the technical specifications of the material - ranging from the detailed level of the cell to the larger scale of the building and its surrounding environment.

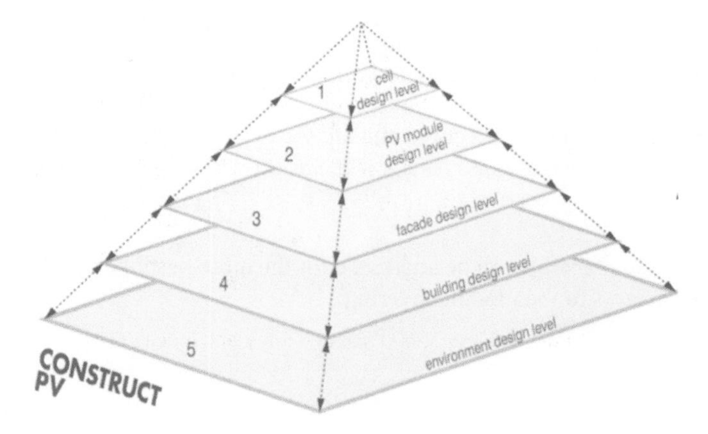

Figure 2-1 Range of architectural scale levels for the Construct-PV design research

2.2 State of the Art Photo-Voltaic modules

The "Multi Wire" PV-module technology from Swiss manufacturer Meyer Burger was chosen as the basis for creating the design layering. This module has particularly good characteristics in terms of pairing high energy yield with the intended additional design layers. In addition it has been designed to reduce the visible impact of the main bus-bars by replacing these with a grid of smaller wiring that blends in more seamlessly with the PV cell behind.

The advantage of the finely branched wiring is that any negative effects of shading created with the additional design layers are largely mitigated. Although some shading is created by the elaborate designs introduced to integrate the PV appearance within the larger design, the network of wiring allows flexibility concerning which area of the cell can be electrically active.

An additional goal was to use the new design insights derived from this product to improve future PV production lines that Meyer Burger will introduce. This equipment is aimed at European as well as other manufacturers and is intended to further establish the local production of customized BIPV products that will make it possible for this specialized form of production to be carried out regionally and close to building sites in Europe and elsewhere.

2.3 Layering Concept

The principles of mass customization and production taken from the building industry were used to come to the aesthetic requirements that should be set for the PV modules. Each production step allows a finite number of design changes and, once properly planned, this allows improvements and cost reductions in the manufacturing process. In the production of glass-glass PV modules each separate manufacturing step needs to be considered to understand which design principle can be applied. This includes steps such as preparing the cells, cell placement and finally the assembly of the modules. Within these standard steps additional layers of architectural finishing can be planned and prepared using pre-fabricated elements such as glass panels with various finishes that are fed into the production line.

The following list of design parameters that can be addressed for the more bespoke panels include the following enhancements to the module layering:

- glass surface finishing
- cell colours
- cell form and placing
- back panel colour
- module framing

The two techniques chosen for glass surface finishing are silk screen printing and sandblasting, creating flexibility in the outer and inner layers of the external glass panel respectively. This design disruption of the basic layering setup of the module requires planning changes but no physical adjustment to the production line. The more bespoke approach would be dependent on the project budget and needs. The balance between aesthetic freedom of choices in design layering and actual costs is a balance between cost and effect.

Experience gained from the Construct-PV research shows that using a coloured back panel, together with a dense cell placement and a subtle pattern on the front sheet of the glass leads to the best results. This would lead to a relatively minor loss in terms of electricity yield from the PV module, while adding to the intended design effect and making use of standard materials in a pre-fabricated alteration. The integrated PV module becomes a strong visual component capable of adding high end material quality to the building's envelope. The separate production steps can be outsourced to specialized suppliers with competitive pricing benefits and layered in the production line of the PV module manufacturer using a standard production line. Finally, the mounting and framing of the modules can be considered either directly in the production facility or by a different, specialized façade subcontractor from the building industry.

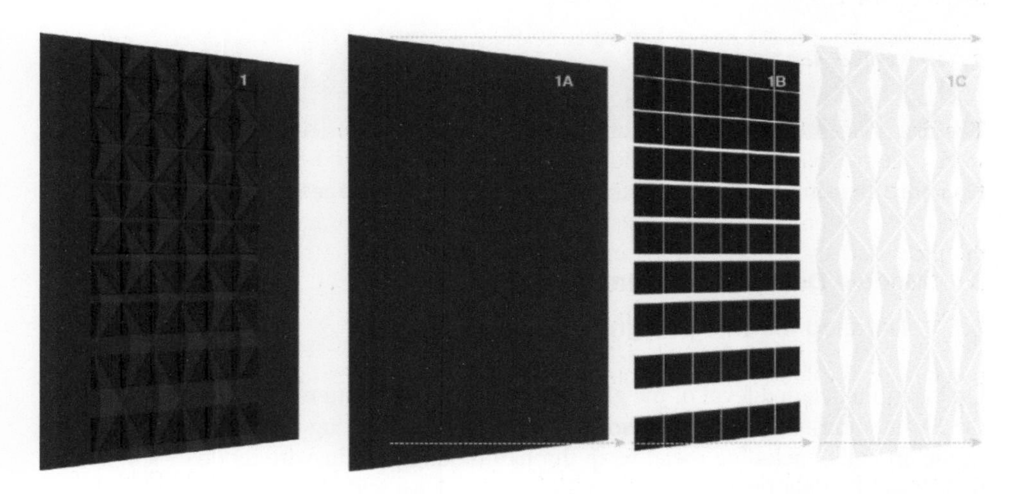

Figure 2-2 The various design layers of the module.

2.4 Glass Finishing

Given that the transparency of the outer layer of glass has a direct impact on the functioning and electricity output of the PV cells behind, UNStudio has applied a design approach whereby the coverage of the pattern is reduced. Using design software to simulate a coverage which can be perceived by the human eye as visible geometries instead of a pattern with actual 100 % coverage, the negative effects of the shading are reduced. The effective coverage using this design software approach is limited to approximately 10-15 % of the glass surface of the module.

The pattern is adapted to the cell strings to provide a similar output throughout the module. In the setup of the pattern, particular attention is given to positioning the coverage of the pattern primarily in the areas where no cells are placed. Together with specialized manufacturers, the production techniques were tested to compare the advantages of finishing methods such as silk screen printing and sand blasting in terms of quality of the finish, operation and maintenance, light transmission to the PV cells and the visual clarity of the external appearance.

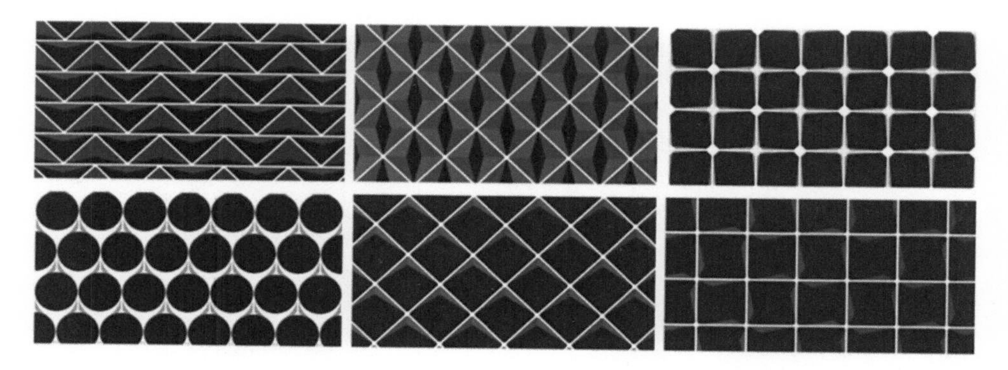

Figure 2-3 Patterns enhancing the rigid orthogonal repetition of the cells.

3 Module Design Development

3.1 Detailed Module design

On the more detailed level of the glass design, a range of pattern variations was tested in the PV-modules. Ranging from modules focused on displaying solely the technical state of the art cells, including metal wrap-through modules, multi-wire modules and the experimental Fraunhofer-ISE mosaic module, to modules with design layering. The design modules intend to showcase a range of example patterns that are adapted to various local contexts.

The starting point of these patterns was the existing Z3 façade, from project partner Züblin. This façade has vertical wooden lamellas in a weaving and undulating geometry. These are reflected in a more detailed and refined angular pattern on the PV module. The scale of this angular pattern reflects both the geometry of the larger shading elements as well as the placements of the PV cells. To the left of the initial Z3 pattern a second angular pattern is introduced with a more playful approach showing various opening sizes which enable different apertures for the cells behind.

Two additional patterns were proposed to showcase a further range of possibilities that adapt to the placement of the cells. The patterns shown are populated in design software and can be parametrically adapted to changes in the underlying cell designs. These four examples shown in Figure 3-1 show how the glass surface can be manipulated by applying a pattern design using various application techniques, in this case using sandblasting.

The advantage of sandblasting is that only the top surface of the glass is altered, resulting in a different refraction of light and eliminating the need for additional opaque surfaces. For larger quantities silk screen printing would be a more viable and cost-effective alternative, providing durability of the design for a prolonged period of time and one that matches the lifetime of a modern façade.

3.2 Module Design as displayed within the Mock-Up

The detailing in the mock-up is engineered to allow seamless integration of the PV module designs within the various possible façade geometries. The PV is placed in the opaque parts of the façade, mainly in the spandrel panel and the window breast area, leaving the glazed window areas transparent and available for a visual connection from inside to outside. The modules are displayed within the following actual facade types.

Figure 3-1 Photos of the design modules in the mock-up on display at the Energy Forum in Bern.

4 Future Projects

4.1 Integration of PV forms in Architectural Projects

The intention is to disseminate the described design principles through conferences and trade fairs such as the Glasstec, educative lectures to students, information sessions to influence policymakers and other stakeholders, as well as directly to clients and developers. The aim is to use these channels of communication to strengthen both the visibility of the Construct-PV project and the design ambitions and principles behind the research and in this way encourage the use of high quality PV. In this way PV design will be promoted and it will be possible for these new forms of PV to become accepted and built in the European Union and in a wider international context.

For UNStudio as an architectural practice this is an important element of the Construct-PV research and forms a viable basis for new projects in a range of design articulations, depending on requirements of future clients and the local context for the intended designs. The innovation within these new forms of PV design is driven by further integrating existing technology into façades through the use of architectural design strategies. This includes the mapping of the surrounding context and local influences, detailed solar radiation analysis and ensuring the optimization of functionality and aesthetics.

For this purpose clients will need to be well informed and encouraged to consider investing additional resources for the functionality of energy generation through PV in the building envelope. These Energy Forms can then be utilized to complement the appearance of this PV layer with the aesthetics and customization that will allow a level of expression and premium design that fits the future ambitions of both architects and their clients.

5 References

[1] B. van Berkel, C. Bos, 2006. UNStudio: Design Models – Architecture, Urbanism, Infrastructure. Thames and Hudson, London T. Herzog, R. Krippner, W. Lang, 2004. Facade construction manual, Birkhauser, Basel.

[2] T. Soderstrom, P. Papet, Y. Yao, J. Ufhell 2014 Smartwire connection technology, Whitepaper SWCT.

[3] G.Z. Brown, M. Dekay, 2001. Sun, Wind and Light – Architectural design strategies. John Wiley & Sons, New York.

[4] B. Weller, C. Hemmerle, S. Jakubetz, S. Unnewehr, 2010. Detail Practice: Photovoltaics – Technology, Design, Construction. Birkhauser, Basel.

[5] European Commission, 2010. Roadmap for moving to a low-carbon economy in 2050, Brussels (http://ec.europa.eu/clima/policies/roadmap/).

[6] A. Watts, 2010. Modern construction envelopes. Springer, Vienna.

[7] S. Roberts, N. Guariento, 2009. Building integrated photovoltaics a handbook. Birkhauser, Basel.

Construct-PV: The research leading to these results received funding from the European Community's Seventh Framework programme (FP7/2007-2013) under grant agreement no 295981 [www.constructpv.eu]

Nursery +e kita in Marburg – solar architecture at its best

Dieter Moor[1]

1 ertex solar, Peter Mitterhofer Straße 4, 3300 Amstetten, dieter.moor@ertex-solar.at

This document contains information about the solar facade and roof at the Nursery +e kita in Marburg Germany. The facade is well known in the architecture business related to the homogenous design. The cells are not visible at a fist view. The article will consider the possibilities in the photovoltaic sector to meet the needs of the architects.

Keywords: photovoltaic, façade, BIPV

1 Available solutions

In the photovoltaic business we have to decide between crystalline technologies and so called thin film technology. There was a big growth in the thin film sector started in 2008 related to the bottleneck in the crystalline technology. Architect loved the design of thin film because of the more or less homogenous design compared to the crystalline cells.

End of 2011 some supplier for thin film disappeared and that one who survived were in economic depression. This was a very difficult situation for Architects who are willing to integrate such new technology. Opus Architekten from Darmstadt Anke Mensing and Andreas Sedler tried to find a very homogenous design compared with high power output.

Which approach you have to undertake to find such a solution, means what is the problem and are the possibilities?

2 Contrast

First of all if you see a so called standard module (Figure 2-1), you see a lot of different materials which are from the color point of view not homogenous.

- Metallic frame:
  It's basically made from anodized aluminum, the cheapest solution for corrosion protection.
- Solar cells:
  They are usually in a color which consume as much light as necessary to have the best efficiency, dark blue, grey, black, the older versions very shine as you can see in Figure 2-1.

Engineered Transparency 2016. Glass in Architecture and Structural Engineering. First Edition.
Edited by Jens Schneider, Bernhard Weller.

- Soldering tapes:
  The tapes are necessary to interconnect one cell to the next. This is the biggest contrast, because the tapes are reflecting the sunlight according to the angle of the sun.
- Backsheet foil or glass:
  The rear side of a standard panel is usually protected with a plastic sheet, its available in different colors, but normally not in the exact color of the cell.

Figure 2-1 This is a typical (old) so called standard module.

3 Solutions

According to the list above, we try to match the materials to reach the best appearance. This is what the Architect expected for the +e kita in Marburg.

3.1 Frame

Basically, you don´t need a frame if you integrated a PV panel in to a building. On one hand it depends on the mounting system on the other hand there are some helpful arguments for a frame.

In this case, we combined the helpful arguments with the mounting system and a nice appearance. Therefore, the frame is black anodized and the glass is glued and sealed together with the frame. You can see in the sketch at Figure 3-1 how it was built.

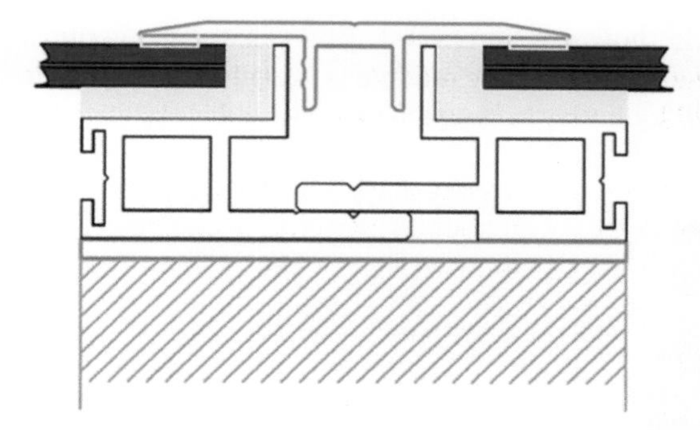

Figure 3-1 Section shows the mounting system.

3.2 Cells

As mentioned above thin film technology was in trouble at that time, so we decided together with the architect to go on efficient mono crystalline cells.

This was important also to cover some of the triangular areas related to the zig zag Design of the façade and the roof. The "pixel" of a monocrystalline cell is 156 x 156 mm that one for thin films is between 700 x 1000 mm. If you cut them they will be electrically damaged and you can use it only as a decorative element, or so called dummy.

3.3 Tapes

There are some solutions on the market how you can avoid this annoying reflection.

- Print on glass at #2:
  You can print at level 2 (inside) of the glass a similar color like the cell. But, this solution is on one hand very expensive and on the other hand not very practicable. You have to match the strings with the printed tapes very exactly otherwise the tapes will maybe look thru and the shine effect is the same as before.
- Colored tapes:
  Some manufactures of tapes produce black, or white colored tapes. This is available for the cell area but not for the interconnections between the individual strings. That means you have to find a second solution either print on glass as mentioned above or some other colors.
- Covered tapes:
  We decided to invest in a special printer with a careful selected ink. This printer could color the tapes in an inline process during soldering. With a second printer we can cover also the string connection in the same color.

A sample of such an unified module is shown in Figure 3-2. You can see the contrast related to the flashlight and a very close distance. If the module is mounted in the façade and you see it from 5 m distance, it will look very homogenous.

Figure 3-2 Sample of a homogenous module design.

3.4 Backsheet

In the standard module business the backsheet is only to protect the PV module from the electrically point of view.

By the way, I always have to note that single glass modules only laminated with one backsheet are not suitable building products. If the tempered glass is broken, it will collapse and only a very special mounting system or glue could hold it in the structure.

Nevertheless, if the color behind the cell comes in a similar color code, the contrast will be reduced again and you will have at the end a more or less homogenous design.

Important in any case, is that this solution is certified by the so called IEC norms to avoid any efficiency losses or reaction between the components.

In Figure 3-3 and 3-4 you can see the result of the development between architecture and industry.

Figure 3-3 South view at the +e kita in Marburg.

Figure 3-4 Side view at the +e kita in Marburg.

Finally I´d like to cite the architecture critic Arne Winkelman who said: "*I was really exited about the taylor made façade and rooftiles, which were discovered only at the second view.*" [1].

4 References

[1] Winkelmann A.; db deutsche bauzeitung 09.2015 Schwerpunkt: Dächer page 52.

Small scale demonstration of Building-Integrated-Photovoltaic (BIPV)

Robert Hecker[1], Tom Minderhoud[2]

1 Ed. Züblin AG, Central Technical Division – Business Unit of Façade Engineering, Albstadtweg 3, 70567 Stuttgart, Germany, robert.hecker@zueblin.de

2 UNStudio, Stadhouderskade 113, 1070AX Amsterdam, The Netherlands, t.minderhoud@unstudio.com

Opaque surfaces on buildings represent massive large-area spaces of untapped harvesting potential for renewable energy production. Photovoltaic modules are due to the lack of moveable parts an emission-free source of renewable electricity which can be integrated in the building skin, roof and façade, like conventional building products. Since 2013, the European Commission funds the cooperative research project "Construct PV – Constructing buildings with customizable size photovoltaic modules integrated in the opaque part of the building skin" coordinated by the central department of façade engineering at Ed. Züblin AG. In this research project partners from the whole value chain of building integrated photovoltaics collaborate like the architect, photovoltaic module producer, building product manufacturer, supplier of the electrical components, the general contractor and different research partners. As tangible output of the project three small-scale and two large-scale demonstration sites with integrated photovoltaic modules were constructed. One of these demonstrators, the life-size small-scale façade demonstrator with dimensions of 6 m x 6 m x 2.5 m (length x height x width) is exhibited in parallel at the stand of TU Dresden at the Glasstec trade fair. In this article the different types of façade integration of photovoltaic modules exhibited in the small-scale façade mock-up are presented. Furthermore the development of this Mock-Up as well as the concept and intention is presented.

Keywords: BIPV, demonstration, photovoltaic, façade, mockup, Construct-PV

1 Introduction and Motivation

The Construct-PV research project is now running since 2013 (February 2012?). The project aims at developing and demonstrating customizable, efficient and architect-friendly cost BIPV (Building Integrated Photovoltaic) modules for façade and roof integration. This aim is in line with the current European energy efficiency legislations, thus it is apparent that renewable sources of energy such as photovoltaic systems are one of the fundamental elements that will enable European Union meeting the three key objectives commonly known as „20-20-20 targets". In this context, such cost-effective and customizable BIPV-modules are needed on European level in order to reach the EU 2020 and 2050 goals concerning the reduction of the CO2-footprint of the building stock by converting large parts of the building skin (namely roof and façade) into energy harvesting surfaces. Indeed, opaque building surfaces have been selected because they represent

Engineered Transparency 2016. Glass in Architecture and Structural Engineering. First Edition.
Edited by Jens Schneider, Bernhard Weller.

massive wide-area spaces of untapped harvesting potential across Europe. Nevertheless, a high market potential can only be achieved when the modules are cost-effective, easy to use from a construction point of view and when they also correspond with the aesthetic requirements of architects, owners and building users.

2 Small scale façade demonstrator

The demonstration activities are the key-activites in this project and will demonstrate the output and results of the research activities in this project. The aim is to demonstrate the steps to integrate different photovoltaic technologies in different façade and roof types. As a typically procedure in construction projects first of all different MockUps where built and installed to verify and ensure the quality of the developed design. ZÜBLIN planned and built up a Façade MockUp, based on a UNStudio design. Such MockUps are typical for big construction projects. They present all stakeholders, as for example the owner, the architects and also the construction companies to get a first impression of the planned building and the planned façade.

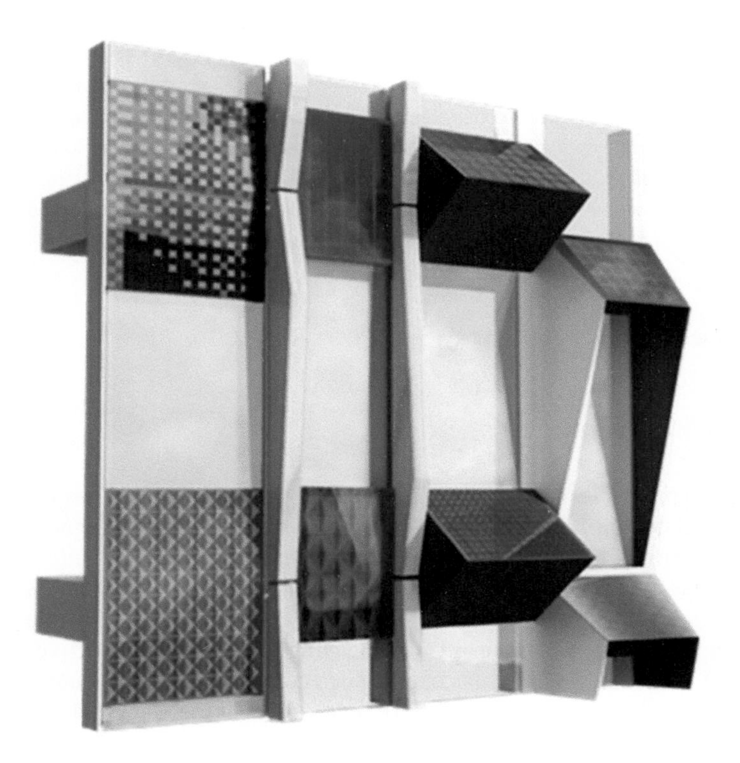

Figure 2-1 First Design of the façade demonstrator by UNStudio.

Figure 2-2 Realized BIPV-Façade-MockUp – Copyright by Construct-PV-Consortium.

Figure 2-3 Realized BIPV-Façade-MockUp – Copyright by Construct-PV-Consortium.

3 Different façade types

The Demonstrator presents four different façade types, which are common in Europe. Based on the selected techniques a number of designs were prepared and adapted to existing design conditions. With the aim of linking these test designs to actual built conditions, four current façade designs were selected with typical geometries to prepare and to adapt to future PV functionality. The design principles formulated within the Construct-PV research where then applied to these different façade types and geometries. The façade mockup setup consists of the following façade designs:

1. Standard curtain wall, with Mosaic PV technology prepared by Fraunhofer-ISE integrated into the opaque spandrel panel.

Figure 3-1 Typical transom-mullion façade on the left hand side, and the correlated integration of PV in the MockUp – Copyright by Construct-PV-Consortium.

2. A retrofit of the Züblin Z3-building façade, which represents a ventilated façade. This office building is used by Züblin in Stuttgart, Germany.

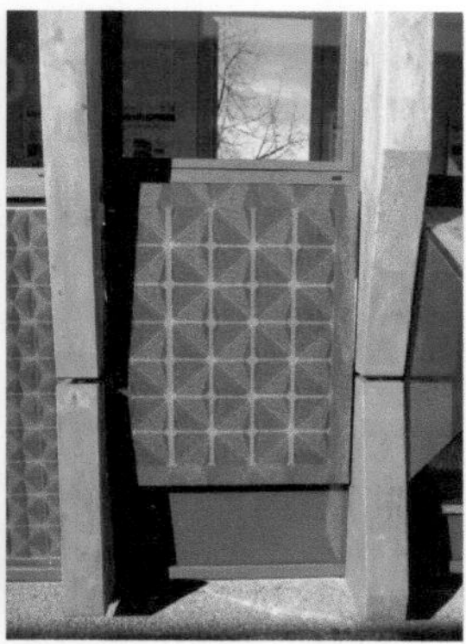

Figure 3-2 Typical ventilated façade with glass cladding on the left hand side, and the correlated integration of PV in the MockUp – Copyright by Construct-PV-Consortium.

3. The Education Executive Agency and Tax Office façade in Groningen is displayed as the third façade. The horizontal shading element of this building from UNStudio was adapted to incorporate PV capability.

Figure 3-3 Façade designed by UNStudio with horizontal fins on the left hand side, and the correlated integration of PV in the MockUp – Copyright by Construct-PV-Consortium.

4. The Hanwha Headquarter façade designed for leading environmental and PV technology manufacturers. This UNStudio design is currently being installed for an office building retrofit in Seoul, South Korea.

Figure 3-4 Design of the Hanwah Headquarters in Seoul with metals frames on the left hand side, and the correlated integration of PV in the MockUp – Copyright by Construct-PV-Consortium.

In addition the Mock-Up is designed to be demountable and mobile, so that it can be exhibited in different locations. This mobile platform allows transportation to trade fairs and exhibitions and will be exhibited at a number of fairs in the coming year. The back structure consists of steel tubes and incorporates an actual office space to give visitors an experience of the façade from the interior perspective, looking outward through the façade.

4 Conclusions

The small scale façade Mock-Up should present the possibility of integrating PV in different façade types. The small scale demonstration activity was a success and will be transferred in a large scale demonstration site.

5 Acknowledgements

Construct-PV: The research and demonstration activities leading to these results received funding from the European Community's Seventh Framework program (FP7/2007-2013) under grant agreement no 295981 [www.constructpv.eu].

6 References

[1] Zentrale Technik (2013): Konzernbroschüre Research, Development Innovation 2012/13, S. 72-75.

[2] Zentrale Technik (2015): Konzernbroschüre Research, Development Innovation 2014/15, S. 120-121.

[3] Construct PV (2014): Publishable Summary.

Laminated BIPV glass: approaches for the integration in the building skin

Erika Saretta[1], *Pierluigi Bonomo*[2], *Francesco Frontini*[3]

1 University of Applied Sciences and Arts of Southern Switzerland (SUPSI), Trevano Campus, Canobbio CH-6952, Switzerland, erika.saretta@supsi.ch

2 University of Applied Sciences and Arts of Southern Switzerland (SUPSI), Trevano Campus, Canobbio CH-6952, Switzerland, pierluigi.bonomo@supsi.ch

3 University of Applied Sciences and Arts of Southern Switzerland (SUPSI), Trevano Campus, Canobbio CH-6952, Switzerland, francesco.frontini@supsi.ch

When integrating PV cells in glass based building elements to form a laminated BIPV glass, the requirements on construction safety and the mechanical behaviour are particularly significant and higher than in conventional PV plants. When installed in facades, laminated BIPV glass can be installed with different mounting structures and fixing configurations that should ensure mechanical safety and adequate operational conditions (e.g. adequate resistance preventing excessive deflections) along with the safety of users indoor or outside the building. However, so far, the BIPV glass elements are mainly tested as conventional PV products in accordance with the IEC 61215:2005. In detail, in order to assess the mechanical ability to withstand wind and/or snow loads, a standard load test is performed with a pass/fail criteria based on the absence of electrical damages rather than on the assessment of the static behaviour and the limit states for safety and/or operation. For example, the deflections are not evaluated in this approach even though they are an important parameter to check, especially where glass is used as primary building skin such as in case of curtain walls. This paper includes the evaluation of the deflections of laminated BIPV glass elements resulting from the IEC mechanical testing, including the comparison with the limit states prescribed in the Eurocode and other recommendations for laminated glass to be used in buildings.

Keywords: laminated BIPV glass, mechanical reliability, deflections, fixing configurations

1 Introduction and motivation

Laminated BIPV glass elements are assemblies made of two glass panes with PV cells inside, jointed together with a transparent encapsulant. Depending on the PV design and arrangements, laminated BIPV glass captures part of the sunlight for producing electrical energy and can let a part of the natural light to pass through. As a consequence, laminated BIPV glass elements seem particularly suited for architectural uses in semi-transparent facades, skylights and balustrades if they are able to replace conventional building elements and to fulfill their specific requirements. Indeed, the recently approved EN 50583-1:2016 “Photovoltaics in buildings” [1] states that the properties of a BIPV module should be evaluated with respect to both relevant building requirements as set in the European

Engineered Transparency 2016. Glass in Architecture and Structural Engineering. First Edition.
Edited by Jens Schneider, Bernhard Weller.

Construction Product Regulation CPR 305/2011, and electro-technical requirements as specified in the IEC standards. In the case of laminated BIPV glass, the relevant building requirements are specified for laminated glass. Specifically, the standard EN 14449:2005 [2] represents a reference for the evaluation of conformity of laminated glass introducing the standard EN ISO 12543-1:1998 [3] for the specific product characteristics. With regard to the design criteria of laminated glass elements in buildings, only draft codes, such as prEN 13474-2:2000 [4] and prEN 16612:2013 [5], and national codes are available. However, a pre-normative document "Guidance for European Structural Design of Glass Components" [6] has been developed as a preliminary document of the Eurocode on design rules for structural glass that is still under development. With regard to relevant electro-technical requirements, the reference standard for laminated BIPV glass made of monocrystalline silicon cells is the IEC 61215:2005 [7]. Such a standard sets to evaluate the electrical reliability, the durability and the mechanical characteristics. Moreover, in the case of laminated BIPV glass, a non-mandatory standard already exists – ISO/DIS 18178:2014 "Glass in building. Laminated solar PV glass" [8], in order to take into account specific requirements such as appearance and dimensions, as well as durability and safety (e.g. impact resistance and electrical insulation).

When laminated BIPV glass is used as façade element, it can be subjected to severe and prolonged wind loads in real operating conditions. Therefore, both resistance and deflections should be verified in order to provide safety and adequate operational conditions. With regard to the deflections, on which the paper is focused, it is important to note that they depend on several aspects, such as the glass thickness, the load value, the duration of the load and the module temperature, as well as the fixing configuration. Nowadays, there are several BIPV building envelopes that are provided with laminated BIPV glass fixed with punctual clamping systems, as arises from the ISAAC-SUPSI BIPV database (www.bipv.ch) that collects more than 100 BIPV examples in Switzerland and abroad. Specifically, this research deals with such punctual clamping system since it is the mounting typology identified for the demonstration facade of the Smart-Flex project [9], within which the framework of this investigation is based. Indeed, the demonstration project will be a double skin façade assembled with large frameless laminated BIPV glass and fixed to the floor slabs on the short edges with two point fixings. However, the research activities have considered, in this first stage, frameless standard size BIPV glass elements with the goal to focus the investigation on the mechanical behavior of the module configuration that represents the widest market of "low-cost" and "not customized" conventional modules.

Hence, this paper describes the deflections of laminated BIPV glass elements subjected to the IEC mechanical test, with the purpose to compare the results with the requirements and the approach prescribed in the Eurocodes for construction elements. The reason of this investigation is to point out the gap – in terms of mechanical performances in integration scenarios – between some qualification procedures developed for conventional PV elements and the performance-based approach of the façade engineering.

2 Investigation on the deflections of laminated BIPV glass

As previously introduced, this research activity is aimed at evaluating the deflections of a laminated BIPV glass resulting from the IEC mechanical test and at comparing such approach with the one of the Eurocodes. As a first step, different experimental mechanical load tests have been performed in order to evaluate deflections in different static conditions. Afterwards, with reference to the performance-based design approach set in the Eurocodes, preliminary results and considerations have been described.

2.1 IEC approach

The standard IEC 61215:2005 represents the reference for evaluating the electro-technical conformity of monocrystalline PV modules. Concerning mechanical safety, it prescribes a mechanical load test aimed at evaluating the ability of the specimen to withstand the snow or the wind load without presenting visual and/or electrical damages. In accordance to the IEC procedure, the specimen shall be loaded in the fixing configuration prescribed by the manufacturer (if multiples, the worst one) with an uniform load of about 2400 Pa in 3 cycles. Each cycle consists of a 1 hour pressure load and 1 hour suction load. At the end, the test is passed if there are not visual damages, intermittent open-circuit faults during the test, a degradation of the output power or a loss in the insulation resistance but there is not a reference to limit-state conditions (resistance, breakage mechanisms, etc.). In detail, deflections do not represent an evaluation criteria, so that theoretically any deflection of the module could be accepted if compliant with the above mentioned criteria. However, deflections of building elements are a significant parameter for ensuring safety, durability and reliability in operative conditions. So, in order to investigate the deflections of a laminated BIPV glass, the mechanical load tests have been performed on several specimens at the SUPSI PV Lab.

The test setting included: the preparation of the sub-structure mechanically anchored to the main load-bearing structure of the tester and the installation of the module by means of aluminum or stainless steel clamps to reproduce the real static configuration, the positioning of displacement sensors in representative points and the positioning of an electroluminescence (EL) camera below the laminated BIPV glass in a dark environment created with black tents to monitor the PV cells status during the mechanical load.

The specimens are frameless laminated BIPV glass and they measure 170 cm in length, 100 cm in width and 0.54 cm in thickness (2+2 mm of glazed panes). Each specimen has 60 cells within two PVB encapsulant foils that bond together the two glass panes. All the specimens have been mounted in the mechanical load device with a different punctual clamping system. In some configurations the interaction with the frame of the sub-structure in relevant as reported in comments. Then, the 3 load cycles have been performed and the deflections have been recorded. Table 2-1 shows the test conditions, the resulting maximum displacement for the points considered and the emerging aspects in the comments.

Table 2-1 Mechanical tests conditions and main results for different mounting/fixing configurations.

Configuration	Max Deflections	Comments
2 clamps on the long side, orthogonal bars 	Sensor1: 2.36 cm; Sensor2: 0.87 cm; Sensor3: 2.31 cm; Sensor4: 3.11 cm; Sensor5: 1.40 cm; Sensor6: 0.93 cm	The test passed even though a clamp had a significant plastic permanent deformation. The deflections are wider in the suction load than the pressure load due to the interaction of the module with the underlying mounting bars.
2 clamps on the short side, parallel bars 	Sensor1: 4.34 cm; Sensor2: 1.20 cm; Sensor3: 4.43 cm; Sensor4: 4.87 cm; Sensor5: 1.31 cm; Sensor6: 1.52 cm	The test failed because the module fell out from the clamps due to an excessive deflection. The module fell out at the second suction load.
2 clamps on the long side, parallel bars 	Sensor1: 2.47 cm; Sensor2: 0.80 cm; Sensor3: 2.44 cm; Sensor4: 2.87 cm; Sensor5: 2.22 cm; Sensor6: 0.74 cm	The test passed but during the 1 hour cycles the initial deflections continued to increase. Indeed, in such a configuration there is not the contribution of the mounting bars. However, if the time load is longer than 1 hour the module could fall out from the clamps.
2 clamps on the long side, parallel bars 	Sensor1: 4.33 cm; Sensor2: 0.53 cm; Sensor3: 4.24 cm; Sensor4: 4.78 cm; Sensor5: 2.50 cm; Sensor6: 0.67 cm	The test passed but during the 1 hour loads the deflections increase. Indeed, in such a configuration there is not the contribution of the mounting bars. However, if the time load is longer than 1 hour the module could fall out from the clamps.
3 clamps on the long side, orthogonal bars 	Sensor1: 0.99 cm; Sensor2: 1.42 cm; Sensor3: 1.01 cm; Sensor4: 2.53 cm; Sensor5: 1.42 cm; Sensor6: 1.48 cm	The test passed. The deflections are wider in the suction load than the pressure load because of the absence of the underlying mounting bars.

From such preliminary results, the following critical issues arise:

- the deflections are limited in the pressure load by the underlying mounting bars, indeed they are wider in the suction load (e.g. the depression load of the wind);
- in most of the observed tests the deflections are not stabilized after 1 hour of suction load, but they slowly increase. This could be related to a progressive weakening of the clamps potentially involving a fall of the laminated BIPV glass out from the clamps in case of load with long duration;
- from the EL measures, it arises that there are not electrical damages of the PV cells;
- the limit state has been reached due to the clamp deformation and rotation, involving higher deflections and safety issues. This phenomena pointed out that the fixing/mounting system is an essential part of the resistance chain. Accordingly, using different clamps for glass made of stainless steel (punctual and not-deformable connections) in further experimental tests on type „A" and type „B", the mechanical behavior showed the module breakage (fig. 2-1). Moreover, the position of the junction box influences the module breakage mechanism because, depending on its dimensions and position, it can stiffen a part of the laminated BIPV glass.

Figure 2-1 Laminated BIPV glass fixed with punctual stainless steel clamps. The breakage is due to the more punctual and not-deformable connections than the traditional clamps.

2.2 Eurocode approach

In the framework of this investigation, a particular focus on the case-study of Smart-Flex project motivated the choice of the configurations used in the experimental tests. In particular, since a double skin façade was the design solution of the demonstrative building, also a consideration on the design approach linked to this façade typology has been included. Since double skin façades are considered a typology of curtain walls, they shall be designed in accordance with the EN 13830:2003 [10]. As a consequence also laminated BIPV glass – when used as the outer glazing element of a double skin façade – should be designed in accordance with the design principles set in the Eurocode EN 1990 and the actions determined in accordance with the Eurocode EN 1991.

In detail, the design principles for laminated glass are specified in the prEN 16612:2013 that sets to verify the resistance of the glass at the ultimate limit state (ULS) as well as the deflections at the serviceability limit state (SLS). Such draft version represents an evolution of the prEN 13474-2 that specifies the design methods for particular laminated glass solutions, depending on their shapes and fixing configurations. However, analytical solutions are available only for rectangular or triangular glass with linear supports, whereas rectangular glass with punctual clamping systems are not envisioned. As a result, numerical solutions can be used in order to determine the maximum deflections and the maximum tensile stress. Thus, such values can be compared and verified with the specific limit values set in national codes and standards in order to design safety glass elements.

In accordance to the Eurocode approach for the calculation of actions, the laminated BIPV glass is supposed to be subjected only to the wind load due to its use as vertical outer glazing element of a double skin façade. Therefore, the calculation of the wind pressure load that affects a building shall be determined in accordance with the Eurocode 1 „Actions on structures – General actions – Part 1-4: Wind actions" [11]. However, the specific coefficients for the determination of the wind pressure load are developed at the national level.

Only for the purpose of next considerations, the laminated BIPV glass is supposed to be installed as glass elements of a double skin façade of a 20 m tall building located in Lugano, Switzerland. Hence, the wind load is calculated in accordance with the Swiss standard SIA 261:2014 „Actions on Structures" [11], thanks to the following equation:

$$q_{ek} = q_{p0} \cdot C_h \cdot C_p \tag{2.1}$$

where „q_{ek}" represents the wind pressure. In accordance with the data of Table 2-2, the worst wind load is a suction load in the amount of 0.891 kN/m^2. It is clear that the design load value depends on the specific aspects as well as on safety coefficients and load combinations. In conclusion, when PV modules (qualified according to IEC) are integrated in building, it is important to assess the coherence of the mechanical requirements and performances (of modules, clamps and the whole system) to establish if the use in building is compliant with the relevant codes.

Table 2-2 Wind pressure coefficients in accordance with SIA 261:2014.

Coefficient	Input	Value	Reference
Geographical location, q_{P0}	Lugano	0.9 kN/m^3	SIA 261:2014 – Annex E
Building height and terrain category, C_h	Building height = 20 m and large-scale urban areas (category IV)	0.9	SIA 261:2014 – Table 4
Building form, C_p	Building form n. 38	Front facade: -1.1; Side facades: -0.95, Pressure on the side façade: +0.85	SIA 261:2014 – Table 38

2.3 Preliminary results and considerations

Figure 2-2 shows an overview of the deflections resulting from the IEC mechanical tests and reports the limit values defined by the Italian CNR-DT-210:2012. In detail, such Italian code defines the maximum deflections at the SLS for monolithic glass (or equivalent) when installed by means of specific punctual fixing systems. From such a comparison, it arises that the deflections are wider than the limit values for each fixing configuration.

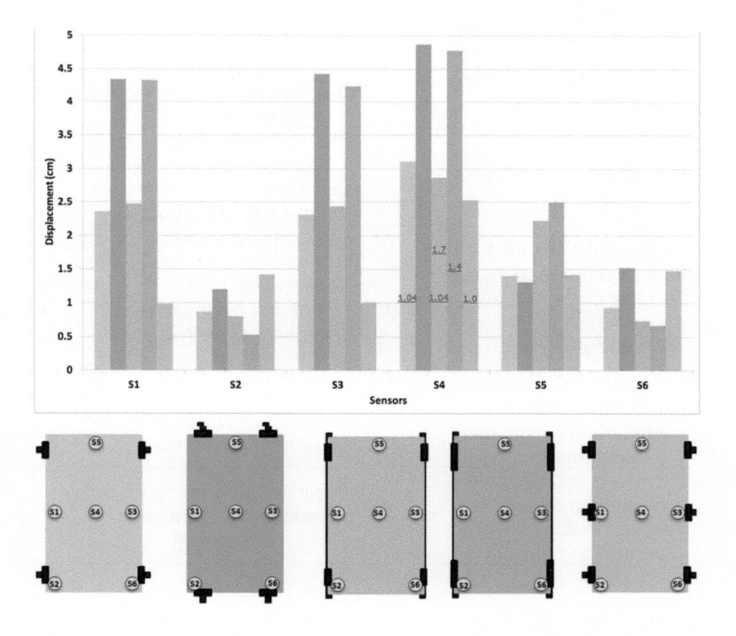

Figure 2-2 Deflections resulting from the IEC mechanical tests performed at ISAAC-SUPSI labs and comparison with the limit values (red lines) defined in the Italian CNR-DT-210:2012.

However, it is important to note that the load used in the IEC mechanical load test represents a standard load that is applied independently from the location where the module will be installed only for 1 hour, and it is not calculated in accordance with the local-based approach of the Eurocode. For instance, the load applied to the laminated BIPV glass in the mechanical load tester corresponds to 2400 Pa that is lower than the "calculated" load in accordance with SIA 261, taking also into account the safety coefficients. Indeed, further mechanical tests to be performed at SLS conditions could be eventually envisioned to take into account some aspects that can significantly represent the mechanical capacity of the laminated BIPV glass according to relevant building codes. Moreover, the presence of the PV cells and the consequent overheating due the electricity production can involve a loss of the shear modulus of the PVB-encapsulant, increasing the deflections so that further investigations on the temperature of laminated BIPV glass should be included in technical procedures.

Therefore, it arises that the IEC approach is addressed to assess the product conformity in order to design a „low cost" and „non-customized" PV system to be applied on buildings, whereas the Eurocode approach has significant differences (as shown in Table 2-3) which can lead to a better design of laminated BIPV glass elements.

Table 2-3 Main results of the comparison between the IEC approach and the approach defined in the standards for laminated glass to be used in building.

	IEC 61215:2005	**Standards for laminated glass to be used in building**
Approach	Test with pass/fail criteria to verify electrical/thermal requirements	They set design principles in accordance with the performance-based design approach of Eurocodes
Load	Independent from the location. The wind load (correspondent 800 Pa) has been increased with a safety factor of 3 for gusty winds.	Calculated in accordance to Eurocode EN 1991, taking into account the location, the building height and form, the terrain category, as well as safety coefficients when required
Shapes/Fixing configurations	The configuration prescribed by the manufacturer shall be evaluated (if multiples, the worst one)	Numerical solutions are required for specific shapes and fixing configurations
Deflections	Deflections are not considered as an evaluating criteria and they can be influenced by the mounting bars	Deflections are considered time-dependent and temperature dependent

In addition to this, from the experimental tests it arises that there are different ways of collapse of the laminated BIPV glass installed with point fixings in facades, that could be schematically summarized through the following reference limit states of capacity for static load conditions:

- Ultimate limit states for mechanical/electrical safety: glass breakage (fragile), clamp/fixing breakage (fragile/plastic), sub-structure limit capacity (fragile/plastic), frame breakage (fragile/plastic), module fallen due to expulsion from clamp (e.g. excessive deformation), electrical wiring breakage for excessive deformation/module collapse, junction box breakage;
- Serviceability limit state: power loss (e.g. for cell's breakage due to excessive deformation), limit of deformation/tension (e.g. for a proper interaction with sub-structures or for avoiding further mechanical consequences).

Such ways of collapse are just a first setting of possible limit states for BIPV elements, but they highlight the need for a performance-based approach that should be „adopted" during the whole process: ranging from the design phase to the qualification phase.

3 Conclusions

As previously introduced, when laminated BIPV glasses are installed as glazing elements of building facades, it should guarantee both the building and the electro-technical requirements. With regard to the safety requirement, the reference standard IEC 61215:2005 prescribes a mechanical load test based on prescriptive approach that does not take into account some important aspects for safety and serviceability in buildings such as the deflections of laminated BIPV glass as an evaluating criteria. On the other hand, the relevant standards for laminated glass to be used in buildings (Eurocodes, standards for curtain walls, standards for laminated glass) define the principles and the evaluation methods for the design of a laminated glass according to a performance-based design approach. Nowadays, some main industries producing laminated BIPV glass for curtain wall and/or ventilated façade already design and provide special tests and procedures for developing and qualifying BIPV glass according to CPR 305/2011 and for CE marking. However, there is still a difficulty in recognizing the „added value" of BIPV products with a high quality (also by the normative entities), which risk to be not competitive with other solutions proposed for building integration without a specific qualification for construction (e.g. due to possible higher costs linked to specific tests or design/production/installation procedures). Starting from the new „integrated" approach introduced by EN50583, an effort of harmonization and definition of performance reference and procedures for BIPV products is today necessary and should be referred to all the design and building process. This not only calls into question the reference to building codes but also, in some cases, the definition of new procedures specifically developed for laminated BIPV glass and their peculiarities.

4 Acknowledgements

This research has been developed in the framework of the project ACTIVE INTERFACES – Holistic strategy to simplify standards, assessments and certifications for building integrated photovoltaics (#153849) founded by the Swiss National Science Foundation SNF. Moreover, the authors would like to thank the module manufacturer ViaSolis for providing the specimens within the Smart-Flex project funded from the European Union's 7th FP (Grant Agreement n. ENER/FP7/322449).

5 References

[1] EN 50583-1:2016, Photovoltaics in buildings – Part 1: BIPV modules.

[2] EN 14449:2005, Glass in building – Laminated glass and laminated safety glass – Evaluation of conformity/product standard.

[3] EN ISO 12543-1:1998, Glass in building – Laminated glass and laminated safety glass – Part 1: Definitions and description of component parts.

[4] prEN 13474-2:2000, Glass in building – Design of glass panes – Part 2: Design for uniformly distributed load.

[5] prEN 16612:2013, Glass in building – Determination of the load resistance of glass panes by calculation and testing.

[6] Feldmann, M., Kasper, R.: Guidance for European Structural Design of Glass Components. Scientific and Policy Report by the Joint Research Centre of the European Commission, 2014.

[7] IEC 61215:2005, Crystalline silicon terrestrial photovoltaic (PV) modules – Design qualification and type approval.

[8] ISO/DIS 18178:2014, Glass in building – Laminated solar PV glass.

[9] SmartFlex European Project. http://www.smartflex-solarfacades.eu/, 2013.

[10] EN 13830:2003, Curtain walling – Product standard.

[11] EN 1991-1-4:2005, Eurocode 1: Actions on structures – Part 1-4: General actions – Wind actions.

[12] SIA 261:2014, Actions on Structures.

Solar energy gains and thermal loads at large scale transparent building envelopes in the presence of indoor solar ray tracing

Nina Penkova[1], Kalin Krumov[1], Ivan Kassabov[1], Liliana Zashkova[1]

1 University of Chemical Technology and Metallurgy, blvd. "Sv. Kliment Ohridski" 8, Sofia, Bulgaria, nina_ir@mail.bg

The transparent facade elements on the external walls with different spatial orientation may occur simultaneously on the path of the sun rays at certain times of day. The sunlight enters the room through the first transparent layers, crosses the non-opaque indoor environment and strikes over the next transparent elements where parts of it are transferred, absorbed and reflected back. That indoor solar ray tracing is not taken into account at the assessment of the solar energy gains and thermal loads at transparent building elements with relatively small areas and surface to surface view factors. But it may influence on the temperature fields at large scale curved and flat glass panes, close to see each other. The ignoring of the internal solar ray tracing and surface to surface radiation at such geometrical configurations leads to inaccuracy in the prediction of the solar gains and thermal comfort in the buildings. The transient solar ray tracing in the indoor environment and its impact on the energy gains and thermal loads at large scale glass flat and bent facades are discussed on the base of numerical experiments and analysis.

Keywords: solar energy gain, bent glass, large scale facades, solar ray tracing

1 Introduction

Since transparency is an important element of the architecture, the modern large-scale buildings often are equipped with a high proportion of areas with transparent flat or bent façades. This can lead to major problems from an energetic and comfort point of view, especially in high-rise buildings, where conventional external overheating protection cannot be used due to the wind loads. On the one hand, high solar radiation during summer can result in overheating of the building without a proper shading device. On the other hand, the large advantages of a transparent façade in winter are the solar gains – heating demands can be compensated and the energy balance of the building can be improved by allowing solar radiation to enter the building.

As the sunlight is transferred into the room it is absorbed and reflected in different directions by the opaque indoor environment, depending on the orientations and the roughness of the sunlit surfaces. If there are transparent elements on the room walls with different spatial orientation, penetrating into the building solar energy can strike on them where it is reflected back, absorbed or transmitted to the outdoor environment.

Engineered Transparency 2016. Glass in Architecture and Structural Engineering. First Edition.
Edited by Jens Schneider, Bernhard Weller.

That solar energy distribution is named "*indoor solar ray tracing*" in the present paper.

The absorbed by the interior energy is emitted in the infrared spectrum due to relatively small temperature of the indoor environment (usually accepted as equal to the room temperature). In that spectrum the glass is accepted as nearly opaque material – it omits to the outdoor environment only the energy in the solar spectrum. The fraction of that "loosed" energy is ignored in the energy balance of buildings with relatively small areas of the windows in comparison with the opaque solid walls.

The solar distribution can be different at large scale flat and bend transparent facades, seeing each other (Figure 1-1). Possible impacts of that process on the solar energy gains, thermal loads on the transparent glass elements and thermal comfort in the indoor space, are discussed below.

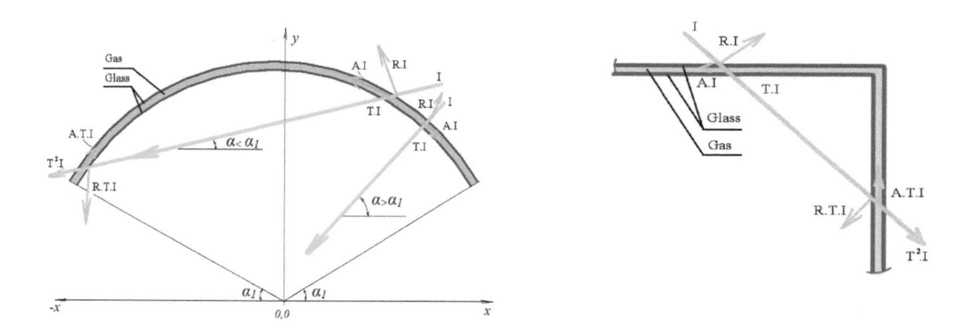

Figure 1-1 Solar ray tracing at cylindrically bent (left) and flat (right) double glazed facades.

2 Indoor solar ray tracing at flat large scale transparent facades

An example of one stage of a building, situated in Sofia with sizes $a = b = 8$ m, height $h = 4$ m, nearly 100 % transparent vertical façade and a lack of interior walls is considered in order to demonstrate the processes, object of the paper. The way of the solar beams is visualized on Figure 2-1 till 2-6 at the middle of December and July (the mounts with the maximal and the minimal solar energy in the chosen region). The solar altitude, elevation angles and other sun position parameters at clear sky are computed and systematized in Table 2-1 according well known methods (Ashrae [1]).

Table 2-1 Sun position parameters at Sofia (longitude 23.33 and latitude 42.69).

	December			July		
Local time, h	9.00	13.00	16.00	9.00	13.00	16.00
Local solar time, h	8.50	12.50	15.50	7.46	11.46	14.46
Hour angle, °	-52.47	7.53	52.53	-68.06	-8.06	36.94
Sun elevation angle, °	8.43	23.89	8.40	30.31	67.83	52.73
Sun azimuth angle, ° (South = 0°; East = - 90°; West=90°)	-47.52	7.58	47.57	-87.80	-20.21	67.27
Sun inclination angle at south oriented vertical surface, °	47.99	24.98	48.21	91.88	69.21	76.50
Solar beam intensity I, Wm^{-2}	366.38	700.94	365.05	767.70	929.24	892.23
Diffuse solar irradiation on vertical surface I_{dR}, Wm^{-2}	31.26	76.19	31.15	90.51	178.42	146.63
Direct solar irradiation on south oriented vertical surface Id, Wm^{-2}	244.91	635.56	243.82	26.24	330.22	209.74
Global solar irradiation on south oriented vertical surface Is Wm^{-2}	276.17	711.75	274.97	116.75	508.63	356.36

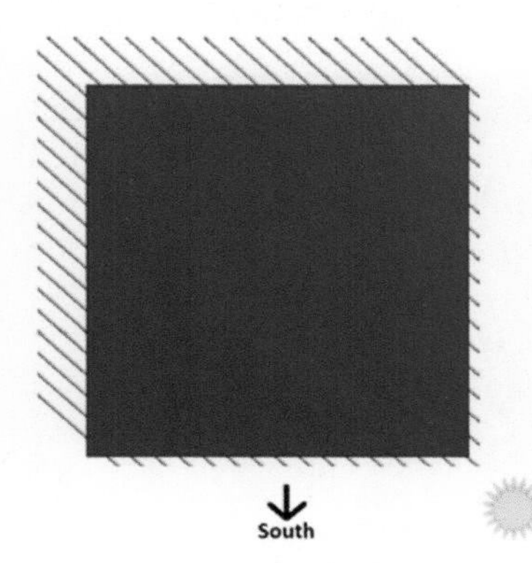

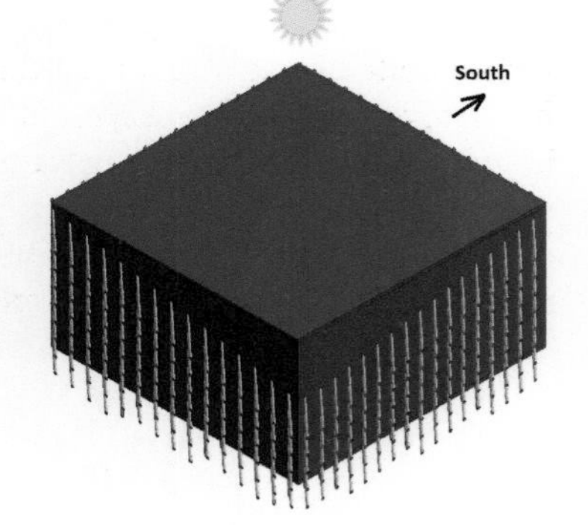

Figure 2-1 Solar beams at 9.00 h on December 15.

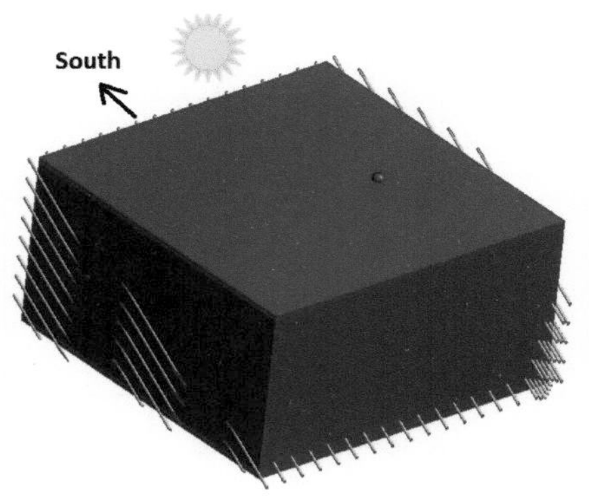

Figure 2-2 Solar beams at 13.00 h on December 15.

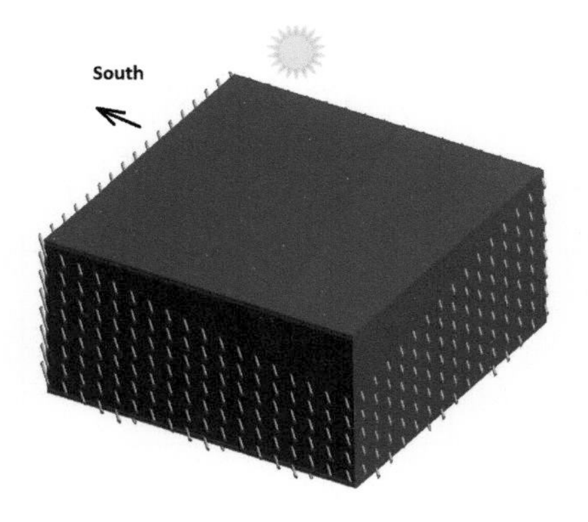

Figure 2-3 Solar beams at 16.00 h on December 15.

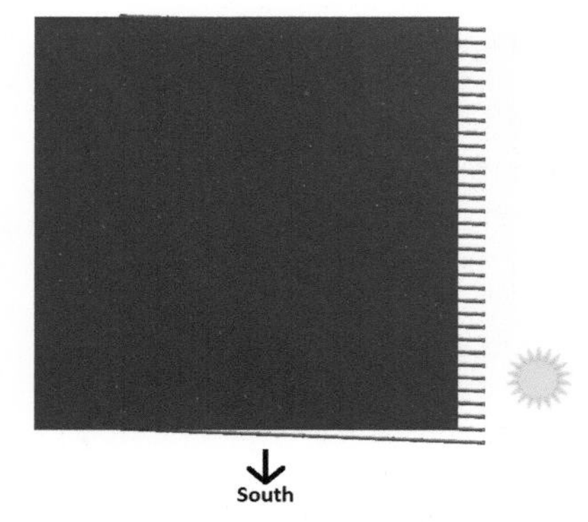

Figure 2-4 Solar beams at 9.00 h on July 15.

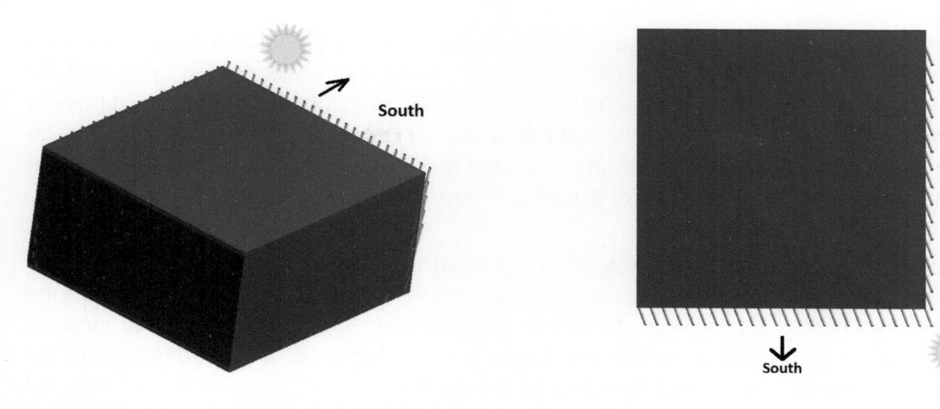

Figure 2-5 Solar beams at 13.00 h on July 15.

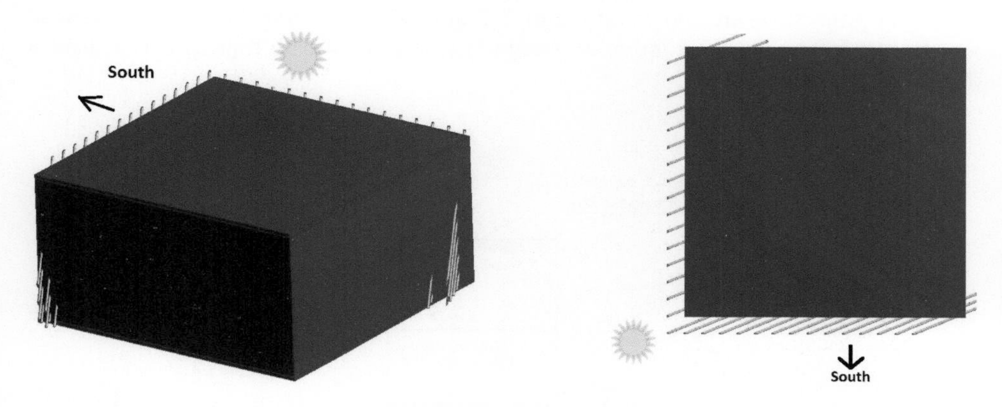

Figure 2-6 Solar beams at 16.00 h on July 15.

The diffuse radiation is accepted as equal for all verticals facades regardless of their orientation. The direct solar irradiation (beams) is visualized on the figures above. It is seen that parts of the direct solar energy, entering thought a façade can leave the room at the neighbor perpendicular and opposite parallel transparent facades at some positions of the solar beams (Figures 2-2 till 2-4). The sun rays can cross the perpendicular façade only (Figures 2-5 till 2-6) if

$$b > h\frac{\cos|\gamma - \varphi|}{tg\beta} \tag{2.1}$$

where b = distance between the parallel walls; β = solar altitude angle; φ = solar azimuth angle; γ = surface azimuth angle, h = distance between the upper edge of the transparent element and the floor. In the discussed example h is the height of the room.

Condition (2.1) is met at all illustrated cases at July in Sofia. Then small parts of the solar energy, entered in the example of building will leave it at areas near the edges of the unlighted facades. That effect is negligible at 13.00 h on July 15 due to the high solar elevation angle. As smaller is the right term in (2.1) as smaller is the part of the solar energy, leaving the room. Condition (2.1) is not met in the examined moments at December – there are higher possibilities for solar energy losses at those cases.

The reflected by the internal glass surface parts of the solar beams are not shown on the figures above, but that reflection is according the laws of the optics and the reflected solar energy is directed primarily at the indoor environment. It is also important to mention here that the transmittance of the glass units is much higher than the reflectance of the internal glass surface so the proportion of the reflected to the indoor energy is expected to be smaller in comparison with the transmitted one.

Different projections are shown on Figure 2-7 in order to illustrate better the process of entering and out coming of the solar energy in empty room with transparent facades at fixed morning time.

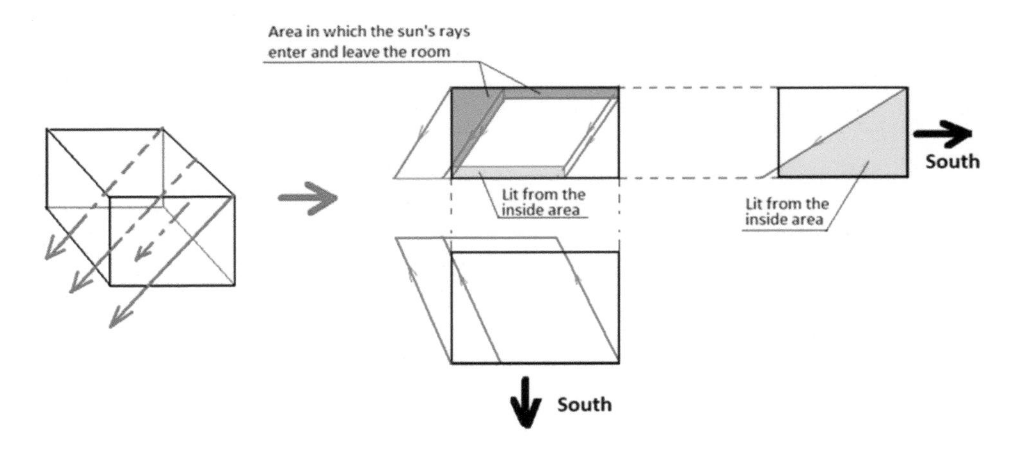

Figure 2-7 Solar beams distribution on the vertical surfaces.

It is obvious that parts of the direct solar energy, entered through the trapezoidal and rectangular areas on the south facade can leave the room via the west and north ones at the chosen sun position parameters. The areas in which the sun beams enter and leave the room are changed with the time of the year and the day.

All that processes are happening if the solar beam doesn't meet opaque interior surfaces at their way. At the real situations there are possibilities for shadowing of the solar energy by the interior.

3 Indoor solar ray tracing at bent transparent facades

In principle the solar load on bent external surfaces is not equal due to the variation of the solar incidence angle with the radius of curvature. That variation leads to subsequent change of the angle of light refraction and the optical properties of the glass. The solar flux, the absorbed solar energy and the subsequent glass temperature are higher at the places on the bent surface with the smaller incidence angle.

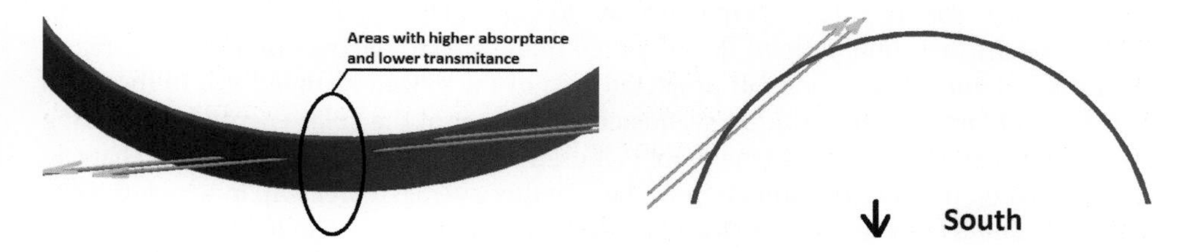

Figure 3-1 Solar beams position according a transparent vertical cylindrically bent façade at 16.00 h on December 15.

A thermal load analysis of cylindrically bent insulating glass unit, reflecting the mentioned variations at a fixed position of the solar beams, crossing ones the transparent layers is presented (Penkova et al., 2014 [3]). Such analysis, taking into account the indoor solar ray tracing at bent facades is not reported to now.

A visualization of solar rays, crossing twice a cylindrically bent façade is presented on Figure 3-1. The illustrated situation is highly probable due to difficulty to fill the indoor space near the curved transparent facades. That results in loses of solar energy on the one hand and uneven temperature field of the transparent layers on the other hand. At vertical stripes, where the solar beams are nearly tangential on the curved surface, the sun incidence angle and glass absorptance are higher, and the transmittance and heat flux are lower. Those stripes are “moving” with the daily variation of the sun position.

4 Possible results of the indoor solar ray tracing

4.1 Overestimation of the solar energy games

The method for estimation of solar energy gains through transparent façade elements in the international standard ISO 13790 [1] is based on the assumptions that solar rays cross ones the transparent building envelopes. The method is valid for flat transparent elements and probably is applicable for bent facades with relatively high radius of curvature. The heat flow entering the room at transparent facade elements is computed by:

$$\Phi_s = F_{sh,ob} A_s I_s, \ W \tag{4.1}$$

where $F_{sh,ob}$ = shading reduction factor for external obstacles for the solar effective collecting area of the surface; A_s = effective collecting area of surface with given orientation and tilt angle, m^2; I_s = solar irradiance, W/m^2. The effective solar collecting area of a glazed envelope element (e.g. a window) is:

$$A_s = F_{sh,g} F_w g_{g,n} (1 - F_F) A_{w,p} \,,\; m^2 \tag{4.2}$$

where $F_{sh,g}$ = shading reduction factor for movable shading provisions; $g_{g,n}$ = solar energy transmittance for radiation perpendicular to the glazing; F_w = correction factor; F_F = frame area fraction, ratio of the projected frame area to the overall projected area of the glazed element; $A_{w,p}$ = overall projected area of the glazed element, m^2. In the cases of large scale transparent facades, seen each other, part of the solar energy, transferring directly through the area $A_{w,p}$ in the room is loosed at the next transparent layers. That can be reflected on the base of estimation of the monthly averaged areas $\bar{A}_l$, in which the sun rays enter and leave the rooms (given in orange on Figure 2-7). The loosed monthly averaged solar energy flux can be determined by:

$$\Phi_{s,l} = F_{sh,ob} F_{sh,g} F_w^{\,2} (1 - F_f) \overline{A_l} I_s \tau_{1g,n} \tau_{2g,n},\; W \tag{4.3}$$

where $\tau_{1g,n}$ and τ_{2gn} are the total transmittance of the first and the second transparent system on the way of the solar beams. Additional coefficient, reflecting the interior shadowing can be added in (4.3).

4.2 Non uniform thermal loads on the transparent façade elements and radiation asymmetry in the rooms

The temperature differences between the transparent and non-transparent internal surfaces, surrounding a person, lead to radiation asymmetry and can cause corresponding local discomfort. ASHRAE Standard 55 sets limits on the allowable temperature differences between various surfaces: the walls may be up to 23 K warmer than the other surfaces. That limit can be exceeded in the hot seasons - the glass temperature is higher than the temperature of non-transparent surfaces due to the lower thermal mass of the single or multilayer glass system in comparison with the massive walls. The presence of coatings can increase additionally the glass temperature.

At the case of sunlit parts of the internal surfaces (Figure 2-7) parts of solar radiation on those areas are absorbed and the glass temperature, and view factors of warmer internal surfaces, participating in the indoor surface to surface radiation are increased. That process exists near the edges even at relatively high sun elevation angles (Figure 2-5).

The solar energy, reflected to the indoor environment by the last transparent system on its way, can also contributes to the thermal discomfort.

5 Conclusions

The indoor solar ray tracing at large scale flat and bent facades, seeing each other may influence on the energy balance, thermal comfort of the buildings and cause additional thermal loads on the glass elements. The estimation of that effect can be implemented by mathematical modeling and numerical simulation of the heat transfer at the glazed systems, taking into account the transient sun position, proportion, construction and optical properties of transparent elements, facades orientation and geometry.

Acknowledgments: The investigations are implemented with the financial support of project DFNI E 02/17 ”Parametric analysis and estimation of the efficiency of transparent structures in solar energy utilization systems”, funded by Bulgarian Ministry of Education.

6 References

[1] ASHRAE, Handbook of fundamentals, American Society of Heating, refrigeration and Air conditioning Engineers, Atlanta, W.S.A, 1981.

[2] EN ISO 13790:2008, Energy performance of buildings - Calculation of energy use for space heating and cooling, 2008.

[3] Penkova, N., Iliev, V., Zashcova, L., Neugebauer, J., Thermal load analysis of cylindrically bent insulating glass units, Challenging glass & COST Action TU0905 Final Conference, Lausanne, Proceedings, 2014, 123-132.

Photovoltaic glass in building skin. A tool for customized BIPV in a BIM-based process

Francesco Frontini[1], Pierluigi Bonomo[1], Robert Hecker[2]

1 SUPSI, University of Applied Sciences and Arts of Southern Switzerland, Trevano Campus, Canobbio CH-6952, Switzerland

2 Ed. Züblin AG, Central Technical Division – Business Unit of Facade Engineering, Albstadtweg 3, 70567 Stuttgart, Germany

Building Integrated Photovoltaics (BIPV) means today the possibility for the building skin to produce renewable energy in a safe, reliable and affordable way. Façade engineering and planning plays a fundamental role in this challenge and BIPV, along with other innovative technologies and solutions, opens a concrete opportunity to address the target of nearly-Zero or Plus-Energy Building. Most of the BIPV applications in façades concern glazed systems where PV cells are included within a laminated glass (e.g. in curtain walls or structural façade). Integrating photovoltaic in surfaces of the building envelope, involves a strong integration of energy, electrical, architectural and construction requirements during the whole process, from early-design phase till to manufacturing and operation. A collaborative and integrated planning approach by architects, engineers and manufacturers becomes essential, so that the development of methods, models and tools oriented to optimally support an integrated planning and construction process (i.e. the BIM approach) is a crucial aspect of growing interest. In the framework of the European project Construct PV, the authors collaborated in developing a web-tool aimed to support, since the early design phase, the integrated design of a customizable BIPV component for the integration into the building skin, within an interoperable process based on Building Information Modelling (BIM). The tool allows the customization of both physical (light transmittance, power,..) and constructive (dimensions, layers,..) features with the goal to cover all the main possibilities in terms of product design, such as the module's layering, the use of different materials (such as different glass types), the shape, the cell's arrangement, etc. with the result to define a 3D geometry of the component in a realistic graphic environment. The second step focused on developing a plug-in to create interoperability with a BIM-based process. The research is expected to cover a first step for supporting the design of advanced systems of building envelope including customized BIPV elements for transparent façade, in the perspective of an integrated approach ensuring quality, transparency and time/cost benefits.

Keywords: BIPV, building skin, semi-transparent photovoltaics, BIM, façade, tool

1 Introduction and motivations

1.1 BIPV and glazed façade: between product and process innovation

BIPV has always had a special relation with glass since its early usage. Conventional PV module requires transparent flat glasses in order to ensure strength, rigidity, environmental stability and high light transmission and also, with ultra-thin glasses, to reduce the

Engineered Transparency 2016. Glass in Architecture and Structural Engineering. First Edition.
Edited by Jens Schneider, Bernhard Weller.

module weight and the cost since glass is a moderately large part of the final cost for a solar module (Burrows & Fthenakis, 2015). But whereas research and manufacturing in conventional solar panels are mainly aimed to increase electro-technical requirements and energy performances, glass systems for BIPV are expected to evolve from relatively crude systems to a true integration of PV in glazed parts of the building skin (skylights, facades, spandrels, curtain walls and atrium roofing). BIPV glasses have been used so far almost entirely for prestige projects but today a market evolution is forecasted to shift towards a larger market of ordinary buildings (e.g. commercial, residential), with the challenge to integrate an attractive panel design (advanced optics, cell's arrangements, shape, etc.), to fit with new PV technologies (thin films, DSSC, organic PV, etc.) and to address cost-effectiveness (Gasman, 2012). In BIPV, the solar module becomes a constructive component of the building skin, language and material of its architecture. Such a different approach opens further issues in the use of glass, especially for façade, since aesthetics and building performances become an essential part of the expected requirements along with energy production, electrical safety and reliability (Frontini, Scognamiglio, Graditi, Pellegrino, & Polo Lopez, 2013). Moreover when BIPV glass is used for semitransparent elements, the solar façade acts as a key-element enabling the indoor comfort (solar gains, solar control and shadowing, daylighting along with acoustical and visual comfort) and the global building energy demand for heating and cooling (Lai & Hokoi, 2015) (Scognamiglio & Røstvik, 2012). In this framework BIPV urgently requires to complete the path of "technological transfer" in the building sector and, consequently, to overcome some barriers in terms of awareness, supporting tools, normative compliance and cost-effectiveness (Bonomo, Chatzipanagi, & Frontini, Overview and analysis of current BIPV products: new criteria for supporting the technological transfer in the building sector) (Bonomo, Frontini, De Berardinis, & Donsante, 2016). A different approach is required since the early Design Phase (EDP): PV becomes a "fundamental" of the design concept and of the envelope engineering, not only a multi-functional physical part of the building skin but also more and more part of an integrated and complex building process, both real and digitized, involving design, construction and information management. In this framework, the European project Construct PV (www.contructpv.eu) is aimed to develop and demonstrate customizable, efficient and low cost BIPV by means of a collaborative and integrated approach involving architects, engineers and manufacturers.

Innovative methods, models and tools oriented to optimally support an integrated design play a fundamental role in BIPV, since integrating PV involves a strong integration of energy, electrical, architectural and construction requirements during the whole process, from early-design phase till to manufacturing and operation. One of the practical reasons for which BIPV still today remains a niche market, along with other possible barriers (such as costs, normative framework, feed-in-tariffs), can be identified in the absence of supporting tools capable to overcome the gap between PV and building sector and to effectively integrate PV along the whole design and construction process. Goal of the work presented in the paper is to offer a versatile, flexible and effective software environment for design and evaluate a BIPV customizable component where energetic, architectural and construction aspects are jointly taken into account. The software platform, object of this document, has the general goal to facilitate the main stakeholders during the "process" of creation and implementation of a BIPV customized component, since the Early Design in connection also with BIM technology. Furthermore the web-tool allows a real time visualization of the BIPV object created by the user considering the different module's materials. Different glass types and finishing are implement using different UNITY assets: transparency, reflectance and refraction are considered and represented and the relative influence on the final power is introduced.

1.2 Building Information Modeling (BIM) for design and engineering of BIPV glazed façade

Building Information Modeling is a methodology for the planning, construction and operation of a building. The continuous model-based process aims a continuous use of the digital representation of a product or a building component from the first planning phase to the construction phase and finally for the use of the product in the phase of the facility management. The key word for this procedure in the construction industry is Building Information Modeling or Building Information Management (BIM). STRABAG/Züblin developed this method further and defines internal processes based on these methods as 5D. 5D means: → 3D three dimensions: geometry; → 4D the fourth dimension: time; → 5D the fifth dimension: all other connected processes like estimating and logistics. So first of all BIM is a method to use the product data or building component data for the planning process. The data represents on the one hand geometrical data such as for example the height and the width of a BIPV-Element. On the other hand quality data is represented. This could be the material and detailed properties of an element. This data has to be used in different phases of the planning and construction process. Furthermore the data can be used by different stakeholders. The façade is the envelope of and for a building and it is responsible for the protection of the interior, for the representation, and thanks to newest developments it is also responsible for energy harvesting. Due to these functionalities and further different circumstances façades and façade-technologies become more and more complex. BIPV is such a new and highly complex technology. Why? In most cases BIPV is built by Glass. Glass has static properties, which makes this product highly regulated at least in Germany. Now, due to the integrated PV Cells it becomes also an electrical component. For the traditional construction industry a new interfaces between the trades of the façade engineering and the electrical engineering. To support the planning and engineering of BIPV, newest planning technology is necessary. In this perspective the motivation of the presented work, as described in the next sections, was to create interoperability between the BIPV design phase and a BIM-based process, allowing BIPV to enter a digital environment assisting the stakeholders (architects, owners, contractors, etc.) to make better-informed and shared decisions throughout the building lifecycle.

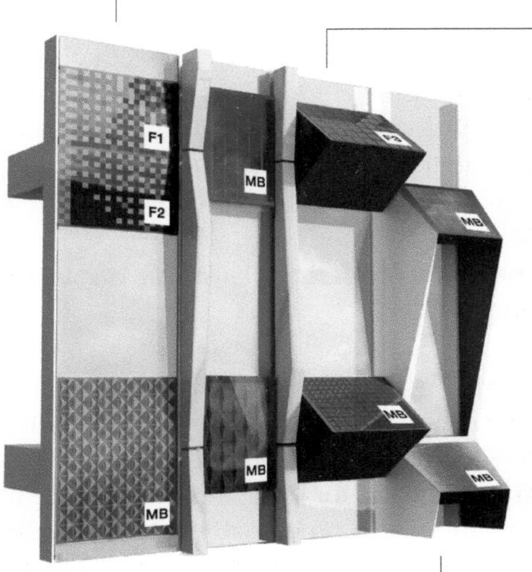

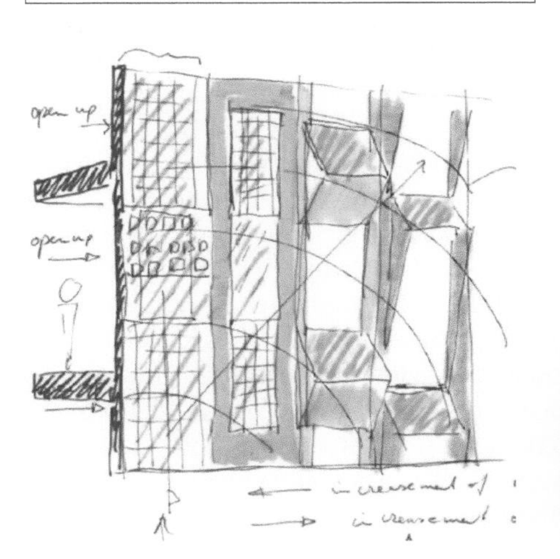

Figure 1-1 A mobile façade mock-up prepared by the coordinator of the project Züblin AG, will demonstrate in real-scale innovative BIPV glass-based technologies for the building envelope developed in Construct PV. The mockup, originated by the collaboration of Züblin with UN Studio, Fraunhofer ISE and Meyer Burger, demonstrates variants and possibilities in terms of architectural and technological design/manufacturing of BIPV systems for the building skin, showing different façade systems for glazed BIPV modules. It will be exhibited around Europe in the next years so that it will become a mobile demonstrative facility (more information at www.constructpv.eu).

2 Approach and methodology

The software platform, object of this document, has the general goal to facilitate the main stakeholders during the "process" of design, evaluation and implementation of a BIPV customized component within the design and building process. Since the architect is often the first key-player of the building process, the first step was the development of a platform with the function to flexibly support the **early design phase** for a customizable BIPV element. A **BIPV design web-tool** has been developed at this purpose as discussed in 2.1.1. Once that the configuration of the BIPV component has been established, in terms of architectural and constructive general layout, the development can move from a conceptual phase to the "design development" adding more details and information. The transition to a BIM-based process becomes strategic in order to really support the penetration of BIPV within the real Architectural, Engineering and Construction (AEC) process. Thus, the shift from a static component (such as an image, a CAD element, etc.) to a BIM parametric object was the second milestone of the work, as discussed in 2.1.2. The **Interoperability**, that is the interaction and information exchange of the design tool with Revit interface, has been defined through the development of a plug-in allowing effective and efficient exchange between the two software. As further discussed in 2.1.3, native **BIM objects** (Revit families) with specific features, parameters and behaviours (BIM objects adaptively working and customizable in BIM environment) have been predisposed to support this "transformation".

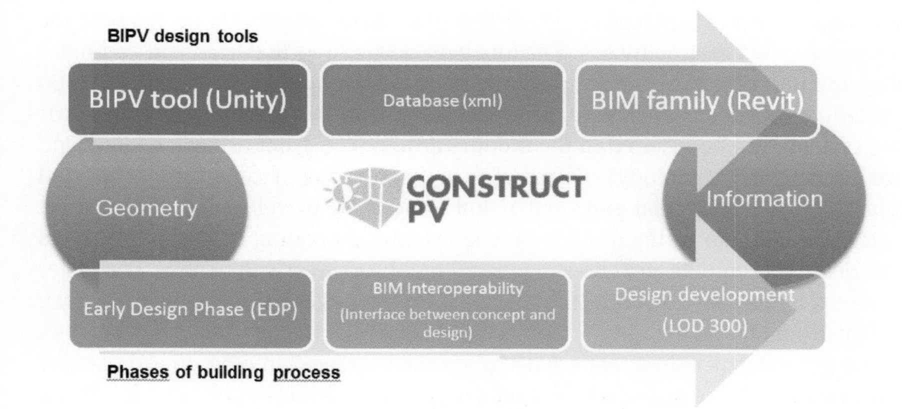

Figure 2-1 The possible levels of transfer of BIPV tools in the building process.

2.1.1 BIPV design web-tool for Early Design Phase

The main goal of the BIPV design web-tool is the development of an attractive web-platform with the function to support architects in creating customizable photovoltaic element to be used as part of the building skin. The tool has been developed on Unity platform that is a flexible and powerful development platform for creating multiplatform 3D games and interactive experiences. This software was chosen for the rapid iterative editing and the high visual fidelity, rendering power and ambience of the design environment (Real-time Global Illumination and physically-based shader) able to create an evocative dynamic platform. Thanks to this highly-flexible platform creating life-like images on the screen in real time, a high level of interactivity with the user (mainly architects, designers) is possible and BIPV design is transformed from a technical phase to the design concept of an architectural component in a scene looking relatively realistic. All the main architectural/constructive design possibilities for crystalline BIPV are available (module shape, layering, materials, type and material of solar cells, cell dispositions within the module, etc) thus offering a wide range of design opportunities to architects. In this “conceptual design” phase, even though the design creativity is an important goal, it’s essential to take into account all the rules of PV technology, taking care of a good final energy performance and considering, in each single case, the compatibility of the prototype with the industrial manufacturing possibilities. The main advantage is that a web-based software can be used over the internet with a web browser without installing any CDs, download any software, or worry about upgrades. Data are stored on secure, always-updated, backed-up servers and all data are centralized and accessible over the web from any computer at any time. The tool is placed on a web-link that can be directly connected with the project website (www.constructpv.eu). The tool generally allows to design and configure a crystalline module. It is possible to select the main design options at cell level. Mono or polycrystalline cells are available, in different colours and sizes. Also different electrical contact systems can be inserted in the module design (e.g. front or back contact and smart wire connection technology). Different grid dispositions of cells can be selected (rectangular uniform, rectangular exponential and hexagonal) as well as disposition type, spacing, etc. The layering of the module can be established adding or removing layers through the dedicated buttons and choosing the material and layer’s thickness. The material list provided in the database collects some main typologies and possibilities. For each material of the front layer is assigned a “Reduction Factor” parameter, linked to its energy transmittance (affected by colour, pattern and glass transparency) in order to estimate the final module power that is displayed in real time during the module design. The following correlation between the transparency/colours/patterns of the glass and the power has been introduced:

$$P_{max} = n \cdot PP_{cell} \cdot RF \qquad (2.1)$$

P_{max} = Peak Power of module in STC (W) n = number of cells; PP_{cell} = peak power of a single cell in STC (W); RF = reduction factor (0-1)

The Reduction Factor (*RF*) takes into account how the transparency, colored filters/glass and/or patterns affect the power.

$$RF = RF_{tr} \cdot RF_{co} \cdot RF_{pat} \tag{2.2}$$

RF_{tr} = RF for transparency of layers in front of the cells (0 - 0.95); RF_{pat} = RF for glass surface pattern = pattern light transparency (0 - 1); RF_{co} = RF for layer's color (0 - 1)

In order to simply estimate these factors, considering the target of the tool for EDP, the following assumptions have been done:

- in case of white module (obtained by the diffused colored reflection of a colored coating behind the glass), where all the visible part of the light is reflected, the PV efficiency decrease of 40 % (Perret-Aebi, 2015);
- Light Reflectance Value (LRV) is the total quantity of visible and useable light reflected by a surface when illuminated by a light source. (ref. British Standard BS 8300:2001/A1:2005). LRV runs on a scale from 0 % to 100 %. Zero assumed to be an absolute black and 100 % being an assumed perfectly reflective white (Diamond Vogel, 2016).

Mismatching at cell level/cell's string is not considered as well as the influence of thickness of glass on transparency. The LRV value, linked to each colour (http://encycolorpedia.it/), can be used for simply estimating the reduction factor:

$$RF_{co} = 1 - (0.40 \cdot LRV) \tag{2.3}$$

Table 2-1 Material/Patterns used both in the Design tool and Autodesk Library with the relative value of the Reduction Factor (*RF*). The patterns are considered applied onto an extra-clear glass.

Glass Material	$RF_{,tr}$	$RF_{,co}$	$RF_{,pat}$	RF
name	0-0.95	0-1	0-1	0-1
ExtraCLEAR	0.93	1	1	0.93
Micro-structured	0.94	1	1	0.94
Matt	0.80	1	1	0.80
Frosted	0.75	1	1	0.75
colored_meringue	0.92	0.89	1	0.82
colored_skilla	0.92	0.92	1	0.85
colored_can can	0.92	0.94	1	0.86
colored_zen retrat	0.92	0.95	1	0.87
colored_starfish	0.92	0.88	1	0.81
1_Pattern	1	1	0.88	0.88
2_Pattern	1	1	0.66	0.66
3_Pattern	1	1	0.50	0.50
4_Pattern	1	1	0.45	0.45
CPV_Pattern				0.80

Table 2-2 NCS (Natural Colour System), R G B and Light Reflectance Value (*LRV*).

RGB	NCS	*LRV (%)*	$RF_{,co}$
RGB 0, 148, 124	NCS S 2060-B70G Meringue	27	0.89
RGB 0, 122, 190	NCS S1565-B Skilla	20	0.92
RGB 189, 52, 107	NCS S 2060-R30B Can can	14	0.94
RGB 95, 95, 95	NCS S1565-B (grey scale)	13	0.95
RGB 255, 55, 36	NCS S 0580-Y80R	28	0.88

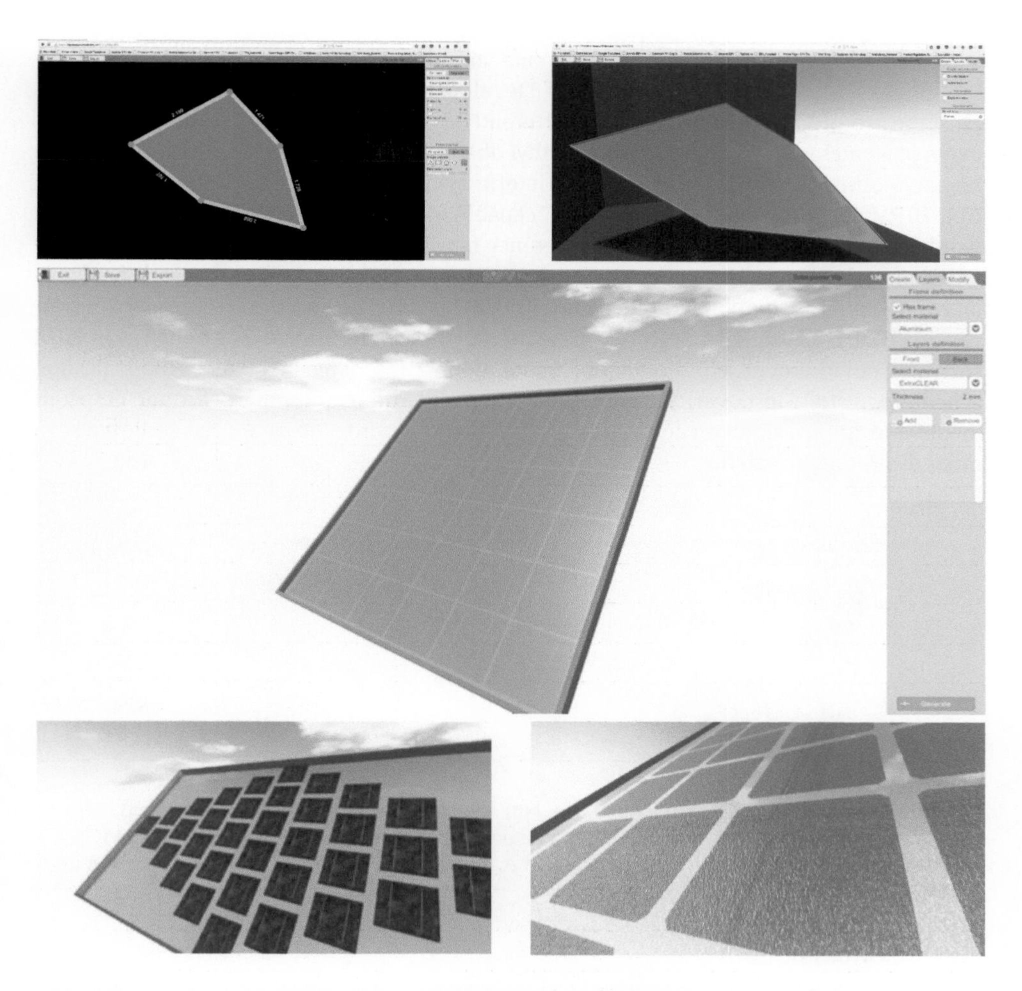

Figure 2-2 A screenshot of the software window for defining the layout of the BIPV glazed component in terms of layering, material, shape and transparency.

2.1.2 BIM interoperability and BIPV-BIM objects

The interoperability between the design web-tool (Unity) and the BIM design platform (Revit) is relevant in order to transfer the pre-defined BIPV object within a BIM-based process. Therefore, a plug-in has been developed in order to create an "exchange platform" able to transfer the parameter's values from Unity database (an xml file) to Revit Database, by establishing a direct communication between the two software. Briefly it allows to apply "in real-time" the design configuration defined in the web-tool to the BIPV object selected in the Revit project, thus allowing to "transform" the BIPV module pre-configured in the tool in a Revit family both in terms of geometry and parameters.

This step is very important in order to create an object completely operating in BIM environment with a certain LOD (Level of Development). The adopted approach is based on some prototype BIM-families that, differently by current BIM objects, are not components of a catalogue but rather adaptive BIM objects with parametric attributes which can be parametrically adapted. Thus a specific effort has been dedicated at the development of a BIPV adaptable “reference family”, capable to be flexibly adapted and customized thanks to a parametric behavior. Two reference families have been developed: an adaptive BIPV module (for project in design development phase). It can be used on wall/roof/curtain wall/skylight components and a curtain BIPV panel pattern based (for project in conceptual/EDP). These BIM objects can be used and directly customized by users in BIM software environment as autonomous stand-alone BIM objects for modelling, design, scheduling, etc. Moreover, through BIM software editing interfaces, further developments and improvements can be achieved by users in order to adapt the prototypes to other design requirements.

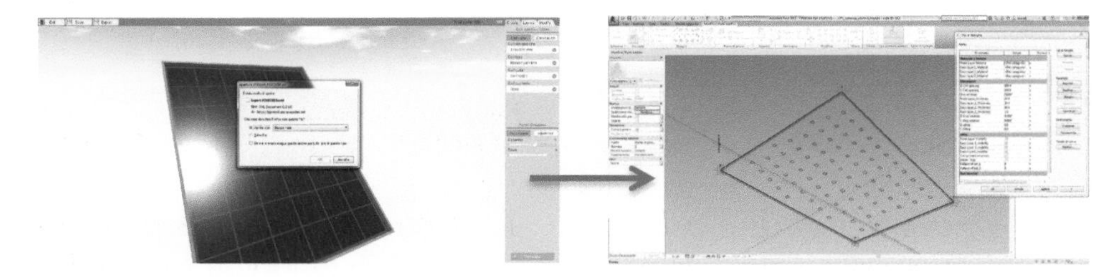

Figure 2-3 Export window in the tool (above, left) and the xml file generated (above, right). The xml file is used by Revit plug-in to import parameters in the “reference” family with the goal to update the parameter's values. Family prototype A: adaptive BIPV module with parametric features (below).

2.2 Conclusions

Integrating photovoltaic glasses in building skin involves a strong integration of energy, electrical, architectural and construction requirements during the whole process, requiring a multidisciplinary and integrated approach by architects, engineers and manufacturers since the initial design stages. The development of methods, models and tools oriented to optimally support an integrated process is a crucial aspect of growing interest. As discussed in the paper, starting from the development of a design web-tool that is mainly conceived for designing a BIPV component in the EDP, thanks to interoperabiliy loadable BIM families have been obtained with the purpose to transfer BIPV in a BIM-based process. The web-tool is available at the Construct-PV project website. Further possible evolutions of the tool can be planned and they could include additional features and properties to establish a more effective relation with the whole building process and the building life-cycle (4D, 5D, 6D and 7D).

3 Aknowledgments

Construct-PV: The research leading to these results received funding from the European Community's Seventh Framework programme (FP7/2007-2013) under grant agreement no 295981 [www.constructpv.eu].

4 References

[1] Burrows, K.; Fthenakis, V. Glass needs for a growing photovoltaics industry. Solar Energy Materials and Solar Cells. 2015, pp. 455-459.

[2] Gasman, L.; BIPV glass market and strategies. Solar business Focus. 2012.

[3] Frontini, F. et al.; From BIPV to multifunctional component.; Paris, 2013. 28th European PV Solar Energy Conference.

[4] Lai, Chi-Ming and Hokoi, S. Solar façades: A review. Building and Environment. 2015, Vol. 91 , pp. 152-165.

[5] Scognamiglio, A. and Røstvik, H.N.; Photovoltaics and Nearly Zero Energy Buildings: a new opportunity challenge for architecture. 2012, Progress In Photovoltaics: Research And Applications.

[6] Bonomo, P., Chatzipanagi, A. and Frontini, F. Overview and analysis of current BIPV products: new criteria for supporting the technological transfer in the building sector. VITRUVIO – International Journal of Architectural Technology and Sustainability.

[7] Bonomo, P. et al. BIPV: building envelope solutions in a multi-criteria approach. A method for assessing life-cycle costs in the early design phase. Advances in Building Energy Research. 2016.

[8] Perret-Aebi, L.-E. (2015). New approaches for BIPV elements: from thin film terra-cotta to crystalline white modules. CSEM.

[9] Diamond Vogel. (2016). *Color Finder*. Tratto il giorno November 2015 da www.diamondvogel.com : http://www.diamondvogel.com/colorfinder.

Development of a Composite PV Module with Phase Change Material (PCM)

Bernhard Weller[1], Sebastian Horn[1], Julia Seeger[1], Leonie Scheuring[1], Franziska Rehde[1]

1 Technische Universität Dresden, Institute of Building Construction, George-Bähr-Straße 1, 01069 Dresden, Germany, Julia.Seeger@tu-dresden.de

Opaque façade areas feature a great potential for generating energy by using PV modules. Their use in the façade sector has so far been limited mainly to cold fronts because the performance of these modules drops with increasing temperatures of the PV cells. In contrast to cold fronts, a rear ventilation to cool the modules is not possible in standard mullion and transom façades. To ensure low module temperatures and thus a high performance, the PV modules are to be combined with rear-mounted phase-change materials (PCM). When the temperature in the façade reaches the (freely adjustable) melting point of the PCM, it results in a phase transition from solid to liquid. The material absorbs thermal energy from the PV-module and the rising of the module temperature can be buffered. At night, the thermal energy latently stored in the PCM is released back into the ambient air.

Keywords: building envelope, latent heat storage, mullion-transom façade, photovoltaic, solar

1 Introduction

In recent years, beside the endeavour to save energy, attention has also increasingly been paid to the aspect of producing energy in the building sector. Photovoltaics (PV) can contribute notably to achieve these objectives and specifications. PV modules can be installed on the building in various ways and convert a part of the incident solar radiation into electrical energy. The most common method so far is to mount the PV modules onto flat or slanted roofs. The useable mounting space, however, is limited which increasingly puts forward the potential capacity of façade areas. Nonetheless, new design-related complications emerge which must be taken into consideration.

The efficiency factor of a PV module indicates how much solar radiation is converted into electrical power. In addition to a variety of external impacts, the temperature in particular affects this conversion rate. Rising temperature decreases the efficiency and the power output recedes. This effect is displayed by the temperature coefficient of a PV module. Integrated into a warm façade, the module temperature can be up to 55 K higher than the ambient temperature. For crystalline modules, this can reduce the annual energy gain by up to 10.5 % (Weller [1]). The implementation of PV modules in façades is therefore mostly limited to ventilated curtain walls because in conventional warm façades, like the widely-used mullion and transom construction, a rear ventilation of the modules is not possible.

Engineered Transparency 2016. Glass in Architecture and Structural Engineering. First Edition.
Edited by Jens Schneider, Bernhard Weller.

To reach lower module temperatures and, therefore, higher conversion rates even in mullion-transom systems, the possibility of cooling PV modules with rear-mounted phase change materials (PCM) is to be analysed. PCMs can absorb thermal energy from the PV module by melting from solid to liquid in the daytime, thus lowering the core temperature of the PV cell. As a result, an opaque insulated façade panel consisting of a PV module and PCM, which can be incorporated into mullion and transom façades, is to be created (Fig. 2-1). In the process, a series of design-related questions need to be solved to obtain a structurally functional façade element.

2 Construction principle and functioning

The façade panel with an integrated photovoltaic module (PV) and phase change material (PCM) is insulated on the rear side and closed up towards the interior with a metal sheet (Fig. 2-1). It can be incorporated into the skeletal grid structure of the mullion-transom construction like a conventional opaque façade element. The panel is fastened by pressure plates that are bolted to the mullion-transom system and therefore secure the panel against wind suction. The dead weight of infilling elements is transferred to the horizontal transom through setting blocks and passed on to the vertical mullion. In a warm façade, the developed front panel also serves the purposes of weather protection and thermal insulation, in addition to power generation.

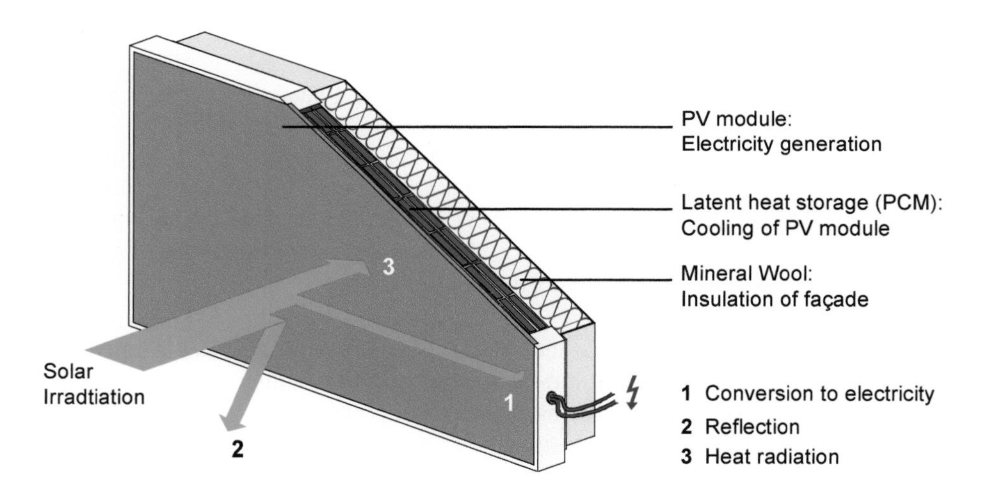

Figure 2-1 Functioning of the PV façade panel with rear mounted PCM capsules for cooling. The efficiency of the PV module drops as the temperature increases.

By integrating PCMs, often known as latent heat storage, the maximum temperatures of the PV module are supposed to be reduced in the façade panel. Through phase transition, latent heat storage units can retain thermal energy with minimal losses for a long period of time. During the phase change (melting), heat energy is bound without increasing the temperature of the PCM. When the ambient temperature drops below the melting point of the PCM, the phase change of liquid to solid begins and the latent heat energy is released again. For instance, the thermal energy will be withdrawn from the PV module during the day and returned back to the environment at night. Thus, the temperature in the PV module remains largely constant over a longer period of time so that there is no temperature-related reduction of efficiency. Salt hydrates or paraffins are used as PCMs. The melting point can be varied through the chemical composition and is adapted to the respective application.

For implementation in a façade panel, a PCM based on salt hydrates was chosen for use in the building envelope as it is less flammable than the alternative of paraffin. Considering previous temperature measurements on an identical façade panel without PCM (Chapter 4), a salt hydrate with a melting temperature of 31 °C and a storage capacity of 220 kJ/kg is used (Rubitherm [2]). Enclosed in capsules (macro-encapsulated), it is placed on the back of the PV module in two layers. The capsules are made of 18 mm thick extruded aluminium profiles, which are divided by a central bar into two chambers. The flat surface of the aluminium capsules and the good thermal conductivity of the material also allows for a high heat transfer between PV module and PCM.

The PV module is a thin film PV module and consists of the semiconductor material cadmium telluride (CdTe). Compared to a crystalline PV module, this material offers a very homogenous visual appearance and its efficiency loss due to rising temperatures is lower. The module efficiency of 11.1 %, however, is lower than for a commercial crystalline PV module. Since the electrical connections in commercial PV modules are located in a box attached on the back of the PV module, the aluminium capsules could initially not be applied to the entire module surface. To avoid this, a junction box was designed which is placed on the edge of the PV module and fits into the seam of the mullion and transom construction. This allows full-surface application of aluminium capsules over the entire back of the PV module.

3 PCM integration

3.1 Connection of PCM capsules and PV module

In order to keep the temperature increase in the PV module as low as possible the highest possible heat dissipation from the PV module into the PCM capsule should be achieved. This approach uses the space available for PCM capsules on the back of the module to full capacity. For a sufficient temperature buffering in the PV module, a total of two consecutive levels with 5 ½ PCM capsules each are necessary. The melting point of the PCM has to be adjusted accordingly. At this point reference is made to section 4.

For the connection of the aluminium PCM capsules to the PV module an adhesive bonding was chosen. Concerning the selection of a material especially the following requirements for the bond line have to be considered (Fig. 3-1).

- Compensation of different thermal expansions between the glass of the PV module and the aluminium of the PCM capsule
- Good adhesion properties on glass and aluminium
- Durability of the compound; in particular, resistance to large and frequent temperature changes from -40 °C to +85 °C according to DIN EN 61646 [3]
- Good heat transfer between PV module and PCM capsule
- Small effort for the process of application

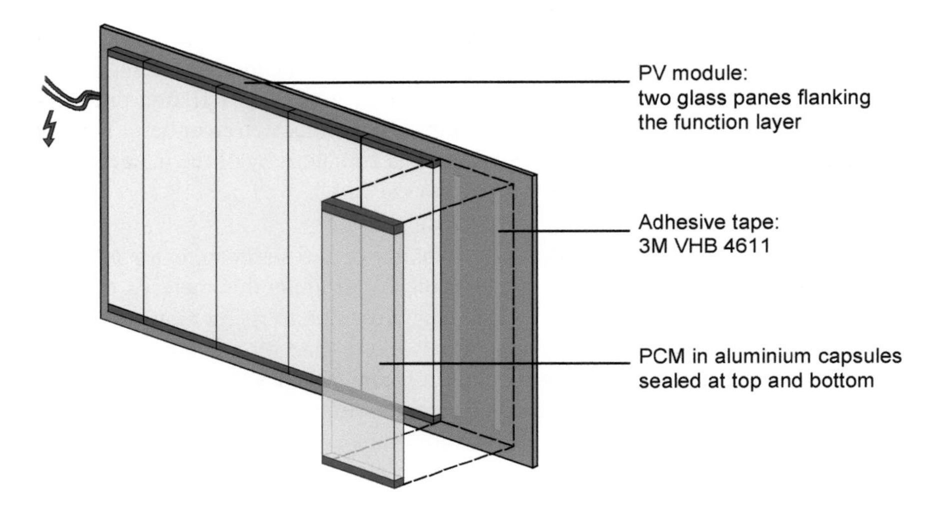

Figure 3-1 Rear view of PV module: PCM capsules are attached to PV module via adhesive tape.

Based on extensive research and preliminary investigations with different adhesive systems (e.g. silicones) finally a double-sided adhesive tape was selected: 3M VHB 4611. It is a single layer tape made of foamed acrylate. The tape features a very high thermal stability and is particularly suitable for the bonding of high-energy surfaces such as metal and glass. Moreover, foamed acrylate tapes show a viscoelastic material behaviour and can compensate varying linear expansions of the adjacent joining parts by up to 300 % of their thickness (Data sheet 3M [4]). The selected tape VHB 4611 is 1.1 mm thick and therefore allows an opposing horizontal translation of the aluminium capsule and the glass (PV module) of 3.3 mm. This corresponds to a shear strain tan γ of 3.0. The acrylate adhesive tape VHB 4611 is glued as two vertical strips on to each capsule. The bond line geometry has a high dimensional accuracy and is easy to produce. Therefore low production costs can be reached. However, a readjustment of the joining parts on short notice is not possible.

3.2 Experimental studies on shear behaviour

For the selection of a suitable adhesive the specifications provided by the producer were not sufficient, so that a closer examination of the material properties became necessary. In a first step a preselection of adhesives was made based on the respective data sheet. For these materials the shear strain and strength of unaged and artificially aged specimens as well as the thermal conductivity were examined. Of all experimental tests only the results of the preferred option with the double-sided adhesive tape 3M VHB 4611 are presented in the subsequent section. The determined material properties also serve to verify previous theoretical assumptions, calculations and numerical simulations.

Using a universal test machine (UPM) a shear test was performed on small specimens as shown in Figure 3-2. The small specimens consist of glass (backside PV module) and aluminium (PCM-capsule) connected with the double sided adhesive tape. In the process the two joining parts are shifted parallel to each other until the adhesive bond line fails. The geometry of the glass-aluminium specimens was developed in accordance with the ETAG 002-1 [5]. Likewise, the experimental setup, its implementation and the number of test specimens conform to the requirements of this directive. After the compulsory curing time of 28 days, the samples were preconditioned for 24 hours at temperatures of + 23 °C, +80 °C and -20 °C and then tested under shear while maintaining the respective temperature levels. The experiment was conducted on 5 test specimens per temperature level; at room temperature on 10 specimens each.

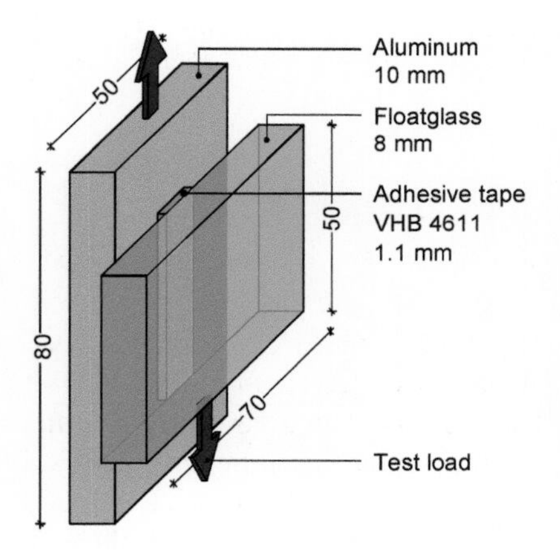

Figure 3-2 Geometry of the small adhesive specimens for the shear test.

Figure 3-3 Examplary test specimen with acrylate adhesive tape VHB 4611 after the shear test.

During the test, the load and crosshead travel was measured. In addition the axial displacement of the two assembly components was recorded by a video extensometer (non-contact measurement). Based on these values, the shear stress in the adhesive joint (failure load divided by bond line area) was calculated and the shear strain tanγ was determined (axial shifting of the joining parts relating to the bond line thickness). Figure 3-3 exemplarily shows a tested specimen after the adhesive joint has failed.

Besides, also the residual shear strength and strain after artificial aging of the specimens were tested. With the results the durability of the bond between PV module and PCM capsule can be assessed. On the basis of DIN EN 61646 [3], which regulates the design qualification and type approval of terrestrial thin-film PV modules, three artificial aging scenarios were defined: Temperature cycling (-40 °C to +85 °C, 200 hours), Humidity-Frost cycling (-40 °C to +85 °C, 85 % relative humidity, 240 hours) and Damp Heat Test (85 °C, 85 % relative humidity, 1000 hours). Each aging scenario is performed with 10 specimens. Afterwards the differently aged specimens are tested under shear at room temperature in order to determine the residual shear strength and shear strain. The experimental setup is unchanged compared with the unaged specimens.

Figure 3-4 shows the mean shear strain tan γ of the unaged and artificially aged samples (light green tones) at room temperature (+23 °C). Additionally the results for the other two temperature levels -20 °C (blue) and +80 °C (red) are presented. Likewise, the values for the shear strength of unaged and aged specimens are displayed in Figure 3-5.

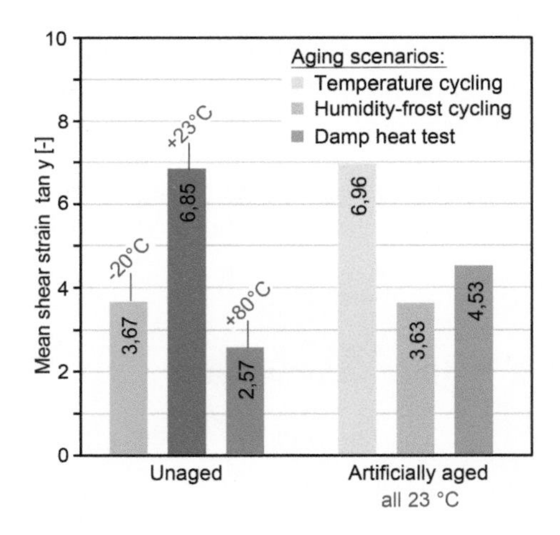

Figure 3-4 Mean shear strain of unaged specimens at temperatures of -20 °C (blue), +23 °C (green), and +80 °C (red) in comparison with artificially aged specimens (green tones) at room temperature (23 °C).

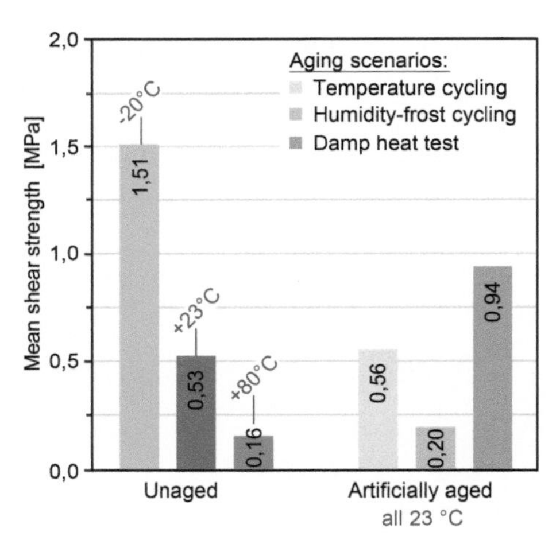

Figure 3-5 Mean shear strength of unaged specimens at temperatures of -20 °C (blue), +23 °C (green), and +80 °C (red) in comparison with artificially aged specimens (green tones) at room temperature (23 °C).

Generally, the shear tests indicate a temperature-dependent material behaviour of the adhesive tape VHB 4611. While the material can compensate displacements of almost

700 % of its thickness at room temperature (equivalent to tan γ= 7.0), this capacity is approximately halved at -20 °C and +80 °C. The temperature-dependent material behaviour is even more significant in terms of the shear strength: At +80 °C the value drops to a third, whereas at low temperatures (-20 °C), the material strengthens so that a shear strength of about 1.5 MPa is achieved.

The results for the aged specimens show that pure temperature cycling has no significant influence on the shear strength and strain. However, the adhesive tape reacts sensitively to humidity influences especially in combination with frost cycles. Here the shear strain decreases by 50 % and the shear strength even by 65 % in comparison to the unaged material at +23 °C. In fact the measured value for the shear strain (tan γ= 3.63) is still higher than the data sheet of the adhesive tape indicates (300 %, equivalent to tan γ= 3.0). The specimens which were exposed to the damp heat test show a moderate loss in shear strain, while the shear strength even increases significantly.

The failure of the majority of specimens show an adhesive failure on the glass surface. In the process, the detachment of the tapes was always limited to partial areas of the adhesive surface and there was no complete separation of the two joining parts. To improve the adhesive properties, a surface pre-treatment of the glass is considered.

Due to the very low shear strength at +80 °C (0.16 MPa) and after humidity-frost-cycling, the use of the adhesive tape VHB 4611 for general load bearing functions is certainly limited. For the connection of the PCM capsules and the PV module in the façade panel it is nevertheless suitable, as the capsules rest directly on the edge band (Phonotherm). Therefore the dead load is not transferred via the bond line. In case of failure of the adhesive tape the PCM capsules basically remain at their position. Only a slight tilting backwards is possible inside the panel, so that the heat transfer between PV module and PCM capsule would deteriorate.

4 Monitoring at the outdoor exposure test rig

4.1 Concept

Despite existing research studies on the connection of PV modules and PCM, there is still a great need for research. In most projects, measurements to improve the performance were either carried out under laboratory conditions or only on certain days. Also, most of the examined systems were not finished façade elements but mere experimental setups (Huang et al. 2004 [6] and Browne et al. 2015 [7]). Furthermore, since the schematic of the complex melting processes in the PCM is very difficult and so far only available for certain paraffin-based PCM calculation models, an accurate quantification of performance improvement through PCM integration is not possible without further investigation.

For that reason, the performance of the newly developed façade panel is to be investigated within the scope of an open air monitoring over a longer period of time. For this purpose, the Institute for Building Construction at the TU Dresden operates an outdoor exposure test rig, on which a mullion-transom façade with PV integration is examined (Fig. 4-1). The monitoring provides the data logging of different measurement categories, following the requirements of IEC 61724 [8], which describes the monitoring of the operating behaviour of photovoltaic systems.

Figure 4-1 Outdoor exposure test rig with PV façade panels in a mullion transom construction.

Regarding the meteorological variables, mainly the recording of global radiation or the irradiation at module level is important as it allows a quantification of the energy available for the photovoltaic process. In the façade, the module temperature and the temperatures over the cross section of the front panel are recorded. Hereby the module temperature is measured at the back of the PV module in accordance with method A described in DIN EN 61829 [9]. As electrical parameters the electrical power in direct current circuit (P_{DC}) as well as in alternating current circuit (P_{AC}) are recorded. For further investigations only the electric power in the direct current circuit (P_{DC}) was used, since it is not influenced by inverter losses. All measurement parameters are logged at an interval of 10 seconds and converted to five-minute averages to improve further processing.

In order to make a statement on the impact of the PCM integration, PV façade panels without PCM capsules were installed in a first step. For a period of 10 months the module temperatures are recorded under different weather conditions. The measurement data obtained are used in particular for adjusting the melting point of the PCM. In a second step, the façade panels are installed with and without integrated PCM capsules at the same time and are also meteorologically recorded on the outdoor exposure test rig over a longer period of time. This setup allows a direct comparison of the performance improvement and the temperature development in the PV module by integrating PCM capsules.

4.2 Analysis

The measured data of the system without PCM shows that the generated electric power as well as the development of the module temperature mainly depend on the irradiance at module level (Fig. 4-2). This, in turn, is strongly dependent on the orientation of the building. In the case considered here, the PV panels are aligned vertically to face south. Consequently, the maximum irradiation under ideal conditions without the influence of clouds occurs at noon. The performance curve evidently follows the radiation course. The module temperature also follows this course, though time-delayed. This phase shift is mainly based on the inertia of the building structure.

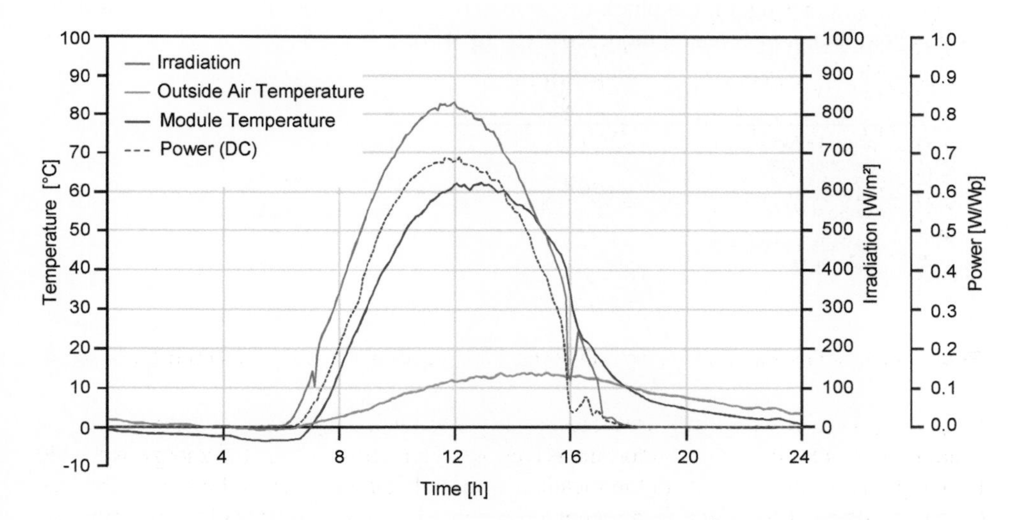

Figure 4-2 Course of irradiation at module level and module temperature without PCM on a sunny winter day.

The analysed monitoring data for a whole year shows a distinct difference between summer and winter. Thus, the maximum irradiance at module level is higher on a sunny winter day than on a sunny summer day. The reason for this difference is the cycle of the sun and the seasonal fluctuations of solar altitude angle. In winter, the solar radiation at noon impinges nearly perpendicularly on the module surface. Due to the high solar altitude angle in summer, the angle of incidence is sharper and hence reflection losses increase. However, the earlier sunrise and later sunset in summer effectuate a significantly longer duration of irradiance than in winter. The fact that high module temperatures also occur during the winter months shows that the integration of PCM is of great importance, not just for summer application, but also the whole year.

When comparing the measured data of systems with and without PCM, the effect of PCM integration becomes apparent (Fig. 4-3). While the curve of the module temperature without PCM (dark blue) follows the course of the irradiation at module level, the curve of the module temperature with PCM is nearly horizontal during the melting of the PCM (light blue). The incipient phase change from solid to liquid prevents a further increase of the module temperature. That this straight line can not always be kept at the melting point of 31 °C is mainly because the thermal energy is not passed onto the PCM quickly enough. Nevertheless, it can be seen that the power of the modules with PCM is higher than without PCM. However, any influences from age-related degradation of the PV modules are not taken into account here. After all, the PV modules without PCM have already been operating about 1 year longer than the PV modules with PCM. In the evening and night hours, the module temperature without PCM drops relatively quickly. This is not the case for the modules with PCM, though. Here, the latent heat in PCM has to be discharged first. Once again, the phase of the material changes from liquid to solid and the module temperature decreases more slowly (Fig. 4-3).

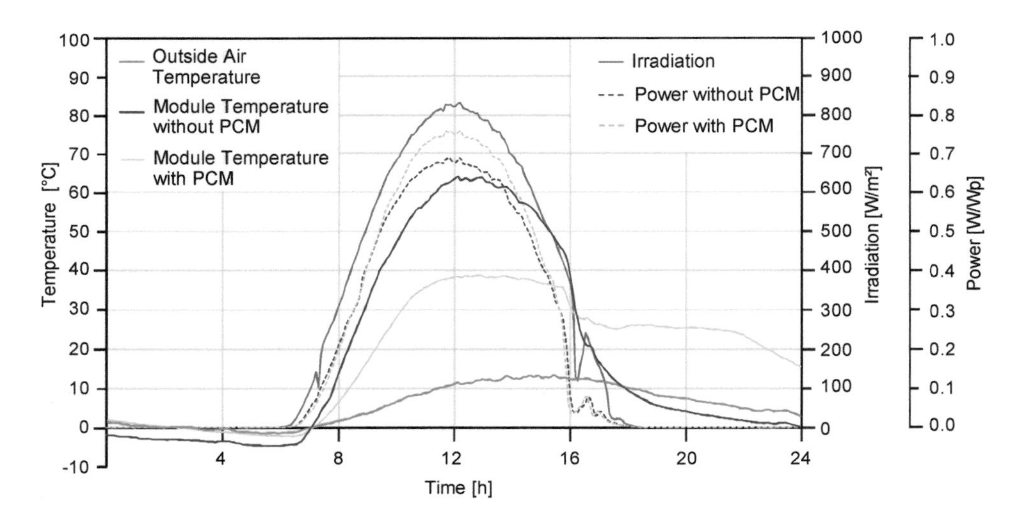

Figure 4-3 Example of the effect of PCM integration in terms of module temperature and power.

A critical factor to the performance improvement in each case is the complete solidification of the PCM in the cool night hours to enable the positive effect of temperature buffering on the following day. Especially on a series of hot days with hot nights, it may happen that no complete discharge occurs. The implications of such events are still under investigation by the means of the continuous monitoring. Such weather conditions could not yet be recorded at the test rig. When analysing the measurement data, it has to be taken into account that the findings apply only to the arrangement on south-facing walls and to the climatic conditions at the test rig. By choosing a new location and different orientation (e.g. east façade), the results and also the optimal melting point of the PCM can vary. However, these considerations are not part of the article.

5 Summary and outlook

The previous evaluation of the monitoring has shown that the module temperature for insulated PV façade elements is occasionally very high, even in winter. Thus there are great losses of performance throughout the whole year. Contributing to an improvement in this situation, the integration of PCM capsules which buffer the temperature rise inside the PV module was investigated. With the aim of developing a composite PV panel for the use in mullion transom façades, a suitable connection between the PV module and the rear mounted PCM capsules was investigated.

Based on preliminary research and experimental tests with different adhesives finally the double-sided adhesive tape 3M VHB 4611 was selected. The acrylate tape distinguishes especially by the capability to compensate great expansion differences between the PV module and the aluminium capsule. Shear tests on small specimens indicate that the tape can compensate varying expansions of up to 690 % of its thickness at room temperature. However, the material behaviour is strongly temperature-dependant and the adhesive is aging resistant to only a limited extent. Nevertheless the material meets the constructive and structural requirements for the use in this designed façade panel.

In a further step, the PV-PCM façade panel and its performance was investigated within the scope of an open air monitoring on an outdoor exposure test rig over a longer period of time. Simultaneously, façade panels without PCM integration are monitored at the same test rig. Through direct comparison the improvement in performance of the newly developed façade panel with PCM capsules can be quantified. Initial analysis of the measurements show good results. For instance, on sunny days in spring the module temperature is up to 25 K lower than in the façade panel without PCM. For a temperature coefficient of 0.25 %/K this corresponds to an increase in performance of 6.25 %.

At that time, no absolute statement concerning the effect of the PCM can be made yet, as the amount of measured data is still too small. An accurate assessment can only be submitted after the results of a whole year under varying seasonal conditions are evaluated. In addition to the monitoring the façade is also to be examined concerning its behaviour in case of fire. Therefore a single burning item test as well as an inflammability test will be conducted.

6 Acknowledgements

This work originated from the research project "Insulated thin film photovoltaic panel (PV) with integrated latent heat storage (PCM)", funded within the scope of the Central Innovation Programme for SMEs (german: Zentrales Innovationsprogramms Mittelstand, ZIM) of the Federal Ministry of Economic Affairs and Energy (grant number VP2050212MF4). The authors are grateful to all project partners for their excellent co-operation and wish to thank Lisa Schoeberlein, student assistant, for her support with the revision of this article.

7 References

[1] Weller, B.; Hemmerle, C.; Jakubetz, S.; Unnewehr, S.: DETAIL Practice: Photovoltaics: Technology, Architecture, Installation. Institut für internationale Architekturdokumentation, Munich: Birkhäuser, 2010.

[2] Rubitherm Technologies GmbH: Technical data sheet SP31. Berlin: 2016.

[3] DIN EN 61646:2009 (IEC 61646:2008): Thin-film terrestrial photovoltaic (PV) modules – Design qualification and type approval. Deutsches Institut für Normung e.V., Berlin: Beuth, 2009.

[4] 3M Deutschland GmbH: Technical data sheet 3M VHB Hochleistungs-Verbindungssysteme 4511, 4613, 4646, 4655. Neuss: 2004.

[5] ETAG 002-1: Guideline for European technical approval for structural sealant glazing kits (ETAG) – Part 1: Supported and unsupported systems. European Organisation for Technical Approvals (EOTA), Brussels: 2012.

[6] Huang, J. M.; Eames, P. C.; Norton, B.: Thermal regulation of building-integrated photovoltaics using phase change materials. In: Internaional Journal of Heat and Mass Transfer. Vol. 47, 2004, pp 2715-2733.

[7] Browne, M. C.; Norton, B.; McCormack, S. J.: Phase change materials for photovoltaic thermal management. In: Renewable and Sustainable Energy Reviews. Vol. 47, 2015, pp. 762-782.

[8] IEC 61724: Photovoltaic system performance monitoring – Guidelines for measurement, data exchange and analysis. Deutsches Institut für Normung e.V., Berlin: Beuth, 1998.

[9] DIN EN 61829: Crystalline silicon photovoltaic (PV) array – On-site measurement of I-V-characteristics. Deutsches Institut für Normung e.V., Berlin: Beuth, 2012.

Solar Panel Breakage During Heavy Rain Caused by Thermal Stress

Reijo Karvinen[1], Antti Mikkonen[2]

1 Tampere University of Technology, PL 527, 33101 Tampere, Finland, reijo.karvinen@tut.fi

2 Tampere University of Technology, PL 527, 33101 Tampere, Finland, antti.mikkonen@tut.fi

Solar panels and thermal collectors are increasingly popular. There is practical experience of large numbers of solar panel glasses being broken during heavy rain. The present paper studies the role of mean heat transfer between rain and the glass on the breaking. Thin tempered glass is preferred for its low weight, durability, and good optical quality. However, thin glass tempering is expensive and by understanding relevant stresses costs can be avoided. The heat transfer between a solid surface and rain is studied experimentally using a hot copper block and free falling drops. The thermal stresses are solved using a one-dimensional theory and the measured mean heat transfer coefficient. The thermal stresses depend on rain rate, surface inclination, glass thickness and temperature difference. The results show that, expect for word record approaching rain rates, the thermal stresses are below 10 MPa. A non-heat treated soda-lime glass should withstand this stress without breaking. The used rain rates were R = 1100, 340, 110 mm/h and the maximum mean heat transfer coefficients h = 600, 250, 140 W/m^{2}K, respectively. All else being equal, the maximum mean heat transfer was observed for surfaces that were inclined 15° from horizontal. Based on the results in the present paper the mean rain heat transfer causes no need to temper soda-lime glass to be use in solar panels. However, one should remember that thermal stresses must be added to all the other stresses.

Keywords: thermal stress, heat transfer, rain, experimental, one-dimensional

1 Introduction

Experience has shown that sometimes solar panel glasses break during heavy rain. A heavy rain in a warm environment can cause a rapid change in the glass surface temperature. Rapid cooling causes tensile thermal stresses at the glass surface. Being a brittle material, glass is susceptible to fractures from such stresses. Fractures initiate at surface flaws such as scratches and their accurate prediction is difficult. Glass can be protected against tensile tresses with heat treatment which produces compressive stress on the glass surfaces. The cost of heat treatment for thin glass increases sharply with increasing surface stress as explained by Rantala [1]. Thin glass is preferred for its lower weight and better optical quality. In the present paper the thermal stress resulting from rain are studied during very heavy rain fall. Mean heat transfer is measured and stress is calculated using a one-dimensional theory.

Natural rain drop size distribution is produced by large drops breaking into smaller drops after they reach their maximum stable velocity. The process is elegantly explained by

Engineered Transparency 2016. Glass in Architecture and Structural Engineering. First Edition.
Edited by Jens Schneider, Bernhard Weller.

Villermaux and Bossa [2]. Most of the rain drops are small and the largest stable drop diameter is about $d = 6$ mm. Drop size distribution is governed by $n(d) = n_0 e^{-d/d_{\text{ref}}}$, where n_0 is a constant and $d_{\text{ref}} = 41 R^{-0.21}$. This correlation was first found experimentally by Laws and Parsons [3] and then explained theoretically by Villermaux and Bossa [2].

The rain fall rate is usually discussed in the units of millimeter per hour. Rain fall rates vary globally and only the heaviest rain conditions are studied in the present paper. The world record for most rain in one minute is 31.2 mm (1872 mm/h) according to World Meteorological Organization [4]. Other heavy rain world records are for one hour 305 millimeters in 42 minutes (435 mm/h) and 1825 mm for 24 hours (76 mm/h) [4]. What passes for heavy rain in casual conversation in Finland is tens of millimeters per hour for a short time. The rain rates studied in the present paper are 1100, 340, and 110 mm/h and far exceed those of normal conditions.

When the initially hot glass is subjected to high speed drops impinging on its surface a high local heat transfer occurs at the glass surface. Depending on the rain rate, surface inclination, impurities in the water, and surface material details the rain water may wet the whole surface or form small streams down the surface like water on a train window. Such details cause the heat transfer to be non-uniform both in space and time. In the present paper, only mean heat transfer over the surface is considered.

The mean heat transfer is measured by impinging free falling water drops on a copper block. The block temperature change is measured and a constant heat transfer coefficient is fitted to the data. The details are explained in section 4. Experimental Methods. A sketch of the setup is shown in Fig. 1-1.

The copper block is large enough to cool slowly under the rain and a lumped heat transfer model is used in the heat transfer coefficient fitting. To solve the thermal stresses, an accurate temperature profile is needed and it is solved with a series based one-dimensional

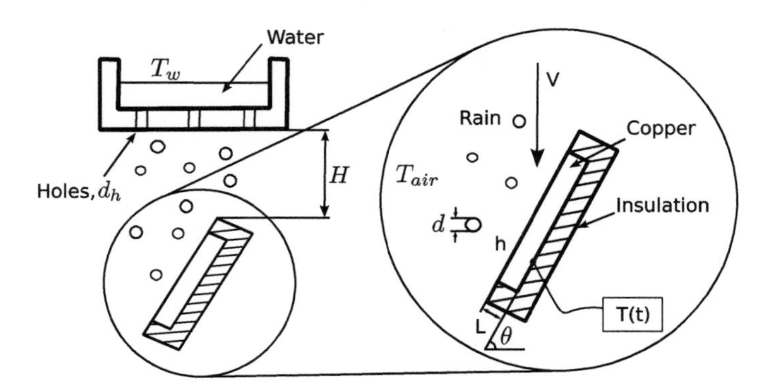

Figure 1-1 Experimental setup.

method. The thermal stresses are then solved using a one-dimensional elastic model. The details of the calculations are given in sections 2. Heat Transfer and 3. Thermal Stress. The largest stresses are found for thick glass with large heat transfer near the cooled surface a short time after the cooling starts.

2 Heat Transfer

The copper block used in the experiments can be assumed to be in a uniform temperature. When compared with the accurate solution in Eq. 2.2 there is a few percentage of difference. The assumption of uniform temperature leads to a lumped model of heat transfer

$$\frac{T-T_w}{T_0-T_w} = e^{(\frac{ht}{\rho Lc})} \tag{2.1}$$

Where T is the solid temperature, T_w is the water temperature, T_0 is the solid initial temperature, h is the mean heat transfer coefficient, t is the time, ρ is the solid density, L is the solid thickness, and c is the solid heat capacity. An analytical solution to the one-dimensional heat transfer problem is given in a textbook by Boley and Weiner [5] as

$$\frac{T-T_w}{T_0-T_w} = \sum_{n=1}^{\infty} A_n e^{-\lambda^2 \tau} \cos(\lambda_n Z) \tag{2.2}$$

$$A_n = \frac{4\sin(\lambda_n)}{2\lambda_n+\sin(2\lambda_n)}, \quad \lambda_n \tan(\lambda_n) = \mathrm{Bi}$$

$$\mathrm{Bi} = \frac{hL}{k}, \quad Z = \frac{z}{L}, \quad \kappa = \frac{k}{\rho c}, \quad \tau = \frac{\kappa t}{L^2}$$

where k is thermal conductivity of the solid and z is distance from the insulated surface, see Fig. 2-1. In the present paper a few hundred of λ_n values were included and they were solved by noting that $\lambda_n \tan(\lambda_n) = \mathrm{Bi}$ has discontinuities at $\lambda = (n - 0{,}5)\,\pi$, is a genuinely growing function between the discontinuities and has exactly one solution in each bracket.

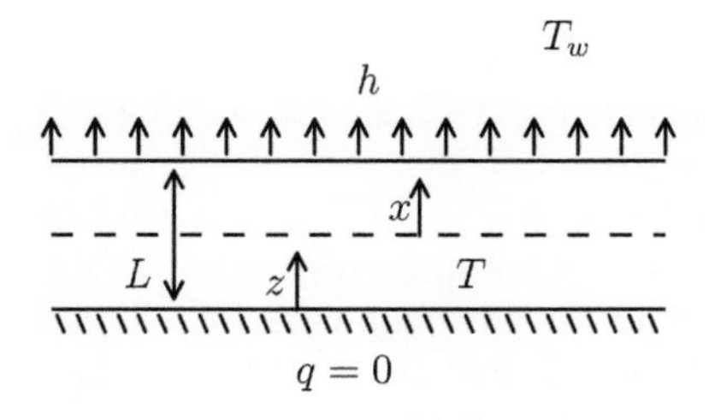

Figure 2-1 Coordinate systems.

Convection from the solid surface is assumed to be dominated by convection to water and heat transfer coefficient is defined as $h = q/T_s - T_w)$ where T_s is the solid surface temperature. All modes of heat transfer are included in this single coefficient. If convection to the water was truly the only mode of heat transfer, the upper limit of heat transfer could be calculated from the rain rate and heat capacity as

$$h_{ideal} = \frac{Q}{A\Delta T} = \frac{\dot{m} c_w \Delta T}{A\Delta T} = \frac{\rho_w RA\cos(\theta)\, c_w \Delta T}{A\Delta T} = \rho_w R c_w \cos(\theta) \tag{2.3}$$

where Q is total heat flux, A exposed solid surface area, c_w water heat capacity, R (m/s) rain rate (often mm/h), $A\cos(\theta)$ is the horizontal projection of solid surface area, and ρ_w water density. As is explained in 5. Results, heat transfer coefficients larger than h_{ideal} do occur.

3 Thermal Stress

Assuming plane stress the thermal stress can be solved using a one-dimensional model from Boley and Weiner [5]

$$\sigma = \frac{\alpha E}{1-\nu}\left(-T + \frac{1}{L}\int_{-\frac{L}{2}}^{\frac{L}{2}} T\mathrm{d}x + \frac{3x}{8L^3}\int_{-L/2}^{L/2} Tx\mathrm{d}x\right) \tag{3.1}$$

Where σ is tension, α thermal expansion coefficient, E Youngs modulus, ν Poisson's ratio, and x distance from center line, see Fig. 2-1. A useful result is that for an infinite heat transfer coefficient there exist a finite upper limit for the thermal stress

$$\sigma_{(h=\infty)} = \frac{\alpha E(T_0 - T_w)}{1-\nu} \tag{3.2}$$

For an infinite heat transfer coefficient, the maximum stress is reached at the surface immediately after the cooling starts.

4 Experimental methods

The rain was generated by drilling 1 mm holes to a plastic box with a 2 cm spacing and keeping the water level in the box constant. The free fall distance of the drops was H = 540 cm ± 1 cm measured from the horizontal solid surface. The plastic surface was rough and produced drops with a diameter of about d = 6 mm corresponding to the upper limit of natural rain drop size [2]. The drops did not have enough velocity to break into smaller drops. There probably is a difference in the heat transfer for different drop size distributions as individual large drops cause large variation in the surface conditions whereas mist like rain would be highly uniform. The plastic box with developing drops is shown in Fig. 4-1. The test was performed for three rain rates R = 110, 340, 1110 mm/h. The rain rates were measured between each measurement with a measuring glass and the mean value used as reference. There was considerable variance over short time scales but they averaged over the long measurement. Measurement time was about half an hour.

The drops were impinged on a copper block with dimensions of 23 x 26 x 6 cm ± 1 mm. The bottom was insulated with a 4 cm thick layer of Glavaflex insulation and the sides with a 2 cm thick layer. The measurement was started from a horizontal position and then repeated for 5 different inclinations: θ = 0, 15, 30, 45, 60, 75°, see Fig. 1-1. A fully wetted surface of the copper block is shown in Fig. 4-1. For the lowest rain rate R = 110 mm/h the surface was not always fully wetted. K-type thermocouples were used for the temperature measurements. There was one thermocouple for the air temperature, one for the water temperature and 11 for the copper block bottom temperature. At equilibrium, all the thermocouples gave the same temperature with less than half a degree deviation from the mean. During the measurement, all the bottom thermocouples gave the same temperature with a few percentages of deviation from the mean. The mean value was used as the measured copper temperature. Temperatures were measured once a second.

During an experiment the copper block was heated to a temperature of 70 – 80 °C, allowed to stabilize, subjected to the rain and was cooled to a temperature of 30 – 40 °C. Some data was always cut from the beginning to allow the heat transfer to stabilize before fitting the constant heat transfer to the results. The minimum sum of deviations for each measured point was used as a fitting criteria. For the larger rain rates an almost perfect fit was observed. For the smaller rain rates the heat transfer coefficient was slightly larger in the begin and lower in the end. This is presumed to be caused by evaporation. Even as the solid temperature is well below the boiling point a higher temperature still causes more evaporation than a low temperature. This would also be more visible for a smaller rain rate.

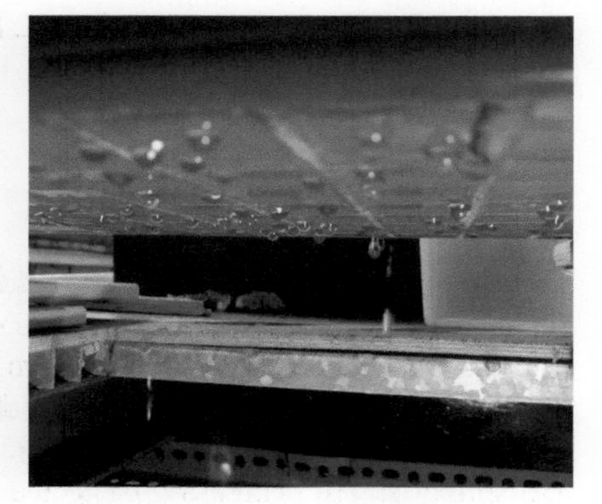

Figure 4-1 Copper block and plastic box.

5 Results

The results for the measured heat transfer and calculated stress are given below. In Fig. 5-1a the measured heat transfer coefficients for each of the rain rates and the inclinations are given. It is observed that larger rain rate causes larger heat transfer coefficients. This effect would most likely become less important with even larger rates as a smaller proportion of the water would be in contact with the solid. The largest heat transfer coefficient for each of the rain rates is observed near the horizontal orientation with small inclination. For large inclinations, the heat transfer drops. The drop is likely caused by the smaller solid area projected to horizontal direction and therefore a smaller amount of rain hitting the surface. Heat transfer, however, drops much slower than would be apparent from the decreasing projected surface area. This is probably caused by the water falling down the inclined surface. The maximum heat transfer somewhere between $\theta = 0 \ldots 30°$ is probably caused by these two opposing processes.

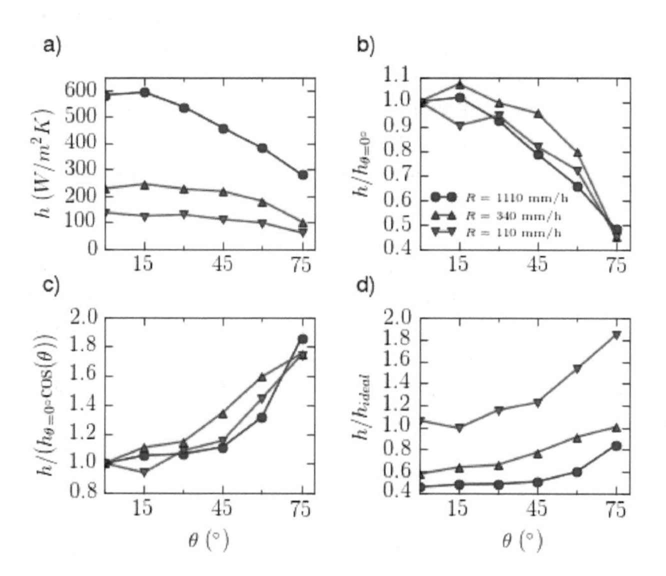

Figure 5-1 Effect of inclination and rain rate on heat transfer coefficient.

To better capture the effect of inclination on heat transfer, each of the heat transfer coefficients is divided by the heat transfer coefficient for a horizontal orientation with the same rain rate, $h/h_{\theta=0°}$. From Fig 5-1b it is now easily observed that the heat transfer coefficient responds similarly to inclination for each rain rate. It is assumed that the local minimum for rain rate R = 340 mm/m at θ = 15° is a random error. In order to visualize the comparison of decreasing projected surface area and water flowing down the surface the results in Fig. 5-1b were divided by cos(θ), see Fig. 5-1c. It is clear the downward motion of the water has a large effect as values are larger than unity.

In Fig. 5-1d, the measured heat transfer is compared with the maximum value of pure convection to water, see Eq. 2.3. For the rain rates R = 340, 1110 mm/h the pure convection with water could explain all heat transfer as $h/h_{ideal} < 1$. For R = 110 mm/h the ratio is $h/h_{ideal} > 1$ and some other forms of heat transfer must also be present. Convection to air and radiation are not enough to explain the magnitude of the extra heat transfer. Natural convection and radiation usually have an effect of less than 10 W/m² and most of the surface was wetted. Therefore it is assumed that extra heat is converted to phase change heat. Some mist was also visible during experiments indicating the presence of water vapor. Some amount of evaporation is probably presence for R = 340, 1110 mm/h even though it is not visible in the measured data.

Using the maximum heat transfer coefficient for each of the rain rates, h = 600, 250, and 140 W/m²K, and solving the maximum tensile stress near the glass surface for various glass thicknesses results in the plots in Fig. 5-2a. The initial glass temperature is T_0 = 80 °C and water temperature is T_w = 0°. Soda-lime-silica glass material properties are used. For all but a few world record approaching rain rates thermal stresses resulting from the mean heat transfer are below 10 MPa and usually much lower. This is below the usually cited allowable non-heat treated glass tensile stresses but of the same magnitude. It therefore seems unlikely that rain would break even non-heat treated glass but the possibility can not be ruled out. It must be noted that the thermal stresses should be added to those caused by other reasons such as gravity or wind. Too rigid a support will also cause higher thermal stresses.

In Fig. 5-2b, the maximum stresses resulting from given heat transfer coefficients are shown. The same material properties and temperatures as in Fig. 5-2a are used. The most interesting detail is that for an infinite heat transfer corresponding to a step change in the surface temperature there is a finite value of stress, namely $\sigma_{max} \approx 65$ MPa. This amount of surface compression stress is easily available in tempered glass. It should also be noted that all the stresses respond linearly to temperature differences.

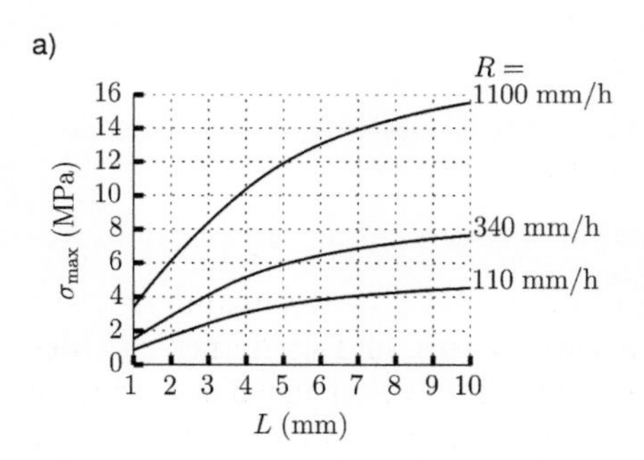

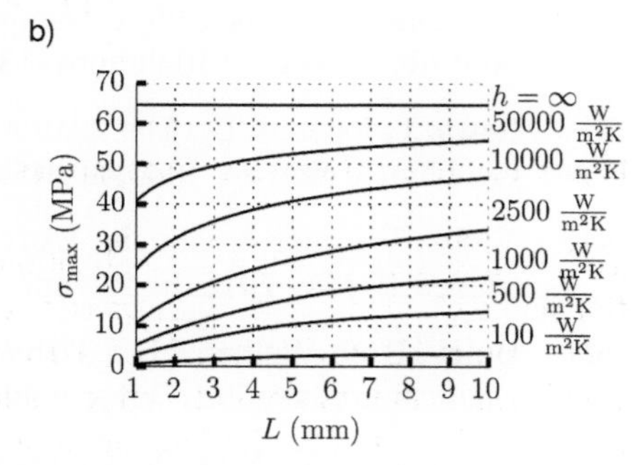

Figure 5-2 Effect of heat transfer on maximum tension for a given rain rate in a) and given heat transfer coefficient in b).

A few attempts to break 2 mm and 4 mm glasses by placing them on the heated copper block and letting rain fall them were also made. Rain rates up to $R = 3000$ mm/h were tested. None of the glasses broke. The effect of locally varying heat transfer on thermal stresses should be studied. Wind effects should also be studied. According to the measurements made by Rabadiya and Kirar [6] heat transfer resulting from wind with no rain is an order of magnitude smaller than rain heat transfer measured in the present paper.

6 Conclusion

Heat transfer between a solid surface and very heavy rain was measured using a copper block and free falling drops. The maximum surface stresses were then calculated for various glass thicknesses. The rain rates were $R = 1100, 340, 110$ mm/h and the corresponding maximum heat transfer coefficients were $h = 600, 250, 140$ W/m^{2}K. Glass surface stress depends on the glass thickness and temperature difference but was observed to be below 10 MPa for all but the heaviest rain rates. This is below but of the same order as usually cited allowable non-heat treated soda-lime glass tensile stresses. In the absence of other forms of stress there is no need to temper soda-lime glass exposed to rain. A theoretical maximum stress corresponding to an infinite heat transfer coefficient is $\sigma_{max} \approx 65$ MPa. The maximum value is easily achievable in tempered glass. The present paper did not study the effect of local heat transfer resulting from individual drops impinging on the solid surface. Heat transfer is likely to be much higher near the impinging drops.

7 References

[1] Rantala, M.: Heat Transfer Phenomena in Float Glass Heat Treatment Process. Phd thesis, Tampere University of Technology. 2015.

[2] Villermaux, E., Bossa, B.: Single-drop fragmentation determines size distribution of raindrops. In: Nat Phys. Nature Publishing Group. Vol. 5, 2009, pp 697-702. http://dx.doi.org/10.1038/nphys1340.

[3] Laws, J., Parsons, D.: The relation of raindrop-size to intensity, ii. In: Trans. Am. Geophys. Union 24, 1943, pp. 452-460.

[4] Global Weather & Climate Extremes. World Meteorological Organization. Retrieved 16 May 2016. http://wmo.asu.edu/#global.

[5] Boley, B.A., Weiner, J.H.:. Theory of Thermal Stresses. In: Dover Civil and Mechanical Engineering. Dover Publications, 2012.

[6] Rabadiya, A.V., Kirar, R.: Comparative Analysis of Wind Loss Coefficient (Wind Heat Transfer Coefficient) For Solar Flat Plate Collector. In: Int. Jour. of Emerging Technology and Advanced Engineering. Vol. 2, 2012. ISSN 2250-2459.

Mosaic Modules for Improved Design Options in Building Integrated Photovoltaic Modules

Matthieu Ebert[1], Max Mittag[1], Tobias Fellmeth[1], Alma Spribille[1], Tom Minderhoud[2], Ger Gijzen[2], Karoline Fath[3], Robert Hecker[3], Ulrich Eitner[1]

1 Fraunhofer Institute for Solar Energy Systems ISE, Heidenhofstr. 2, 79110 Freiburg, Germany

2 UNStudio, Stadhouderskade 113, PO Box 75381, 1070AJ Amsterdam, Netherlands

3 Züblin Zentrale Technik, Albstadtweg 3, 70567 Stuttgart, Germany

The integration of photovoltaics in façades requires the development of new techniques for aesthetical and efficient photovoltaic module designs. Architectural demands cannot be satisfied with industrially available solar modules due to strict design limits [1, 2]. Industrial solar cells are usually square and available with an edge length of 156 mm. Cell interconnection of commercial solar modules requires strings of solar cells that strictly limit possible designs. Transparent areas or non-regular positioning of solar cells is only possible by cost intensive manual module assembly. We present a novel module concept to overcome design limitations by cell size and electrical interconnection that allows higher degrees of freedom. The module concept is based on an electrical conductive back sheet and back-contact solar cells.

Keywords: building integrated photovoltaic modules, semitransparent PV modules, design freedom, back contact PV modules

1 Design concept

Overcoming the limitation in PV module design through new technological concepts is a key improvement for a higher acceptance of photovoltaic components in the building industry [1, 2]. The current PV module design is strongly limited by the solar cell size and appearance. The mosaic module concept breaks up the typical 6 by 6 inch pattern by using solar cells flexible in size in the range from 2 x 2 cm² up to 156 x 156 cm² full wafer size. This concept allows new design possibilities making the PV module a façade design element. Figure 1-1 displays the concept with one design example.

Engineered Transparency 2016. Glass in Architecture and Structural Engineering. First Edition.
Edited by Jens Schneider, Bernhard Weller.

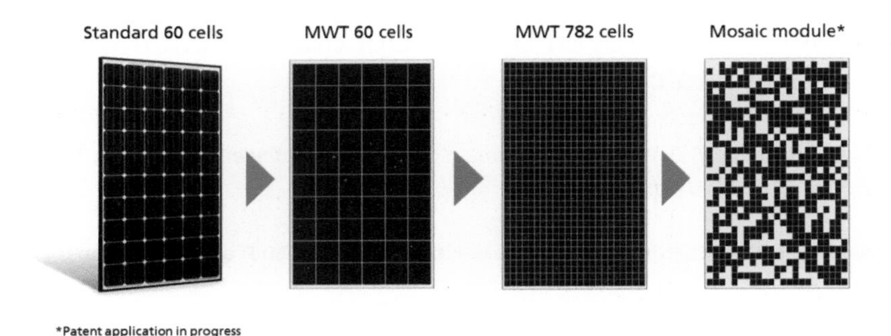

Figure 1-1 Representation of mosaic module design concept.

Additionally to the flexible cell sizes, the interconnection layer (copper) and the backsheet (BS) layer can be used as design elements. They can be coloured and patterned as desired. In Figure 1-2 the different layers are shown.

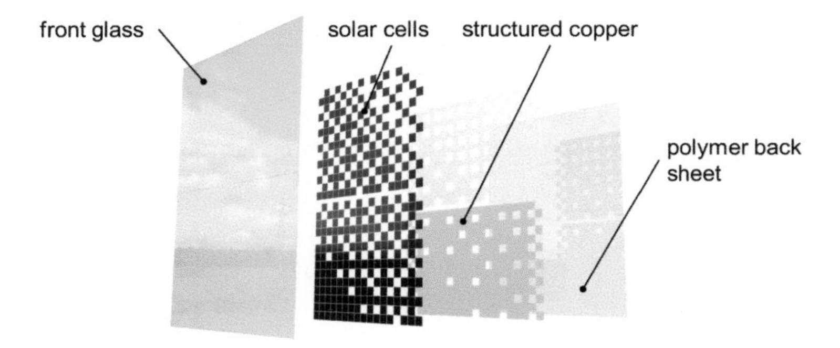

Figure 1-2 Layers in a mosaic module: 1. Front glass, 2. Solar cell matrix, 3 Structured copper layer, 4. Polymer backsheet.

2 Solar Cells

AP-MWT Cell concept

The mosaic module concept is built on back contact metal wrap through solar cells [3, 4]. This solar cell concept allows higher cell efficiency due to reduced front side shading by interconnection and maximum flexibility regarding cell interconnection [3]. In addition, the AP-MWT concepts features the possibility to cut the 156 x 156 cm^2 sized MWT-cells into subcells with a multiple of the unit cell format of 2.25 x 1.0 cm^2.

Mosaic module cells

The All-Purpose cell concept used within the mosaic module is based on 4-by-4 unit cells. The size of one cell is thus 70 mm x 67.5 mm. The front and rear cell layout is displayed in Figure 2-1.

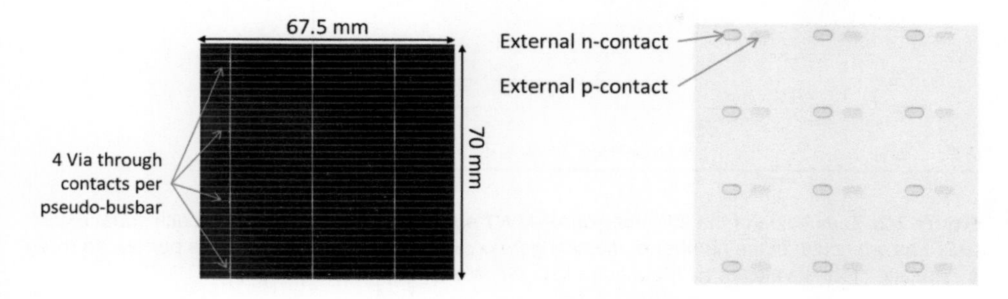

Figure 2-1 Photograph of All-Purpose Metal Wrap Through solar cells in the Construct PV design. Left: Front side of an AP-MWT solar cell, which features 4 via through contacts per pseudo-busbar and 3 pseudo-busbars. Right: Rear side of an AP-MWT solar cell, which features 4 external n-contacts underneath the pseudo-busbar.

In order to adapt the AP-MWT cells for façade applications with reduced irradiance in comparison to standard field PV, a simulation of the cell efficiency as a function of the metal fingers on the front was performed using GridMaster [5].

The simulation is based on an illumination of 700 W/m^2 for vertical façade integration. The optimal metallization features 37 fingers on the front side. The optimal metallization layout was applied for the fabrication of the AP-MWT solar cells of the demonstration modules. The resulting metallization with a reduced numbers of silver fingers leads to a solar cell with a more homogeneous blue appearance.

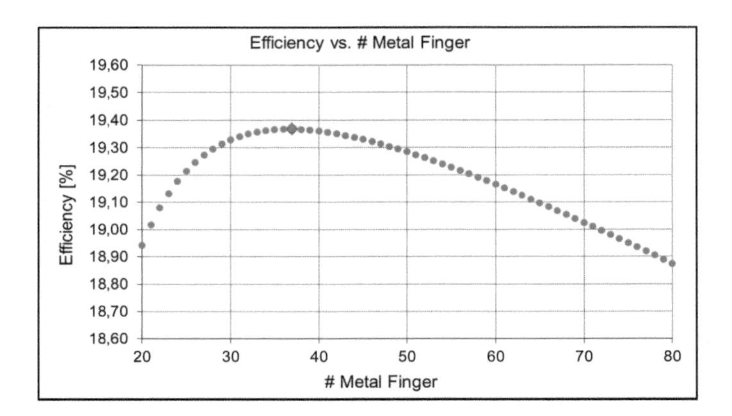

Figure 2-2 Simulation of the efficiency of AP-MWT solar cells with an aluminum back surface field (BSF) as a function of the number of metal fingers on the front. The simulation was performed using a simulation tool developed by Fraunhofer ISE: GridMaster [5].

2.1 Electrical interconnection

To obtain a useful current and voltage range, the solar cells are interconnected in series and in parallel in the module. Several different interconnection methods have been reported for back contact MWT solar cells in the past [6-10]. In the mosaic module the cells are interconnected with an electrically conductive back sheet. The copper layer on the polymer backsheet is structured according to the electrical design. Figure shows a structured backsheet with a couple of cells placed during manufacturing.

Figure 2-3 Photograph of a copper coated structured backsheet used to interconnect the MWT solar cells.

2.2 Manufacturing concept

The concept is designed to allow a fast and flexible module production process based on industrial standard pick-and-place (PnP) processes. In the Fraunhofer ISE ModuleTEC module prototypes are manufacture by using a six axis robot for positioning the solar cells on the patterned conductive back sheet. The pick-and-place process is assisted by a computer vision system to increase the positioning precision.

Figure 2-4 Photograph of the automated manufacturing facility in the Fraunhofer ISE Module Tec.

Separate soldering of the solar cells is not necessary since the production process is optimized for an inline soldering during the lamination of the module. A low temperature melting solder paste is applied on the rear contact pads of the solar cell before the PnP process. This solder paste melts during the module lamination. The quality of the solder joint is examined with X-Ray imaging. Figure 2-5 shows a X-Ray image of two soldered solar cells and a close up of one solder joint.

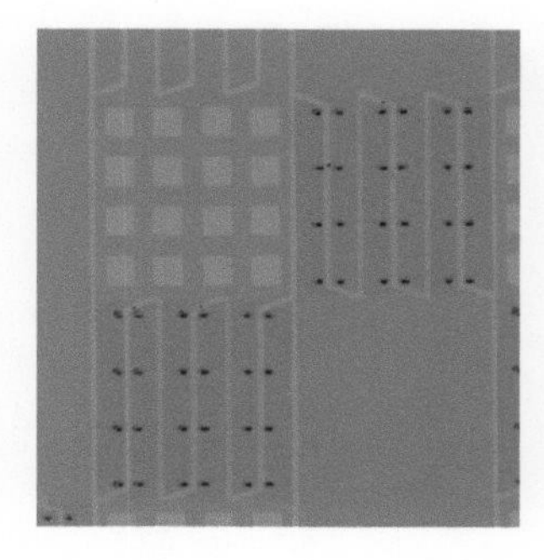

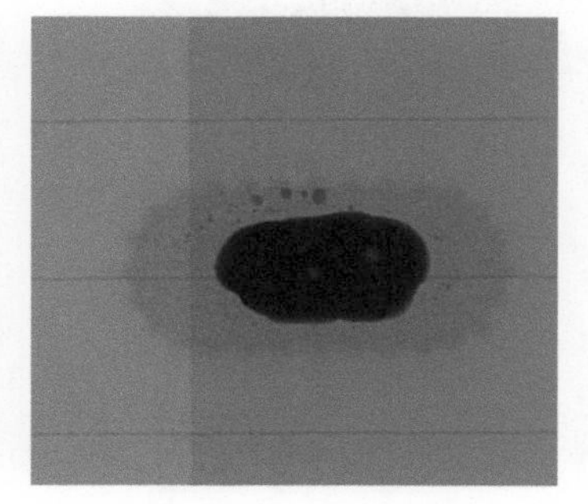

Figure 2-5 Left: X-Ray image of a small scale a mosaic module sample showing the solder connection. Right: X-Ray image of a single solder joint.

3 Demonstration Mock up

In the framework of the European project Construct PV [11] several demonstration modules are built and integrated in a full size façade mock up for demonstration purpose. Figure 3-1 shows a photograph of one of the modules built for the mock up. This module has three design layers where colours and patterns can be varied: Solar Cell, Copper layer and backsheet transparency.

Figure 3-1 Photograph of the a mosaic module for the façade mockup.

Figure 3-2 shows the façade mockup with the two mosaic modules on the top left on the Advanced building skin conference in fall 2015 in Bern.

Figure 3-2 Photograph of the façade mockup built by the company Zublin in the project Construct PV.

4 Façade design

The mosaic module concept allows more than just design on module level. With a combination of several mosaic modules, an entire façade design is possible. In the demonstrated mock up the pattern designed by UNStudio uses two mosaic modules.

5 Acknowledgement

This work is funded by European Commission through Grant agreement no 295981 in the project Construct PV [11].

6 References

[1] SUPSI, Institute for Applied Sustainability to the Built Environment (ISAAC), Lugano, Switzerland, *FROM BIPV TO BUILDING COMPONENT.*

[2] UNStudio, *DESIGN INNOVATION FROM PV-MODULE TO BUILDING ENVELOPE: ARCHITECTURAL LAYERING AND NON APPARENT REPETITION.*

[3] T. Fellmeth *et al, Eds, Industrially feasible all-purpose metal-wrap-through concentrator solar cells.* Photovoltaic Specialist Conference (PVSC), 2014 IEEE 40th, 2014.

[4] B. Thaidigsmann, M. Hendrichs, and S. Nold, "P-type MWT solar cells: current status and future expectations," in *Proceedings of the 28th European Photovoltaic Solar Energy Conference and Exhibition*, 2013.

[5] T. Fellmeth, F. Clement, and D. Biro, "Analytical Modeling of Industrial-Related Silicon Solar Cells," *Photovoltaics, IEEE Journal of,* vol. 4, no. 1, pp. 504-513, 2014.

[6] U. Eitner, D. Eberlein, and M. Tranitz, "Interconnector-Based Module Technology for Thin MWT Cells," in *Proceedings of the 27th European Photovoltaic Solar Energy Conference and Exhibition*, 2012, pp. 3461-3464.

[7] M. Francis, "Conductive Adhesives for Back Contact Applications," in *Proceedings of the 5th Workshop on MWT Solar Cell and Module Technology*, 2013.

[8] B. Verschoor, "Backcontact module fabrication," in *Presentation at the 4th International Conference on Crystalline Silicon Photovoltaics,* 's-Hertogenbosch, Netherlands, 2014.

[9] Ali, "Conceptual comparison between standard Si solar cells and back contacted cells," in *Proceedings of the 4th International Conference on Crystalline Silicon Photovoltaics*, 2014.

[10] G. Beaucarne, "Study of compatibiliy of silicone-based electrically conductive adhesives and conductive backsheets for MWT modules," in *Proceedings of the 4th International Conference on Crystalline Silicon Photovoltaics,* 2014.

[11] Construct PV Website, *Construct PV Website.* [Online] Available: http://www.constructpv.eu/project/.

TPEDGE: Glass-Glass Photovoltaic Module for BIPV-Applications

Max Mittag[1], Tobias Neff[2], Stephan Hoffmann[1], Matthieu Ebert[1], Ulrich Eitner[1], Harry Wirth[1]

1 Fraunhofer Institute for Solar Energy Systems, Heidenhofstr. 2, 79110 Freiburg, Germany

2 Bystronic Glass, Karl-Lenhardt-Str. 1, 75242 Neuhausen, Germany

With TPedge we present an advanced frameless, polymer free encapsulation concept for silicon solar cells which addresses several disadvantages and significant cost factors related to conventional solar modules. TPedge represents a gas-filled, edge sealed, glass-glass module without polymeric encapsulation foils that requires less module production time. The cost calculation indicates 15.3 % lower module material costs for TPedge production compared to the standard module production due to savings for encapsulation foils and frame. The results from successful and extended module testing such as 400x thermal cycling (ΔP_{STC} = 0 %), 4000 hours damp-heat (ΔP_{STC} = -1.3 %) or 5400 Pa (ΔP_{STC} = -0.9 %) mechanical load testing show that several critical IEC tests are passed.

Keywords: PV module, module manufacturing, durability, reliability, cost reduction, building integrated PV (BIPV), encapsulation, façade, polymer film

1 Introduction

Conventional photovoltaic modules use polymeric foils like ethylene-vinyl acetate (EVA) as solar cell encapsulation. Several effects are known that cause failure or power loss of the solar module and are directly related to the encapsulation material or incomplete protection of the solar cells from environmental influences [1]. Degradation effects such as discoloration, degradation and corrosion that is increased by the generation of acetic acid, make lifetimes longer than 30 years difficult to reach [2]. Additionally a lamination process is needed during module production with polymer foils. This process takes 8-15 minutes and is known to represent a bottle neck in solar module production. The costs of the foils are a substantial part of the module production costs. Furthermore conventional module concepts rely on aluminum frames to ensure mechanical stability. Those frames add again a significant cost factor to PV module production [3].

Various innovations have been proposed to overcome the disadvantages of polymeric encapsulants including other material groups [4], multi stage laminators or glass-glass-laminates with improved aging stability. None of these measures were able to completely eliminate the intrinsic disadvantages of the encapsulation foils at competitive cost levels.

In this work we show that with the TPedge approach a significant improvement of long term stability and simultaneously a decrease of module production costs can be realized.

Engineered Transparency 2016. Glass in Architecture and Structural Engineering. First Edition.
Edited by Jens Schneider, Bernhard Weller.

The TPedge-module concept applies an edge sealing process, well known from the manufacturing of double glazing insulation windows. The sealing consists of a thermoplastic spacer (TPS) filled with drying silicates and a silicone which renders the mechanical stability of the module. The glass spacing is filled with air. A double side coated ARC front glass is required to minimize reflection losses. Small pins consisting of an UV-curing adhesive, glue the solar cells to the rear side glass pane. Glass spacing is provided by a set of transparent distance pins on the front side of the solar cells that cover approximately 0.02 % of the cell area and provide additional mechanical stability. Metal frames or similar additional supporting constructions are not necessary. A wide range of advantages such as simple recycling, reduced fire load and a potential for the use of larger formats results from neither using foils nor lamination. Figure 1 shows a schematic drawing of the cross section of a TPedge-module with the positions of the adhesive pins on the front and backside of the solar cell.

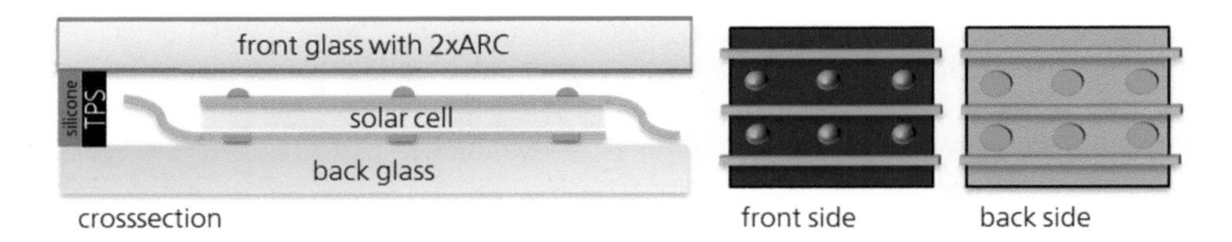

Figure 1-1 TPedge-module sketch with position of adhesive pins and double layer edge sealing.

2 Experimental

2.1 Module Production

The production of TPedge-modules is a combination of well-manageable processes that enable reliable and fast manufacturing: glass washing, dispensing of fixation pins (back side), string layup, UV-curing of fixation pins, dispensing of distance pins (front side), UV-curing of distance pins, TPS application (primary seal), sealing press, silicone application (secondary seal).

TPS application with an industrial applicator is shown in Figure 2-1 (left). Figure 2-1 (right) shows dispensing of the adhesive pins with a semi-automated robot system installed at Fraunhofer ISE.

Figure 2-1 TPedge-module production (primary sealing with a Bystronic TPS-applicator, left) and at Fraunhofer ISE (dot dispensing, right).

Cycle times of less than one minute are expected. A production line has been projected using several existing machines that are already used for prototype manufacturing.

Different module setups are produced for detailed module testing and qualification. The modules of setup A contain 60 solar cells. To prove flexibility of the concept several commercially available monocrystalline solar cells by different manufacturers are used. The full size modules are built with 3 mm thick, hardened float glass sized 1640 x 1000 mm. For mechanical testing some modules are built with 2 mm thick glass.

The modules of setup B are customized BIPV modules made of 42 MWT-HIP-back contact solar cells produced at Fraunhofer ISE [5] and connected by structured interconnectors [6]. Glass sizes are 1240 x 1005 mm. A 4 mm thick front glass is partially black enameled on the module edge for architectural purposes. The back glass pane is 5 mm thick. Junction boxes are applied on the back side covering a drilled hole.

2.2 Module Qualification

Module qualification is performed at Fraunhofer ISE TestLab PV modules. Accelerated aging as well as mechanical tests, PID stability and hot spot endurance tests are conducted. Critical test sequences from IEC 61215 / 61730 are performed and extended on several TPedge modules. The test sequence is completed with a two year outdoor exposure in a BIPV façade at Fraunhofer ISE.

Each module type is tested for at least 1000 h damp-heat (85 °C, 85 % r.h.) and 200 thermal cycles (-40 °C / 85 °C). Modules of setup B are also tested for a combined 1000 h damp-heat and 200 thermal cycles. Tests are accompanied by electrical safety measurements. The mechanical stability of the modules is assessed by performing mechanical load tests with uniform loads up to 5400 Pa according to IEC 61215 / 61730.

Three different commercial available module clamps and one backrail system are tested in different configurations (four or six clamps per module and suspended mounting or fixed substructures). Hail and mechanical load tests are additionally performed on modules with 2 mm thick glasses.

A building façade at Fraunhofer ISE is equipped with 70 TPedge-modules of setup B. The first ten modules have been installed in August 2013 and are electrically monitored. Additional modules were installed in September 2015. Figure 2-2 shows the building and the TPedge modules.

Figure 2-2 BIPV-installation of TPedge-modules at Fraunhofer ISE.

2.3 Cost Analysis

A cost of ownership calculation is performed for three different module setups. An industrial standard module, a glass-glass-laminate and a TPedge-module are compared. The calculation takes material and process costs as well as yield rates, productivity and other significant factors into consideration. The input parameters for TPedge are based on the projected module production line. The calculation is performed for an annual production of approximately 200 MWp. All module concepts are calculated with four busbar, 6" solar cells, a cell efficiency of 18.5 % and a cell price of 0.30 €/Wp.

3 Results

3.1 Thermal Cycling and Insulation Resistance

Thermal cycling is performed with different TPedge-module setups. Aside from some finger interruptions due to failure of the front side metallization no module related failure mechanism can be observed. For an extended test (400 thermal cycles) module B2 (for BIPV, 42 cells) is used and shows no additional micro cracks after the test. The structured interconnectors are successfully tested in combination with the TPedge-setup.

Power measurements are performed before and after the test and show a power loss within measurement uncertainty. Detailed results are listed in Table 3-1.

Table 3-1 Results of power measurements before and after thermal cycling tests (-40 °C / 85 °C).

Module No.	Thermal cycles	Initial power [W]	Power after test [W]	Change [%]
A4	200	219.6	221.2	+0.7
B2	400	171.1	171.1	±0.0
B3	200	169.1	167.4	-1.0

3.2 Damp Heat

Several modules are tested for extended periods under damp-heat conditions (85 °C, 85 % r.h.). Tests are hardened by performing combined tests to modules of setup B. The test of module B3 is performed after a thermal cycling test. The results show an excellent damp-heat resistance of TPedge-modules.

Table 3-2 Results of power measurements before and after damp-heat tests (85 °C, 85 % r.h.).

Module No.	Test length [h]	Initial power [W]	Power after test [W]	Change [%]
A8	4000	262.6	260.7	-0.7
B1	2000	163.5	164.6	+0.7
B3	1000	169.1	167.4	-1.0

3.3 Mechanical Load test

The results of the mechanical load test show strong deflection of all tested modules during the test. Figure 3-1 shows the deflection of a TPedge-module built with 2 mm glass panes under a load of 2400 Pa mounted with backrails.

Modules mounted with clamps face the risk of modules slipping out of the clamps due to high deflection which would result in test failure. Nonetheless feasible solutions could be identified that allow a safe module mounting for loads up to 5400 Pa with commercially available clamp systems. Also backrail systems are successfully tested. With backrail mounting the module is glued to the mounting structure and therefore the deflection is reduced and no slipping is possible. A module with 3 mm glasses has been successfully tested for extended loads up to 5400 Pa.

Figure 3-1 TPedge-module with 2 mm glass panes, backrails and supported mounting during mechanical load test (2400 Pa).

Table 3-3 shows the results of power measurements performed before and after the mechanical load tests. No significant change in the module power output can be observed after mechanical load test. On module B1 the mechanical load test is performed after 1000 h of damp-heat. Power measurements indicate no power loss within measurement uncertainties before and after the mechanical load test procedures.

Table 3-3 Results of power measurements before and after mechanical load tests.

Module No.	Initial power [W]	Power after test [W]	Change [%]
A2	221.0	219.4	-0.9
A3	220.4	221.6	+0.6
B1	163.5	164.6	+0.7

3.4 BIPV Application

Ten modules have been installed and are electrically monitored since August 2013. Additional 60 modules were installed in September 2015. Figure 3-2 shows the results of performance measurements during outdoor installation.

After two years of service a module is taken off the façade for inspection. Electroluminescence images show no degradation or cell damage. A power measurement shows no power loss within measurement uncertainty. Results of extended accelerated aging, monitoring and module inspection after two years of outdoor installation show a good resistance of the TPedge module concept against environmental stress factors.

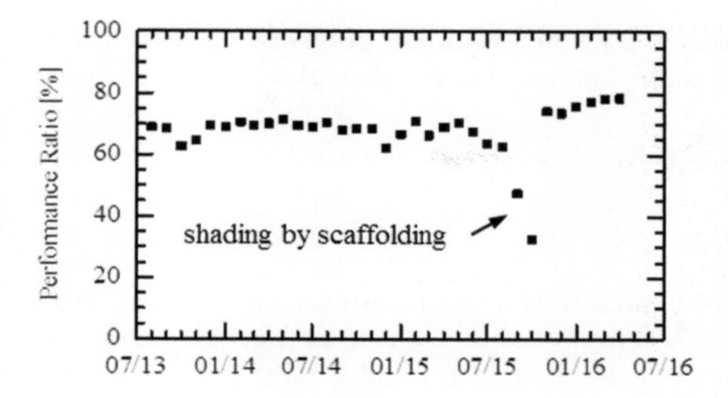

Figure 3-2 Performance Ratio of façade modules.

3.5 Cost Analysis

A cost calculation is performed for three module setups. Results show a significant advantage of the TPedge-concept compared to the industrial standard as well as to a glass-glass-laminate. Saving the metal frame is the main factor for the glass-glass-module's price advantage compared to the standard module. The TPedge-module's material costs are 15.3 % lower than the standard module's and 8.5 % below the glass-glass laminate. Figure 3-3 (left) shows the material cost structure of different module setups.

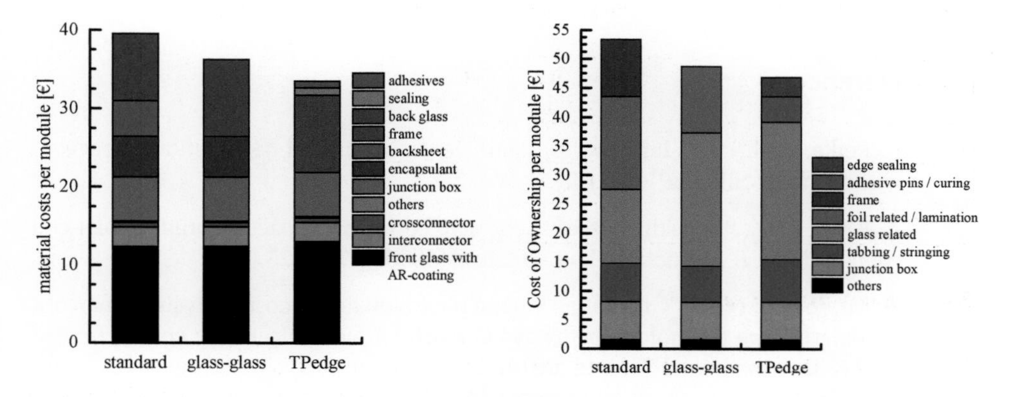

Figure 3-3 Material cost structure for different module setups (without solar cells) (left), Cost of Ownership analysis for different module setups (without solar cells) (right).

Considering production steps and equipment, TPedge is able to compete with the other module concepts. While TPedge module production has more production steps (adhesive dispensing, UV-curing, TPS application, pressing, secondary sealing) than standard module production they are faster (cycle time < 1 minute). Figure 3-3 (right) shows the cost of ownership structure for different module setups.

The use of the TPedge module concept leads to a total cost of ownership reduction of 12.3 % compared to the standard module. Compared to the glass-glass-laminate TPedge saves 8.8 %.

4 Summary & Outlook

The TPedge-module concept has been intensively tested on critical stress factors. Tests have been successfully passed with several full size modules. TPedge has passed the damp-heat-test (4000 h), the thermal cycling test (400 cycles), mechanical load tests (5400 Pa), hail stability tests, PID stability tests, a hot spot endurance test, UV stability tests and an outdoor exposure for more than two years. Accompanying electrical safety tests have been passed as well.

The maturity of the TPedge module concept has been demonstrated. Customized BIPV modules as well as full size modules with different commercially available solar cells have been manufactured and innovative features such as back-contact cells and structured interconnectors have been used. Possibilities of an automated production have been successfully demonstrated and a module production line was projected.

A cost analysis shows that TPedge offers advantages in material costs and costs of ownership compared to industrial standard modules as well as glass-glass-laminates.

5 References

[1] C. Peike, et al., "The influence of laminate design on cell degradation", Proceedings of the 3rd SiliconPV, 2013.

[2] A. Kraft, et al., "Investigation of acetic acid corrosion impact on printed solar cell contacts". IEEE Journal of Photovoltaics Vol.5 No. 3, 2015.

[3] D. M. Powell, et al., "Crystalline silicon photovoltaics: a cost analysis framework for determining technology pathways to reach baseload electricity costs", In: Energy & Environmental Science, 2012.

[4] M. Poliskie, "Solar Module Packaging – Polymeric Requirements and Selection", CRC Press, 2011.

[5] Drews, A., et al., "HIP-MWT solar cells-pilot-line cell processing and module integration", Proceedings of the 27th EU PVSEC, 2012.

[6] U. Eitner, et al., "Interconnector-based module technology for thin MWT-cells", Proceedings of the 27th EU PVSEC, 2012.

Thin film and crystalline glass/glass modules in BIPV – desiccated edge seals for improved durability and output

Christoph Rubel[1], *Lori Postak*[2]

1 Edgetech Europe GmbH, Quanex Building Products, Heinsberg, Germany, christoph.rubel@quanex.com

2 Quanex IG Systems, Quanex Building Products, Solon, Ohio, USA, lori.postak@quanex.com

Manufacturing of thin film PV modules is very much like producing laminated glass with a coated layer to the inside of one glass sheet. However, the precious and sensitive PV coating on the glass needs more protection against moisture intake than an ordinary laminated glass would require since moisture degrades PV cell activity coatings that are critical to module performance (anti-reflective coating and transparent conductive oxide). Maintaining the PV output efficiency demands better protection against moisture intake through the glass edge. This can be done with a thermoplastic butyl-based sealant that has a fragment of the moisture permeation compared to PVB. Adding desiccant to the butyl formulation critically extends the barrier properties of the edge seal extending the expected service life and cumulative output efficiency of the PV module.

Keywords: thin film, crystalline, glass/glass PV module, output efficiency, durability, BIPV

1 Photovoltaic cells and modules

1.1 Silicon solar cells and thin film cells / modules

The „classical", this means the most commonly used photovoltaic modules consist of silicon cells, which in itself group different types of solar cells, such as polycrystalline or monocrystalline cells, or amorphous crystalline cells. These cells are all made out of silicon as the photoelectric or semi-conductor base substrate. Solar cells are differentiated by its thickness, or clustered into „development generations". Both differentiations are good enough for the level of detail that is described in this paper. [1]

The most-thick solar cells, or the first generation solar cells, conventional and predominantly used as of today, are the crystalline-, mono- or polycrystalline silicon wafer-based solar cells. The second generation, thinner solar cells, that include amorphous silicon cells, CdTe and CIGS solar cells. These are the thin-film solar cells. The third generation would be a number of different, yet not really commercialized, thin film technologies and organic cells technologies, which are in the different stages of development until nearly ready for commercial application. [2]

The first generation solar cells are typically built into modules that have a top sheet out of glass, the wafers are bonded in different methods to the top sheet glass, and with the electrical connections then sealed up on their back sides against environmental impacts

Engineered Transparency 2016. Glass in Architecture and Structural Engineering. First Edition.
Edited by Jens Schneider, Bernhard Weller.

by a back sheet made out of a polymeric film. This paper is mainly looking at the second generation thin film solar cells, which are not really cells, as the photoelectric circuits are chemically vapor deposited layers of different substrates onto one side of a glass sheet, that is then, with an interlayer of EVA or PVB, laminated to a second sheet of glass. These are then called glass/glass modules. The external behavior and the assembly of the glass/glass laminate is very much like the production process of a laminated (safety-) glass for the building construction industry. This is also what makes this photovoltaic technology so suitable for the use as building integrated photovoltaic system, besides the fact that by arranging the cell structure in a certain way, the efficiency can be maximized without the need to have always optimal orientation of the PV-module towards the sun radiation, or maintaining PV-output when the sun radiation is diffused by cloudy or foggy weather situation. On the other hand, the available building envelope area in west to south facing orientation plus rooftop area is often times big enough to justify the use of BIPV in a façade or building skin, especially when other targets, such as sustainability, environmentally friendly image, etc. play a role for a building or its owner, user, too.

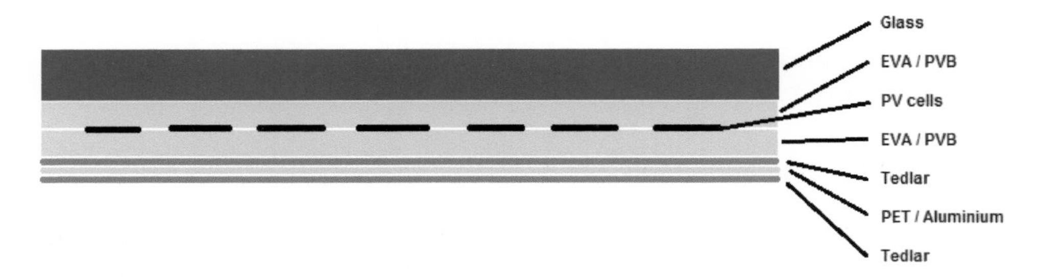

Figure 1-1 Schematic build-up of a 1st Generation PV module based on crystalline cells [3].

1.2 BIPV modules with thin film aSi, Cd/Te or CiGS technology

Figure 1-2 Opaque and semi-transparent BIPV silicone modules [4].

Figure 1-3 BIPV façade with thin film modules: Kulturhaus Milbertshofen, Germany [5].

1.2.1 Schematic assembly process

The major difference between standard first generation PV modules and thin film modules, is that the PV cells in a thin film module are in the coated layers of chemically vapor deposited substrates on the so-called „super strat" top glass sheet. This top glass sheet carries a TCO (transparent conductive) layer, on which different other coatings have been applied, and in parts etched away again, to make the semiconductor and the doped layers, the connections between the cells and the backside contacts.

The photoelectric generation is often times based on a Cadmium-Telluride (CdTe) or a Copper-Indium-Gallium-Diselenid (CIGS) semiconductive technology. This is why these thin film modules are often called CadTel or CIGS modules.

Behind this multi-layer, yet very thin chemical vapor deposited solar cells and its two back contacts, there will be an interlayer of EVA (ethylenvinylacetate) or sometimes PVB (polyvinylbutyral) and the back-side glass sheet.

As mentioned, these TFPV module technologies have an often times lower output efficiency (max. Wp) compared to wafer cell based modules, but can be manufactured in a more automated, therefore more efficient process, and have often times a better efficiency in utilizing solar power over more indirect sun radiation.

So having the two glass sheets and the interlayer, the assembly process of this thin film PV module is then very much like that of a laminated glass, and the handling of these modules are very similar to laminated glass panes. Therefore mounting and application of these glass/glass modules can be very similar to laminated glass in building construction.

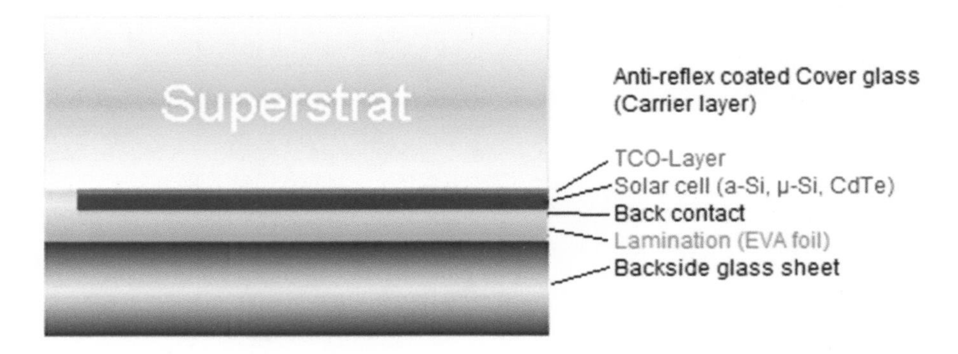

Figure 1-4 schematic build-up of a 2nd generation thin film PV module [6].

As the semiconductive coatings in the thin film module are very sensitive to environmental impact and may degrade, it is important for the thin film PV module to seal the perimeter, the only area that can allow moisture to enter into the module. One way to do this effectively is to use a perimeter seal made out of a butyl-based polymer formulation. A strip, either pre-formed and wound on tape reels or directly extruded as a rectangle from a drum, is applied at the edge area of the one glass sheet, and the pre-cut interlayer is placed within the applied edge seal. In the matching station, the one glass sheet with the edge seal and the interlayer is married with the other glass sheet, slightly compressed and prepared for the laminator. Thick, heat resisting and non-sticking interlayer foils are used as separators when a number of modules are stacked for the laminating process. Using heat and vacuum in a fully automatically controlled process, very similar to the ordinary laminated glass, the glass/glass modules are then laminated, and both, the interlayer and the edge seal, are tightly adhering to the glass sheets as one module.

Using an edge seal that is thermoplastic, based on butyl and contains desiccant in its formulation, has three advantages for the glass/glass module:

1. It is thermoplastic, so it is a mono-component material that changes its viscosity in certain temperature ranges. So it is easy to apply, no mixing needed and can be dosed very precisely.
2. The base of the formulation is butyl, one of the polymers with the highest barrier properties, best weathering and chemical dissolution resistance.
3. It contains desiccant to even further delay the ability for moisture to permeate into the module.

The formulation of the edge seal using butyl and desiccant in the mixture leads to measurable improvements in durability helps maximize the lifetime power output per module.

The following study shows the difference in moisture penetration when using no edge seal, a butyl-based edge seal and a desiccated, butyl-based edge seal in a glass/glass module. The first two test specimens were aged at maximum relative humidity at 60 °C for 5 weeks. One test specimen was a glass/glass laminate with just the EVA interlayer and no edge seal. The other test specimen had a butyl-based and desiccated edge seal. Both specimens were prepared with dots of cobalt chloride, a substrate that changes color from blue to pink when the moisture level increases over a certain amount. In just five weeks moisture penetrated 64 mm into the laminated glass without edge seal. With desiccated butyl, no moisture penetration can be measured.

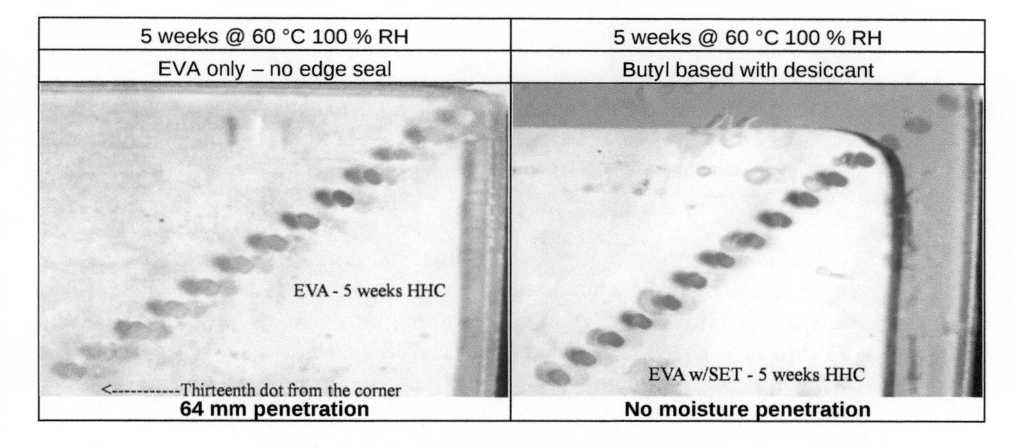

Figure 1-5 Glass/glass module, EVA without edge seal (left) vs. EVA with edge seal (right) [7].

The second part of the study compares a butyl-based edge seal with and without desiccant in the formulation. Again cobalt chloride dots were used to indicate moisture ingress. The test specimen were aged at maximum relative humidity at 60 °C for 13 weeks. In thirteen weeks, moisture penetrated 64 mm into the laminated glass with butyl edge seal without desiccant. With desiccant, only 8 mm of penetration occurred.

Butyl edge seal alone extended time before significant moisture penetration by 2 - 3 times (from 5 weeks to 13 weeks). With desiccated butyl, the extension time was many times greater, not even measurable in this study.

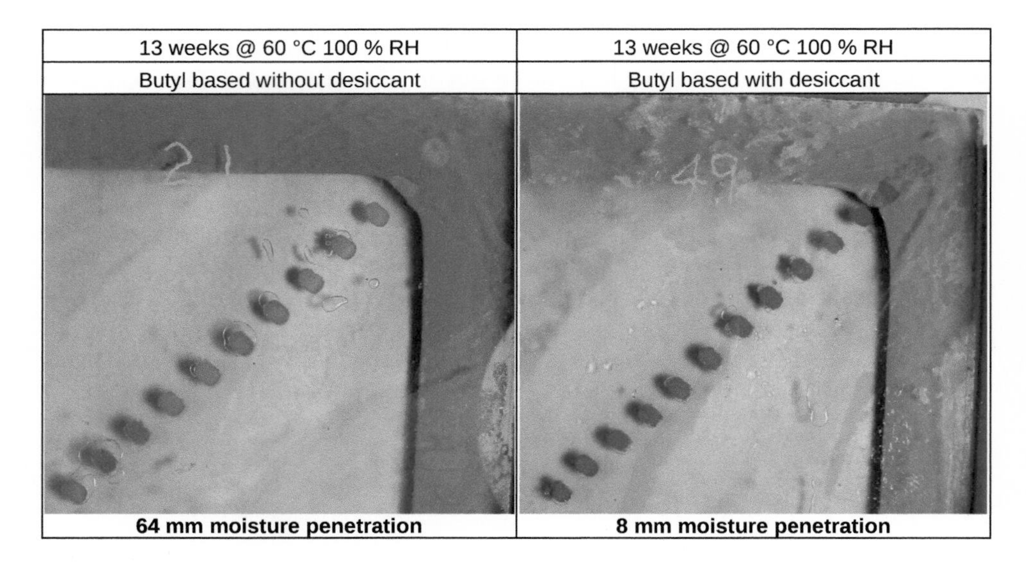

Figure 1-6 Glass/glass module edge seal without desiccant [8] (left) and glass/glass module edge seal with desiccant (right).

These two studies indicate on a real data basis that the durability of a glass/glass PV module can be increased by using a butyl based edge seal, and can be even more increased by using a desiccated, butyl-based edge seal. Although this study is performed using EVA, since PVB has a higher MVTR than EVA, similar results are expected with PVB. Shockingly, BIPV modules installed with PVB laminate today are not using an edge seal because they pass the PV design and safety standards. However, IEC 61730, IEC 61215 and IEC 61646 assess the initial qualification/rapid failure and do not assess module „wear out" meaning longevity.

Now, how can this effect of significant time delay for moisture ingress when using a desiccated butyl be explained? Encapsulants such as EVA or PVB have a moisture penetration rate of around 15 to 50 g/m² and day, depending on how this is measured.

An edge seal, like a desiccated butyl-based solar edge tape, has a moisture permeation rate of below 1 g/m² and day, sometimes just a fracture. This shows already that it is significantly more difficult for moisture molecules to penetrate into a laminated glass module when there is an edge seal around the perimeter of a glass/glass module. And the longer the path length (=wider the edge seal) the more difficult it is for the moisture to penetrate. If this edge seal is then also loaded with desiccant to capture one moisture molecule at a time, one can imagine that it becomes even harder for moisture molecules to pass this barrier loaded seal.

Graphic of moisture transfer through a film until steady state MVTR is reached.

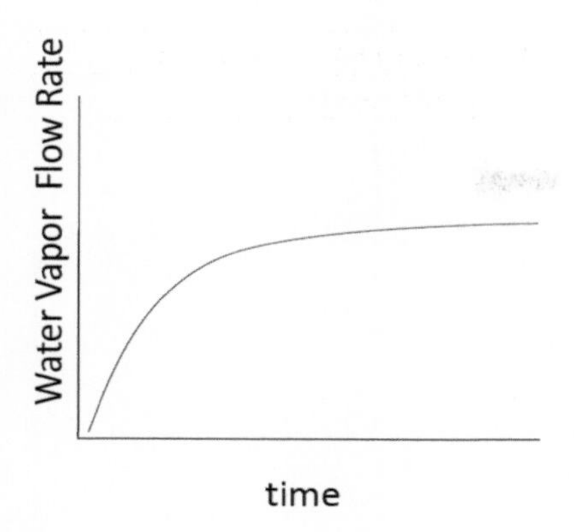

Figure 1-7 schematic explanation of moisture molecules to pass through a conventional film steady state MVTR.

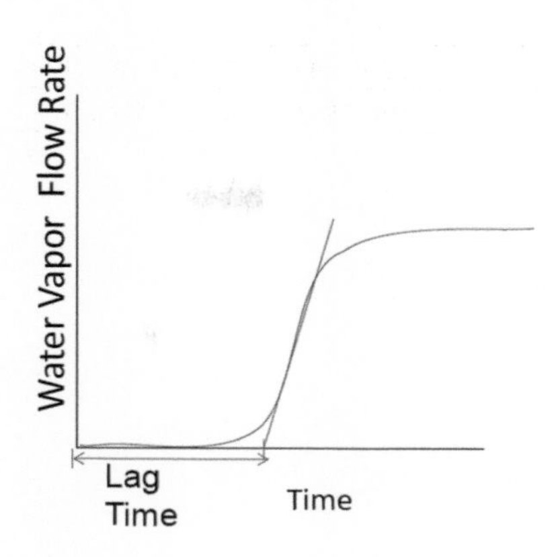

Figure 1-8 schematic explanation of moisture molecules to pass through a desiccated barrier edge seal.

A PV module, especially BIPV with thin film modules, has many similarities to a conventional insulating glass curtainwall. Long lifetime is expected. Costs to repair or replace are extremely high. Designing and installing high quality, durable curtainwall is key to successful, affordable commercial building operation. Just as you wouldn't install argon filled IGU with PIB primary seal but no desiccant, the same goes for PV. If a curtainwall design needs to keep moisture out for a building lifetime, then both PIB (butyl) and desiccant are necessary to meet that lifetime. Upgrading curtainwall to BIPV is a big investment. Adding desiccated butyl edge seal is a small cost to the bill of materials of BIPV for securing the known benefit of significantly delaying moisture entry into the active PV area.

2 References

[1] https://de.wikipedia.org/wiki/photovoltaik 23.3.2016.

[2] https://de.wikipedia.org/wiki/solarzelle 23.03.2016.

[3] Photos are properties of Quanex Building Products.

[4] Photos are properties of Quanex Building Products.

[5] Arnold Glas 2009.

[6] http://www.photovoltaiksolarstrom.de 23.3.2016.

[7] Postak, L.: How desiccants protect TFPV. Photos are property of Quanex Building Products.

[8] Paul, D.R. & Kemp D.R. „The diffusion time lag in Polymer Membranes Containing Adsorptive fillers“ Journal of polymer science, Vol. 41, pp 79 - 93 (1973).

Energetic prospects of glass-polycarbonate composite panels

Univ.-Prof. Dr.-Ing. Thorsten Weimar[1], M.Sc. Sebastián Andrés López[2]

1 Universität Siegen, Institute of Building Structure, 57068 Siegen,
weimar@architektur.uni-siegen.de

2 Universität Siegen, Institute of Building Structure, 57068 Siegen,
andres-lopez@architektur.uni-siegen.de

For architectural design, the glass expresses lightness and modernity. By increasing glass surfaces, especially in buildings with high safety requirements, the glazing must be provided with special security glass. Thereby, the requirement on glazing as a room closure material grows constantly. On the one hand, laminated safety glass offers a passive safety by protecting against injuries caused by broken glass, because the glass fragments adhere after fracture to the interlayer of the composite. On the other hand, it also offers an active safety, which means a resistance against attack. These security glazing can be specified in resistance against manual attack according to EN 356, bullet attack according to EN 1063 and explosion pressure according to EN 13541. The common build-up of security glazing is a laminate of two glass panes with a polymeric film as interlayer. The result of such laminates is an increase in weight; transport and installation are also more complex. Due to climate change and the need to save energy, the building sector responds to these challenges by using insulation glass to achieve thermal protection. In this context, double glazing demonstrates the state of the art, but triple glazing becomes more and more important. The thickness of standard triple glazing reaches about 36 mm without any other specifications. In combination with a resistance against attack, build-ups increase by 54 mm and more. Security glazing consisting of glass-polycarbonate composite panels reduces these effects. Both materials combine the advantageous properties of glass with high material stiffness and hard surface as well as those of polycarbonate with high impact strength and low weight. Due to these various properties, the use of the composite enables an individual design and optimal approach of the glazing to the required specifications, based on tests for transmission and thermal conductivity.

Keywords: laminated glass, glass-polycarbonate composite, transmittance, thermal conductivity

1 Introduction

The properties of laminated glass are continuously improved and developed. Meanwhile, products can be made economically and to high quality. Composite panels of the brittle material glass are already used in safety related applications such as horizontal, walk on or barrier glazing. In this context, laminated safety glass with a sufficient bonding of fragments of broken glass panes as well as a high post-breakage performance is necessary. Compared to common laminated safety glass, the resistance of glass-polycarbonate composite panels under impact loads is much higher. Those laminates combine the positive

Engineered Transparency 2016. Glass in Architecture and Structural Engineering. First Edition.
Edited by Jens Schneider, Bernhard Weller.

properties of glass with high material stiffness and of polycarbonate with high impact strength and low weight. Thereby, the cross section and dead load of glass-polycarbonate composite panels compared to common laminated glass can be reduced. The combination of both materials to prevent from attack and to optimise thermal conductivity by developing specific insulation glazing, represents a significant enhancement to efficient security glazing. Properties are maximum protection combined with low cross-section in consideration of current requirements for thermal protection (Weimar [1]).

Generally, heat will always achieve a temperature balance and flows from the warm to the cold side of a material until both sides reach the same temperature. In this context, three types of heat transmission are described: thermal radiation, thermal conduction and thermal convection. This study on glass-polycarbonate composite panels is referring to the thermal conduction and thermal radiation. The transmittance and heat conductivity in particular will be analysed as two parts of the thermal transmission. Various coatings are tested in different positions of the composite section to analyse effects of different coating types. The test method for thermal conductivity depends on the guarded hot plate method according to (EN 674 [2]). (EN 673 [3]) describes a calculation method for determination of thermal transmittance.

2 State of the art

2.1 Thermal radiation

Solar radiation produces heat, generates electrical power by photovoltaic cells and induces biological reactions. Therefore, solar radiation is a natural resource of energy. Table 2-1 shows the spectrum of different radiations with classification and wavelength.

Table 2-1 Classification of the electromagnetic frequency spectrum (Willems [4]).

<table>
<tr><th colspan="2">Classification</th><th>Wavelength</th><th colspan="2">Classification</th><th>Wavelength</th></tr>
<tr><td colspan="2">Gamma rays</td><td>< 0.005 nm</td><td>Visible light</td><td>VIS</td><td>380 nm - 780 nm</td></tr>
<tr><td colspan="2">X-rays</td><td>0.005 nm - 10 nm</td><td rowspan="3">Infrared rays</td><td>IR-A</td><td>780 nm - 1,400 nm</td></tr>
<tr><td rowspan="4">Ultra-violet rays</td><td>Extreme UV</td><td>10 nm - 100 nm</td><td>IR-B</td><td>1,400 nm - 3,000 nm</td></tr>
<tr><td>UV-C</td><td>100 nm - 280 nm</td><td>IR-C</td><td>3,000 nm - 1 mm</td></tr>
<tr><td>UV-B</td><td>280 nm - 315 nm</td><td colspan="2">Microwaves</td><td>1 mm - 1 m</td></tr>
<tr><td>UV-A</td><td>315 nm - 380 nm</td><td colspan="2">Radio waves</td><td>1 m - 10 km</td></tr>
</table>

A first barrier to restrict solar radiation is the ozone layer and only a limited range of the spectrum, ultra-violet rays, visible light and infrared rays, transmits the atmosphere. Thermal radiation is defined in transmission τ, reflection ρ and absorption α. Equation (2.1) describes the summation and the correlation of the different parts of thermal radiation (Willems [4], Schröder [5]).

$$\tau + \rho + \alpha = 1 \tag{2.1}$$

The non-visible ultra-violet and infrared rays influence material properties. The ultra-violet rays result in fading of materials and induce biological reactions. Due to the absorption of short wave radiation, these rays can cause browning of the skin as well as cell damage and destruction. Infrared radiation is noticeable by feeling of heat (Willems [4]).

2.2 Thermal transmittance

High absorption in the long-wave spectral range in combination with thermal conduction and thermal convection results in a high heat transfer of basic glazing. Single glass panes possess values of heat conductivity about 0.8 W/(m·K) to 1.5 W/(m·K).

Thermal transmittance, also described as U value, is based on heat conductivity. (EN 673 [3]) includes a calculation method to determine the thermal transmittance of glass. In addition to the calculation method experimental tests are given according to (EN 674 [2]) with the guarded hot plate method and to (EN 675 [6]) with the heat flow meter.

$$\frac{1}{U} = \frac{1}{h_e} + \frac{1}{h_t} + \frac{1}{h_i} \tag{2.2}$$

where	h_e, h_i	[W/(m²·K)]	exterior respectively interior heat transfer coefficient
	h_t	[W/(m²·K)]	total conductivity coefficient
	U	[W/(m²·K)]	thermal transmittance coefficient

U values of single panes are about 5 W/(m²·K) to 6 W/(m²·K) and of common insulation glass nearly reduced by half. The heat loss is based on two-thirds of radiation, and one-third of convection and conduction. An important aspect for thermal radiation is the emissivity of glass of about 85 %. Emissivity means the ability of glass to lose energy as thermal radiation. The subsequent processing of single glass to insulation glass or by using low emissivity (lowE) as well as solar protection coatings (sun) influence the heat transfer significantly. Coated glass panes used in insulation glass represent the state of the art. A thin metal ion layer is applied with few nanometres of the surface of glass. By using low emissivity coatings, it is possible to reduce the emissivity but it provides no protection against ultra-violet rays. Solar protection coatings possess high transmission in the visible range, reduction of the emissivity as well as a high reflexion of thermal radiation (Wörner [7]). Another way to reduce the heat transfer is to prevent convection by filling the gap between inner and outer pane with air or an air-gas mixture.

2.3 Laminated glass, laminated safety glass and security glazing

Enhanced applications placed for example on security, sound protection or design require laminated glass. Generally, the definition in accordance with (EN ISO 12543-1 [8]) differs in laminated glass and laminated safety glass. Laminated glass consists of at least two glass panes and an inner adhesive layer. Laminated safety glass ensures additional the bonding of fragments of broken glass panels, limitation of the opening angle, post-breakage capacity as well as protection from injury at the glass. Layers of the composite mostly consist of polyvinyl butyral films. The production of laminated glass can be provided in two different methods. On the one hand, the autoclave process with an automatic lamination and on the other hand manual production methods with cast-in place resin instead of interlayers (Weller [9]).

Security glazing describes laminated safety glass with advanced properties. (EN 365 [10]) regulates the resistance against manual attack, (EN 1063 [11]) defines demands to resist bullet attack and (EN 13541 [12]) shows a classification of resistance against explosion pressure. With increasing requirements for security in the building envelope, especially regarding attack resistance, commonly used glasses are more and more inefficient. Common security glazing consists of a laminated safety glass with several glass panes. Depending on the resistance class, relatively thick cross-sections with a high dead weight might result.

In order to optimise the usage characteristics for security glazing, composite panels with glass and polycarbonate are used. Glass has a high material stiffness and polycarbonate (PC) has a characteristically high impact strength while being lightweight. The transmission of polycarbonate shows similar the value of glass. However, the heat conductivity is five times lower than of glass (Table 2-2). This results in an advantageous thermal protection of the glazing.

Table 2-2 Selected properties of polycarbonate and glass.

Material	Density ρ	Transmittance τ_V	Elastic modulus E	Heat conductivity λ
Polycarbonate	1.2 g/cm³	89 % (t = 3 mm)	2,350 MPa	0.2 W/(m²·K)
Glass	2.5 g/cm³	91 % (t = 10 mm)	70,000 MPa	1.0 W/(m²·K)

Security glazing from glass-polycarbonate composite panels consists of two outer glass panes, generally made of annealed glass (AG), and one or more inner polycarbonate panels (Figure 2-1). The material combination provides high resistance against external attacks while featuring a slim cross-section and a low dead weight. Glass-polycarbonate composite panels are for example as security glazing of the highest resistance class P8B against manual attack according to (EN 356 [10]), 33 % thinner and 50 % lighter than a common laminated glass. This enables a thinner design for frames and support structures.

Retrofitting into existing buildings and further processing to insulation glass will be easier (Weimar [1]).

Figure 2-1 Security glazing with the resistance class P8B against manual attack consisting of laminated safety glass (left) and glass-polycarbonate composite panel (right).

3 Experimental studies

3.1 General aspects of the testing method

In the context of this research project, the test specimen are examined for light transmission and heat conductivity. At first, the focus has been on the effect of coatings with low emissivity and solar protection coating, both with an emissivity of 1 % or 3 %. The coatings are placed in different positions of the cross section.

The glass-polycarbonate composite panels consist of two outer glass panes (AN) according to (EN 572-1 [13]) and (Ipawhite [14]), one inner polycarbonate sheet (PC) defined in (Lexan [15]) and two interlayers of thermoplastic polyurethane (TPU) described in (Weimar [16]). All composite panels fulfil the resistance class P8B.

3.2 Transmission tests

The transmittance of specimen is measured between 210 nm and 2,150 nm with MCS 600 series instrument from Carl Zeiss GmbH in accordance to (EN 410 [17]). Table 3-1 shows the several cross-section of the specimen analysed. The gaps are filled with about 95 % of krypton.

Table 3-1 Cross-sections of specimen for light transmission tests.

No	Cross section
1	10 mm AG
2	6 mm AG | 2 mm TPU | 6 mm AG
3	6 mm AG | lowE 3 % | 2 mm TPU | 6 mm AG
4	6 mm AG | lowE 3 % | 2 mm TPU | lowE 3 % | 6 mm AG
5	10 mm AG | Sun 1 %
6	6 mm AG | 2 mm TPU | 4 mm PC | 2 mm TPU | 6 mm AG | 10 mm gap | Sun 1 % | 6 mm AG
7	6 mm AG | 2 mm TPU | 4 mm PC | 2 mm TPU | 6 mm AG | Sun 1 % | 10 mm gap | Sun 1 % | 6 mm AG

3.2.1 Effectiveness of low emissivity coatings

Figure 3-1 illustrates the transmission curve of a single glass pane and laminated glass as well as the effect of low emissivity coating with 3 % regarding to transmission. The coated glass surface is always placed to the interlayer. Single glass panes show from a wavelength of 300 nm a transmittance above 90 % and in comparison to laminated glass a continuous transmission level. Laminated glass absorbs more ultra-violet radiation than a single pane. Laminated glass without coating shows 5 % lower transmittance in the visible and infrared range up to 1,700 nm in comparison to single glass. At the beginning of the range of visible light, the transmission of laminated glass without coating is about 40 %, with one coating about 30 % and two coatings about 25 %.

It is important to reduce the transmission of ultra-violet rays, because polycarbonate and thermoplastic polyurethane embrittles under this radiation. The thermoplastic polyurethane interlayers contain specific blockers to protect the polymers against ultra-violet radiation. By using low emissivity coatings the maximal transmittance in the visible light spectrum is above 75 % to 80 %. The maximum values of transmittance start at a wavelength of about 500 nm, subsequently in the infrared range the transmittance decreases continuously until under 10 % at a wavelength of 1,700 nm for coated glazing. Low emissivity coatings on this position of the cross section of laminated glass reduce the transit of heat rays from the long-wave infrared spectrum.

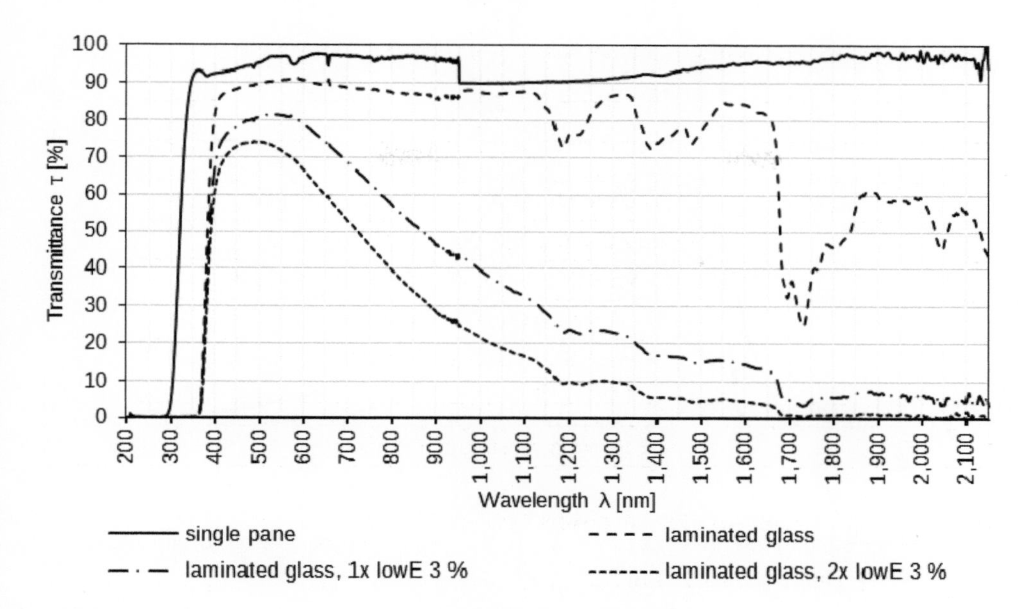

Figure 3-1 Transmittance τ in according to wavelength λ of single glass pane in comparison to laminated glass and low emissivity coatings.

3.2.2 Effectiveness of solar protection coatings

Figure 3-2 shows the transmission curve of glass with solar protection coatings. A single glass pane with a solar protection coating is compared to an attack-resistant double glazing with one or two solar protection coatings of 1 % placed to the gap.

Ultra-violet radiation transmit extensively through the single glass pane in contrast to double glazing with blockers in the thermoplastic polyurethane interlayers of the glass-polycarbonate composite panels. The trend of transmission above the infrared radiation do not differ between single glass panes and the double glazing with one solar protection coating of 1 %. The transmittance at the visible range is also reduced in comparison to a single pane and double glazing with solar protection coatings. However, double glazing with two solar protection coatings reduce the transmission of infrared radiation of about 10 %.

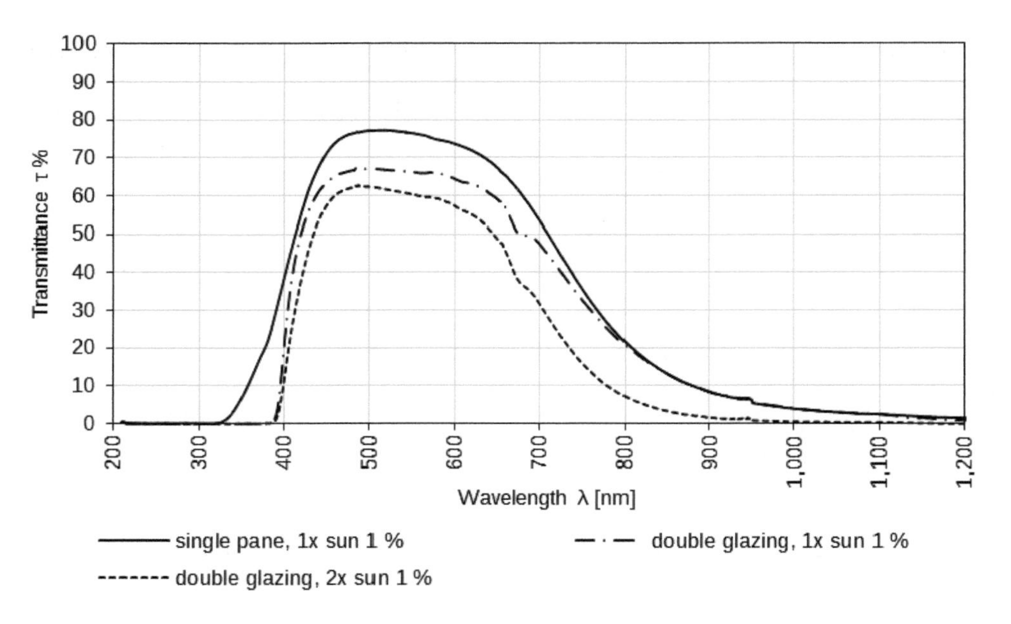

Figure 3-2 Transmittance τ in according of wavelength λ of single glass pane in comparison to double glazing with one or two solar protection coatings of 1 %.

3.3 Heat conductivity tests

Table 3-2 shows various thermal transmittance coefficients of double glazing with laminated safety glass and glass-polycarbonate composite panels calculated in accordance to (EN 673 [3]).

Table 3-2 Thermal transmission coefficient of double glazing with common laminated safety glass and glass-polycarbonate composite panels.

No	Type	Cross-section	Nominal thickness	Thermal transmittance
1	Laminated glass [18]	31 mm - 16 mm Ar - 6 mm	53 mm	1.1 W/(m^2·K)
2	Laminated glass [19]	27 mm - 15 mm Ar - 6 mm	48 mm	1.1 W/(m^2·K)
3	Laminated glass [20]	36 mm - 8 mm Kr - 6 mm	50 mm	1.2 W/(m^2·K)
4	Laminated glass [21]	32 mm - 16 mm Ar - 6 mm	54 mm	1.1 W/(m^2·K)
5	Glass-polycarbonate	24 mm - 12 mm Kr - 6 mm	42 mm	0.9 W/(m^2·K)

All cross-sections achieve resistance class P8B. The gap between the inner and the outer layer is filled with argon (Ar) or krypton (Kr). Cross-sections with laminated safety glass provide a nominal thickness of 48 mm to 54 mm with a thermal transmittance of about 1.1 W/(m^2·K). Glass-polycarbonate composite panels reach a reduced nominal thickness of 42 mm with a thermal transmission coefficient of 0.9 W/(m^2·K) in consequence of higher impact strength and lower heat conductivity of the polycarbonate sheet.

The heat conductivity of specimen is analysed with TPL 800S from TAURUS instruments GmbH according to the guarded hot plate method defined in (EN 674 [2]). Various cross-sections of security glazing as insulation glass with glass-polycarbonate composite panels with dimensions of 800 mm in length and width are measured. The coated glass surface is always placed to the gap, which is filled with at least 95 % of krypton. Table 3-3 shows the cross-sections of the specimen analysed for heat conductivity. Difference between nominal and effective thickness are caused in material tolerance of the build-ups.

Table 3-3 Cross sections of security glazing as insulation glass with glass-polycarbonate composite panels for head conductivity tests.

No	Cross-section	Nominal thickness	Effective thickness
1	6 mm AG | sun 1 % | 10 mm gap | 6 mm AG | 2 mm TPU | 4 mm PC | 2 mm TPU | 6 mm AG	36.0 mm	36.7 mm
2	6 mm AG | sun 1 % | 10 mm gap | sun 1 % | 6 mm AG | 2 mm TPU | 4 mm PC | 2 mm TPU | 6 mm AG	36.0 mm	36.7 mm
3	6 mm AG | lowE 1 % | 10 mm gap | lowE 1 % | 6 mm AG | 2 mm TPU | 4 mm PC | 2 mm TPU | 6 mm AG	36.0 mm	36.0 mm

The ratio of thickness and thermal conductivity results in total thermal resistance expressed by equation (3.1).

$$R = \frac{t}{\lambda} \tag{3.1}$$

where R [m^2·K/W] total thermal resistance
t [m] effective thickness
λ [W/(m·K)] thermal conductivity

Equation (3.2) describes the reciprocal value of thermal transmittance in accordance to (EN 674 [2]) with exterior heat transfer coefficient set to 25.0 W/(m^2·K) and interior heat transfer coefficient set to 7.7 W/(m^2·K).

$$\frac{1}{U} = R + \frac{1}{h_i} + \frac{1}{h_e} \tag{3.2}$$

where $1/U$ [m²·K/W] reciprocal value of thermal transmittance
h_e, h_i [W/(m²·K)] exterior respectively interior heat transfer coefficient
R [m²·K/W] total thermal resistance

Table 3-4 shows the results measured of heat conductivity, total thermal resistance and thermal transmittance.

Table 3-4 Heat conductivity, total thermal resistance and thermal transmittance in according to (EN 674 [2]).

No	Effective thickness	Heat conductivity	Total thermal resistance	Thermal transmittance
1	36.7 mm	0.042 W/(m·K)	0.874 m²·K/W	0.958 W/(m²·K)
2	36.7 mm	0.038 W/(m·K)	0.966 m²·K/W	0.880 W/(m²·K)
3	36.0 mm	0.047 W/(m·K)	0.766 m²·K/W	1.069 W/(m²·K)

Specimen of build-up 1 with one solar protection coating shows a thermal transmittance of 0.958 W/(m²·K) and specimen of build-up 2 with two solar protection coatings provides the lowest value with 0.880 W/(m²·K). The thermal transmittance of specimen of build-up 3 with two low emissivity coatings amount to 1.069 W/(m²·K), which represents the most unfavourable value for the insulation glazing analysed.

Glazing with low emissivity coatings induces lower thermal insulation in comparison to solar protecting coatings, because the emissivity of glazing is reduced, but not transmission of infrared radiation. In contrast, solar protection coatings in the cross-section provide lower thermal transmittance. In accordance to the previous transmission tests, infrared rays do not pass through the glazing. In consequence, the glazing with solar protection coatings shows an improved thermal protection.

4 Summary and conclusion

The improvement of energetic performance of insulation glazing with requirements to security is analysed by application of glass-polycarbonate composite panels. Therefore, the transmittance and heat conductivity of glass, laminated glass and glass-polycarbonate composite are compared to analogue specimen with additional coatings for low emissivity and solar protection.

Laminated glass shows a reduced transmittance of about 5 % in the range of visible light and infrared radiation in contrast to single glass panes. Ultra-violet radiation does not pass through the laminated glass in consequence of specific blockers containing in thermoplastic polyurethane interlayers to protect the polymers. By using low emissivity coatings, the maximum transmittance in the visible light spectrum is above 75 % to 80 % and in the infrared range the transmittance decreases continuously until under 10 %. Tests with single glass panes and glass-polycarbonate composite panels in double glazing each with one solar protection coating do not differ above infrared radiation. However, double glazing with two solar protection coatings reduce the transmission of infrared radiation of about 10 %.

Security glazing with glass-polycarbonate composite panels as insulation glazing shows reduced cross-sections and thermal transmittance compared to common build-ups with laminated safety glass in consequence of higher impact strength and lower heat conductivity of the polycarbonate sheet. Subsequently, insulation glazing with glass-polycarbonate composite panels and low emissivity as well as solar protection coatings are analysed by experimental studies with guarded hot plate method. Low emissivity coatings provide higher thermal transmittance in contrast to solar protection coatings, because the emissivity of a low emissivity glazing is reduced, but not transmission of infrared radiation.

5 Acknowledgments

The research project founded by the German Federal Ministry for Economic Affairs and Energy is carried out in cooperation with SILATEC Sicherheits- und Laminatglastechnik GmbH and managed by AiF Projekt GmbH.

6 References

[1] Weimar, T.: Experimental research on glass-polycarbonate composite panels. Bautechnik 92 (2015). Issue 4.

[2] EN 674: Glass in building – Determination of thermal transmittance (U value) – Guarded hot plate method. Berlin: Beuth, 2011.

[3] EN 673: Glass in building – Determination of thermal transmittance (U value) – Calculation method. Berlin: Beuth, 2011.

[4] Willems, W.: Lehrbuch der Bauphysik. Wiesbaden: Springer Vieweg, 2013.

[5] Schröder, G.; Treiber, H.: Technische Optik. Würzburg: Vogel, 2007.

[6] EN 675: Glass in building – Determination of thermal transmittance (U value) – Heat flow meter method. Berlin: Beuth, 2011.

[7] Wörner, J.-D.; Schneider, J.; Fink, A.: Glasbau. Grundlagen, Berechnung, Konstruktion. Berlin: Springer, 2001.

[8] EN ISO 12543-1: Glass in building – Laminated glass and laminated safety glass – Part 1: Definitions and descriptions of component parts. Berlin: Beuth, 2011.

[9] Weller, B.; Krampe, P.; Reich, S.: Glasbau-Praxis: Konstruktion und Bemessung Band 1: Grundlagen. Berlin: Beuth, 2013.

[10] EN 356: Glass in building – Security glazing – Testing and classification of resistance against manual attack. Berlin: Beuth, 2002.

[11] EN 1063: Glass in building – Security glazing – Testing and classification of resistance against bullet attack. Berlin: Beuth, 2001.

[12] EN 13541: Glass in building – Security glazing - Testing and classification of resistance against explosion pressure. Berlin: Beuth, 2012.

[13] EN 572-1: Glass in building – Basic soda lime silicate glass products – Part 1: Definitions and general physical and mechanical properties. Berlin: Beuth, 2004.

[14] Ipawhite: Product data sheet. Interpane Glas Deutschland, 2016.

[15] Lexan® 9030: Product data sheet. SABIC Innovative Plastics, 2014.

[16] Weimar, T.: Research on glass-polycarbonate composite panels. Dissertation. Technische Universität Dresden. 2011.

[17] EN 410: Glass in building - Determination of luminous and solar characteristics of glazing: Beuth, Berlin, 2011.

[18] SG-Sanco Safe: Product data sheet. Stader Glas GmbH & Co KG, 2016.

[19] SGG STADIP Protect: Product data sheet. Saint-Gobain Glass Deutschland, 2012.

[20] ALLSTOP P8B B-27: Product data sheet. Flachglas Markenkreis, 2015.

[21] Ipasafe P8B: Product data sheet. Interpane Glas Industrie AG, 2016.

Critical evaluation of chemically strengthened glass in structural glazing applications

Guglielmo Macrelli[1]

1 Isoclima SpA – R&D Department, via A.Volta, 14 – I-35042 Este (PD) Italy, gmacrelli@finind.com

Chemically strengthened glass by ion exchange (CSG-IX) below glass transition temperature is becoming a candidate in structural glazing applications. When compared to thermally strengthened or heat strengthened glass (TSG/HSG), CSG-IX exhibits superior optical quality (example: absence of anisotropies) and no limitations in thickness and shape. These features make CSG-IX an appealing material in some architectural and transportation structural applications. Examples in structural applications, both architectural and transportation will be presented. Recent standard literature (example: pr EN 16612:2013) is considering this material in load resistance calculations practically without any specific limitation. The introduction of CSG-IX in standards related to load resistance of glass and glazing will be critically discussed. Safety aspects will be discussed considering post breakage behaviour related to fragmentation patterns of CSG-IX. Characteristic strength values for Soda-Lime float glass and Sodium Alumino-Silicate glass will be discussed on the basis of the particular residual stress profile generated by the ion exchange process. Issues related to glass mechanical strength will be identified in both design and process control in order to provide arguments to be considered when using CSG-IX in structural applications.

Keywords: glass chemical strengthening, ion exchange, glass strength, structural glazing

1 Introduction

Chemical strengthening of glass by ion exchange below glass transition temperature (herewith we will use the acronym CSG-IX) is a known process since several decades (A. K. Varshneya [1]). Reviews have been published (R. Gy [2], S. Karlsson et al. [3]) where ion exchange process fundamentals and application to glass strengthening are introduced and discussed. In spite of the applications in several and different fields like aerospace windows, marine glazing, automotive and railway glazing, consumer electronics, pharmaceutical and medical devices and architectural special projects, there are still some fundamental issues under discussion in the glass science community (A. K. Varshneya et al [4], A. K. Varshneya [5]). Applications in structural glazing have been discussed (G. Macrelli and E. Poli [6]) where it has been clearly identified how the strengthening effect is depending from parameters related to the glass article to be strengthened: original glass chemical composition, surface quality (statistics of surface flaws distribution), glass thermal history. Parameters related to ion exchange process and expressed in terms of residual stress distribution (see Figure 1-1) are also determining the strengthening effect namely: compression layer depth (C_d), surface compression (S_C) and central tension (S_T).

Engineered Transparency 2016. Glass in Architecture and Structural Engineering. First Edition.
Edited by Jens Schneider, Bernhard Weller.

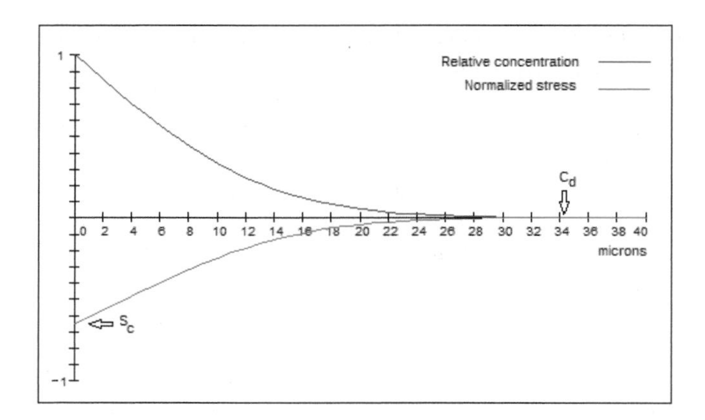

Figure 1-1 Concentration and Residual Stress distributions in the considered CSG_IX glass; positive values are normalized concentration, negative values are compressive stress with relaxation normalized at un-relaxed surface compression (600 MPa).

Because ion exchange is performed below glass transition temperature, the process does not introduce additional deformations to the glass article. This results in a superior optical quality when compared to other thermal strengthening processes, that require glass article heating above glass transition temperature. Additionally, as ion exchange is a process that involves compression layer depths – C_d – of the order of tens of microns, it can be considered a surface process where the minimum glass article thickness is not practically an issue. For these two last features and because, with this process, glass strength can be considerably increased (G. Macrelli and E. Poli [6]), CSG-IX is a material of great interest for structural applications. The most famous example are structural glass stairs [7] where most of the load is taken by curved laminated CSG-IX glass through sealed holes (see Figure 1-2).

Figure 1-2 Self-bearing CSG-IX glass stairs.

International standardization is moving towards the consideration to this type of strengthened glass [8]. While the motivations for this standardization process of CSG-IX are quite understandable, it shall be pointed out that this type of strengthened glass articles have a far different residual stress characteristic in comparison to conventional TSG/HSG due to the reason that their compression layer depth – C_d – even when larger than the deepest surface flaw, is anyway of the same order of magnitude. This last evidence requires consideration for all those situations when, for any reason, the deepest surface flaw may be larger than compression layer depth. The present study is relaying on results presented in a previous publication (G. Macrelli and E. Poli [6]) and it is based on the first order strength prediction model initially introduced in H. Aben and C. Guillemet [9] and successively completed (G. Macrelli and E. Poli [6]) with the introduction of the central tension stress term. The strength prediction model has the purpose to identify critical effects due to deeper surface flaws already present before ion exchange, not detected during production steps and surface flaws generated after ion exchange during the glass article lifetime. In figure 2-1 the strength prediction model has been used to evaluate strength reduction for original not strengthened glass and CSG-IX as a result of increased critical flaw depth. Residual stress profile reported in figure 1-1 have been evaluated according to the model outlined in G. Macrelli and E. Poli [6] according to equation (1.1) considering a complementary error function distribution for concentration and a KWW relaxation function R(t).

$$\sigma(x,t) = \frac{B \cdot E \cdot Co \cdot V}{(1-\upsilon)} \cdot \left[\frac{1}{d} \sqrt{\frac{D}{\pi}} \int_0^t \frac{R(t-t')}{\sqrt{t'}} dt' - \frac{2}{\sqrt{\pi}} \int_{\frac{x}{2 \cdot \sqrt{D \cdot t}}}^{\infty} R\left(t - \frac{x^2}{4 \cdot D \cdot p^2}\right) e^{-p^2} dp \right] \quad (1.1)$$

$$R(t) = e^{-\left(\frac{t}{\tau}\right)^{\gamma}} \quad (1.2)$$

In equation (1.1) E is the glass Young modulus, v is the Poisson ratio, Co is surface concentration of incoming ions, B is the Cooper coefficient (A. K. Varshneya et al [4]) and V is the Varshneya factor (0< V< 1, introduced by the author after A. K. Varshneya et al [4] and A. K. Varshneya [5]), d is the glass thickness and D is the interdiffusion coefficient. The first term of equation (1.1) represent glass relaxed central tension while the second term represent the relaxed built-up compression component. In equation (1.2), $R(t)$ is the KWW relaxation function, τ is the relaxation time while γ is the non Maxwellian factor for structural relaxation (γ = 3/7).

2 Strength characteristics of soda-lime CSG-IX

Strength of soda-lime CSG-IX has been determined by the author (G. Macrelli and E. Poli [6]) by four point bending test according to EN 1288-3. Results are summarized in Table 2-1.

Table 2-1 Characteristic bending strength (5 fractile at 95 % lower confidence limit) of soda-lime float glass before and after IX.

Statistical parameters	Annealed Glass	CSG-IX C_d = 35 μm ; S_C = 380 MPa
Weibull parameter θ (MPa)	76.5	296.6
Weibull parameter β (MPa)	8.5	18.7
Expectation value <σ> (MPa)	72.2	288.3
Characteristic bending Strength (MPa)	42.3	226.9

These results are referred to a soda-lime float glass where a specific control based on visual optical inspection has been put in operation in order to identify surface visible scratches or defects and where edge finishing (grinding and polishing under effective flow of cooling water) has been controlled by careful final optical visual inspection. This result has been evaluated (G. Macrelli and E.Poli [6]) as generated by a surface flaws population typically distributed around a value of c =27 μm that, with a C_d of 35 μm, generates a strengthening effect of $\Delta\sigma$= 193 MPa with an expected final strength of 265 MPa in a substantial good agreement with the test result of 288 MPa. Following the same first approximation model outlined in G. Macrelli and E.Poli [6], it can be concluded that final strength is depending from the original surface quality of the glass that can be represented by the original surface flaws population distribution.

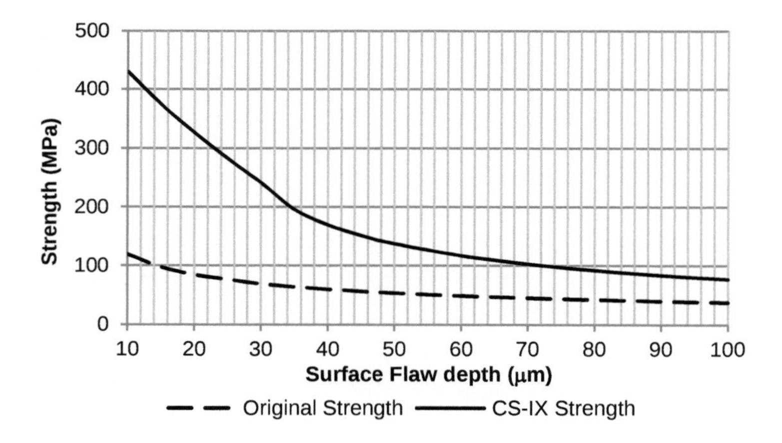

Figure 2-1 Strength as function of critical surface flaw depth for original not strengthened glass (dotted line) and CSG-IX (continuous line).

In Table 2-2 an evaluation is presented based on the same model for glasses with different surface flaws values: 15 μm (excellent initial optical quality), 25 μm (good initial optical quality) and 50 μm (poor initial optical quality) submitted to the same IX process (C_d = 35 μm; S_C = 380 MPa).

Table 2-2 Influence of initial glass surface quality to final strength of CSG-IX.

Glass surface conditions	Surface flaw depth c (μm)	Original glass strength (MPa)	Final glass strength (MPa) C_d = 35 μm; S_C = 380 MPa
Excellent surface quality	15	109	385
Good surface quality	25	76	283
Poor surface quality	50	53	140

The same argument leading to Table 2-2 can be considered for in-service surface damages that may induce surface flaws deeper than the original case depth. If a surface flaw is induced during glass article lifetime such that, as an example c = 100 μm, glass original strength is expected to be accordingly reduced to 37 MPa and CSG-IX strength is expected to result 77 MPa. This last argument lead to a second conclusion that glass local strength is depending on the evolution of surface quality during the glass article lifetime.

3 Design value of strength for CSG-IX

Characteristic bending strength values ($f_{g;k}$ for annealed glass and $f_{b;k}$ for prestressed glass) are the input values for the determination of the design values of strength $f_{g;d}$. The transfer of characteristic bending strength values into design values is normally a matter of standardization. As already mentioned, a draft has been published [8] that includes both annealed and pre-stressed glass and, in this last category, chemically strengthened glass. Following [8], for annealed glass, the calculation of design strength requires the value of $f_{g;k}$, and the value of the following parameters: $\gamma_{M;A}$ (material partial factor), k_{sp} (factor for glass surface profile) and k_{mod} (factor for load duration). Considering values of Table 2-1 for $f_{g;k}$ and $\gamma_{M;A}$ = 1.8, k_{sp} = 1 and k_{mod} = 1 (wind action, single gust), it results for annealed glass a design value for strength $f_{g;d}$ = 23 MPa. If we perform the same calculation for CSG-IX considering the factors for pre-stressed glass: k_v = 1 and $\gamma_{M;v}$ = 1.2 we obtain, according to equation (3.1) taken from [8], a design strength $f_{g;d}$ = 177 MPa for CSG-IX. Even using the characteristic bending strength values indicated in [8] for annealed glass ($f_{g;k}$ = 45 MPa) and for CSG ($f_{b;k}$ = 150 MPa), we arrive to a design value of strength for CSG-IX of $f_{g;d}$ = 112 MPa.

$$f_{g;d} = \frac{k_{mod} \cdot k_{sp} \cdot f_{g;k}}{\gamma_{M;A}} + \frac{k_v \cdot (f_{b;k} - f_{g;k})}{\gamma_{M;v}} \tag{3.1}$$

It has been shown in section 2 that expected strength values, from initially poor surface quality to damaged surfaces during lifetime, are in the range between 70 MPa to 140 MPa.

This means that, accepting design strength values within this range according to [8], may lead to critical situations. The obvious conclusion is that standards about design strength values of CSG-IX should consider carefully the specific features of this kind of pre-stressed glass. Additionally, as achievable case depth values are at the same order of magnitude of surface flaws, also limitations in case depth (C_d) and surface compression (S_C) should be considered when thinking of a standards covering this glass product. A classification standard for CSG-IX in terms of case depth and surface compression has been introduced [10] without any correlation to strength values or strength determination methods. This standard [10] clearly states that it does not purport to address end-use performance. It can be concluded that, from the standardization point of view, something is missing or it is not properly completed for end-use performance definition of CSG-IX.

Safety issues related to CSG-IX are due to the limited values of central tension levels in thick glass as a consequence of ion exchange. Surface compression stress are developed as a result of the constrained molar volume expansion at the surface due to the larger volume of the invading ions, this surface compression is balanced by internal tension because of elastic compatibility. The depth of this compression layer (C_d) is normally far smaller than glass thickness so a very small central tension is generated as a result of ion exchange. The weak central tension is responsible of a coarse fragmentation of CSG-IX after breakage with large and sharp glass shards, this means that CSG-IX cannot be considered a safety glass as, for example, thermally strengthened glass (TSG). Safety issues with CSG shall always to be solved by using laminated glazing constructions. An interesting development in CSG-IX is the possibility to use glass with different chemical compositions that allows a far deeper compression layer depth. This can be achieved with alkali alumino-silicate glass both Sodium and Lithium alumino-silicate. With these glasses case depth of the order of hundred microns can be achieved in similar processing conditions of soda-lime glass.

4 Conclusion and outlook

The main conclusions achieved can be summarized as follows:

1. Final strength of CSG-IX is strongly depending from the original surface quality of the glass that can be represented by the original surface flaws population distribution.
2. Glass local strength of CSG-IX is depending on the evolution of surface quality during the glass article lifetime.
3. Limitations in case depth (C_d) and surface compression (S_C) should be considered when thinking of standards covering CSG-IX.
4. Safety issues with CSG-IX shall always to be solved by using laminated glazing constructions.

On the basis of the above conclusions future research work should be based on following directions:

5. Testing of glass with different surface and edge quality (Flaws population) after ion exchange.
6. Testing of CSG_IX submitted to different lifetime surface damages (abrasion, scratches).
7. Perform same testing schedule for glass with different chemical composition namely: Sodium Aluminosilicate and Lithium Aluminosilicate.

5 References

[1] Varshneya, A.K.; Fundamentals of inorganic glasses 2nd Edition. Sheffield: Society of Glass Technology, 2006.

[2] Gy, R.; Ion exchange for glass strengthening. In: Materials Science and Engineering. Vol. 149, 2008, pp 159-165.

[3] Karlsson, S.;Jonson, B.; The technology of glass chemical strengthening – a review. In: Glass Technology: European Journal of Glass Science and Technology. Vol. 51, 2010, pp 41-54.

[4] Varshneya, A.K.; Olson, G.A.; Kresky, P.K.; Gupta, P.K.: Buildup and relaxation of stress in chemically strengthened glass. In: Journal of Non-Crystalline Solids. Vol 427, 2015, pp 91-97.

[5] Varshneya, A.K.; Mechanical model to simulate buildup and relaxation of stress during glass chemical strengthening. In: Journal of Non-Crystalline Solids. Vol 433, 2016, pp 28-30.

[6] Macrelli, G.; Poli, E.;Chemically strengthened glass by ion exchange: residual stress profile and strength evaluation. In: Engineered Transparency. International conference at Glasstec, Düsseldorf (Germany), 2014, 231-240.

[7] http://www.austrian-tis.at/seele/retail_store.php.

[8] prEN 16612 Glass in building – Determination of the load resistance of glass panes by calculation and testing. CEN 2013.

[9] Aben,H.; Guillemet, C.; Photoelasticity of glass. Berlin Heidelberg: Springler-Verlag, 1993.

[10] ASTM C 1422/C1422M -15 Standard Specification for Chemically Strengthened Flat Glass.

Strength testing of thin glasses

Louisa Blaumeiser[1], Jens Schneider[2]

1 Technische Universität Darmstadt, Institute of Structural Mechanics and Design, Franziska-Braun-Straße 3, 64287 Darmstadt, Germany, blaumeiser@ismd.tu-darmstadt.de

2 Technische Universität Darmstadt, Institute of Structural Mechanics and Design

This paper deals with the modification of the coaxial double ring test for the testing of thin glasses. By substituting the loading ring with a silicone pad in a finite element model, a constant stress field in the central region of the glass panel could be generated. For this, compression tests on silicone specimens of different Shore-A hardnesses were carried out to analyze the material behavior and to fit a material model. In a parametric study on the Finite Element Model the different material models and different geometries of the silicone printing pad were investigated.

Keywords: thin glass, bending strength, coaxial double ring test, constant stress field

1 Introduction

The coaxial double ring test according to DIN EN 1288-5 [1] is a common testing method for determining the bending strength of flat glass. During bending, the concentrically arranged rings (supporting and loading ring) induce a constant tensile stress field in the glass panel in the area circumscribed by the loading ring. In this case, the radial stresses σ_{rad} as well as the tangential stresses σ_T are equal (Figure 1-1). Thus, the surface strength is determined with this method by excluding the influence of the edges. However, concerning thin glasses with a thickness of 2 mm and smaller that are, for example, used as cover glasses of PV modules, the large deflection relative to the glass thickness activates membrane forces and generates a highly non-linear stress field, especially in the radial stress distribution (Figure 1-2). These glasses always break beneath the loading ring due to the local tension stress maximum.

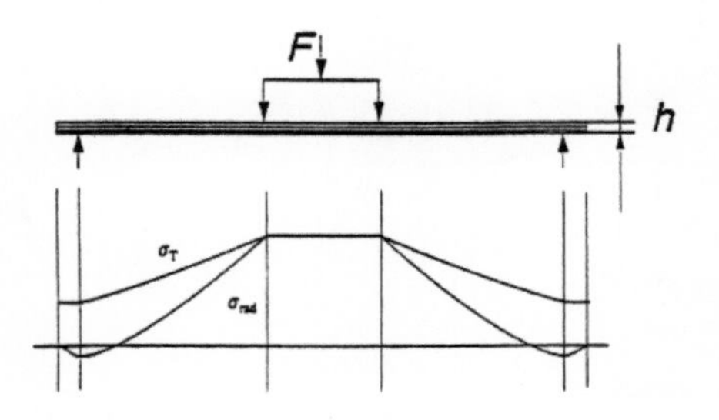

Figure 1-1 Stress distributions at small deflections (DIN EN 1288-1 [2]).

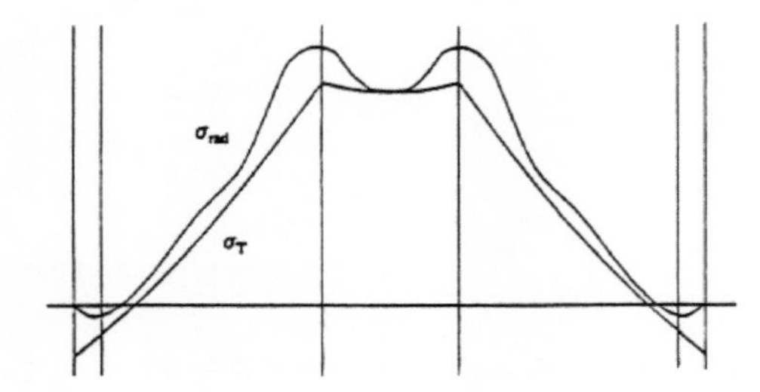

Figure 1-2 Stress distributions at large deflections (DIN EN 1288-1 [2]).

Engineered Transparency 2016. Glass in Architecture and Structural Engineering. First Edition.
Edited by Jens Schneider, Bernhard Weller.

The DIN EN 1288-2 [3] tries to face this phenomenon with the application of an additional gas pressure below the loading ring, which has to be in a specific relation to the loading ring force. The experimental setup gets, however, significantly more complex due to the need of a non-linear controlling of testing speed and is rarely used.

Similar to the concept of applying a uniform gas pressure we developed the idea to apply a distributed load via a polymeric body that can match the deflection of the glass panel. Silicone samples with different Shore-A hardnesses usually used as printing pads were found suitable for this.

2 Simulation of the coaxial double ring test

For testing PV module glass panels with a size of about 1700 x 1000 mm, it is necessary to scale the geometry of the coaxial double ring test of DIN EN 1288-5 [1]. By scaling the geometry, the possibility of the critical surface damage lying within the area of the loading ring increases whereas the value of the fracture stress decreases. A factor of five has been chosen for scaling both the loading ($r = 5 \cdot 9 = 45$ mm) and the supporting ring ($r = 5 \cdot 45 = 225$ mm). The effect of different glass thicknesses was investigated by a finite element simulation using the rotational symmetry. Figure 2-1 shows the increasing non-linear effect in the center area resulting from membrane stresses activated in thinner glass plates. The stress level of 120 MPa at the loading ring approximately corresponds to the breaking stress of heat-strengthened glass often used in PV modules.

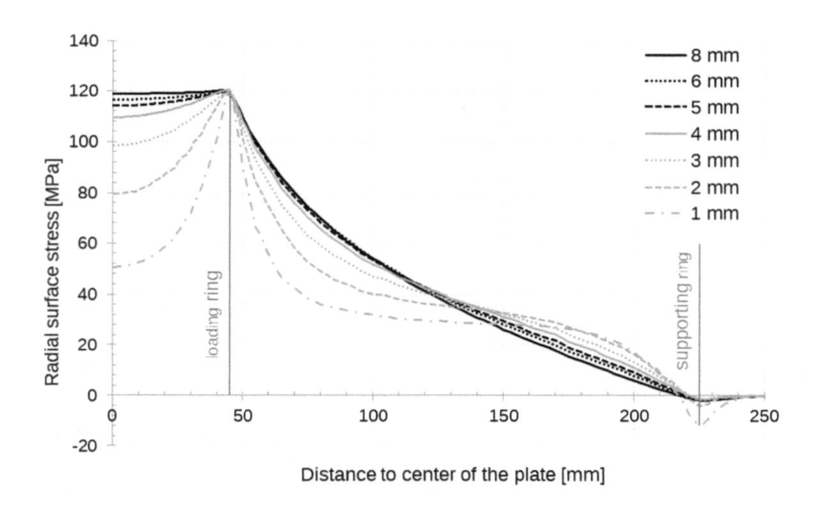

Figure 2-1 Stress distribution of different glass thicknesses.

3 Material model of the silicone pads

3.1 Introduction

For a finite element simulation of the coaxial double ring test it is essential to state the material behavior of all components used in the model. The linear-elastic material behavior of the steel rings and the glass plate is well known. In contrast to this, the silicone as an elastomer shows a hyper-elastic material behavior, which can be formulated by various material models described for example in Ali et al [4]. By means of stress-strain curves determined in mechanical load tests, a suitable material model can be fitted.

3.2 Uniaxial compression tests

For determining the hyper-elastic material behavior of an elastomer, it is necessary to perform various tests i.e. simple extension, equi-biaxial extension, uniaxial compression and shear (see Treloar [5]). In good approximation, a uniaxial compression test following ISO 7743 [6] in this case is sufficient to obtain the stress-strain behavior since the load condition of the silicone in a uniaxial compression test is nearly identical to the intended coaxial double ring test.

Figure 3-1 Universal testing machine type M.A.N. 100 kN. **Figure 3-2** Sample geometry.

In a first step, the initial loading was compared with several re-loadings each within a time interval of about five minutes (Figure 3-3) because in the coaxial double ring test the silicone pad would be used several times without being replaced. Furthermore, various loading rates (10 mm/min, 5 mm/min and 2 mm/min) have been investigated to determine a possible strain-rate dependency. The scattering in both series was very low especially in the range up until a distance of 6 mm which corresponds to about 50 % strain. Thus, for the following series the specimens were tested only once each with a mean loading rate of 5 mm/min.

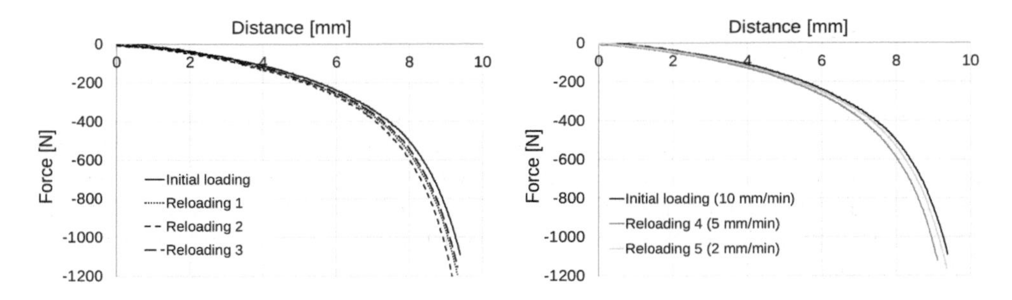

Figure 3-3 Force-distance diagram of initial loading and reloading, 7.8 Shore-A.

Figure 3-4 Force-distance diagram of different loading rates 7.8 Shore-A.

Four different series have been performed. The series differed in the Shore-A hardness of the silicone samples (4, 10, 16 and 22). Prior to the compression tests, it has been necessary to measure the diameter and thickness of the samples in order to generate the engineering stress-strain diagram from the force-distance diagram. The engineering stress-strain diagram of all samples is displayed in Figure 3-5. By tendency an increasing stiffness of the material can be assessed with an increasing Shore-A hardness.

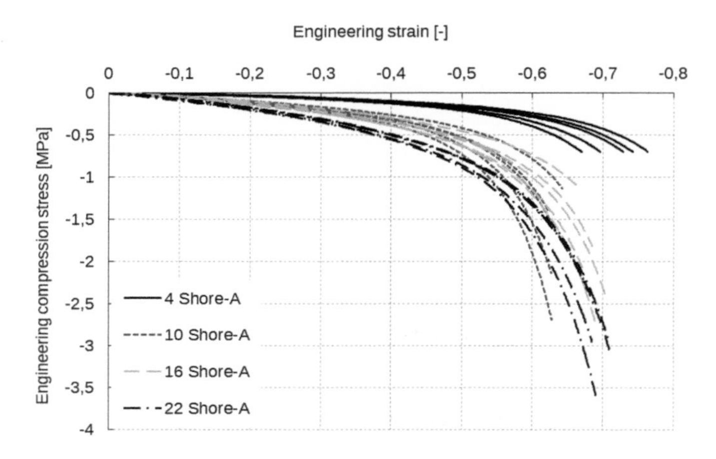

Figure 3-5 Engineering stress-strain diagram of all samples, loading rate 5 mm/min.

3.3 Curve Fitting

Subsequently, the stress-strain curves were fed into the finite element software ANSYS Workbench [7] in which a curve fitting feature is implemented. In Hanich [8] different material models (Neo Hooke, Mooney-Rivlin, Ogden and Yeoh) were investigated and a criterion in the form of the minimal standard deviation S was set for a comparison of the different material models.

$$S = \sqrt{\frac{1}{n-1}\sum_{i=1}^{n}\left(\sigma_{fit,i} - \sigma_{test,i}\right)^2} \tag{3.1}$$

As a result the *Yeoh*-material model for incompressible materials with three parameters C_{i0} turned out to be the model best fitting the measured data for all Shore-A hardnesses. In Figure 3-6 the measured data of sample five with a Shore-A hardness of 16 as well as the fitted data with *Yeoh 3* are depicted. For a uniaxial stress state with the principle stretch λ and for small strains, the strain energy density $W(\lambda)$ and the consequent stress $\sigma(\lambda)$ are as follows:

$$W(\lambda) = \sum_{i=1}^{3} C_{i0}\left(\lambda^2 + \frac{2}{\lambda} - 3\right)^i \tag{3.2}$$

$$\sigma(\lambda) = \sum_{i=1}^{3} C_{i0}\, i\left(\lambda^2 + \frac{2}{\lambda} - 3\right)^{i-1}\left(2\lambda - \frac{2}{\lambda^2}\right) \tag{3.3}$$

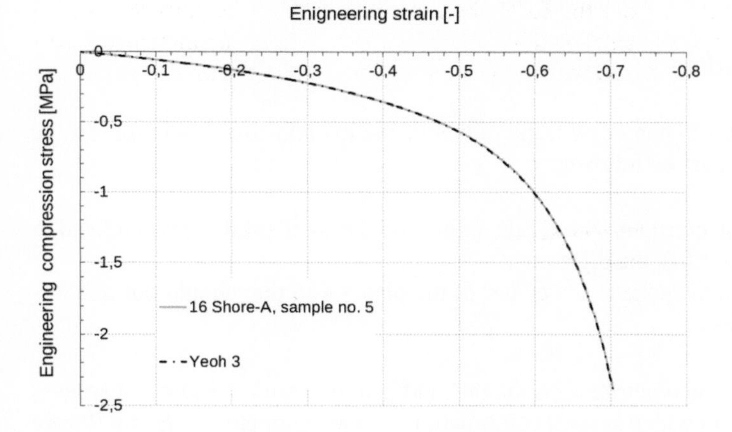

Figure 3-6 Engineering stress-strain curves of measured and fitted data.

4 Modification of the coaxial double ring test

4.1 Finite element model

In Hanich [8], a ten angular degree circular segment finite element model of the modified coaxial double ring test has been generated. Instead of the steel loading ring, a silicone pad with a steel plate on top has been inserted for applying the load on the glass plate (2 mm). Elements of type SOLID187 have been used. To simplify the model, a frictionless bearing on the supporting ring and a frictionless contact between glass and silicone have been chosen. The steel plate and the silicone pad are modeled as fully bonded.

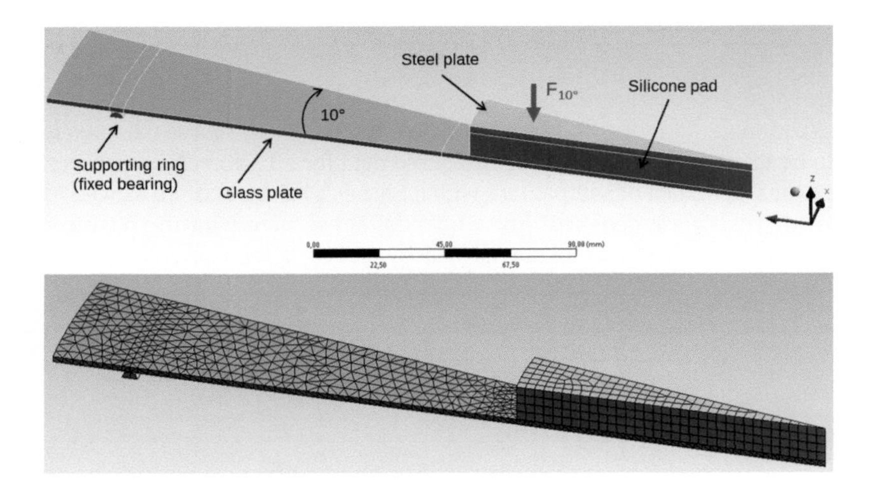

Figure 4-1 Finite element model with mesh of the modified coaxial double ring test.

4.2 Parameter studies

In the scaled test set-up of chapter two, the radius of the loading ring is 45 mm. So the targets of the simulation are as follows:

- The stress should be constant within this radius and the standard deviation related to the mean value quantifies the difference.
- The stress beyond a distance to the center of the plate of 45 mm should not rise further but decrease fast.

First of all, two different geometries („rectangle" and „circle") with a Shore-A hardness of 7.8 were investigated in which several combinations of the parameters *V*, *H* and *R* were used.

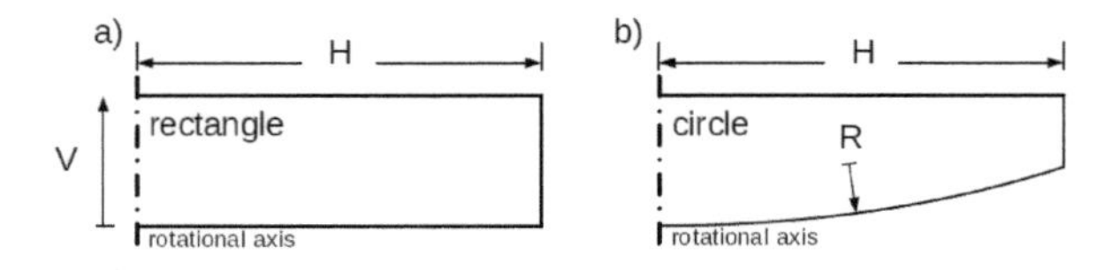

Figure 4-2 Geometries a) rectangular and b) circle of the silicone pad using the rotational symmetry.

There are several parameter combinations for both geometries that match the above-mentioned criteria. For example, a height $V = 14.5$ mm and a width $H = 100$ mm for the rectangular silicone pad are appropriate dimensions for creating an almost constant stress distribution in the defined area.

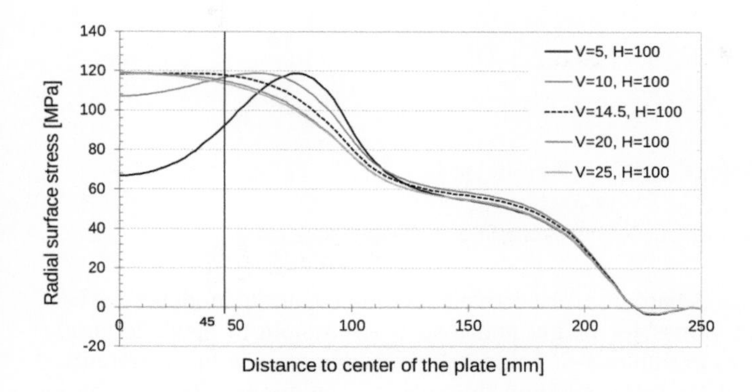

Figure 4-3 Parameter study with silicone pad „rectangle".

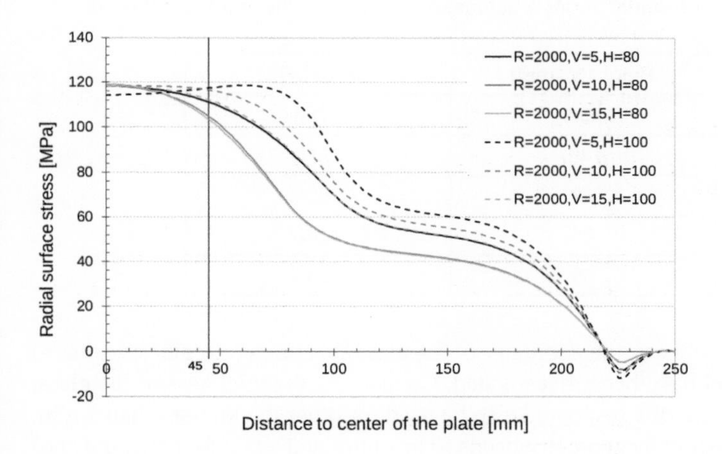

Figure 4-4 Parameter study with silicone pad „circle".

As both geometries nearly exactly yield the same stress distributions (compare Figure 4-5), the rectangular geometry was chosen for further studies because this geometry would be easier to manufacture.

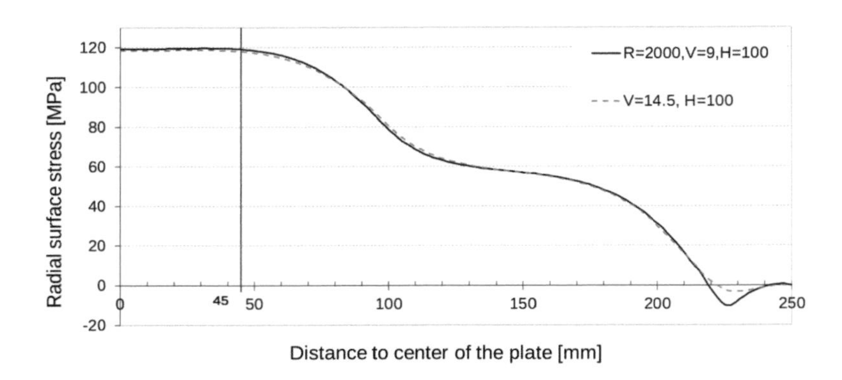

Figure 4-5 Optimized stress distribution with geometries „rectangle" and „circle".

In a second stage, the Shore-A hardness, respectively the fitted material models have been varied. In using different values for height and load it is possible to generate similar curves for each hardness as in Figure 4-5. Table 4-1 gives an overview of the optimized parameters whereas the width is $H = 100$ mm throughout. The applied force in the FE model $F_{10°}$ is distributed over an area of $A = \pi H^2\ 10° / 360°$ at this.

Table 4-1 Parameter sets for different Shore-A hardnesses to optimize the stress distribution.

Shore-A hardness	Height V [mm]	$F_{360°}$ [N]
4	13.5	3780
7.8	14.5	3816
10	22	3996
16	25	4140
22	35	4356

Additionally, the glass thickness, the geometry of the glass specimen (circular vs. square) and the applied load level have been investigated. Concerning the thickness of the glass, the modified coaxial double ring test could also be used for other thicknesses than 2 mm, although in that case the silicone geometry needs to be optimized again to match the chosen criteria. The influence of the glass plate geometry is negligibly low in regard to the center area of the glass plate. With rising distance to the center, the stress in a square specimen increases in comparison to the stress in a circular specimen due to the stiffening effect of the overhanging edge. Finally, the effect of different load levels has been investigated. With decreasing load the area of a constant stress distribution increases which afterwards has to be considered in a stochastic evaluation (compare Figure 4-6).

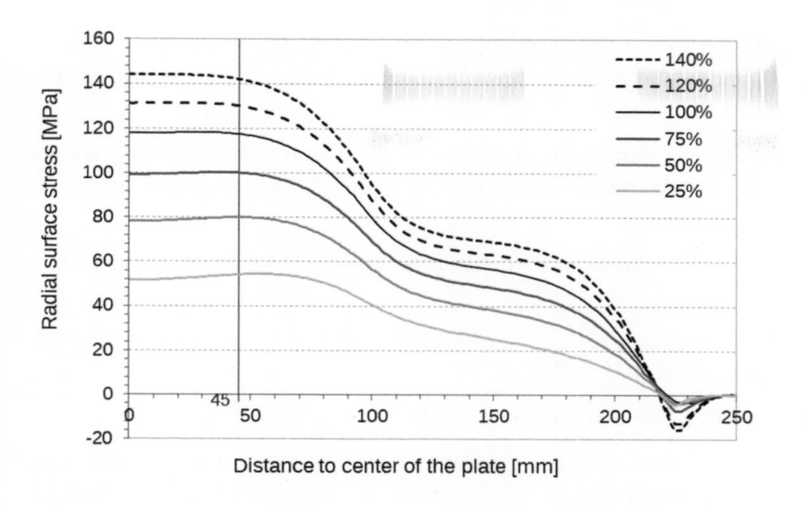

Figure 4-6 Variation of the load level; glass thickness = 2 mm, silicone geometry = rectangle.

5 Conclusion and outlook

In this paper a possible modification of the coaxial double ring test for the testing of thin glasses has been presented. By replacing the loading ring with a silicone pad, a constant radial stress distribution in the center of the glass plate could be achieved. Various geometries, Shore-A hardnesses, glass thicknesses and glass geometries have been investigated. In the next step, the results have to be proofed in an experimental test set-up, e.g. by using strain gauges on the glass surface. If the results can be confirmed, a functional correlation between the applied load and the resulting surface stress needs to be established.

6 Acknowledgement

This research emerged within the scope of the research project "Tempergy Glass", supported by the Federal Ministry for Economic Affairs and Energy.

7 References

[1] DIN EN 1288-5: Glass in building – Determination of the bending strength of glass – Part 5: Coaxial double ring test on flat specimens with small test surface areas. September 2000.

[2] DIN EN 1288-1: Glass in building – Determination of the bending strength of glass – Part 1: Fundamentals of testing glass. September 2000.

[3] DIN EN 1288-2: Glass in building – Determination of the bending strength of glass – Part 2: Coaxial double ring test on flat specimens with large test surface areas. September 2000.

[4] Ali, A.; Hosseini, M.; Sahari, B.: A review of Constitutive Models for Rubber-Like Materials. In: American Journal of Engineering and Applied Sciences. Vol. 3.1, 2010, pp 232-239.

[5] Treloar, L. R. G.: The Physics of Rubber Elasticity. Third Edition. Oxford University Press. 1975.

[6] ISO 7743: Rubber, vulcanized or thermoplastic – Determination of compression stress-strain properties. May 2008.

[7] ANSYS Inc.: ANSYS 16.2 Workbench. 2015.

[8] Hanich, F.: Modification of the coaxial double ring test to investigate the bending strength of tempered thin glasses. Bachelor Thesis. TU Darmstadt – Institute of Structural Mechanics and Design. 2016.

Study on Glass Roof – Spannglass Beams as an Option?

Bernhard Weller[1], Jens Oman[1], Michael Engelmann[1]

1 Technische Universität Dresden, Institute of Building Construction, George-Bähr-Straße 1, 01062 Dresden, Germany, jens.oman@tu-dresden.de

Recent research covers the topic of improving conventional glass beams by reinforcement and post-tensioning techniques. The literature states numerous promising results: enhanced load-bearing capacity, improved post-fracture behaviour and optimised material utilisation. However, building projects require spatial facades and full roof systems to cover openings. Therefore, the development of reinforced and post-tensioned glass beams for roof applications is considered promising to connect research results with real-life structures. This paper shows the efficiency of those structural options in a numerical study comparing an exemplary reference glass roof with novel options. The results give a preliminary estimation of the possible performance. They will contribute to build full glass roof systems with enhanced material effectivity while keeping the transparency in the future. Specifically, the paper discusses the question whether Spannglass beams are a realistic option in forthcoming glass roofs.

Keywords: glass beam, glass roof, roof structure, load bearing, post-tensioning, reinforced girder

1 Introduction

1.1 Long span glass beams: reinforced, post-tensioned and interconnected

Numerous projects showed a trend for transparency during the last decades. Those structures include glass beams for long-spans. Hereby the design needs to cover the brittleness of glass. Oversizing conventional glass girders to reduce the failure probability of the vulnerable glass edge is a common but rather material inefficient approach. Advanced structural options include a reinforcement to provide a redundant load path and post-tensioning techniques to pre-compress the glass edge that is loaded in tension.

The promising results in [1-6] and Figure 1-1 include:

- enhanced load-bearing capacity as the total load is split between the glass cross section and the additional (post-tensioned) reinforcement material,
- improved post-fracture behaviour as the reinforcement carries the tensile part of the total load and
- an optimized material utilization making use of advanced adhesive connections and removing unnecessary sacrificial layers from the laminate.

Engineered Transparency 2016. Glass in Architecture and Structural Engineering. First Edition.
Edited by Jens Schneider, Bernhard Weller.

Figure 1-1 Spannglass Beams with post-tensioned unbonded reinforcement (left, [7]) and adhesively bonded steel tendons (right, [3]).

Production, handling and transportation set technical limits in sizes of glass. For spans larger than available standard sizes of maximum 9.0 to 15.0 m structural connections are necessary. Figure 1-2 shows common examples.

Figure 1-2 Connections for large-span glass beams. Steel plates (Helsinki Music Hall, left), pointwise connections (Opera House Oslo, centre, [8]) and adhesive splice connections (All-glass enclosure IFW Dresden, right, [9]).

1.2 Glass Roofs

The state of the art includes several glass roofs and systems including glass beams. Common spans in representative buildings or extensions of existing building (Figure 1-3) range from 6.0 m to 9.0 m and consist of a combination of single glass beams. Thus, the promising characteristics of reinforced and post-tensioned single girders need to be applied to full structures in order to make use of their properties.

Figure 1-3 Glass roof "Alte Mensa" Dresden [10]. A promising choice to extend floor space while keep the natural illumination and remaining cautious out of respect for the existing building.

1.3 Resulting questions and approach

Thus, we summarise that reinforcing and post-tensioning glass beams are a meaningful option to create a higher state of redundancy, a material appropriate design and an optimised material utilisation. However, all numerical and experimental studies are limited to single girders so far.

We asked whether a use of novel post-tensioned glass beams – Spannglass Beams – is a reasonable option in a modular glass roof system. Furthermore, we suppose that the advantageous properties of single glass beams are transferable to full glass roof systems.

The approach is to perform a numerical design of a chosen glass roof. Therefore, a finite element model of a glass roof covering a courtyard of 13.8 x 9.0 m^2 was created. Variations of the model include a reference of a regular all-glass system, an un-reinforced system with the cross section concept of [4] and a post-tensioned option. The following section 2 will explain the method used. Afterwards, chapter 3 lists all relevant results for a subsequent comparison and discussion in chapter 4.

2 Numerical Analysis

2.1 System

Within the framework of the present paper, flat roof systems based on biaxial structures are analysed. The structures are subdivided into main and secondary beams that are further subdivided into segmental beams. Due to the biaxial structure, the perpendicularly ordered main and secondary beams span rectangular horizontal facade fields. Each facade field is filled with non-load-bearing infill glass element.

According to the defined covered courtyard, the roof is 9.0 m in width and 13.8 m in length. The roof structure is subdivided into three 3.0 m long segments in width and into six 2.3 m long segments in length. An isometric view of the glass roof structure without infill glass elements is given in Figure 2-1.

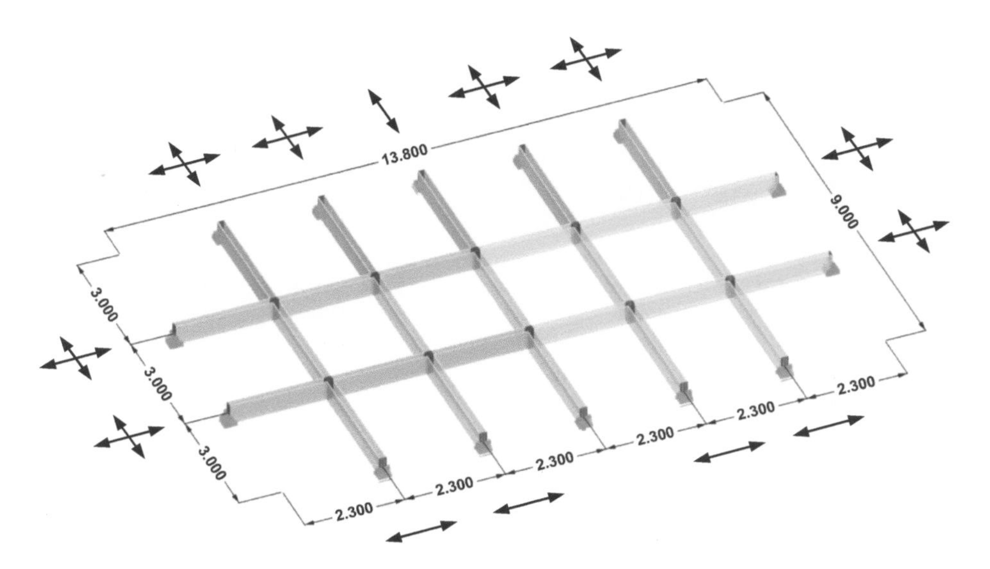

Figure 2-1 Isometric view of the analysed glass roof structure and support situation [mm].

Both the main and secondary beams are supported at the roof edges. Each end is supported in direction of gravity. To respond to the needs of expansion, the structure is supported free of constraint forces. For details of the support construction situation, please see Figure 2-1.

In order to compare different systems, the joint stiffness of intersections and glass girders is rigid. Based on rigid joint stiffness, the glass fins are designed as standard double laminated safety glass made of two heat-strengthened glass panels of 12 mm each and an ionomer interlayer. The height is fixed to 500 mm for all glass beams. The cross section of the reinforced Spannglass structure (System A) is shown in Figure 2-2.

The post-tensioning steel cable runs in a parabolic shape in the middle of the two glass girders of each Spannglass beam. It is fixed at both ends of the segmented main beam at the height of the centreline of the beam. The cable is bypassed and relocatable in longitudinal direction at each intersection. In the numerical study, the reinforcement cable is considered as a spiral strand made of stainless steel with a diameter of 16.6 mm [11]. The value of pre-tensioning is adjusted to the structure deflections.

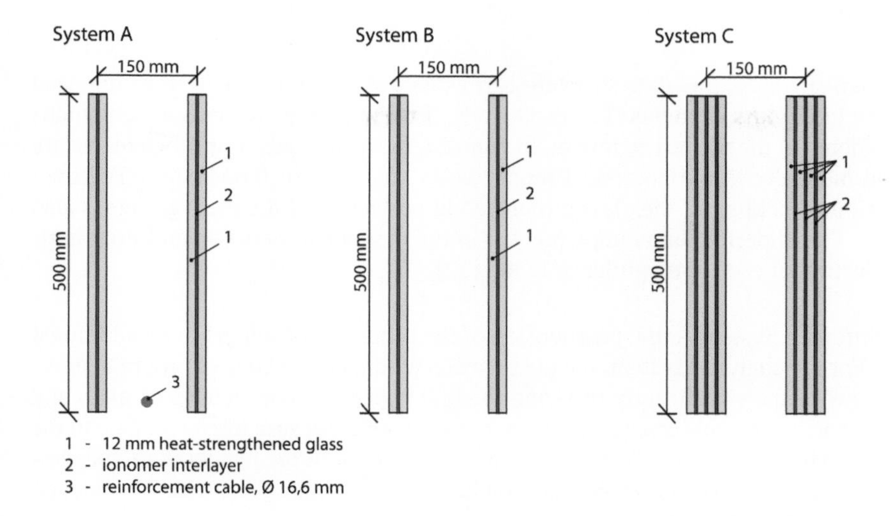

Figure 2-2 Reinforced cross section of System A (left) and unreinforced cross sections of System B (middle) and System C (right).

Additionally, two reference systems are analysed. To keep the comparability of the three systems, the glass girder height, the roof length and roof width as well as the joint stiffness are fixed. The only difference between the Spannglass structure (System A) and the reference structures (System B and System C) are the cross sections of the beams.

The first reference system (System B) varies from System A in the missing reinforcement cables. That means that the structure is unreinforced, and the full loads are to be exclusively bore by the glass girders itself.

The second reference structure (System C) is also unreinforced. In addition to that, the cross section is made of quadruple laminated safety glass. This oversized cross section is built up on girders including two outer glass layers as sacrificial layers. This quadruple laminated safety glass meets the demanded level of safety requirements in Germany. In intact condition, the two outer glass layers are part of the load bearing structure. Should both outer glass layers fail, the two remaining inner glass layers can still bear the full design loads. The quadruple laminated safety glass represents the regular cross section of common glass beams for roof structures, e.g. "Alte Mensa" of Technische Universität Dresden, Germany [10]. The cross sections of System B and System C are shown in Figure 2-2.

2.2 Loads

In the numerical study, the glass structure carries its own dead load as well as the dead loads of the infill glass elements. The dead loads of the structure elements are automatically considered in the numerical model. In contrast, the dead loads of infill elements are considered manually in the numerical model as an area load of 0.60 kN/m². To cover lateral torsional buckling of the glass girders, an imperfection of the glass girders is also considered. The imperfection is implemented in the numerical model as an initial horizontal deflection of each glass girder of $L/300$ [12].

For the reinforced System A, the post-tension of the cables is considered in an additional load case. For the analysed system, the post-tension amounts to 90.0 kN at room temperature. In view of the significantly differing thermal expansion coefficients of glass and steel, the temperature loads can have impact on the Spannglass structure [13]. Due to the complex behaviour of hybrid roof structures under influence of the temperature, the temperature loads are not taken into consideration in the analysis since this would be beyond the scope of the present paper.

But the environmental loads from wind and snow have to be taken into account. The characteristic loads are assumed to be +0.88 kN/m² for snow and +0.60 kN/m² for wind pressure.

2.3 Limit state design

Load bearing glass structures have to be checked in three types of analyses: first, the analysis for ultimate limit state (ULS), second the analysis for serviceability limit state (SLS) and, finally, the analysis for residual load-bearing capacity, which is needed for glass structures. The load cases are combined according to DIN EN 1990 [14]. Due to the reduced spectrum of load cases, the decisive load case combination consists of Dead Loads (DL), Imperfection (IMP), Snow Load (SL) as well as Wind Pressure Load (WP) and Post-Tension (PT) in case of System A. As second variable action load, wind pressure is reduced by a ψ_0-factor of 0.6.

The decisive load case combinations are as follows:

ULS: 1.35 DL + 1.00 PT + 1.00 IMP + 1.50 SL + 0.6 x 1.50 WP

SLS: 1.00 DL + 1.00 PT + 1.00 IMP + 1.00 SL + 0.6 x 1.00 WP.

As described in Chapter 2.1, the analysed glass girders are made of laminated safety glass built up on heat-strengthened glass panels. Under those conditions, the design resistance amounts to 51.3 N/mm² according to DIN 18008-1 [15]. To keep the serviceability of the roof structure, the deflections need to be limited to a maximum value. The analysed systems are based on segmented glass beams, which are connected to each other via intersections. In real life, the intersections and their joint stiffness are key aspects of the deflection resistance of systems. This is why the deflections of segmented systems can be larger than the deflections of unspanned systems made of continuous glass beams. Thus, the deflection limit of the segmented system is assumed as $L/300$.

3 Results

To find out the benefits of an innovation like Spannglass roof systems, it is necessary to compare the novel system to common systems currently in use. Hence, the Spannglass roof system (System A) will be compared with two unreinforced reference systems (System B and C) in terms of stress (ULS) and deflections (SLS).

The reinforced System A is compared to the unreinforced System B in order to find out the influence of post-tensioning cables. Furthermore, System A is compared to the common glass beam system C to figure out the benefits of Spannglass roof structures for forthcoming glass roof projects in real life.

The main criteria for the comparison are the tensile stresses and the deflections in initial state as well as the maximum deflection. The results of the analyses are shown in Figure 3-1.

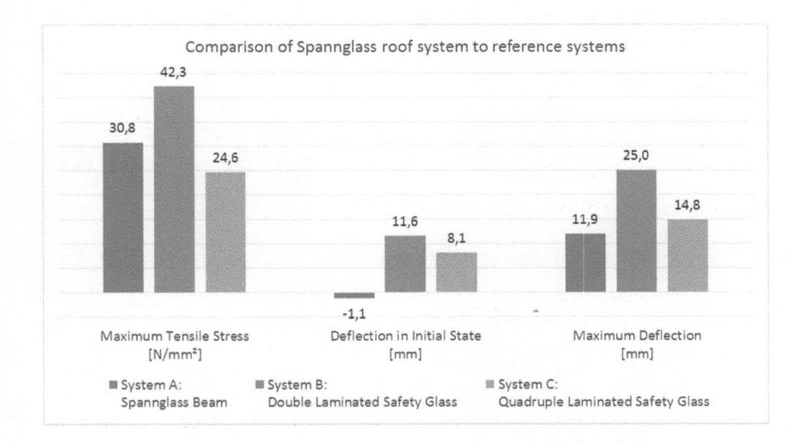

Figure 3-1 Comparison of the Spannglass roof system to the reference systems.

For System A, which is made of Spannglass beams, the maximum tensile stress is 30.8 N/mm^2. The existing maximum tensile stress of the unreinforced System B is significantly higher and amounts to 42.3 N/mm² (+37 %). System C made of quadruple laminates safety glass has the smallest amount of the maximum tensile stress with 24.6 N/mm² (-20 %).

To evaluate the cross sections of the analysed systems, the glass weight of each system is to be related to the maximum tensile stresses. The cross sections of System A and System B have a glass weight of 60 kg/m each. System C has a cross section of 120 kg/m weight. The relative glass consumption of the reference systems relating to System A is 82.4 kg/m (= 60.0 kg/m x 42.3/30.8) for System B and 95.8 kg/m (= 120.0 kg/m x 24.6/30.8) for System C.

There are significant differences in the deflection in initial state between the reinforced System A and the reference systems. System A is the only roof structure that has a cambered shape in initial state. The uplift amounts to -1.1 mm in the centre of the roof (Figure 3-2, left picture). Due to the parabolic shape of the cable, the structure can be uplifted depending on the pre-tensioning of the cable. The reinforced System A can equalise the dead-load deflections of the roofs structure in initial state exclusively.

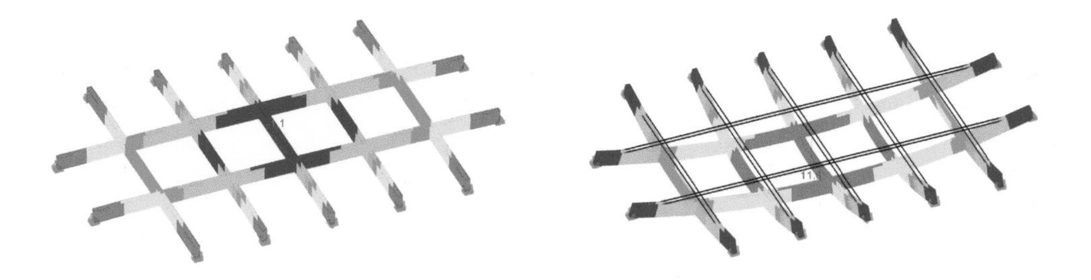

Figure 3-2 Dead-load deflection in initial state, System A: -1.1 mm (left), System B: +11.6 mm (right). Deflection shape of System C is analogue to System B but amounts to +8.1 mm. Scaling factor is 100.

System B and System C do not entail reinforcement cables that can create an uplift. Due to that, the structures of System B and System C deform in the direction of gravity (Figure 3-2, right picture). System B has a maximum deflection in initial state of +11.6 mm. The deflection of System C amounts to +8.1 mm.

The maximum deflection during SLS of reinforced System A is also smaller than the maximum deflections of the unreinforced Systems B and C. The reinforced System A has a maximum deflection of +11.9 mm. In contrast, System B has a maximum deflection of +25.0 mm, which is two times larger than the deflection of System A. Closer to the amount of System A is the maximum deflection of System C with +14.8 mm.

4 Discussion

To compare the novel Spannglass roof with common glass roof structures, varying systems based on the same glass girder height were analysed. Despite the same conditions, the systems differ significantly with regard to their load bearing behaviour. Although the same glass girder height have been provided, the results of tensile stresses and deflections of the analysed systems vary.

As mentioned in Chapter 2.3, the design resistance is 51.3 N/mm². The existing maximum tensile stress of each analysed system is to be less or equal to that amount. Due to the analyses using 2nd order theory and the imperfections of each glass girder, the stress analyses include lateral torsional buckling.

In comparison, the maximum tensile stress of the reinforced System A is lower than that of System B, which has the same glass girder geometry but the post-tensioning cable is missing. In System B, all the loads are bore by the glass girders itself. This load bearing behaviour is reflected in the stresses of System B (42.3 N/mm²) that are higher than the stresses of post-tensioned System A (30.8 N/mm²).

Due to the oversized cross section, the maximum tensile stress of System C is less than the maximum tensile stress of System A. Analogous to System B, the loads are bore by the glass girder itself due to the missing post-tensioning cable. In contrast to the glass girders of System A and System B, the glass girders of System C are built up on quadruple laminated safety glass. That means that the glass girders are two times thicker than the double laminated safety glass girders of systems A and B. As a result of the oversized glass girders and the missing post-tensioning cables, System C has the lowest maximum tensile stress of the analysed systems (24.6 N/mm²).

Related to those maximum tensile stresses, both reference systems inherit a higher relative glass consumption than System A. System B has a relative glass consumption of 82.4 kg/m related to System A, which has a glass consumption of 60 kg/m. That means, that the relative glass consumption of System B amounts to 137 % of the glass consumption of System A. System C with a relative glass consumption of 95.8 kg/m that is 160 % of the amount of System A. That means, that System A has the lowest glass consumption of all analysed systems related to the tensile stress level.

Also with respect to the deflections, System A can be seen as the most beneficial solution of the analysed systems. As has been described in Chapter 3, the post-tensioning cables of Spannglass roof systems support the structures in each deviation point at the intersections. This additional support increases the deflection resistance of the roof system. Thus, the post-tensioning cable significantly reduces the deflections in the direction of gravity. The option of post-tensioning cables can even be used to uplift the roof structure as needed, e.g. to equalise the dead-load deflections in initial state. Furthermore the post-tensioning cables are also an option to improve the residual load-bearing capacity. The cables can provide alternative load paths in case of glass failures.

For System A, the deflection in initial state amounts on -1.1 mm. The negative value represents an uplift of the roof structure which prevents the collection of rain water. On the given conditions of the analyses, the reinforced System A is the only approvable structure mentioned in Chapter 3. In terms of the deflection in initial state, the reference systems B and C are deformed to an irregular grade. System B is deformed by +11.6 mm and System C is deformed by +8.1 mm in the direction of gravity. Both systems can trigger collections of rain water in initial state. Depending on the interaction of deflecting in the direction of gravity due to mass of water and collecting more water due to deflection, the loads increase. In the worst case, the mass of water collected on the roof can cause defects in liquid tightness over time. For real life projects, the cross sections of System B and System C need to be designed with larger glass girders to increase deflection resistance and to avoid excessive rain water collection.

The comparison of System A and System B shows that the reinforcement cables reduce the maximum deflections significantly. Under the same conditions, the reinforced System A has a maximum deflection of +11.9 mm while the unreinforced System B has a two times larger deflection of +25.0 mm. That comparison highlights the advantage that reinforced Spannglass has over unreinforced systems. The maximum deflection of System A is even less than the maximum deflection of System C made of quadruple laminated safety glass (+14.8 mm). According to chapter 2.3, the deflection limit in SLS is L/300. In accordance with the roof width of 9.0 m, the allowable deflection is 30.0 mm. Each of the analysed systems meets that requirement.

Given the minimal glass consumption, the favourable deflection and the redundant load path via the reinforcement, System A made of Spannglass beams is the most beneficial structure of all analysed systems. It generates manageable values of tensile stress and deflection based on the same height of the glass girders as System B and System C. Additionally, the structure of System A can be uplifted by the post-tensioning cables. Transferred to the general type of structure, it can be said that the high design resistance and deflection resistance as well as the low material consumption make the segmented Spannglass roof system a reasonable option for forthcoming glass roofs.

5 Conclusion and summary

In framework of the present paper, varying segmented glass roof systems were analysed and compared to each other. A reinforced glass roof structure made of Spannglass beams was compared to two unreinforced reference systems.

The deflection resistance of the Spannglass structure is significantly higher than the deflection resistance of the unreinforced structures on the supposition that all glass girders have the same height. Unreinforced systems bear the loads exclusively on the glass girders. The post-tensioning cables of the reinforced Spannglass system support the structure at the cable deviation points and thus increase the resistance of the system. In this manner, load-bearing glass roof structures with high design resistance as well as deflection resistance can be generated by Spannglass system with compact sizes of glass girders.

This finding is fundamental for upcoming research activities. Extensive parameter studies are in work and focus on the deflection resistance and load-bearing behaviour of Spannglass roof structures depending on the degree of segmentation. Those studies include load cases that were excluded to the present paper, for example wind suction and temperature loads. At same time the real life stiffness of defined intersections get determined via experimental research activities. The residual load-bearing capacity of Spannglass roof systems will also be examined later on.

6 Acknowledgement

The scientific project is sponsored by the German Federal Ministry of Economics and Technology and is conducted cooperatively with THIELE Glas Werk GmbH (Wermsdorf, Germany) and KL-MEGLA GmbH (Eitorf, Germany) from 2016-2018.

7 References

[1] Bos, F.; Veer, F.; Hobbelman, G.J.; Louter, P.C.: Stainless steel reinforced and post-tensioned glass beams. In Proceedings of ICEM12 – 12. International Conference on Experimental Mechanics. Italy, 2004.

[2] Louter C.: Fragile yet Ductile. Structural Aspects of Reinforced Glass Beams. Dissertation, Delft University of Technology. Netherlands: TU Deft: 2011.

[3] Louter C.; Cupac J.; Lebet, J-P.: Exploratory experimental investigations on post-tensioned structural glass beams. In: Journal of Facade Design and Engineering 2. 2014. Pages 3-18.

[4] Weller, B.; Engelmann, M.: Spannglasträger – Glasträger mit vorgespannter Bewehrung. In Glasbau 2014, Weller B, Tasche S. Ernst & Sohn: Berlin, Germany: 2014.

[5] Weller, B.; Engelmann, M.: An innovative concept for pre-stressed glass beams. In: Proceedings of IABSE Conference. Nara, Japan: 2015.

[6] Engelmann, M.; Weller, B.: Load-bearing Adhesive Connections in Spannglass Beams – Experimental Study with Post-tensioned Reinforcement. In: Book of Abstracts: SEMC 2016, Advances and Trends in Structural Engineering, Mechanics and Computation. Cape Town, South Africa: 2016.

[7] Engelmann, M.; Weller, B.: Post-tensioned Glass Beams for a 9 m Spannglass Bridge. In: Structural Engineering International (26) 2016, Issue 02. Zürich, Switzerland: 2016. Pages 103-113.

[8] Snøhetta, O. A.: Opernhaus in Oslo. In: Detail Konzept, Musik und Theater. Zeitschrift für Architektur (49). München, Germany: Institut für internationale Architektur - Dokumentation GmbH & Co. KG, 2009.

[9] Weller, B.; Döbbel, F.; Nicklisch, F.; Prautzsch, V.; Rücker, S.: Geklebte Ganzglaskonstruktion für das Leibnitz-Institut für Festkörper- und Werkstoffforschung in Dresden. In: Stahlbau Spezial 2010 – Konstruktiver Glasbau (2010). Pages 34-40.

[10] Zschippang, S.; Wies, W.; Weller, B.; Schadow, T.: Glasdach Mensa und Rektorat der Technischen Universität Dresden. In: Stahlbau 75. 2006.

[11] Pfeifer Seil- und Hebetechnik GmbH: European Technical Approval, ETA-11/0160, Pfeifer Wire Ropes. Memmingen, Germany: Pfeifer Seil- und Hebetechnik GmbH, 2011.

[12] Weller, B.; Engelmann, M.; Nicklisch, F.; Weimar , T.: Glasbau-Praxis: Konstruktion und Bemessung – Band 2: Beispiele nach DIN 18008. Berlin: Beuth, 2013.

[13] Engelmann, M.; Bukieda, P.; Weller, B.: Experimental Investigation on Post-tensioned Spannglass Beams during Temperature Loads. In: Challenging Glass 5, Belis, Bos & Louter. 2016.

[14] DIN EN 1990, Dezember 2010: Eurocode: Grundlagen der Tragwerksplanung; Deutsche Fassung EN 1990:2002 + A1:2005 + A1:2005/AC:2010. Deutsche Norm. Berlin: Beuth, 2011.

[15] DIN 18008-1, Dezember 2010: Glas im Bauwesen – Bemessungs- und Konstruktionsregeln – Teil 1: Begriffe und allgemeine Grundlagen. Deutsche Norm. Berlin: Beuth, 2011.

Different methodologies for PVB interlayer modulus characterization

Wim Stevels[1], Pol D'Haene[2], Pu Zhang[3]

1 Eastman Chemical Company, Ottergemsesteenweg Zuid 707, Gent, Belgium, wimstevels@eastman.com

2 Eastman Chemical Company, Ottergemsesteenweg Zuid 707, Gent, Belgium, ppdhae@eastman.com

3 Eastman Chemical Company, 730 Worcester Street, Springfield, MA, USA, ppzhan2@eastman.com

The proper measurement and interpretation of modulus data for glass laminate interlayers can be quite complex. The development of master curves using different deformation modes and the preparation of the samples for measurement can significantly affect the results. The shear modulus of polyvinylbutyral (PVB) materials varies to a great extent, e.g. 1 – 400 MPa, over the temperatures and durations encountered for glass laminates in a building. International standard ISO 6721, determination of dynamic mechanical properties, uses modulus as a primary criterion for method selection. We have evaluated the use of tensile, plate-plate and torsion geometries for a high rigidity ("structural") PVB interlayer material. Datasets from different sources have been compared. This paper will discuss the results we obtained using different methodologies, and explore the effect with regards to positioning of the interlayers in the „stiffness families" and the associated shear transfer coefficients as in draft European norms prEN 16612 and prEN 16613.

Keywords: polyvinylbutyral, structural PVB, interlayer, storage modulus, laminated glass, prEN 16613

1 Introduction

Standard PVB interlayers have been used as an interlayer technology providing safety characteristics in laminated glass applications for decades. They can also provide some transfer of shear stresses between glass panes under specific conditions such as short load durations and/or modest temperatures. This is reflected in proposed or accepted glass standards such as prEN 16612, Determination of the load resistance of glass panes by calculation and testing [1], and ASTM 1300, Standard Practice for determining the load resistance of glass in buildings [2]. For cases where a higher level of structural performance is required, or load durations are longer, or occur at higher temperatures, PVB interlayers with high rigidity ("structural PVB") have recently become available on the market. The rigidity of these products contributes to reduced stresses and deflections in glass laminates for use in applications such as balustrades (Stevels, [3]), façades, and

Engineered Transparency 2016. Glass in Architecture and Structural Engineering. First Edition.
Edited by Jens Schneider, Bernhard Weller.

canopies. The trend towards ever larger glass surfaces is another driver for this development, as deflections need be limited to levels that allow full functionality of the glazing, meet standard requirements if absolute limits are in place, and are compatible with constraints in the framing design. Structural PVB's are available in roll widths up to 3.2 m and allow the use of conventional PVB lamination processing equipment and settings, facilitating standard production processes.

In order for an interlayer to be used in a structural glazing, the modulus of the material as a function of time and temperature must be known in detail. Draft European standard prEN 16613, Determination of interlayer mechanical properties, [4] is proposing a specific method for determination of interlayer modulus properties based on tensile vibration measurements. Whereas this specific method is certainly useful, no comparison to other potentially useful methods was made at the time. Generic standards for plastic materials exist under the ISO 6721 Plastics-Determination of Dynamic Mechanical Properties series [5]. In ISO 6721-1-Part 1: General principles, it is stipulated that different deformation modes may produce results that are not directly comparable. Thus far, limited information on the methodology and measurements of this type is available for structural PVB interlayers. A recent example of dataset that was both transparent and interpreted in terms of prEN 16613 was recently published (Zhang et al., [6]).

From a practical perspective, in terms of the interlayer stiffness family classification of prEN 16612, it is more relevant to review if various methods of interlayer characterization would result in a change of stiffness family as determined by specific Young's modulus values under this standard. Therefore, the properties of a structural PVB interlayer were determined using dynamic mechanical analysis using various deformation modes, and the results have been interpreted in terms of the load scenarios proposed in prEN 16612. In some cases, test reports were obtained from independent laboratories to benchmark results, or have more equipment types available.

2 Experimental

2.1 Materials

Eastman Saflex® DG41 structural interlayer was selected as the representative structural PVB interlayer. Interlayers were obtained from commercial production with a nominal thickness of 0.76 mm. Test specimens were conditioned inside a desiccator at room temperature for a minimum of 48 hours before being tested, thus bringing the moisture content of the specimen to zero.

PVB interlayers are completely amorphous, thermoplastic materials, which undergo no chemical cross-linking or crystalline melting during the normal lamination process (Schneider et al., [7]). More importantly, this process will not change any physical properties of the material. Therefore, measurements on the non-processed interlayer is possible, and was used in this testing.

2.2 Rheological measurement

A TA Discovery HR-3 rheometer, TA Discovery HR-2 hybrid rheometer and Anton Paar Rheometer MCR 702 (independent laboratory) were used for shear mode characterization measurements, using an 8 mm plate/plate geometry for a single layer of, dried, non-processed PVB interlayer (interlayer not put through a lamination and autoclave process) punched by an 8 mm circular die. In order to ensure good bonding between the PVB sample and the metal plates, each test specimen was loaded at 65 °C and heated to 100 °C first under program controlled pressure and then cooled to the test temperature for each frequency scan or temperature sweep.

Measurements in a tensile geometry were performed on two different TA Instruments, either a Q800 Dynamic Mechanical Analyzers (DMA) using a film tension setup (samples were cut to 15 mm x 5.6 mm), or on a TA Discovery HR-3 rheometer operated in tensile mode.

Measurements in a torsion geometry were executed an Anton Paar Rheometer MCR 702 (independent laboratory), using a strip of 21.5 mm (length) x 10 mm (width) x 0.76 mm at a constaint strain of 0.05 % from 0.01 to 10 Hz from -60 °C to 55 °C, with the normal force of -0.1 N to strain the sample. No reliable measurements could be obtained at higher frequencies or temperatures.

3 Results and discussion

The ISO 6721 series and prEN 16613 were taken as leading documents for method selection. Considering the modulus range of interest (1-400 MPa), and making use of the methods that seem to be most wide-spread for the characterization of plastic materials, we used tensile vibration (ISO 6721-4), torsional vibration (ISO 6721-7), and oscillatory plate-plate deformation (ISO 6721-10) for this study. A more elaborate review of method selection and deformation modes can be found in Stevels et al. [8]. The relative ease of sample preparation is an advantage for these methods. Kuntsche et al. [9] have reported on the use of double shear vibration (ISO 6721-6), but reported relatively complex sample preparation and installation.

3.1 Single instrument, single sample, different modes

A temperature sweep measurement on a single structural PVB interlayer sample was executed, in shear mode and in tensile mode, executed on the same TA Discovery HR-3 rheometer, to get a basic understanding of the structural PVB rheological characteristics and differences that may arise as a result of method choice. This has the advantage that differences in sample composition, sample preparation and e.g. oven temperature calibration and control are fully eliminated as causes for potential differences in the measurement. The results are shown in Figure 3-1, with the Young's modulus *E* plotted as divided by 3 (*E*/3) to allow easy comparison with the shear modulus *G*.

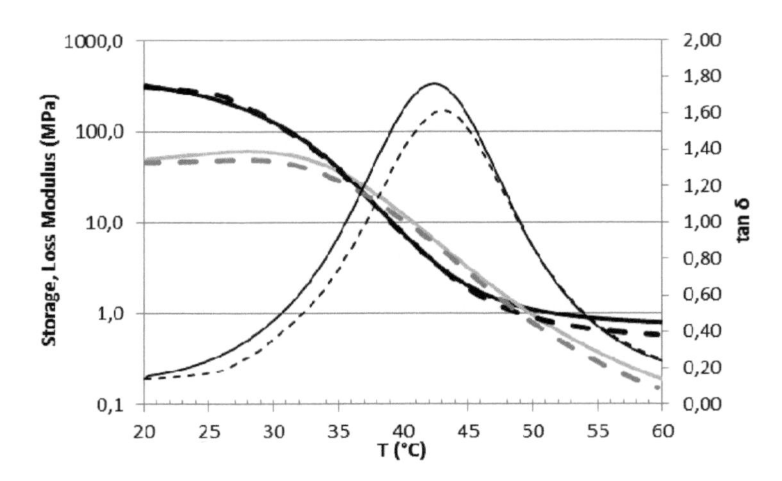

Figure 3-1 Temperature sweep (1 Hz) in shear (plate/plate) (*G*': thick black line; *G*" grey line; tan δ thin line) and tensile mode (*E*'/3 dashed thick black line; *E*"/3 dashed grey line; tan δ dashed).

As expected, the general shape of the curves as determined in either mode is very similar. The Young's Modulus value *E* is related to the shear modulus G through $E \approx 3G$ if a Poisson's ratio of the interlayer close to 0.5 is assumed. This is reflected in the values found for the storage and loss moduli in this experiment. The glass transition temperature, as reflected by the tan δ peak temperature, is slightly higher than as measured in tensile mode, but only by approximately 1 °C.

To review if the consistency between the different modes holds over a wider range of frequencies, frequency sweeps were carried out at 20, 40 and 50 °C. These temperatures represent predominantly elastic behavior at the lower temperature to predominantly viscous behavior at the higher temperature. It is known that the variation in moduli around the glass transition temperature of material is most prone to variation, and therefore the temperature of 40 °C was selected as well. The results are shown in Figure 3-2 for the storage moduli only, and compared to the results in tensile mode again represented as Young's modulus values divided by 3 (*E*/3) to allow easy comparison with the modulus curves measured in shear mode. It can readily be seen that the assumption of the Poisson's ratio is 0.5 is reasonable, given the proximity of the curves. It also shows that very similar results can be obtained in both modes, all else being comparable.

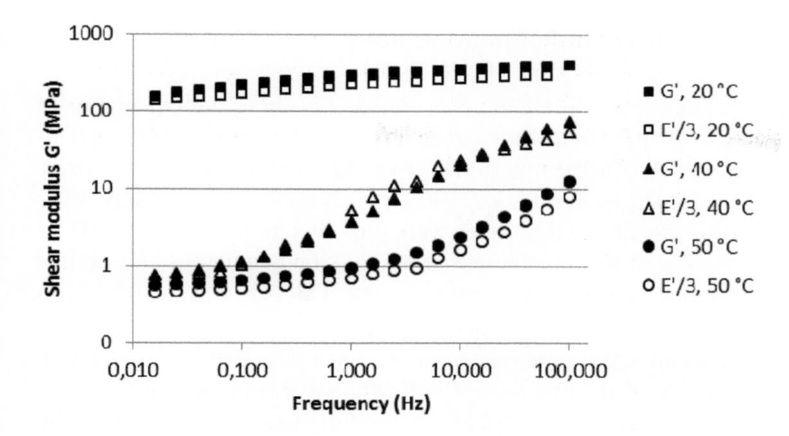

Figure 3-2 Frequency sweeps at 20, 40 and 50 °C in shear (*G*, plate/plate) and tensile mode (*E*). The latter results are represented as *E*/3 for ease of comparison. Different instruments, single sample, different modes.

As an extension of the work in previous paragraphs, the same sample was measured on three different rheometers. A temperature sweep and a frequency sweep at 40 °C were recorded. A short representation of the results is given in Table 3-1, in terms of tan data δ only.

Table 3-1 Key characteristics of a single structural PVB interlayer sample measured on different machines.

Rheometer, mode	Glass transition temperature (°C)	Tan δ peak value
TA Discovery HR-3 rheometer, tensile	43.7	1.7
TA Discovery HR-3 rheometer, shear	42.6	1.8
TA Instruments Q800, tensile	43.1	1.6
TA Discovery HR-2 hybrid, shear	43.0	1.8

The similarity of these results is reflected when the full temperature and frequency sweep curves are compared. This experiment was repeated for a few different samples, with a similar outcome each time. In this case, where sample and sample preparation consistency were ensured, and where instrumentation parameters were chosen from a similar basis, reasonable alignment in results between tensile and shear deformation (plate/plate) modes could be achieved even on different equipment.

3.2 Structural PVB: comparison of methodology

To determine the stiffness family according to prEN 16613, it is not enough to measure a single temperature sweep or a single frequency sweep. Frequency sweeps at multiple temperatures have to be executed to construct mastercurves at different temperatures, as they relate to the different load scenarios. A structural PVB was characterized internally using plate-plate geometry in this manner, and samples were provided to different external laboratories for characterization in plate-plate, tensile and torsion geometries. An outline for the data generation methodology can be found in Zhang et al. [6].

These data were then processed to generate mastercurves, and values extracted pertaining to the load scenarios of prEN 16613 that have a non-zero shear transfer coefficient in stiffness family 3, in order to assess the effect of the measurement methodology on the stiffness family classification. Snow loads of unheated buildings were not included, because mastercurves at 0 °C could not be constructed for all datasets. The results are collected in Table 3-2.

Table 3-2 Comparison of Youngs modulus as determined for Saflex® DG41 interlayer using different deformation modes.

	Plate plate geometry (Det. I)	**Plate plate geometry (Det. II)**	**Tensile mode**	**Torsion mode**
Load case	*E* (MPa)	*E* (MPa)	*E* (MPa)	*E* (MPa)
Wind load (Mediterranean areas)	318	381	85	528
Wind load (other areas)	1023	819	679	1428
Personnel balustrade loads – normal duty	114	180	19	198
Personnel balustrade loads – crowds	20	63	4.3	96
Snow load – roofs of heated buildings	21	12	3.6	15

In contrast to the work in the earlier paragraphs, where consisted practices were adopted, a much wider variation of individual values is now observed. Nevertheless, all these modulus results would put Saflex® DG41 interlayer in stiffness family 3, with the exception of the personnel loads as measured in this particular tensile determination with a very slight margin (< 1 MPa). Given the issues observed in this particular measurement as executed, and the seemingly different results by other authors, it would be recommended to repeat this work with more attention to sample dimension control by pre-stress and/or pre-strain setting, clamping practice and/or interlayer thickness.

4 Conclusion

The rheological properties of PVB interlayer materials were determined using different deformation modes. It was found that in experiments where sample and sample preparation consistency were ensured, and where instrumentation parameters were chosen from a similar basis, reasonable alignment in results between tensile and shear deformation (plate/plate) modes could be achieved.

It was found that this alignment was less apparent when these elements varied, in e.g. external studies, between the methods studied. There was significant variation in the individual values determined for the Young's modulus values E for different load scenarios. This variation tends to be largest around the glass transition temperature of the material. However, the positioning of the interlayer in the stiffness family as proposed in prEN 16612 was generally not affected.

It is recommended that further studies are executed to establish best practices in interlayer rheological property determination to confirm these findings.

5 References

[1] ASTM International: ASTM E1300 -12a: Standard Practice for determining the load resistance of glass in buildings (2012).

[2] European Committee for Standardization: prEN 16612 Glass in Building – Determination of the load resistance of glass panes by calculation and testing (2013).

[3] Stevels, W.: Design and testing of annealed glass balustrade panels using a structural PVB interlayer. Proceedings Glass Performance Days (Tampere Finland), pp. 169-171 (2015).

[4] European Committee for Standardization: prEN 16613 Glass in Building – Determination of interlayer mechanical properties (2013).

[5] International Organization for Standardization: ISO 6721-1 Plastics – Determination of Dynamic Mechanical Properties- Part 1: General principles (2011).

[6] Zhang, P.; Stevels, W.; Haldeman, S.; Schimmelpenningh, J.: Shear modulus measurements of structural PVB interlayer and prEN 16613. In: Proceedings Glass Performance Days (Tampere Finland), 2015, pp. 148-152.

[7] Schneider, J.; Kuntsche, J.; Schuster, M.: Mechanical behavior of polymeric interlayers. In: Proceedings Glas im konstruktiven Ingenieurbau 14 (Munich Germany), 2016, Chapter 16.

[8] Stevels, W.; D'Haene, P.; Zhang, P.; Haldeman, S.: A comparison of different methodologies for PVB interlayer modulus characterization. In Proceedings Challenging Glas 5, Bos, F.; Louter, C.; Belis, J. (eds), Gent, 2016, accepted for publication.

[9] Kuntsche, J.; Schuster, M.; Schneider, J.; Langer, S.: Viscoelastic properties of laminated glass interlayers – theory and experiments. In: Proceedings Glass Performance Days (Tampere Finland), 2015, pp. 143-147.

A deterministic mechanical model based on a physical material law for glass laminates

Dr. Wolfgang Wittwer[1], Thomas Schwarz[2]

1 Kömmerling Chemische Fabrik GmbH, Pirmasens, Germany, wolfgang.wittwer@koe-chemie.de

2 Kömmerling Chemische Fabrik GmbH, Pirmasens, Germany, thomas.schwarz@koe-chemie.de

A new class of polyurethane based materials for liquid lamination is presented. They provide excellent mechanical properties to allow reduction of weight at a required level of stiffness compared to monolithic glass sheets and even the manufacture of mechanically stable cold bent glass laminates. For engineering and design modelling, a material law based on the physical properties model of laminate core is presented. This allows a calculation of the deformation of laminates with high accuracy depending on the applied load, the apparent temperature and the time. The model is based on thermorheologic fundamental principles. The material law obtained from lab scale specimens is proven by correlating the results of tests on full size glass laminates to demonstrate the feasibility of the model's fundamental assumptions.

Keywords: glass laminates, liquid lamination, mechanical model

1 Introduction

GEWE-Composite® is a glass laminate with an organic adhesive core, Ködistruct LG. Ködistruct LG is an aliphatic polyurethane which provides excellent optical and mechanical properties to allow for a reduction in weight at a required level of stiffness when compared to monolithic glass sheets and even the manufacture of mechanically stable cold bent glass laminates as demonstrated in Figure 1-1.

Figure 1-1 Formula 1 Motorhome.

Engineered Transparency 2016. Glass in Architecture and Structural Engineering. First Edition.
Edited by Jens Schneider, Bernhard Weller.

To achieve an official public approval for construction by German DIBT (Allgemeine Bauaufsichtliche Zulassung), we investigated the creep behaviour of the core adhesive to design composites. The target was to establish a material law to describe the development of shear modulus depending on time and temperature.

2 Available Data

From tests performed for German public approval we had creep results from P.L. Geiß (Kaiserslautern, Germany) [1] from small tablet specimen and deformation curves from 4-point bend tests of laminate sheets at various temperatures done by Friedmann&Kirchner (Rohrbach, Germany) [2] (Figure 2-1). They showed a relaxation behaviour, which lacks from similarity and is non-linear.

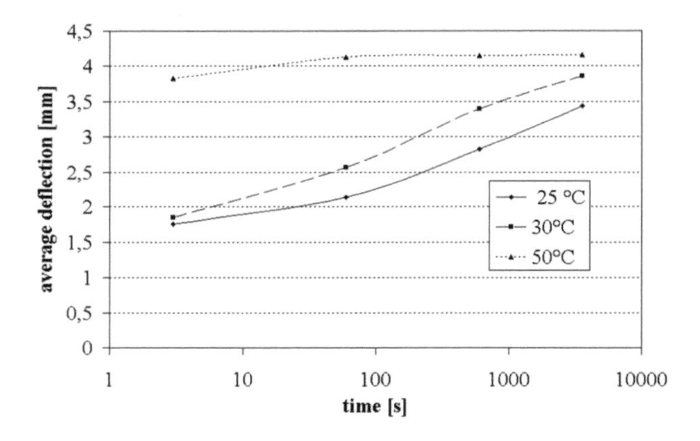

Figure 2-1 Creep results of 4-point bending.

At higher temperatures we see asymptotic creep behaviour of the laminate, ending at a fixed limit whilst at lower temperature a progressive creep is recorded. The target was to identify the creep behaviour and the kinetics of the relaxation process.

3 Investigations on Ködistruct LG relaxation in laboratory scale

To achieve a better understanding the thermorheology of the adhesive resin on small scale lab samples was investigated. To develop an appropriate physical calculation model, three basic investigations were done: Dynamic mechanical thermo-analysis (DMTA) should allow us to determine the Arrhenius activation barriers and to quantify the time-temperature-superposition of relaxation. Determination of the shear modulus at temperature levels above glass transition should provide the entropy elastic modulus. Determination of the creep under permanent load will add data, to be analysed as a base for the material law.

3.1 Dynamic thermo mechanical analysis and time-temperature-superposition of Ködistruct LG

To determine the Arrhenius activation barriers E_a the thermomechanical properties of resin films were measured at six different frequencies f (0.25; 0.5; 1; 3; 10 and 25 Hz) in tear mode. The results for complex elastic modulus E^* and the loss factor tan δ recorded and plotted in figure 3-1.

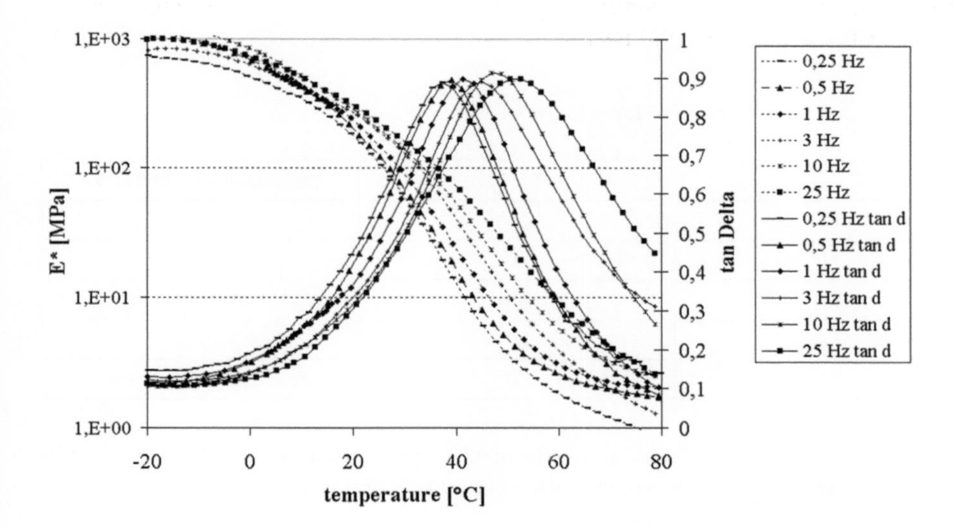

Figure 3-1 DMTA results for Ködistruct.

Analysis was done by determination of the maximum of tan δ at each frequency f as a criterion for glass transition temperature T_g. Using f as velocity in the Arrhenius equation (3.1) [3]

$$k = A \cdot e^{\frac{-E_a}{RT}} \tag{3.1}$$

k = velocity; A = Arrhenius factor; E_a = activation barrier; R = Gas constant; T = absolute temperature.

Regression analysis of the derived logarithmic equation (3.2 provides an Arrhenius activation barrier of 270.22 kJ/mol.

$$ln\, f = -\frac{E_a}{R}\frac{1}{T_g} + ln\, A \tag{3.2}$$

Using the Arrhenius equations (3.3) for different temperatures T_i

$$k_i = A \cdot e^{\frac{-E_a}{RT_i}} \tag{3.3}$$

the influence of temperature to the relaxation speed can be calculated and provide velocity shift factors VF_{12} in reference to different temperature conditions according to equation (3.4) and listed for certain temperatures in table 3-1:

$$VF_{12} = \frac{k_2}{k_1} \tag{3.4}$$

Table 3-1 Superposition factors VF towards 24 °C.

	***T* [°C]**	**Superposition factor *VF***
T_1	24	1.00
T_2	40	$2.69*10^2$
T_3	60	$1.37*10^5$
T_4	80	$3.46*10^7$
T_5	50	$6.69*10^3$

As practical consequence of these factors the mechanical behaviour of the laminate within 1 s at 80 °C is equivalent to approximately 19 months at ambient temperature.

3.2 Determination of entropy elastic shear modulus

For the quantification of shear modulus and its temperature influence the force elongation dependencies of the appropriate lap shear specimen were measured on a standard tensile forward and backward by 0 to 25 % The determined values for shear modulus were as noted in table 3-2:

Table 3-2 Shear modulus depending from temperature.

Temperature [°C]	**Temperature [K]**	**G (N/mm^2)**
60	333	1.404
70	343	1.426
80	353	1.471
90	363	1.494

Regression analysis for equation (3.5) is carried out.

$$G_S = n \cdot k \cdot T \tag{3.5}$$

G_S = entropic shear modulus; n = network density; k = Boltzmann factor;
T = absolute temperature.

To yield in $n \cdot k = 4{,}16 \cdot 10\text{-}3\ \text{N/(mm}^2\ \text{K)}$ or

$$G_S = 4{,}16 \cdot 10^{-3} N/(mm^2 K) \cdot T \tag{3.6}$$

This equation (3.6) is the base for correction of temperature effects on modulus as with increasing modulus creep slide will decrease under a given load.

3.3 Determination of relaxation behaviour of Ködistruct LG

With the same type of lap shear specimens the deformation in the tear machine under constant force at different temperatures (24; 40; 60 °C) was measured and recorded as shown in figure 3-2.

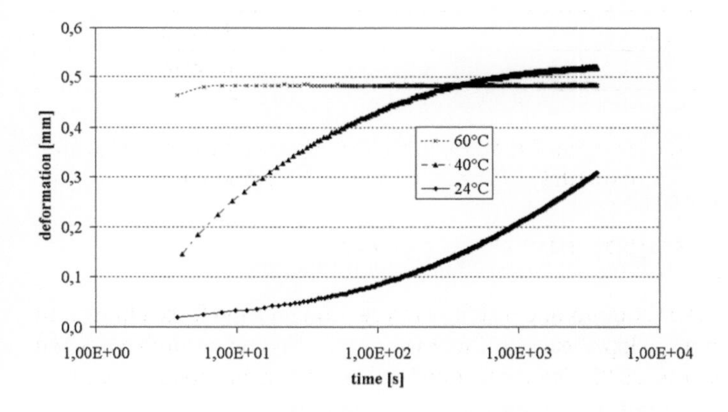

Figure 3-2 Relaxation of Ködistruct LG.

On the first view these curves lack similarity like those recorded in figure 2-1. For analysis the time scale of each curve was recalculated to a constant reference temperature based on the time-temperature-superposition factors of table 3-2. This recalculation to a time scale at fixed temperature provided the chart demonstrated in figure 3-3.

As the entropic modulus is temperature dependent according to equation 3.4, the shear slide is depending from the temperature as well. At a constant load level the maximum creep length for the different temperatures was calculated based on equation 3.5 with equation 3.6:

$$S_{To} = S_T \cdot \frac{T}{T_o} \tag{3.7}$$

S_{To} = Shear slide of reference at T_0; S_T = Shear slide at T; T = absolute temperature; T_0 = absolute temperature of reference.

Using these quantitative physical recalculations for time and creep length all measured data are transformed and normalized to standard ambient conditions. The resulting diagram shows a clear correlation between all three sets of creep data being part of a continuous curve approaching a limit deformation in an asymptotic manner.

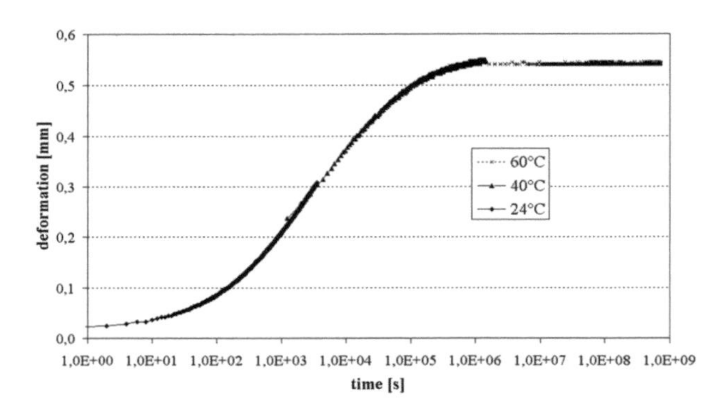

Figure 3-3 Normalized relaxation plot.

This curve can be approximated by a fit based on a Weibull function (3.8).

$$S = S_{max} \cdot (1 - e^{-(\lambda \cdot t)^k}) \tag{3.8}$$

s = shear deformation; s_{max} = limit (maximum) shear deformation;
κ, λ = parameters of Weibull distribution

The maximum deformation is calculated from equation (3.6).

For temperature T = 296 K (= 23 °C) a shear modulus G = 1,223 N/mm² is calculated. Given the parameters for an area A = 300 mm² and a load F = 100 N the limit shear deformation length s_{max} = 0.541 mm. Regression analysis in figure 3-4 with a Weibull function provides the values for λ = 0.000172 and κ = 0.405.

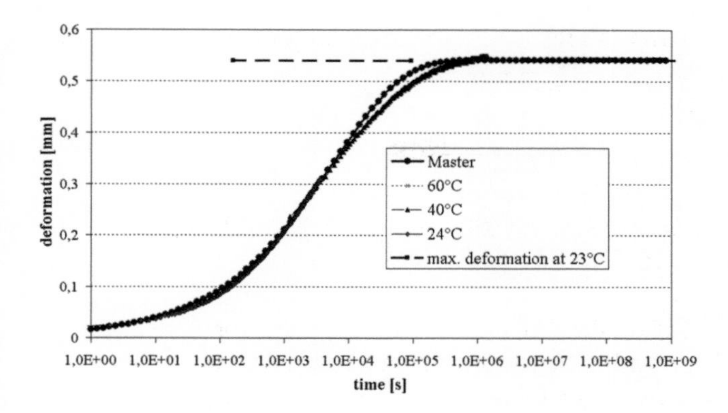

Figure 3-4 Weibull master curve fit.

For relaxation the master curve is representing the measured data in very good accordance. For the shear modulus the following representation can be derived:

$$G_{(t)} = \frac{G_s}{1-e^{-(\lambda t)^k}} \tag{3.9}$$

$$G_{(T,t)} = G_{(T_0,t')} \cdot \frac{T_0}{T} \tag{3.10}$$

$$t' = t \cdot e^{\frac{E_a}{R} \cdot \left(\frac{1}{T_0} - \frac{1}{T}\right)} \tag{3.11}$$

G_s = entropic shear modulus; κ, λ = parameters of Weibull distribution;
E_a = 270.22 kJ/mol.

With this set of equations (3.9), (3.10) and (3.11) the shear modulus and out of that the time and temperature depending deformation of the laminate can be calculated depending from time and temperature by analytical methods.

4 Verification of Physical Model of Polymer Relaxation

4.1 Analytical proof for large scale test results

A first proof for the reliability of the principle is given by use of the time-temperature-superposition factors for normalizing the measured results in figure 3-2 of the 4-point bend trials data. This superposition provides figure 4-1.

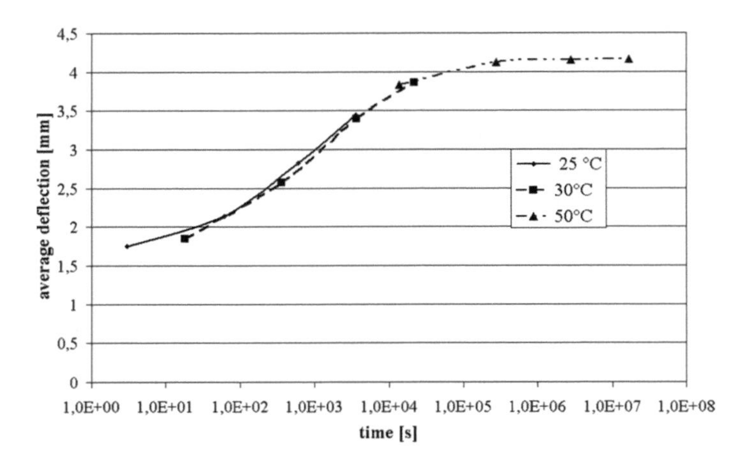

Figure 4-1 Recalculated 4-point bend results of laminates.

Again a continuous curve is the result of recalculation of measured data to a constant temperature axis. Taking into account that the tension distribution within the bent laminates is wide, the superposition factors seem to be feasible at varying shear levels.

4.2 FE-Simulation of Relaxation

Burmeister (Delta-X; Stuttgart, Germany) used the analytical equation of the master curve to do a fit based on Maxwell model using six elements. Finite-element-calculations of the 4-point-bend trials and the tablet creep trials based on this rheologic Maxwell model including the temperature superposition with the above described formalism provide convincing results.

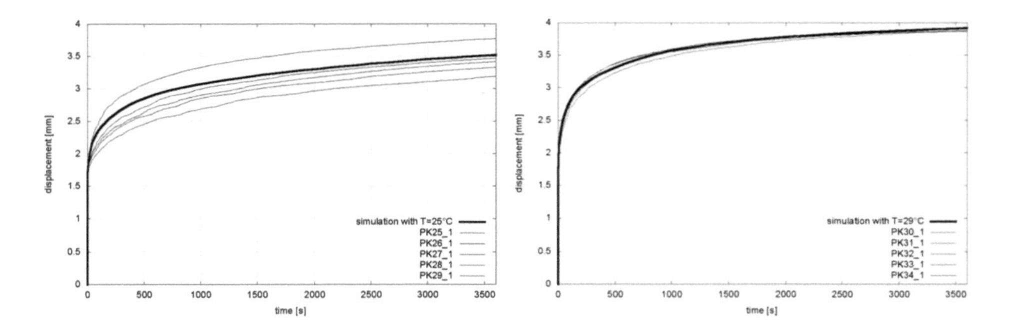

Figure 4-2 Modelling of 4-point bend creep at different temperatures 25 °C and 30 °C measured / 29 °C calculated.

At the 30 °C results a small temperature adaption of 1 °C is required to match experimental and calculated data which is most probably due to the inaccuracy of measured temperature in the oven.

5 Identification of Polymer Creep

To allow predictions at elevated temperatures for longer times the polymeric creep, which results in irreversible deformation has to be modelled as well. For this similar experiments at higher temperatures and for longer times were carried out. As a results out of this a second time-temperature-superposition factor and a equation for the polymer creep is gained (5.1).

$$k_i = A \cdot e^{\frac{-E_{ac}}{RT_i}} \text{ and } VF_{12} = \frac{k_2}{k_1} \tag{5.1}$$

E_{ac} = 134.67 KJ/mol

Table 5-1 Superposition factors VF_c for creep towards 23 °C.

	T [°C]	**Superposition factor *VF***
T_1	23	1.00
T_2	40	1.954*10^1
T_3	60	4.373*10^2
T_4	80	6.833*10^3

For calculation the impacts of both processes, relaxation and creep, have to be combined. Within the Maxwell series this can be done by addition of plastic element representing the creep behaviour. Based on the analytical model a prediction for the change in curvature of a cold bent glass laminate was made and subsequently tested. The measured results are again in excellent conformity with predicted behaviour as depicted in figure 5-1.

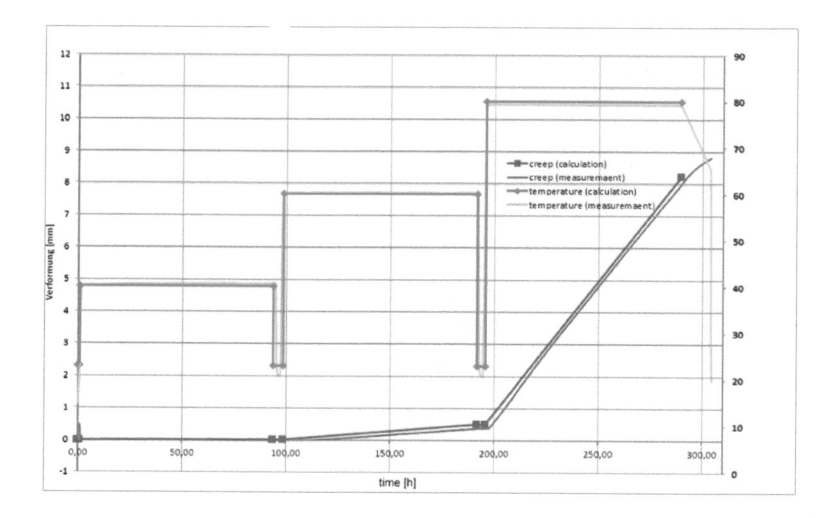

Figure 5-1 Deformation of a cold-bent laminate at heat exposure.

6 Conclusion

A material law represented by a set of analytic equations can be derived from lab scale experimental data. It reflects the material behaviour very well and can be validated with available test results from large scale experiments. The master curve is simulated with the same basic (entropic) temperature dependent elasticity and a time and temperature dependent rheologic model by Maxwell elements. Using these rheologic material parameters for a viscoelastic finite-element-model the experimental results can also be calculated with good relationship to measurements providing a deterministic model for the thermo-mechanical behaviour visualised by Delta-X in figure 6-1.

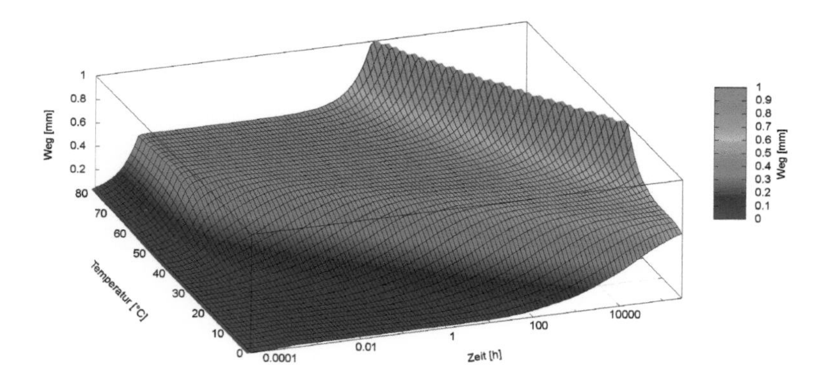

Figure 6-1 Time and temperature dependent model for relaxation and creep.

The model allows shortening investigations of relaxation and creep which are created by the mechanical effects of glass transition and creep. As limits mechanical destruction of the polymer network in case of overload or ageing influences have to be taken into account are not covered by the methodology.

7 References

[1] Paul Ludwig Geiß: Untersuchungsbericht „Prüfungen an GEWE-composite Glasverbunden“, TU Kaiserslautern, Kaiserslautern 2010.

[2] Robert Kirchner: Prüfbericht Nummer: 2009-02-3690-01 „Bestimmung der Steifigkeit von Verbundglas „GEWE-composite““ Dipl.-Ing. Michael Friedmann, Dipl.-Ing. Robert Kirchner, Rohrbach, 2010.

[3] Friedrich Rudolf Schwarzl: Polymermechanik, Springer-Verlag, Berlin, Heidelberg, New York 1990, ISBN 3-540-51965-6.

Investigation of different test set scenarios for determination of ultimate bending stress of thin glass

Jürgen Neugebauer[1]

1 University of Applied Sciences FH Joanneum

Everybody knows thin glass in application as a screen for laptops, tablets or mobile phones. An application of such a glass in building is new and an interesting topic for the future. Glass with a thickness of 0.4 up to 2.0 mm can be defined as a thin glass or even as ultra-light. On the market there are several suppliers, which offer such a thin glass. Not only the design with thin glass causes a totally new kind of thinking, also possible test scenarios for determination of the ultimate bending strength are currently not distinctly regulated in standards. Existing test set-ups described in standards e.g. EN 1288 (four-point bending test or large ring on ring test) cannot be used for the determination of the ultimate bending strength of thin glass without modifications. Different test set-ups published in several papers show possibilities for alternative determination of ultimate bending strength. These different set-ups were investigated for their applicability for determination of bending strength of thin glass. This paper gives a summary of a theoretical investigation.

Keywords: thin glass, ring-on-ring test, pressure pat-on-ring test, four-point bending test, multi-point bending test, bending with constant radius

1 Introduction

Glass with a thickness of 0.4 up to 2.0 mm can be defined as a thin glass or even as ultra-light. On the market, there are several suppliers, which offer such a thin glass. On the one hand there is aluminum silicate glass e.g. "GORILLA GLASS" by Corning Incorporated or "LEOFLEX" by AGC, which are pre-stress by chemical treatment and on the other hand there is soda lime silicate glass, which is pre-stress thermally or chemically.

The design with thin glass causes a total new kind of thinking. This thin glass is very weak against local bending stresses and has a large capacity against membrane stresses. For this reason structures with less portion of local bending stresses and large part of membrane stresses had to be found. Such structures are more or less curved structures. For example, cylindrically or conically shaped geometries of glass are favorable for such transfer mainly by membrane forces [1].

Engineered Transparency 2016. Glass in Architecture and Structural Engineering. First Edition.
Edited by Jens Schneider, Bernhard Weller.

2 Production of thin glass

2.1 Float glass process

Float glass is a sheet of glass made by floating molten glass on a bed of molten tin. After a controlled down process the glass is cut into certain sizes, as typically known as jumbo size. This method gives the sheet a uniform thickness and very plane-parallel surfaces.

2.2 Dawn draw process

The molten glass flows through a small gap at the bottom of the melting tank down and is cooled to ambient temperature by annealing furnaces. After this controlled down cooling process the glass is cut into certain sizes [1].

2.3 Overflow fusing process

The molten glass is poured into an overflow gutter. From this gutter the molten glass flows on both sides down and fuses at the bottom point of the gutter. After a down cooling phase the glass is cut into panels with certain sizes [1].

3 Pre-stressing of glass

3.1 Thermal treatment

Thermal treatment is a typical process of pre-stressing, according EN 12150 [2], of glass in which the glass is moved on rollers forwards into the heating zone and is heated up above the transition point. After this phase of heating, the glass is blown off with air. During the phase of cooling to ambient temperature, glass is permanently moved forwards and backwards on rollers in the furnace. The thinner the glass the bigger so-called roller waves can occur.

For this reason, the Austrian company LISEC has investigated a new process in which the glass is transported on air cushion. This technique gives the possibility to pre-stress thinner glass by thermal treatment without roller waves.

3.2 Chemical treatment – Ionic Exchange

Another possibility to pre-stress the glass is chemical treatment according EN 12337 [3]. The glass is immerged into molten potassium nitrate. At a temperature of approx. 370 - 450 °C the effect of ionic exchange takes place. The smaller sodium ions diffuse from the glass into the liquid potassium nitrate and the larger potassium ions penetrate into the glass matrix. Due to the larger ionic diameter of potassium ions compressive stresses in the close up range of the surface result. The depth of penetration is around 50 - 100 μm. [1]

The values for ultimate bending strength, which are the basis for a structural design, are still missing. Therefore a couple of different test scenarios were investigated for their applicability for determination of ultimate bending strength of thin glass. Due to the application one has to differ between test scenarios with and without the influence of the edge strength (edge quality) – the so called edge effect. In the following a couple of possible test scenarios were investigated with the help of a finite element program and the result are shown.

4 Determination of ultimate bending strength without influence of edge strength

All sides simply supported glass elements e.g. window glass, are good examples for application where the edge effect has not be taken into account. Because the maximum stress arises in the middle of the glass pane. At the edges tiny stresses arises and therefore the influence of the edge strength has not be taken into account.

4.1 Ring on ring test – EN ISO 1288

The test set-up is performed by placing the glass sample on a circular steel reaction ring (supporting ring) and applying on its upper surface a load transmitted through a steel loading ring, until the glass breaks, as shown in figure 4-1 below. The purpose of this test is to achieve a uniform tensile stress field inside of the loading ring that is independent of edge effects. This is described in Blank et.al [4]. Such test set-ups are defined e.g. in EN ISO 1288-1 fundamentals of testing glass [5], EN ISO 1288-2 for large surfaces [6] and EN ISO 1288-5 for small glass samples [8].

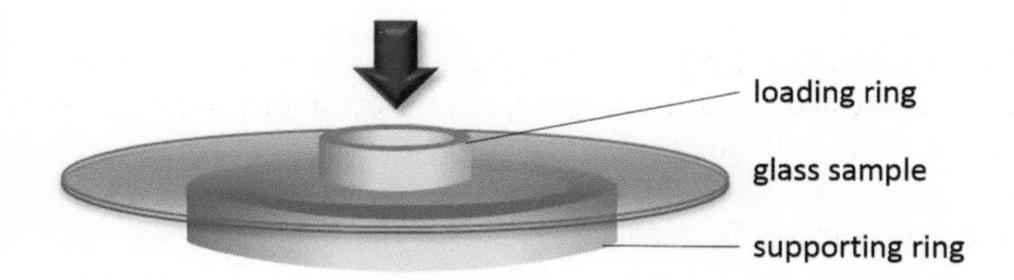

Figure 4-1 Ring-on-ring test set-up.

The test set-up for large surfaces defined in EN ISO 1288-2 is not usable for thin glass, because the deflection of the glass is much too high. The test scenarios (R 30, R 45, R 60 and R 105) defined in part 5 of EN ISO 1288 are more or less applicable for determination of bending strength of thin glass. But effects like as size effect, geometrical non-linearity or imperfections influences the results very much and have to be considered. Due to the thinness of the glass geometrical non-linear effects become dominant in these test scenarios, this is also mention in Wilcox [10].

Another effect, which has to be taken into account for thin glass is, that the stress has no constant distribution inside the loading ring as one assumption for the ring-on-ring test scenario which is stipulated in EN ISO 1288-5 [8]. Figure 4-2 below shows stress distributions for different ring-on-ring test set-ups (R30, R 45, R 60 and R 105) at different levels of stamp forces (0.10 kN up to 0.50 kN). The sizes of the quadratic glass samples are according EN ISO 1288 and the glass thickness is 1 mm.

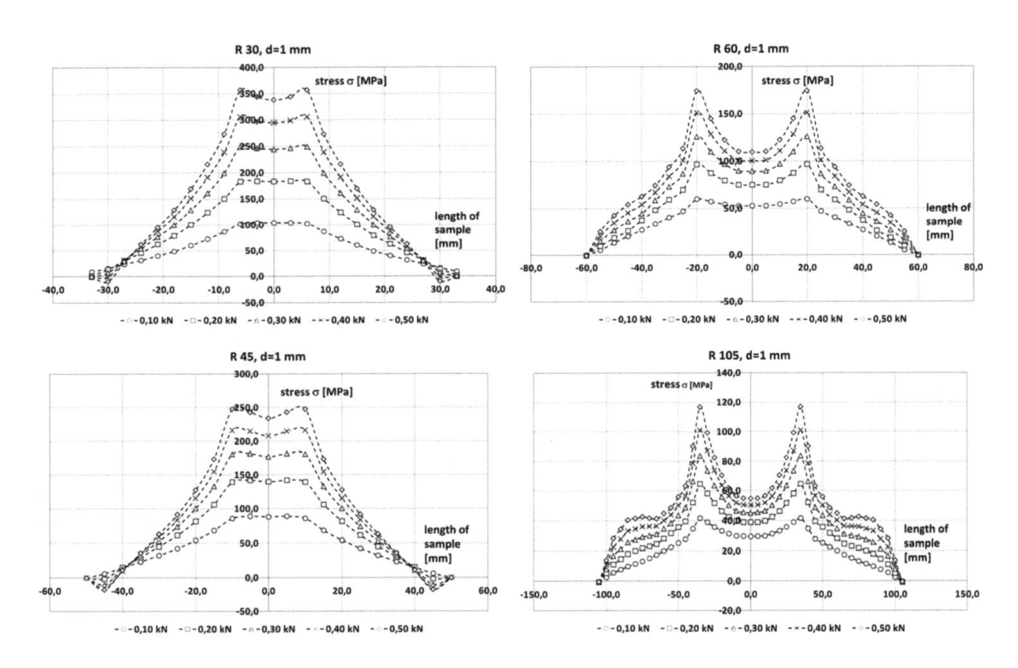

Figure 4-2 Ring-on-ring test – stress distribution along middle axis of glass sample.

Two effects can be summarized. The thinner the glass is the more difference between the stress in the middle of the glass sample and the area below the loading ring arises. The bigger the diameters of the loading ring is the more difference between the stress in the middle of the glass sample and the area below the loading ring arises.

4.2 Pressure pat on ring test

As a possible improvement of the ring on ring test a pressure pat on ring test was investigated. The test set-up is performed by placing the glass sample on a circular steel reaction ring (supporting ring) and applying on its upper surface a load transmitted through pressure pat instead of the loading ring, until the glass breaks, as shown in figure 4-3 below.

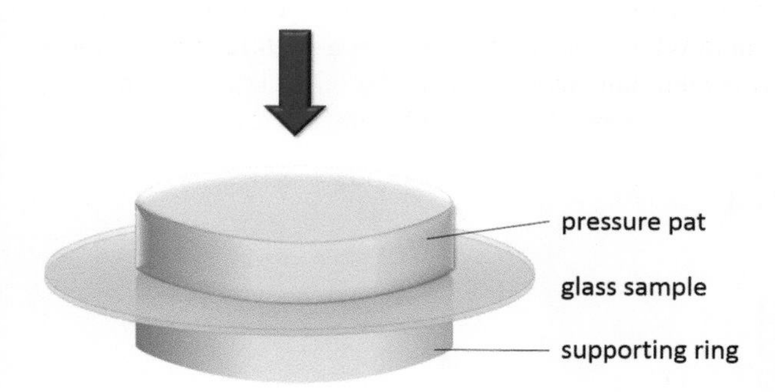

Figure 4-3 Pressure pat on ring test set-up.

The benefit of this scenario is that stability and buckling effects (described later in chapter 9 – effect of imperfections) are minimized and the area in which the stress can be assumed as uniform can be increased in comparison to a ring on ring test. A disadvantage is that the stress inside the supporting ring, as shown in figure 4-4, cannot be assumed as uniformly. Figure 4-4 below shows stress distributions for different diameters of supporting ring (105 and 190 mm) at different levels of stamp forces (0.10 kN up to 0.50 kN). Stress peaks in the area above the supporting ring arise. The thickness of the quadratic glass samples is 1 mm.

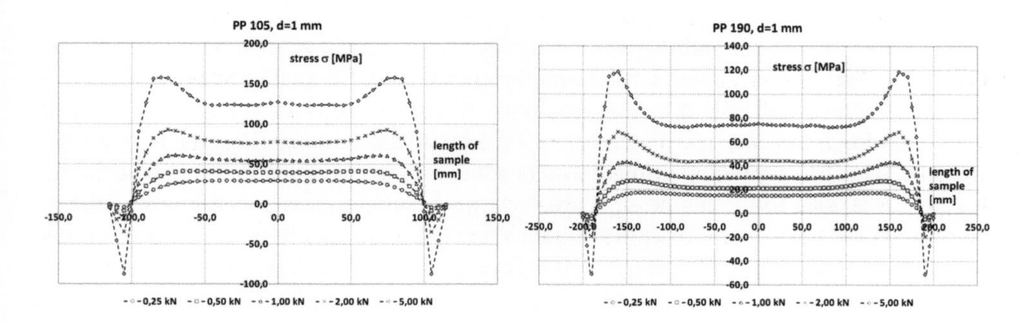

Figure 4-4 Pressure pat on ring test – stress distribution along middle axis of glass sample.

Two effects can be summarized. The thinner the glass is the higher stress peaks in the area above the supporting ring arise. The bigger the diameters of the supporting ring / pressure pat is the higher stress peaks in the area above the supporting ring arise in comparison to the stress in the middle of the glass sample.

5 Determination of ultimate bending strength with influence of edge strength

On two opposite sides simply supported glass elements e.g. room-high façade elements are good examples for application where the edge effect has be taken into account. The reason for this is that in such cases edges are bended and get bending stress at these edges.

5.1 Four-point bending test

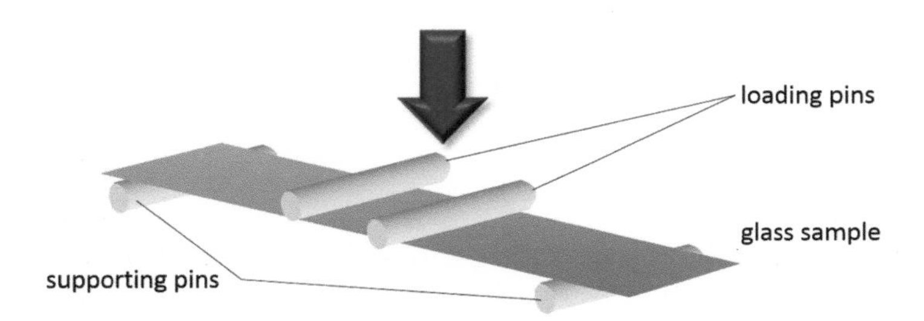

Figure 5-1 Four-point bending principle test set-up.

Figure 5-1 shows the principle test set-up for a four point bending test according EN ISO 1288-3 [7]. The test specimen with a length of 1100 mm and a width of 360 mm is supported on two supporting pins with a distance of 1000 mm. On its upper surface a load transmitted through two additional loading pins is applied until the glass breaks.

Large deflections result and the bearing forces are no longer vertical but inclined. The glass pane distributes its bearing force only by contact and eventually by friction between glass and rubber (EPDM). A simple resolution (breakdown) of the force to vertical and horizontal force shows that with increasing deflection also horizontal components are increasing. This has a growing influence on bending moment and therefore on the bending tensile stress. Due to the thinness no breakage of these thin glass panels can eventually be reached, because of slip from bearing pins due to bowstring effect (distance of pins is constant but end of panes move towards) or on some testing machines reach of maximum piston stroke.

5.2 Multiple point bending

To be able to use the well-known format of 1100 x 360 mm for bending test also for thin glass two possibilities for modifications given for the four-point bending test are possible. On the one hand modify the distance of bearing and eventually loading pins and on the other hand introducing additional pins for bearing as well as for loading, as shown in figure 5-2 below.

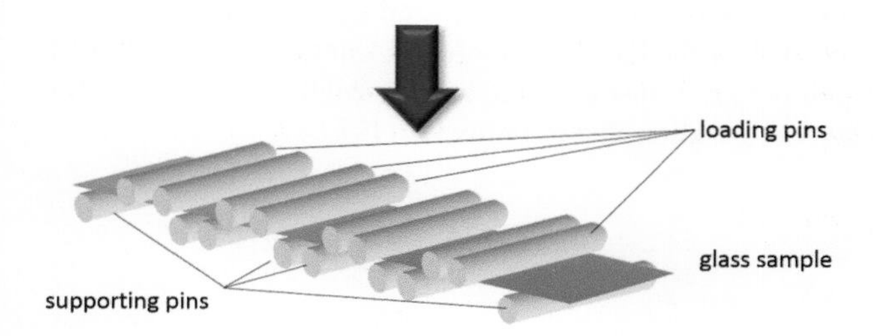

Figure 5-2 Multi-point bending principle test set-up.

Due to disadvantages of tensile stress arises on both surfaces – top and bottom, with the meaning the tensile stress arises in the zone below the pairs of loading pins on the bottom surface and over the supporting pins on the top surface. This set-up induces alternative tensile stress on lower and upper surface [9]. Figure 5-3 below shows the stress distribution along the middle axis of the glass sample for two different test set-ups. The left diagram in figure 5-3 shows the stress distribution for a type a, in which the support pins and loading pins have all the same distance in between them. In the zone at the outer supporting pins higher stresses arises in comparison to the stresses in the middle of the glass sample. The right diagram in figure 5-3 shows the stress distribution for a type b, in which the glass sample is slightly longer and the glass sample is supported at their ends too. The result in type b is better in comparison to test set-up in type a.

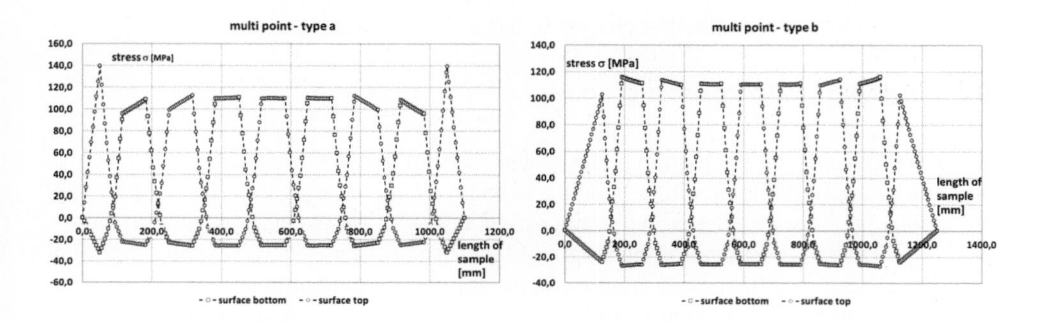

Figure 5-3 Multi-point bending test - stress distribution along the middle axis of glass sample.

Due to the alternative tensile stress distribution (top and bottom surface of the specimen) the determination of the effective area A_{eff} (as described in Siebert [11] or Fink [12]), which represents a homogeneous stress depends on an accurate test set-up and has to be validated by experimental testing.

5.3 Bending by in-plane force

The value for the ultimate bending strength can for example be determined with a kind of a stability test, as shown in figure 5-4 below. With a force F and eccentricity e the maximum stress can be determined according the theory of large deformations. Instead of inducing bending by loading perpendicular to test specimen an alternative concept applies the load in plane of the test pane with bending due to deflection [9].

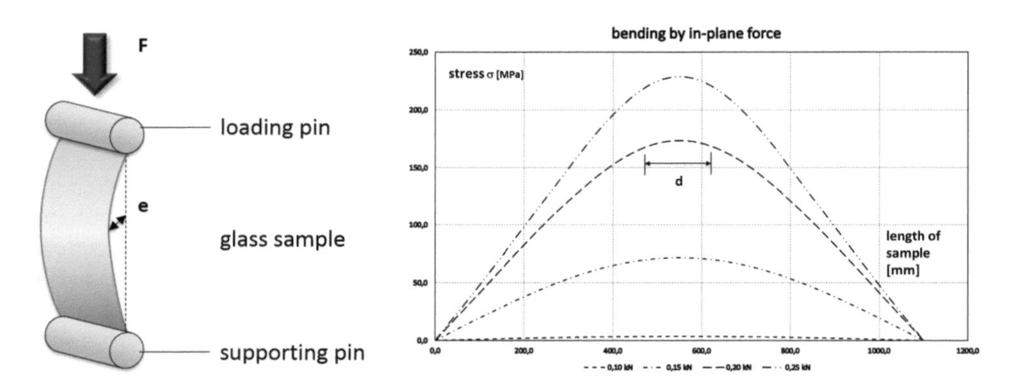

Figure 5-4 Bending by in-plane force principle test set-up, stress distribution along middle axis of glass sample.

Figure 5-4 right above, shows the distribution tensile stress along the middle axis of the test sample at different levels of stamp forces (0.10 kN up to 0.25 kN). The area A_{eff} in which the stress can be assumed as constant, marked with d, is very small and the so-called size effect, as described in chapter 7, has to be taken into account.

5.4 Bending with constant radius

Instead of introducing the load in plane as described in the previous chapter 5.3 it is also possible to apply the load with a bending moment on the straight opposite edges and a reduction of the distance between the supporting hinges, as shown in figure 5-5 left below. With an accurate adjustment of the length of bowstring (distance between the supporting hinges) of the arched bent glass sample and the applied bending moment a constant stress distribution on nearly the whole surface (excluding a small zone at the straight edges where the bending moment is introduced) arises. The stress distribution along the middle axis of the glass sample is shown in the diagram in figure 5-5 below right.

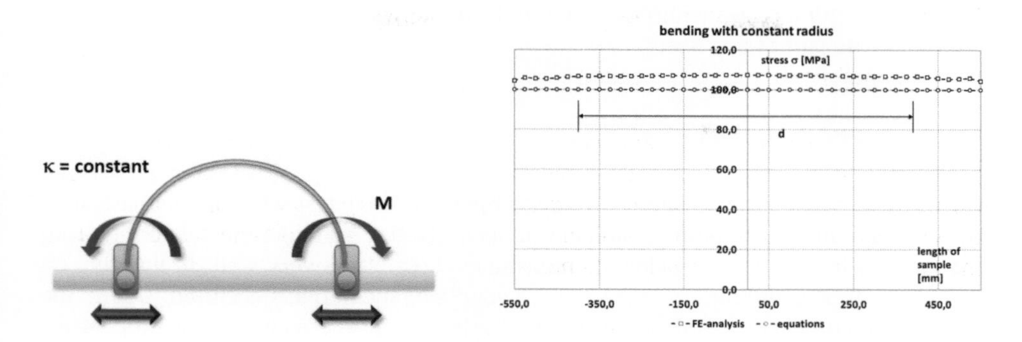

Figure 5-5 Constant bending test set-up, stress distribution along the middle axis of glass sample.

As an approach, the stress can be determined with the differential equation of bending theory including the young´s modulus E, modulus of inertia I and the curvature κ. The tensile stress can easily be computed with the section modulus W, but this approach neglects the influence of the poisson´s ratio ν. The advantage of this test scenario is that area A_{eff}, in which the stress can be assumed as constant, marked with d in the diagram in figure 5-5 above right, can easily be increased and therefore the influence of the so-called size effect can be minimized.

6 Effect of nonlinearity

The effect of geometrical non-linearity e.g. the so-called membrane effect was investigated and published many times, e.g. in Wilcox [10]. For the testing of thin glass, it is of interest whether the determination of the stress in the glass can be determined according the linear theory of small deformations or the non-linear theory of large deformations.

7 Effect of sample size

For the ultimate bending tensile strength, the so-called size effect has to be taken into account. This effect describes the relationship between a measured and statistically evaluated bending strength with a certain size A_0 according the test set-up and the size of the area A_1 of a glass application in which the maximum stress can be assumed as homogeneous, which has the same probability of failure as in the test scenario, as mentioned in Siebert [11] or Fink [12].

8 Effect of load duration

The load history during the lifetime of the glass element an influence on the ultimate bending strength too, as mentioned in Exner [13], but was not investigated in detail and is therefore no part of this paper.

9 Effect of Imperfections

Thin glass is much more sensitive related to imperfections in test set-up in comparison to thicker glass and needs more awareness of such effects. An experimental ring-on-ring test, as shown in figure 9-1 below, demonstrates these issues very well. In this ring on ring test R 105 according EN ISO 1288-5 a couple of such effects occurred. Due to the large deformation in the middle of the glass sample non-linear effects like a stability effect at the edges arose. This effect can be described with the membrane effect, which describes compressive stresses along the edges. At a certain level of loading, a stability effect with large asymmetric deformations of edges was observed, as shown in figure 9-1, below.

Figure 9-1 Ring-on-ring test in laboratory with buckling effects.

In addition to this stability effect a so-called snap through effect at the corners of the sample was observed too. Due to the dead weight of the glass the corners had at the beginning of the test a displacement in direction downward. At a certain level of loading a prompt snap through effect (without breakage of the glass) in direction upwards was observed.

Such imperfections can occur for example due to following reasons.

- imperfections in test set-up
- imperfections in glass samples (e.g. thickness)
- not exact centered load ring

Of course, all these effects of imperfection have to be avoided during determination of the ultimate bending tensile strength of thin glass.

10 Summary

For the determination it is needed to find an accurate balance between size of the effective area, in which the measured stress can be assumed as homogeneous, and sensitivity related to imperfections and non-linear effects. This area has to be increased as much as possible, because in e.g. cold bent glass elements a large area of maximum stress in which the measured stress can be assumed as homogeneous arises, to minimize the size effect. For-ring on-ring tests the in EN ISO 1288-5 given test set-ups have to be improved to minimize the probability of stability effects. The most promising test scenario of bending with constant radius with influence of edge strength shall be investigated much more relating to the applicability of this test scenario.

11 References

[1] Neugebauer J.: Movable Canopy, conference proceedings, Glass Performance Days, Tampere, Finnland, 2015.

[2] EN ISO 12150, ÖNORM EN 12150, Glass in buildings – Thermally toughened soda lime silicate safety glass, 2010.

[3] EN ISO 12337, ÖNORM EN 12337, Glass in buildings – Chemically strengthened soda lime silicate glass, 2004.

[4] Schmitt R.W., Blank K., Schoenbrunn G., Experimentelle Spannungsanalyse zum Doppelringverfahren, Srechsaal – International Ceramics and Glass Magazine, issue 116/5, 1983.

[5] EN ISO 1288-1, ÖNORM EN ISO 1288-1, Glass in building – Determination of the bending strength of glass – Part 1: Fundamentals of testing glass 2014.

[6] EN ISO 1288-2, ÖNORM EN ISO 1288-2, Glass in building – Determination of the bending strength of glass – Part 2: Coaxial double-ring test on flat specimens with large test surface areas, 2014.

[7] EN ISO 1288-3, ÖNORM EN ISO 1288-3, Glass in building – Determination of the bending strength of glass – Part 3: Test with specimen supported at two points (four-point bending), 2014.

[8] EN ISO 1288-5, ÖNORM EN ISO 1288-5, Glass in building – Determination of the bending strength of glass – Part 5: Coaxial double-ring test on flat specimens with small test surface areas, 2014.

[9] Siebert G.: Thin glass elements – a challenge for new applications, Glass Performance Days, Tampere, Finland, 2013.

[10] Wilcox D. et all.: Biaxial stress in Thin Glass during Ring on Ring testing with large deflections, https://www.researchgate.net/.

[11] Siebert G. Maniatis I.: Tragende Bauteile aus Glas, Ernst&Sohn, ISBN 978-3-433-02914-5, 2012.

[12] Fink A.: PhD thesis – Ein Beitrag zum Einsatz von Floatglas als dauerhaft tragender Konstruktionswerkstoff im Bauwesen, 2000, University of Darmstadt, Germany.

[13] Exner G., Erlaubte Biegespannung in Glasbauteilen im Dauerlastfall, Glastechnische Berichte 56 Nr. 11, 1983.

Numerical simulation of residual stresses at holes near edges and corners in tempered glass: A parametric study

Navid Pourmoghaddam[1], Jens Henrik Nielsen[2], Jens Schneider[1]

1 Institute of Structural Mechanics and Design, Faculty of Civil Ambient Engineering, Technical University of Darmstadt, Franziska-Braun-Str. 3, 64287 Darmstadt, Germany

2 Department of Civil Engineering, Technical University of Denmark

This work presents 3D results of the thermal tempering simulation by the Finite Element Method in order to calculate the residual stresses in the area of the holes near edges and corners of a tempered glass plate. A viscoelastic material behavior of the glass is considered for the tempering process. The structural relaxation is taken into account using *Narayanaswamy's* model. The motivation for this work is to study the effect of the reduction of the hole and edge minimum distances, which are defined according to EN 12150-1. It is the objective of the paper to demonstrate and elucidate the influence of the hole and edge distances on the minimal residual compressive stresses at holes after the tempering process. The residual stresses in the area of the holes are calculated varying the following parameters: the hole diameter, the plate thickness and the interaction between holes and edges and corners. Furthermore a comparison between the minimal residual stresses at holes and the residual stresses at other areas of the glass plate (edge, chamfer and far-field stresses) is made.

Keywords: residual stresses, tempered glass, heat transfer coefficient, finite element simulation, edge and hole distance, EN 12150-1

1 Introduction

Over the last couple of decades, glass has gained increasing importance as a construction element of the structural glazing due to its transparency and high resistance to ambient loadings. Glass is a brittle material at room temperature and deformations are linear elastic until fracture. The theoretical strength of the soda-lime-silica glass is about 5000 – 10000 MPa. However, the tensile strength is governed by small flaws in the surface which reduce the actual engineering strength of ordinary cooled float glass to 30 – 100 MPa (Wörner [1]). Furthermore, float glass is unfit for bolted connections due to its time dependent strength (Beason et al. [2]) and high sensitivity to concentrated loads. The strength of glass can be improved by making the flaws inoperative. Due to a compressive residual stress at the surface balanced with an internal tensile stress the surface flaws will be in a permanent state of compression which has to be exceeded by external loading before failure will occur. The residual tensile stresses are carried by the interior part of the material which is almost flawless. A parabolic distribution of the residual stresses along the thickness of the glass plate, as shown in Figure 1-1, can be ob-

Engineered Transparency 2016. Glass in Architecture and Structural Engineering. First Edition.
Edited by Jens Schneider, Bernhard Weller.

tained by the so-called tempering process of the glass. The amount of the surface compressive stress largely depends on the cooling rate and therefore on the heat transfer coefficient between glass and the cooling medium.

The material behavior of solid glass is assumed to be linear elastic. By temperatures above the glass transition temperature T_g the temperature and time dependency of the glass structure and the viscoelastic material behavior of the glass melt is used for the thermal strengthening of the float glass (Narayanaswamy [3]). Considering an "undisturbed" area (no drillings and cuttings) in an endless plate, the maximum stresses in a pure elastic material are achieved when the maximum temperature difference between the surface and the interior is present (Nielsen [4]). For the glass, the maximum stresses are obtained at the end of the process, Figure 1-1.

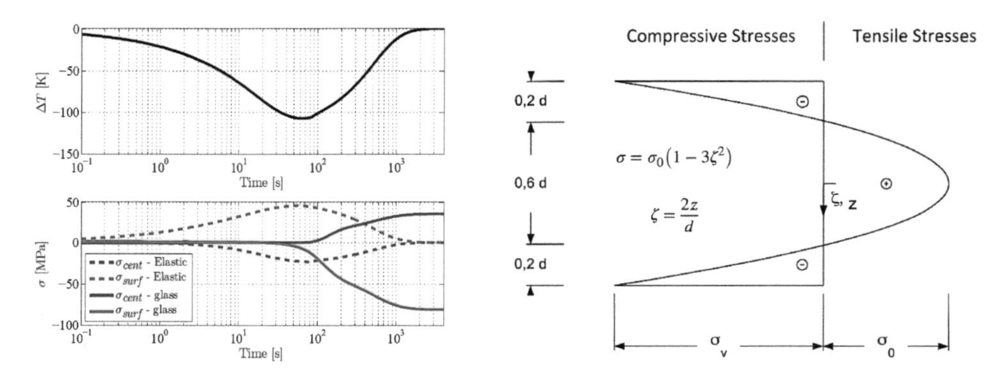

Figure 1-1 Left: Development of temperature difference between the center and the surface during the tempering process and the development of the stresses in purely elastic solid and in glass (Nielsen [4]); right: Distribution of the residual stresses in undisturbed area after the tempering process of glass.

If glass is heated to temperatures above T_g it will become softer and loses the characteristics of a solid body. Glass melt is unable to carry stresses. By cooling the glass melt, due to the temperature distribution along the thickness where the surfaces are cooler than the interior, the surfaces will solidify first. As the outmost layers the surfaces will solidify without generating notable stresses due to the lack of stiffness of the interior. When the interior cools down, solidifies and cools further down it will contract and apply compressive stresses to the surfaces. The temperature of the surfaces is now lower than in the interior and the interior will therefore, in total, contract more during cooling than the surfaces. The surfaces will resist this contraction and thereby end in a state of compression. Due to the equilibration of the surface compressive stresses tensile stresses result in the center of the glass plate. The process line for tempering float glass is sketched in Figure 1-2. If the residual stress state is disturbed sufficiently, the tempered glass will fragmentize completely. Therefore pretreatments like cutting and drilling holes must be done before quenching the glass.

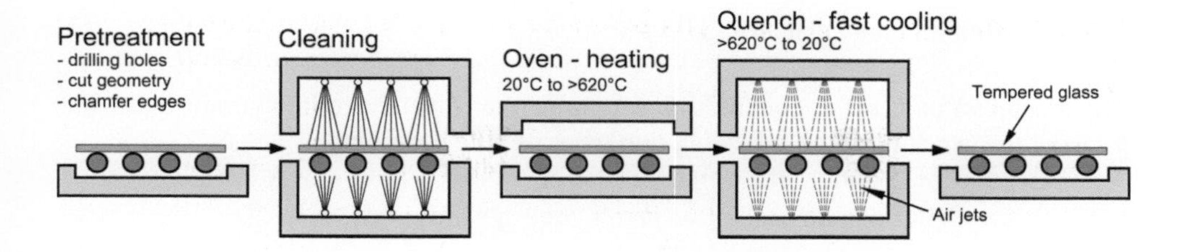

Figure 1-2 Sketch of the process line for tempering float glass.

2 Objectives of the study

Holes are disturbances for the stress distribution and the stress development during the tempering process. Also the hole, edge and corner distances influence the stress development significantly. The position of holes respectively the minimum distances of holes to the edge and the corner and also the minimum distance between two holes in tempered soda-lime-silica glass is defined according to EN 12150-1.

The object of this paper is to calculate the residual stresses at holes numerically and investigate the influence of the different types of the hole, edge and corner distances on the residual stresses at holes. The minimum edge and hole distances, which are defined in EN 12150-1 are reduced from *2t* to the halved value of *1t* (*t* being the thickness of the glass plate), Figure 2-1. For the calculations a new condition is set to consider the diameter of the hole by limiting the minimum distances, which is not a part of the current version of the EN 12150-1.

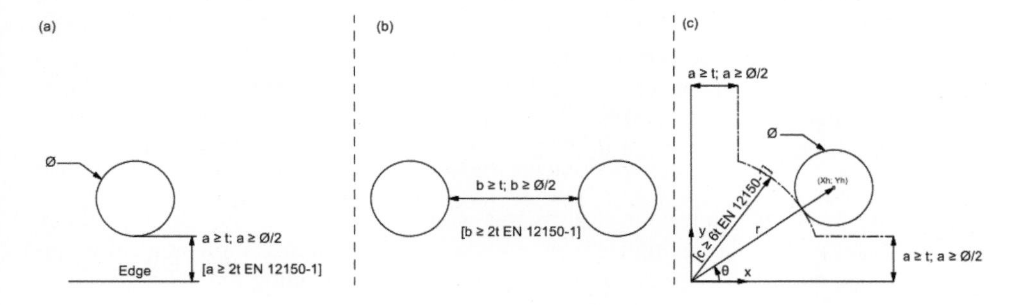

Figure 2-1 Minimum edge (a), hole (b) and corner (c) distances.

The minimum edge- and hole distances have to fulfil the following conditions:

a) Minimum edge distance: $Min\ a = Max\left\{a = t; a = \frac{\varnothing}{2}\right\}$

b) Minimum hole distance: $Min\ b = Max\left\{b = t; b = \frac{\varnothing}{2}\right\}$

The greater value is set as condition to determine the respective minimum distance. The focus of this work is on the following two types of hole positions:

1 Two holes next to each other and in the edge area
   a. t = 6 mm and 10 mm
   b. Ø = 8 mm, 50 mm and 100 mm
2 One hole in the corner
   a. t = 6 mm, 10 mm and 15 mm
   b. Ø = 8 mm, 15 mm, 50 mm and 100 mm

Two types of investigations are carried out by the hole positioned in the corner. First, the angle θ is constant 45° and the distance r between the corner and the hole center varies from the minimum coordinate value of $Min\ x_h = Min\ y_h = Max\{a = t; a = \varnothing/2\}$ to $r = c + \varnothing/2 = 6t + \varnothing/2$. Second, the distance r is constant and the angle θ varies from the minimum coordinate value of $Min\ x_h = Min\ y_h = Max\{a = t; a = \varnothing/2\}$ to $\theta = 45$.

3 FE simulation of residual stresses

3.1 Thermo-mechanical computation

Previous analyses of glass tempering have been concerned with the calculation of residual stresses in glass plates considering the stress, volume and structural relaxation of glass (Narayanaswamy [5]). The computation of residual stresses in a 1D model was carried out in (Aronen [6]). Analyses of glass tempering and the computation of the residual stresses in bored glass plates have been carried out in (Schneider [7]) and (Nielsen et al. [8]). This work presents the numerical calculation of the residual stresses in the hole area based on a 3D FE-model concerning the different heat transfer coefficients of the chamfered holes, edges and the undisturbed far-field area. The influences of the hole distances, as it was shown in Figure 2-1, are studied.

The thermo-mechanical behavior of glass was widely studied in the literature. Kurkjian [9] demonstrated that glass at high temperatures behaves almost thermorheologically simple (TS). Lee et al. [10] introduced a viscoelastic model including the TS behavior of glass. A model for the structural relaxation was proposed by Narayanaswamy [3] . This model is used to calculate the residual stresses in the commercial FE software Ansys. The transient finite element simulation of the residual stresses is carried out in two steps. First, the temperature history during the tempering process is determined in a transient temperature calculation using the 3D-20-Node thermal solid element SOLID90 (Ansys 16.0), which is a thermal element with temperature as the only degree of freedom. In the second step the temperature change over time is put in terms of load steps on a structural mechanical model. Thereby the thermal element SOLID90 is changed in to the 3D-20-Node viscoelastic solid element VISCO89. In Figure 3-1, the temperature time curve and the resulting stress time curve of a 10 mm thick glass plate during the cooling process is shown. The initial temperature T_0 is 943.15 K (equates to 650 °C) and the ambient temperature T_∞ is 293.15 K (equates to 20 °C).

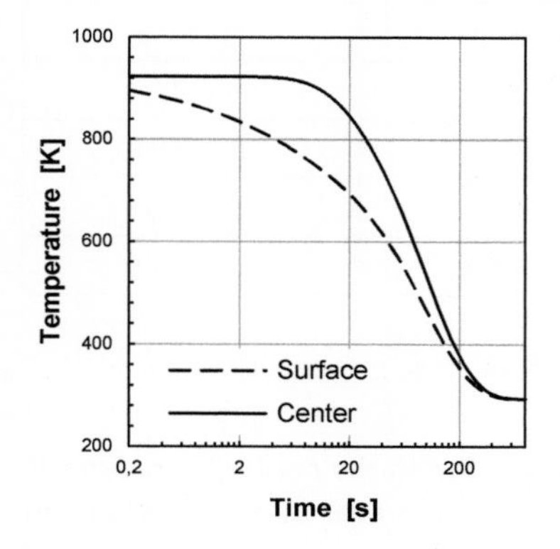

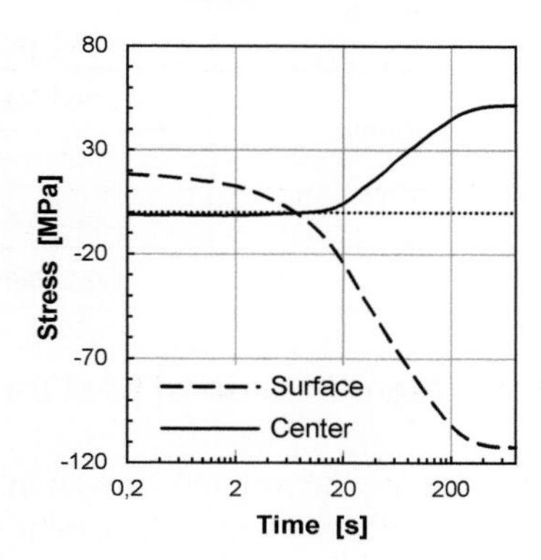

Figure 3-1 Left: Temperature [K] versus time [s]; right: Stress [MPa] versus time [s] – calculated at the surface and at the center of a glass plate with the thickness t = 10 mm – T_0 = 943.15 K (650°C) and T_∞ = 293.15 K (20 °C).

The constitutive equations of the thermo-mechanical behavior and the implementation of a glass tempering 3D-model have been discussed in (Nielsen et al. [11]). The time steps significantly influence the calculated residual stresses.

The whole cooling time is divided into two load steps. In the first load step, t_1, glass is cooled down in fine selected time steps, Δt_1, to a temperature below an assumed glass transition temperature, T_g = 550 K. The fine time stepping is necessary to simulate the rapid cooling of the glass with the initial temperature, T_0. As it is shown in the left diagram in Figure 3-1, the temperature time curve falls steeply at the beginning of the cooling process and afterwards levels off at the ambient temperature, T_∞. The fine time stepping is needed until the model, which was heated by an initial temperature at approximately 100 °C above T_g achieves a temperature below the glass transition temperature at the surface as well as at the center. In the second load step, t_2, glass is cooled down to the ambient temperature, T_∞ in larger time steps, Δt_2, in order to save computing time. In Table 3-1 the time increments according to the thicknesses, which were found by analysis of convergence, are listed. The value of the cooling time for the corresponding load steps as well as the selected time steps vary with the thickness of the glass plate. The thicker the glass plate the longer the cooling time.

Table 3-1 Time steps depending on the thickness.

Thickness	Load step	Cooling time t	Time step
6 mm	Load step 1	t_1 = 10 s	Δt_1 = 0.1 s
	Load step 2	t_2 = 380 s	Δt_2 = 5 s
10 mm	Load step 1	t_1 = 30 s	Δt_1 = 0.2 s
	Load step 2	t_2 = 800 s	Δt_2 = 10 s
15 mm	Load step 1	t_1 = 60 s	Δt_1 = 0.5 s
	Load step 2	t_2 = 1200 s	Δt_2 = 50 s

3.2 Identification of heat transfers

To calculate the residual stresses at holes, the heat convection in the cooling process has to be simulated. Hereby the essential influence parameter is the heat transfer coefficient *h* between glass and the cooling medium. The experimental determination of the heat transfer coefficient is difficult. The convection coefficients in the different area of perforated plates (far away from edges, in the hole, on the straight edges and on the chamfer) have been identified experimentally in (Bernard et al. [12]) using a hollow aluminum model. In Figure 3-2, heat transfer coefficients for the different areas of a plate with a hole, are shown.

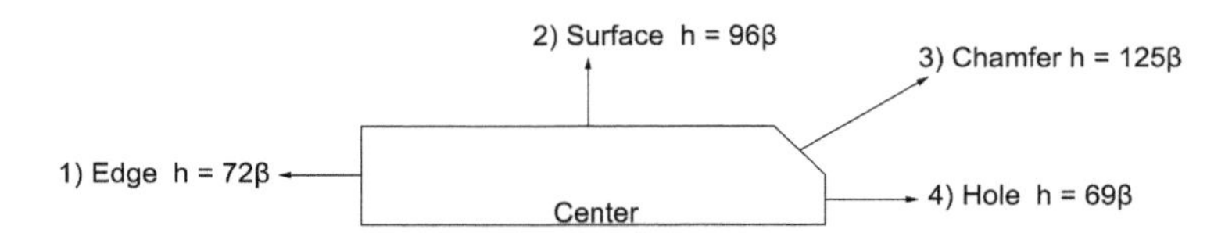

Figure 3-2 Heat transfer coefficients – factor of the different convection areas.

In this work the heat transfer coefficients were calculated iteratively by numerical determination of the variable β (initial value $\beta = 1.0$). The calculations were carried out by means of an "infinite" plate with a hole positioned in the center of the plate. Due to the symmetry, only a piece of the plate under an angle of 5° and half the thickness was modelled, Figure 3-3. Different heat transfer coefficients result in different residual stresses. To have comparable residual stresses for all models of the parametric study the heat transfer coefficients were brought to one stress level. The factor β was varied until a surface compression of approximately 100 MPa was achieved. The heat transfer coefficients were calculated for the different glass thicknesses t = 6 mm, 10 mm and 15 mm, the initial temperature of T_0 = 650 °C and the ambient temperature of T_∞ = 20 °C. The residual stresses after the cooling process are shown in Figure 3-4.

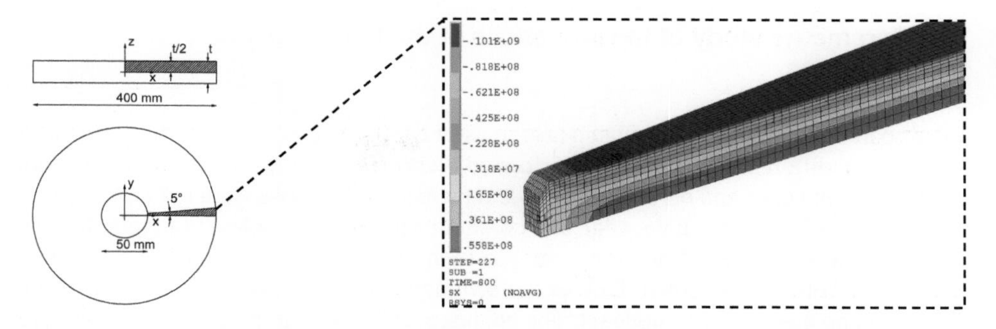

Figure 3-3 Left: Sketch of the plate model; right: Residual stresses in x direction in [N/m²] after the tempering process (t = 10 mm and β = 1.5).

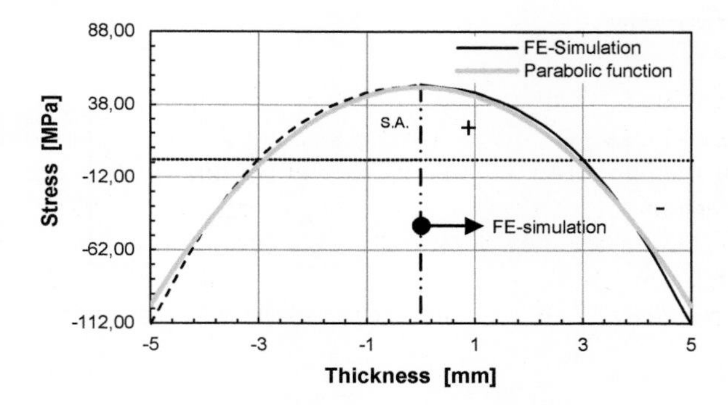

Figure 3-4 The Parabolic function and the FE-simulation of the residual stresses along the thickness of the glass in far-field area (t = 10 mm; β =1.5).

The value of the factor β, which led to the surface compression of 100 MPa in the far-field area of the plate model, was identified as the relevant factor of the heat transfer coefficients of the different convection areas. We have assumed a simple relation of the heat transfer coefficient between surface and the hole area. The results of the calculations are presented in Table 3-2.

Table 3-2 Heat transfer coefficients h [W/(m²K)].

Thickness t [mm]	6	10	15
β	2.7	1.5	1.0
Heat transfer coefficient h [W/(m²K)]			
Edge	194.4	108.0	72.0
Surface	259.2	144.0	96.0
Chamfer	337.5	187.5	125.0
Hole	186.3	103.5	69.0

4 Parametric study of the influences of the hole edge and corner distances

In this part of the study the calculated heat transfer coefficients were set on the convection areas of the different FE models of perforated plates (far away from edges, in the hole, on the straight edges and on the chamfer) and the residual stresses were numerically calculated depending on the hole, edge and corner distances of the holes. In Figure 4-1, the sketch of the FE model of two holes next to each other in the edge area as well as one hole in the corner, are shown. Due to the symmetry of the model only the half of the thickness and one hole was modeled. The boundary conditions at the symmetry axis were considered in the calculations.

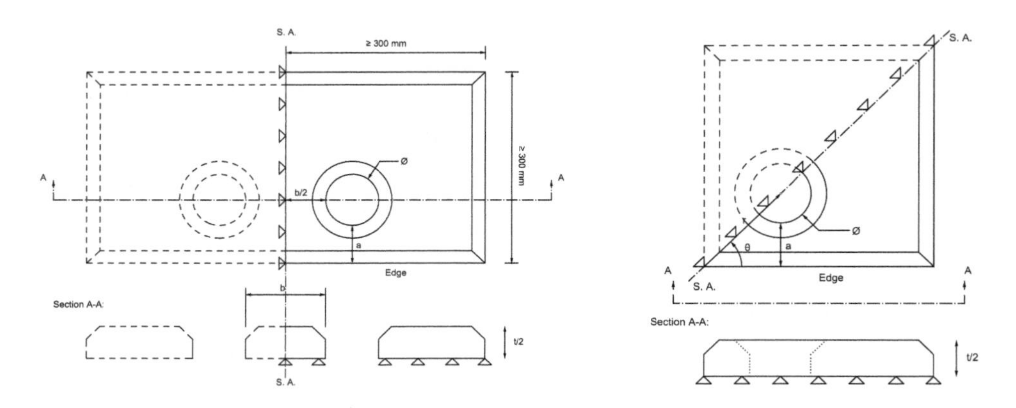

Figure 4-1 FE-Models, left: Two holes next to each other, right: One hole in the corner.

The significant stress at holes is the tangential residual stress, which occurs at the inner surface of the hole and sets it under compression. As it is shown in Figure 4-2, the maximum tangential residual stress, which marks the minimum compression respectively the maximum tension at the inner surface of the hole, occurs at the center of the plate. Negative values are compression and positive values are tension.

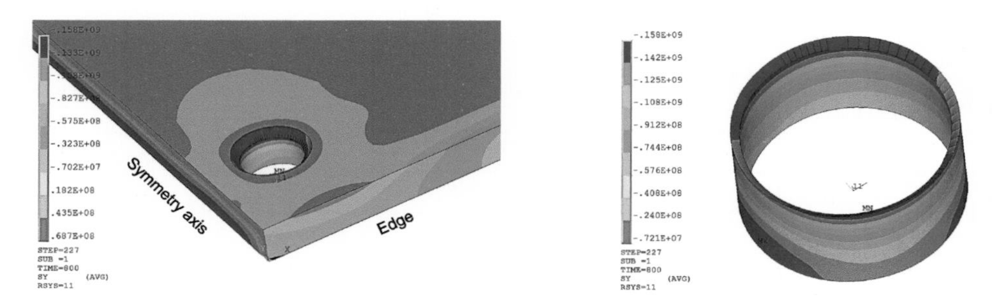

Figure 4-2 Tangential stresses at holes [N/m²] – two holes next to each other and in edge area (Ø = 8mm, t = 10mm, min a = min b = 10 mm).

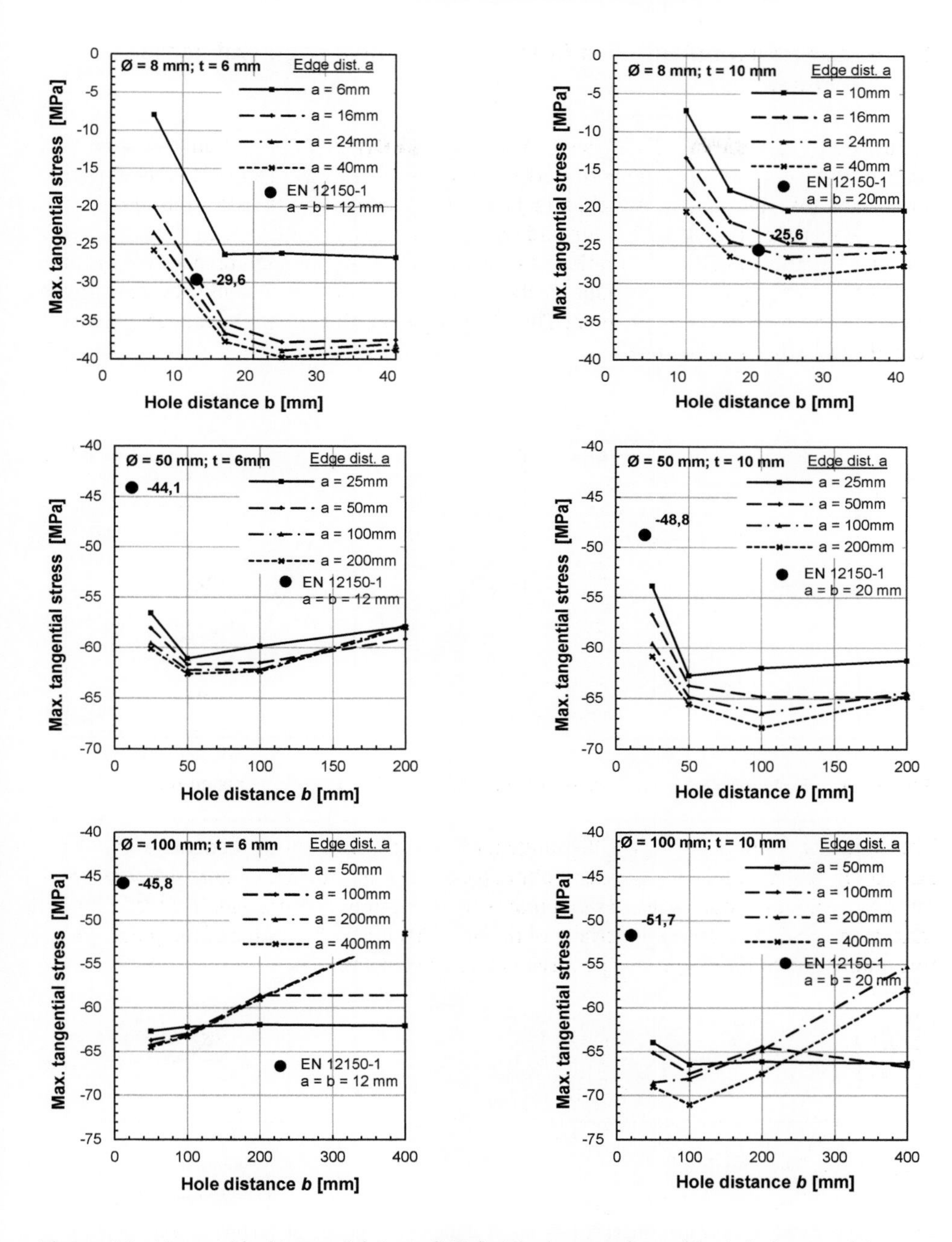

Figure 4-3 Max. residual tangential stress [MPa] at the inner surface of the hole (center of the plate) vs. Hole distance b [mm]; two holes next to each other and in edge area (edge distance a [mm]) – the black point is a result of a calculation with the specification in EN 12150-1 (min a = min b = $2t$).

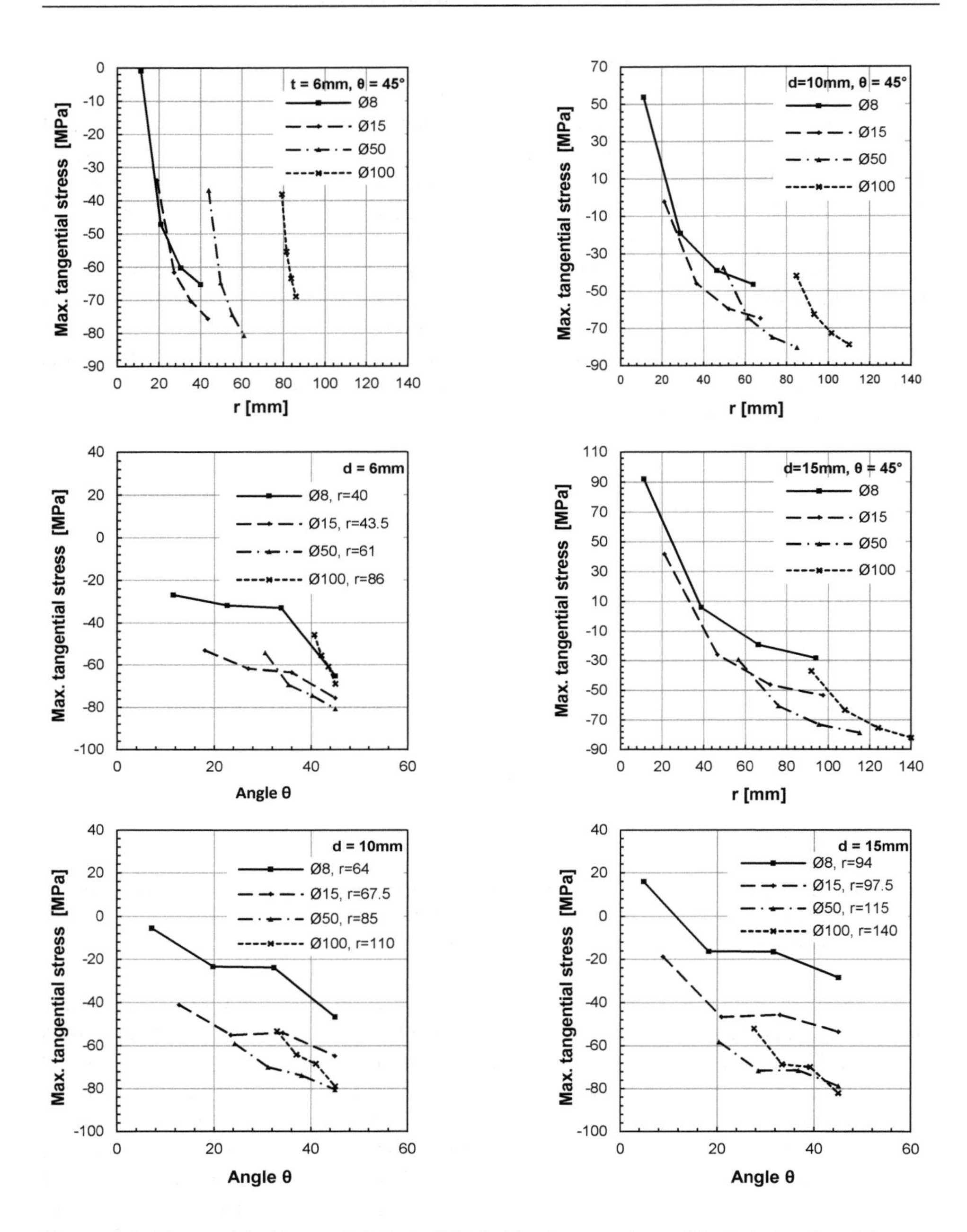

Figure 4-4 Max. residual tangential stress [MPa] at the inner surface of the hole (center of the plate) vs. distance *r* [mm] and angle *θ*; one hole in the corner (edge distance *a* [mm]).

In Figure 4-3, the maximum residual tangential stresses versus hole distance *b* at the inner surface of the hole was shown. Each curve represents one edge distance *a*. It was shown that the hole distance as well as the edge distance significantly influence the development of the tangential stress at holes during the tempering process. The residual compressions sharply fall by decrease of the hole and edge distances. By the small diameter of 8 mm the values of the residual stresses are critical for the assumed minimum distances min *a* and min *b* in comparison to the calculated values with the minimum distances according to EN 12150-1. However, due to the new condition set for the minimum distances, which take the diameter of the hole in to account, the calculations begin with the minimum distances of $a = b = 25$ mm with a hole diameter of 50 mm and $a = b = 50$ mm with a hole diameter of 100 mm. The maximum tangential stress occurs at the center of the plate. Figure 4-5 shows a comparison of the residual stresses calculated at different points of the plate with two holes next to each other and in the edge area.

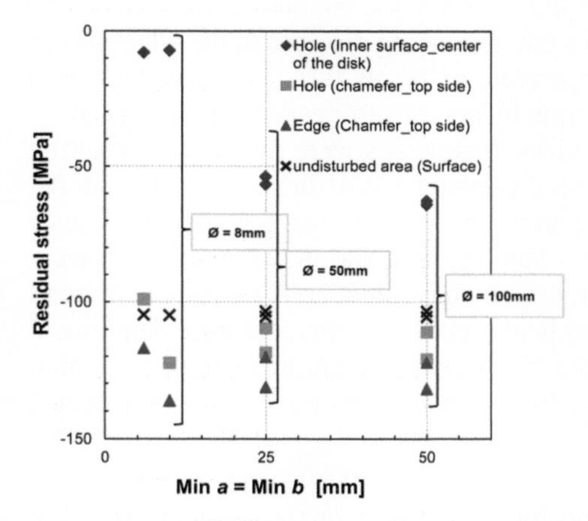

Figure 4-5 Comparison of the residual stresses at different points of the plate for the minimum hole and edge distances by two thicknesses of t = 6 mm and 10 mm and three diameters of Ø = 8 mm, 50 mm and 100 mm – Two holes next to each other and in edge area.

At the top side of the chamfer of the hole and also at the top side of the chamfer of the edge as well as at the surface in the undisturbed area the residual stresses are greater than 100 MPa, Figure 4-5. This is reasonable since such a corner is cooled from two sides and therefore the residual compressive stress is higher. The reduction of the corner distance of the hole to an edge distance of 8 mm ($\theta = 45°$) results in a residual tangential tension up to approx. 90 MPa at the inner surface of the hole in the center of the plate for a hole diameter of $Ø = 8$ mm and a glass thickness of $t = 15$ mm. The critical residual stress by reducing the corner distance to the appropriate minimum edge distance significantly depends on the thickness of the glass. So by the same hole diameter ($Ø = 8$ mm) and a glass thickness of $t = 6$ mm the residual tangential stress is approx. -1 MPa (compression). There is a high influence of the corner distance on the residual stresses. The residual tangential compression at the inner surface of the hole drops strongly by reducing the corner distance. For a diameter of Ø = 8 mm the residual stress difference between the minimum corner distance (a = 8 mm) and the specification in EN 12150-1 is according to amount approx. 65 MPa for t = 6 mm and approx. 120 MPa for t = 15 mm.

5 Summary and Conclusion

The residual stress state in tempered glass is complex and numerical models are needed for estimating the stress fields near edges. However, this study covers several cases for one hole in the corner and two holes next to each other in the edge area which can be used as guidelines. The influence of the hole and edge distance on the residual stress development during the tempering process was calculated using a full 3D Finite Element model. The minimum hole and edge distances according to EN 12150-1 were reduced to consider the flexibility of those specifications. For the simulation of the tempering process the FE-Model of the glass plate with the hole with the corresponding distance to the edge, corner or the adjacent hole was given an initial temperature of T_0 = 650 °C and cooled down to the ambient temperature of T_∞ = 20 °C. The time increments Δt needed for converged results of the tempering process for different glass thicknesses were given. The resulting time-temperature curve was put in terms of load steps on a structural mechanical model to calculate the stress response due to the tempering process. The heat transfer coefficients of the different heat convection areas (edge, chamfer, hole and undisturbed area) were calculated numerically by means of a FE-model of an "infinite" plate with a hole positioned in the center of the plate. The different heat transfer coefficients were determined for yielding a surface compression of 100 MPa. The critical maximum residual tangential stress respectively the minimum tangential compression at the surface of the hole occurs in the center of the plate.

The significant influence of the hole, edge and corner distances on the residual stresses at holes was shown. The reduction of the minimum distances according to EN 12150-1 cannot be proposed in the case of bearing stresses at holes because for all models the critical tangential residual stresses at the inner surface of the hole in the center of the plate (varying from -65 MPa to +90 MPa) were much smaller than the far-field compression stress (approx. -100 MPa). For some of the models tension occurred at the hole after the tempering process. However, for the typical load cases in building constructions (bending moment from wind or snow loading resulting in maximum surface tension stress at chamfers) this means that the small residual tangential stresses at the inner surface of the hole can be acceptable as long as they are compression stresses. Therefore, if the minimum distances according to EN 12150-1 shall be reduced, one has to study the glass resistance based on the geometrical influences from the tempering process as well as the stresses resulting from the external loads.

6 References

[1] J.-D. WÖRNER, J. SCHNEIDER, und A. FINK, *Glasbau Grundlagen, Berechnung, Konstruktion*. Springer-Verlag Berlin Heidelberg, 2001.

[2] W. L. Beason und J. R. Morgan, „Glass failure prediction model", *J. Struct. Eng.*, Bd. 110, Nr. 2, S. 197-212, 1984.

[3] O. S. Narayanaswamy, „A Model of Structural Relaxation in Glass", *J. Am. Ceram. Soc.*, Bd. 54, Nr. 10, S. 491-498, 1971.

[4] J. H. NIELSEN, *Tempered Glass: Bolted Connections and Related Problems*, Nr. PhD Thesis. Technical University of Danmark, 2009.

[5] O. S. Narayanaswamy, „Stress and structural relaxation in tempering glass", *J. Am. Ceram. Soc.*, Bd. 61, Nr. 3, S. 146-152, 1978.

[6] A. Aronen, *Modelling of Deformations and Stresses in Glass Tempering*. 2012.

[7] J. SCHNEIDER, *Festigkeit und Bemessung punktgelagerter Gläser und stoßbeanspruchter Gläser*, Nr. Dissertation. Technische Universität Darmstadt, 2001.

[8] J. H. Nielsen, J. F. Olesen, P. N. Poulsen, und H. Stang, „Simulation of residual stresses at holes in tempered glass: a parametric study", *Mater. Struct.*, Bd. 43, Nr. 7, S. 947-961, 2010.

[9] C. Kurkjian, „Relaxation of torsional stress in transformation range of soda-lime-silica glass", *Phys. Chem. Glas. 4*, S. 128-136, 1963.

[10] E. H. Lee, T. G. Rogers, und T. C. Woo, „Residual Stresses in a Glass Plate Cooled Symmetrically from Both Surfaces", *J. Am. Ceram. Soc.*, Bd. 48, Nr. 9, S. 480-487, 1965.

[11] J. H. Nielsen, J. F. Olesen, P. N. Poulsen, und H. Stang, „Finite element implementation of a glass tempering model in three dimensions", *Comput. Struct.*, Bd. 88, Nr. 17–18, S. 963-972, 2010.

[12] F. BERNARD und L. DAUDEVILLE, „Point fixings in annealed and tempered glass structures: Modeling and optimization of bolted connections", *Eng. Struct.*, Bd. 31, Nr. 4, S. 946-955, 2009.

Benefits of optical distortion measurement – How Moiré technology drives the efficiency of glass production chains

Bertrand Mercier[1]

1 ISRA VISION Albert-Einstein-Allee 36-40 45699 Herten, Germany, bmercier@isravision.com

High-quality glass is facing a significantly growing demand. Especially display glass and automotive applications are boosting the market and the industrial pursuit for highest quality, requiring not only defect free sheets but also flawless optical properties. The technology to be introduced in this paper concerns the measurement of optical power inhomogenities of glass, additional to reliably detecting cords, reams, zip lines and other common optical glass defects. Quality monitoring also comprehensively gives hints for the optimization of the entire float glass process via product and quality data. By relying on the ISRA technology, manufacturers can ensure the delivery of defect-free float glass with certified optical quality regarding Diopter or Zebra values. The paper provides a closer look at ISRA's patented innovative optical distortion measurement, giving an overview about how users can benefit from the inline moiré technology, which ensures highest product quality and enhances global competitiveness.

Keywords: distortion measurement, optical quality, optical power, moiré measurement, float glass

1 Current Situation

Following the current market demands manufacturers are starting to diversify their float glass portfolio by also producing tinted glass and automotive glass, and providing a larger range of thicknesses. Being able to offer different types of products increases the chance of higher added value, but also implies a higher risk of weak optical properties caused by substrate inhomogeneity or affected production processes at the hot end.

Simultaneously, costs rise with increasing process complexity and quality expectations, coming along with these specialized applications. The optical quality of the products has to be controlled reliably – only thus manufacturers can ensure a flawless product and process, resulting in cost effective production and high customer satisfaction.

2 Inspection lab vs automated in-line measurement

Checking glass samples for quality in an offline inspection lab is an alternative, but has several – well known – disadvantages: taking a sample from the current batch to the lab is a time consuming effort, and the controlling capabilities are quite limited. Maintaining a quality lab is rather costly, not only counting technical equipment but also the considerable cost of labor. Besides, measurements still depend on human accuracy, knowledge

Engineered Transparency 2016. Glass in Architecture and Structural Engineering. First Edition.
Edited by Jens Schneider, Bernhard Weller.

and experience. This severely limits the repeatability of results and thus their reliability, especially accurate measurements of optical power and Zebra are very hard to achieve. Beyond that, lab measurements only provide in-formation on distinct aspects of the material resp. the process, and cannot catch punctual optically distorting events during the process, nor give a full picture of the material behavior.

Automated in-line quality assurance offers a powerful alternative to monitor the current process situation by showing the exact optical quality in real time, and by following up the optical quality trend history. This includes searching for and understanding the root cause of optical defects, thus ensuring high process quality and control. Reaction time to reoccurring defects is reduced significantly which as well supports a consistently well-functioning production chain.

By deploying a full circle quality surveillance, manufacturers can monitor their entire process through comprehensive analysis and quality assurance. Thanks to high measurement precision, this also allows an automatic upgrading of the glass quality by automatically and precisely cutting off the non-acceptable optical defects and hence avoiding potential claims.

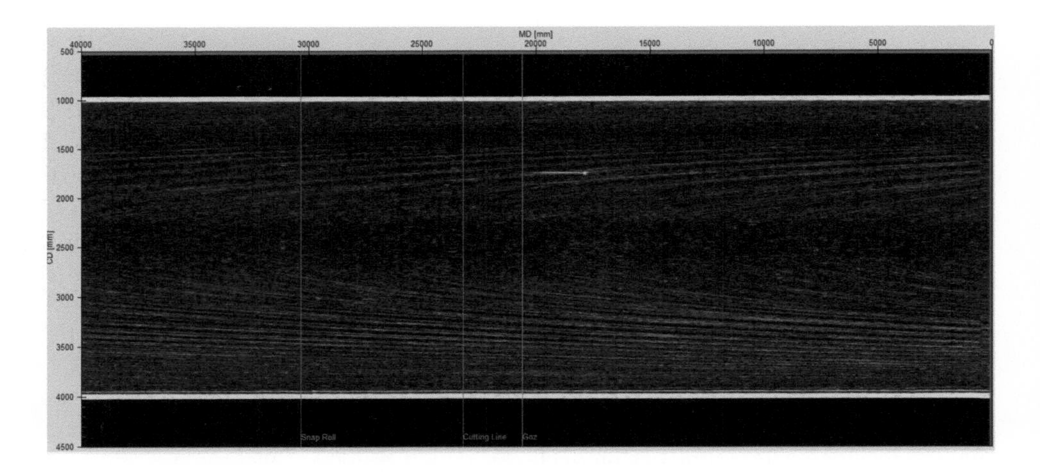

Figure 2-1 Colored coded map of online transmitted distortion realized with FLOATSCAN-5D.

3 Advanced glass inspection using distortion measurement

For achieving such a high-level and unique measurement performance, it has been necessary to develop a patented technology based on the moiré principle.

By using the moiré technology, the glass' optical power becomes measure-able by a direct physical analysis of the deflection of the light through the glass. Measurement results can therefore be expressed in millidiopter value and evaluated with different filters corresponding to the market requirement.

This means a significant difference to optical judging based on grey value monitoring technologies, as it can only try to approach the reality by correlation, and whose visualization is affected by external conditions such as dirt, external light variations and intrinsic glass properties (thickness, transmission, color, etc). In automotive applications, a clear information on optical power measured in Diopter and Zebra has therefore become standard, to ensure the driver's clear view to the street and surrounding environment. The result of this has proven as a strong improvement of the optical quality in the automotive branch during the last years (pls. see figure 3-1).

Figure 3-1 The beginnings of distortion measurement (left, 1999) and the current capabilities (right, today).

4 User benefits

Manufacturers can be fully aware of product and process quality. Opera-tors can derive valuable conclusions for process optimization directly from the visualized inspection results and without additional laboratory analysis. Recognition of cords, reams, draw lines, zip lines as well as monitoring of batch homogeneity via the optical power allows e.g. an optimum setup of the hot end.

As a consequence, the system provides a reliable way to minimize costs while simultaneously maximizing quality, productivity and efficiency. Defective or unsatisfying products can be excluded from further processing.

5 Examples of use

Important glass applications in the automotive industry like e.g. windshields, are nowadays bound to be inspected with the moiré method, with systems like the Faultfinders or Labscans. This has led to a drastical improvement for the driver's clear view on the street with the benefit of an increased quality of the driving experience as well as safety on longer trips.

For that reason, automotive glass manufacturers have also raised their quality exigence, ordering primarily from float glass suppliers who have invested in modern float inspection systems like the FLOATSCAN-5D, which includes nowadays standardly the same moiré technology.

But also display glass producers have increased their requirements to obtain the same satisfying optical properties, also relying on the FLOATSCAN-Thin Superior system. Even constructive glass, although it usually has a higher tolerance regarding the distortion levels, becomes a cleaner and more attractive design element, when a clear view is provided from any perspective. Therefore, manufacturers tend to follow the current trend to higher quality demands by requiring a stronger optical quality control from their suppliers, for the benefit of future higher glass quality products in almost every glass application we can witness in our daily lives.

A systematic methodology for temperature assessment in laminated glass components

Alessandro Baldini[1]*, Lisa Rammig*[1,2]*, Manuel Santarsiero*[1]

1 Eckersley O'Callaghan, 9th Floor, 236 Gray's Inn Road, London WC1X 8HB,
lisa@eocengineers.com

2 Delft University of Technology, Faculty of Architecture and the Built Environment, Architectural Engineering + Technology

In the past decades an increasing demand for transparency in architecture has been observed, leading to an increased use of glass particularly in the building envelope. Together with a development in production capabilities, this has also led to an increased use of structural glass in facades. However, the mechanical behaviour of glass components and particularly laminated components, is often affected by temperature variation. The structural assessment of safety and serviceability performances of glass elements therefore raises the discussion on the design temperature, or more specifically on the design scenario, to be considered. Peak temperatures suggested by design codes occurring simultaneously with peak loads might not be of high probability for many locations worldwide. Specifically the use of viscoelastic interlayer materials with strongly temperature dependant behaviour increases the need for better understanding of temperature impact on glass components. The temperature field distribution is not homogeneous and further to the material properties themselves, dependent on building location, orientation, as well as ambient and environmental conditions such as ambient temperatures and solar incident radiation. Despite the large and increasing use of glass in building envelopes, standards and design codes do not provide a precise methodology on this matter. It is therefore of interest to develop a more systematic approach. This study proposes a methodology to assess temperatures occurring within a glass component over time for any location with available weather data, based on large amount of data, to inform the structural design and lead to an optimised design approach.

Keywords: structural glass, interlayer, temperature, solar radiation

1 Introduction

In the past years an increasing amount of envelopes has been designed and built featuring structural glass elements. The architectural development towards a minimal design as well as the possibility to transfer loads through a stiff interlayer (SGP), has led to structural facade designs being developed without the use of vertical structural elements like solid mullions or glass fins.

This results in very thick glass build-ups spanning between floor and ceiling. The introduction of multiple thick layers of glass with polymer interlayers leads to a reduction in light transmittance and can also lead to a significant reduction of perceived clearness and

Engineered Transparency 2016. Glass in Architecture and Structural Engineering. First Edition.
Edited by Jens Schneider, Bernhard Weller.

transparency of the glass. As transparency is often considered the main driver for the design of these structural glass envelopes, it is of interest to improve these issues. Given that, an optimisation of design criteria can have a significant impact on the visual appearance of the structure, specifically as relates to the assumed design loads. Current standards require the assumption of maximum wind and temperature loads simultaneously, while this might not be a realistic reflection of load cases for many locations globally.

This research aims to understand the impact of ambient and environmental conditions on the temperature in the glass for specific locations, taking into account recorded data (over a significant amount of time). Ambient temperatures as well as incident radiation are the main factors to consider.

This is of particular interest, as the ionomer interlayer considered for the applications discussed in this paper (SGP) shows a significant reduction in its load transfer capacity at temperatures over 50-55 °C, i.e. above its glass transition temperature. This makes it crucial to understand the temperature occurring inside the glass build-up and in particular within the interlayer.

While ambient temperatures cause a considerably linear increase of temperature through the build-up of a glass laminate, the effect of solar radiation is non-linear and highly dependent on the spectral properties of the glass such as reflection, absorption and emissivity. The use of solar and low emissivity coatings can have a substantial effect on these properties, which might increase the absorption of the material notably, leading to significantly higher temperatures in the glass build-up.

The temperature reached by the glazed element is a function of its characteristic optical and thermal properties, of the incident solar radiation and of the ambient temperature, to which the element is exposed.

There are software tools available that allow determining the surface temperature (LBNL Window) as well as the glass build-up temperature (WinSLT) ,based on the ambient temperature and incident radiation.

However, most of the existing tools can perform calculations only for single inputs of specific boundary conditions. In fact, these don't allow assessing temperatures through the glass build-up over time. This is a great limitation as these impede the processing of large sets of data (e.g. hourly data over 50 years) which is crucial for the determination of a detailed critical loading scenario for the assessment of the performance of laminated structural glass components. More specifically, it is essential to better evaluate the correlation between interlayer temperature and applied load cases.

This study therefore aims to develop a simplified methodology for the computation of the temperature within a glass build-up over time due to radiation, convection and conduction phenomena based on a large database of input data. This will allow forming a solid basis for the further development of a systematic method that allows correlating hourly inter-layer temperature with applied load, such as wind pressure. This shall then allow analysing load cases more specifically to the geographic location and might be beneficial to avoid a substantial over-design of the structure due to the conservative assumption of peak wind loads and peak temperature occurring simultaneously over a significant period of time.

To be able to evaluate large sets of data that existing tools do not allow to process simultaneously, a methodology is developed to assess the glass based on the environmental parameters and glass properties.

2 Background

The heat transfer through a solid is a linear function of conduction from one side of the body to the other. If different temperatures occur on either side of a solid, a positive heat flow towards the lower temperature can be observed.

According to the absorption of the material, incident solar radiation may lead to an increase of temperature within the solid, which in turn leads to a non-linear temperature distribution through the solid.

Figure 2-1 (left) shows the linear heat transfer through a solid while Figure 2-1 (right) indicates the effect of radiation on the heat transfer through a solid. (Incropera et al, 2007)

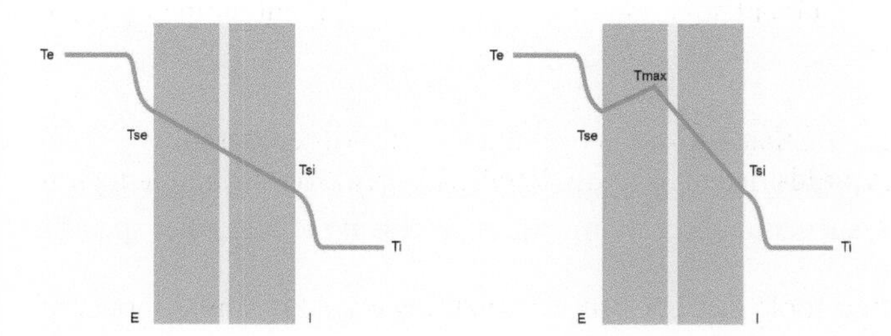

Figure 2-1 Linear heat transfer through a body, based on conduction only (left); heat transfer through a body including increase of temperature due to absorption of radiation (right).

In order to account for the combined effect of solar radiation and temperature on the solid, another parameter was introduced: the sol-air temperature ($T_{s\text{-}a}$). The concept of Sol-Air Temperature expresses the combined action of incident solar radiation and ambient temperature that may lead to a temperature increase in a solid wall. $T_{s\text{-}a}$ is defined as the outside air temperature, which, in absence of solar radiation, would give the same temperature distribution and rate of heat transfer through a solid as exists due to the combined effects of the actual outdoor temperature plus the incident solar radiation. (IHVE Guidebook A, 1970)

The value of T_{s-a} is given by a combination of the glass characteristics, the dry-bulb ambient temperature and the total incident solar radiation on the glass panel.

The Sol-Air temperature is defined as (O'Callaghan, 1977):

$$T_{s-a} = T_0 + R\,(a\,I_T - \varepsilon\,I_l) \tag{2.1}$$

Where:

T_0	Outdoor ambient temperature
R	Thermal resistance of outer surface (from BS EN 6946)
a	Glass solar absorption
I_T	Total incident solar radiation intensity
ε	Glass emissivity
I_l	Long wave radiation intensity from a black body at T_0

The use of this parameter allows a translation of the combined effect of external temperature and solar radiation into resulting surface temperatures, summarising the complex non-linear phenomenon described in Figure 2-1 (right) into a linear problem, making it easier to obtain results over time and understand temperatures for every hour of a theoretical year.

3 Methodology

As outlined in the previous section, the temperature in laminated glass components is a function of several variables, such as ambient temperature, incident radiation and specific glass solar/optical parameters.

The methodology described in this research aims to outline the main steps to investigate the glass temperature variations over time on a laminated single glazed facade. Thanks to its linear nature, the proposed approach is believed to be applicable to a number of conditions in order to isolate the most critical occasions and assess their likelihood of manifestation. Any kind of datasets of environmental conditions can be therefore simultaneously processed with numerical calculation tools (e.g. Microsoft Excel; Matlab) in a timely manner.

The proposed approach is based on the use of $T_{s\text{-}a}$ as an environmental parameter for the calculation of the temperature in the glass build-up. The most common application of $T_{s\text{-}a}$ is the calculation of opaque building fabric heat losses within the field of building services engineering, as described in CIBSE Guide A (2015) and in ASHRAE Fundamentals Handbook (2013). However, although space heat losses and gains through glazed surfaces are generally calculated differently due to the transparent nature of the material, $T_{s\text{-}a}$ may be used to assess the temperature reached by the glass panel itself. Incorporating the effect of the radiation parameter into a comprehensive single boundary temperature condition ($T_{s\text{-}a}$) allows eliminating the non-linear component from the glass temperature calculation, overcoming the limitations of many software simulation tools, which are not able to dynamically model the combined effect of solar radiation and ambient temperature on glazed elements over time.

It is noticeable from eq. (3.1) that $T_{s\text{-}a}$ depends upon outdoor environmental parameters as well as on glass characteristic properties. Worth to mention is that the last term of the equation ($\varepsilon\ I_l$), which refers to the amount of long wave radiation (heat) re-emitted by the glass into the environment, was neglected in the context of this study. This allowed introducing a safety margin into the methodology, as the resultant $T_{s\text{-}a}$ would be higher without this term.

The "simplified" Sol-Air temperature used as the only boundary condition in this study is therefore defined as:

$$T_{s-a} = T_0 + R\ a\ I_T \tag{3.1}$$

It is worth noticing that the external ambient temperature T_0 as well as direct normal radiation and diffuse horizontal radiation data shall be retrieved by a recorded weather dataset. This dataset shall be based on relevant historical data covering several of years in order to include a number of critical environmental conditions reported for the specific location of the study. The accuracy of the study will indeed depend on the coarseness of time-steps at which values are available. However, it shall be considered that the smaller the time-steps will be, the more challenging it would be to obtain and process the dataset.

Once the relevant weather dataset is acquired, values of incident solar radiation shall be calculated according to the methodology described in the ASHRAE Fundamentals Handbook, 2013 (section 14), using the transposition calculation model enables calculating the incident solar irradiance (I_T) on any surface from the specific solar angles (Solar Altitude and Azimuth), direct normal and diffuse horizontal radiations. This method can be implemented into a numerical calculation sheet allowing a great number of conditions to be processed at the same time, providing, for example hourly values for a number of years. Lastly, glass absorptivity (a) can be calculated by modelling the specific build-up with a simulation tool (e.g. LBNL Optics/Window), whereas values of surface heat transfer resistance (R) can be found in ISO 6946:2007 or in CIBSE guide A (2015). All the above variables, can be combined thanks to eq. (3.2) at every time-step available in order to achieve different values of $T_{s\text{-}a}$ over time.

Once the values of $T_{s\text{-}a}$ are determined, the heat exchanged through the laminated glazed element can be calculated assuming a fixed internal temperature. By substituting the outdoor temperature T_O with the Sol-Air temperature $T_{s\text{-}a}$, an appropriate allowance for the solar radiation gain into the heat transfer balance equation is then provided (O'Callaghan, 1977). It is worth mentioning that the glass U-value can usually be extracted from the same simulation tool used for modelling the glass solar absorption in the previous step.

$$Q = U \times (T_{s-a} - T_{int}) \tag{3.2}$$

Subsequently, internal and external surface temperatures can be calculated for every time-step by the use of the following equations:

$$T_{si} = T_{int} - Q \times R_{si} \qquad T_{se} = T_{s-a} - Q \times R_{se} \qquad (3.3, 3.4)$$

Where:

R_{si} Internal surface resistance (from BS EN 6946)
R_{se} External surface resistance (from BS EN 6946)

Lastly, from the determination of the internal and external glass surface temperatures, considering the monolithic homogeneous glass build-up and therefore an almost linear temperature distribution throughout, it is possible to approximate the internal glass temperature as the mean of the two.

$$T_g = \frac{T_{se}+T_{si}}{2} \tag{3.5}$$

The most critical laminated glass temperatures can therefore be identified and their triggering environmental conditions isolated. Further discrete specific analyses can be then carried out with more advanced glass simulation tools if required.

4 Analysis and Discussion

In order to validate the proposed methodology, a simplified case study was carried out. This was based on hourly environmental conditions representing a sample meteorological year for a predominantly hot weather location: the city of Miami. It is worth mentioning that the exemplar conditions used in this analysis shall not be considered representative of the actual worst cases. Much more extended datasets shall be used to assess the risk of high temperature occurrence in structural glass build-ups (e.g. 50 return period conditions). The methodology outlined in section 2 was implemented into a calculation sheet (Microsoft Excel) allowing the determination of the hourly temperature profiles of five hypothetical glass build-ups over the timespan of a year. The panels were all assumed to be vertical and south facing and were simulated under the environmental conditions described by the selected dataset.

Table 4-1 Test panels characteristics.

	Build-up (mm)	Solar coating	U Value (W/m²K)	α (%)	Increase of α
1	12	no	5.5	3.6	-
2	12+12	no	5.12	11.7	325 %
3	12+12+12	no	4.8	19.3	165 %
4	12+12+12+12	no	4.5	23.2	120 %
4	12+12+12+12	yes (face 2)	4.5	35.3	152 %

The build-ups 1 to 4 are characterized by increasing thicknesses based on a 12 mm incremental, representing a commonly used glass layer thickness in structural applications. As a direct consequence, increasing glass absorption and decreasing *U*-values also characterize build-ups 1 to 4. Nevertheless, the presence of a coating in the last build-up (4c) makes it the one with the greater absorption despite featuring the same thickness as build-up 4. As expected, although tested under the same environmental conditions, due to their different thermal and optical characteristics, the build-ups showed dissimilar maximum temperatures. The calculated glass temperatures and the correspondent triggering environmental conditions were then isolated for every build-up and tabled along with the calculated $T_{s\text{-}a}$.

Table 4-2 Worst-case conditions and calculated glass temperature in case study test pan-els

	T_o (°C)	I_t (W/m²)	$T_{s\text{-}a}$ (°C)	T_g (°C)
1	35.6	252	36.15	34.9
2	35	371	37.6	35.8
3	35	371	39.3	36.8
4	31.7	636	40.8	37.4
4	28.3	814	45.4	41.4

The highest calculated glass temperatures in all five build-ups correspond to the maximum calculated sol-air temperature $T_{s\text{-}a}$. However, it was noted that neither the absolute maximum incident radiation level, nor the absolute maximum external temperature leads to the highest temperatures in the glazed panels. The relationship existing between these three variables is governed by the T_{s-a} equation (2) and graphically described in Figure 4-1.

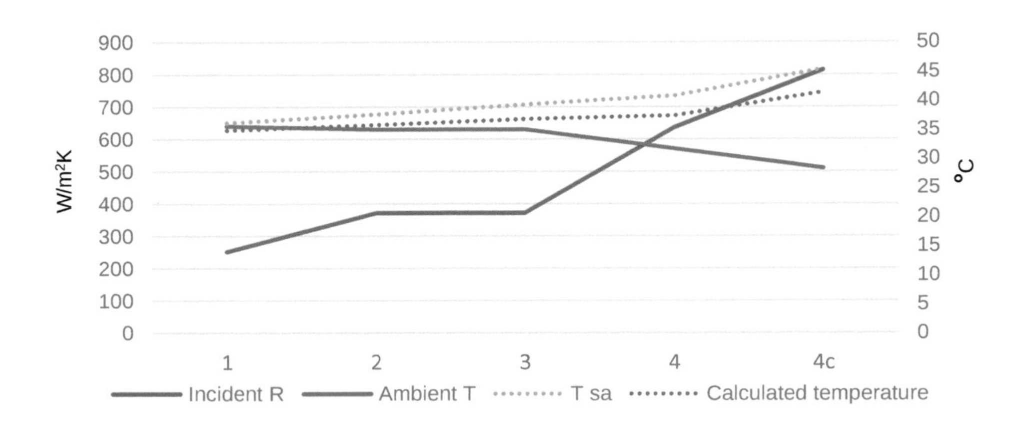

Figure 4-1 Worst-case conditions variation according to panel build-up.

Figure 4-1 shows that for low values of absorption (i.e. build-ups 1-2-3), rather low levels incident radiation levels and simultaneous high external temperatures represent the most critical conditions identified by the study. Conversely, it is shown that the conditions where built-ups with greater absorption (i.e. 4-4c) lead to higher temperatures, are characterized by higher levels of incident radiation and lower external temperatures. This emphasizes the crucial importance of the effect of solar radiation on the temperature of glass build-ups featuring absorption levels greater that 20 %. The identified environmental parameters T_O and I_T (Table 4-1) leading to the maximum glass build-ups temperatures were then input into a glass temperature simulation tool (Win SLT) (Simulation 1) to compare the outcomes of a computer model against those from the proposed numerical calculation. Also, the same modelling tool was used to perform another set of simulations (Simulation 2), this time by replacing the ambient temperature T_O with T_{s-a} and eliminating the I_T parameter, as this was already accounted for in the sol-air temperature equation (2).

Table 4-3 Outcomes comparison matrix.

	Calculated (°C)	Simulation 1 (T_O & I_T) (°C)	Simulation 2 (T_{s-a}) (°C)
1	34.9	31.1	31.4
2	35.8	32.2	32.6
3	36.8	32.7	33.8
4	37.4	32.5	34.6
4c	41.4	33.4	38.1

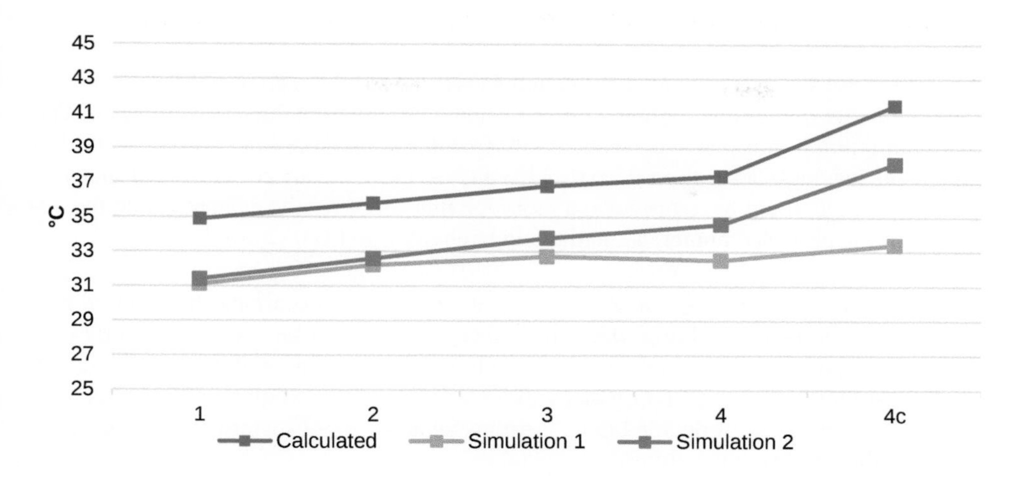

Figure 4-2 Outcomes profiles comparison.

Figure 4-2 shows how the calculated temperatures for every glass build-up are above the corresponding simulated results by at least 3 °C. This is believed to be due to the neglected cooling effect related to the emissivity factor, which was omitted from the calculation to create a safety margin (section 2). In a real case, the glass will in fact be re-radiating heat towards the external space after being heated up, causing a slight decrease in temperature. It is worth mentioning the marginal temperature increase in the Simulation 1 results, despite an increase of the absorption of around 150 % between build-up 4 and 4c (Table 4-1). Conversely, it was noticed a sharp increase of the other two curves based on $T_{s\text{-}a}$ (calculated T and Simulation 2) associated to the same build-ups. The divergence between the results calculated by $T_{s\text{-}a}$ and those simulated by inputs of T_0 and I_T was therefore found to be more substantial for higher levels of absorption. This is related to the non-linear nature of the radiation effect on the glass and to its associated modeling complexity. It should however be noted that the use of $T_{s\text{-}a}$ has lead to more critical results, thus more conservative, mitigating the risk of underestimating the problem.

5 Limitations of the study

The use of the sol-air temperature $T_{s\text{-}a}$ as a single comprehensive boundary condition means assuming the glass panel to be exposed to a completely fictional environment. It shall therefore be stated that the linear temperature distribution generated by the heat transfer balance calculation between a given internal ambient and the described fictional external environment is considered to be only representative of the real temperature profile from the middle of the glass to the internal surface. External surface temperatures are therefore to be modelled with a different approach. A different approach shall be followed also when analysing build-ups including more than one material, such as laminated metal inserts.

The validation analysis presented in Section 4 revealed some uncertainties also related to the complexity of the solar radiation phenomenon on transparent thick constructions. The main limitation of this analysis is the accuracy of the simulation tool (Win SLT) used to model thicker high absorption glazed build-ups. A more advanced non-linear analysis simulation tool would have provided a more reliable set of results against which the accuracy of the proposed numerical methodology could have been verified.

Lastly, it is worth to mention the complexity one may face in sourcing the required historical environmental conditions dataset. To date, this may not be easy to obtain for certain locations and for others it may be subject to charges. However, this is considered as a temporary limitation as currently many of the world's meteorological institutes are collecting large amounts of detailed data, creating reliable sets of historical conditions for future studies of different natures.

6 Future outlooks

The structural behaviour of laminated glass components is dependent on the temperature, predominantly on the temperature of the interlayer. In this work it is shown that the assessment of the temperature in laminated components over a long period of time is dependent on several complex phenomena. In that regard, a systematic simplified methodology is presented for the temperature computation using large amount of weather data. This topic inevitably rises up the discussion on the temperature to be considered when performing the assessment of the mechanical performance of laminated components. The design scenario of maximum action (e.g. maximum gust 3-sec wind speed) to be applied with the highest peak of temperature can be over-conservative. This is because the probability of experiencing the maximum action simultaneously to the maximum temperature is rather limited. In that regards, design codes do not give precise indication on this matter. In the case of structural components with a strongly temperature dependent load bearing capacity, it is therefore of interest to develop a more scientific approach to address this issue. Future work will therefore focus on the development of a systematic methodology on the safety and performance assessment of the structural component considering appropriate safety scenarios.

7 Conclusions

The methodology presented in this research provides an opportunity to reduce a complex phenomenon into linear relationships, including a safety margin. The greatest potential of this numerical model is that it can be applied to a large set of conditions to quickly assess the more critical ones. This, for example, could be used to investigate the worst environmental conditions registered over a relatively long period (e.g. 50 year return period), causing critical temperatures in a specific glass build-up at a specific location. The model's mid-glass temperature outcomes can be taken as preliminary design criteria. Alternatively, the associated critical conditions can be used to further analyse the phenomenon in much detail. The model could in fact be used to isolate the most critical combinations of T_0 and I_T amongst a wide range of conditions in order to subsequently allow, if required, a more detailed non-linear analysis of the characteristic glass' temperature profile to be carried out with advanced simulation software.

This allows overcoming the previously described conservative assumptions design codes in an efficient way, as the methodology allows to process large amounts of data simultaneously. Depending on the severity level of the temperature conditions occurring as well as the duration of temperature load and occurrence over time, further analysis can be carried out to determine the structural implications.

8 References

[1] Incropera, P., et al, Fundamentals of Heat and Mass Transfer, John Wiley & Sons, 2007.

[2] Shelby, J.E., Introduction to Glass Science and Technology, The Royal Society of Chemistry, Cambridge, 2005.

[3] P.W. O'CALLAGHAN, S. D. PROBERT, School of Mechanical Engineering, Cranfield Institute of Technology, 1977.

[4] IHVE Guidebook A, Institution of Heating & Ventilating Engineers, London, 1970.

[5] ASHRAE Handbook Fundamentals, ASHRAE, 2013.

[6] CIBSE Guide A, CIBSE, 2015.

[7] Baldini, A.; Lenk, P.; Rammig, L.; Thermal effects on curved annealed glass, In GPD 2015, Tampere, 2015.

Improvements and Limitations in Fabrication of Structural Glass

Sanmukh Bawa[1]

1 Eckersley O'Callaghan, 9th Floor, 236 Gray's Inn Road, London WC1X 8HB,
sanmukh@eocengineers.com

Applications for structural glass are increasingly affordable and more popular amongst architects and developers are seeking to distinguish their projects from their competitors than ever before. Structural engineers are encountering more aspirational schemes where current engineering knowledge, fabrication techniques and installation conventions are challenged. Engineers writing specifications for structural glass components tend to define strict tolerances to account for possible errors arising during fabrication stage and installation phase. This paper presents findings based on extensive factory site inspections and audits gathered over several years. This paper will introduce and discuss in detail the various typical glass component defects including dimensional tolerances, interlayer overflow, delamination, scratches, blemishes and non-uniform glass pre stress from quenching. The paper will go on to highlight the possible implications of these defects on the reliability and durability of glass components and subsequently the structural design implications. Finally, current state-of-the-art techniques in fabrication and quality control procedures will be described with potential improvements proposed.

Keywords: structural glass, jumbo glass, fabrication, quality control, engineering

1 Introduction

While the use of glass in building envelopes allows for a large amount of transparency, even when used structurally, it can require a large amount of connections, particularly for tall facades and large roof structures. Structural applications with super jumbo glass (height > 9 m) are increasing on trend. The glass edge has a lower strength than the centre of the glass [4] and is yet often subjected to high stress due to bending or other loads. With the desire for maximum transparency, many structures comprising of super jumbo glass opt for structural silicon joints between glazing panels. The importance of clean edges for structural bonding is highlighted later in this paper. Architects and engineers define specifications with very strict requirements which, at times, are simply not achievable by the various fabrication processes or may not be required for project. Most of the check points in quality control take into account the visual aspects and have set acceptable criteria for judgement and normally a structural engineer is not present during the inspection of structural glass to consider the structural integrity.

Engineered Transparency 2016. Glass in Architecture and Structural Engineering. First Edition.
Edited by Jens Schneider, Bernhard Weller.

2 Glass fabrication and implications for structural design

2.1 Dimensional tolerance

Control of dimensional tolerances is very important for structural glass. Panels need to be within the specified dimensional tolerance when arriving on site – otherwise they may not fit into the glazing system or provide enough room for movement. As the glass cannot be trimmed on site, in contrast to other materials such as steel and concrete, the fabricator usually cuts glass panes to be oversized by 1-5 mm, which is then reduced by the edge polishing process. The glass is sent into a tempering furnace for heat treatment, where it is exposed to temperatures in excess of 600 °C. It expands during the heating phase but doesn't contract by an equal amount during the quenching phase. It was observed during inspection audits of jumbo glass panels that an 8 m long, 12 mm thick, low-iron glass expanded by 2.0 mm (0.025 % expansion) and a 15 m long glass expanded by 2.5-3.0 mm (0.016 % expansion) after tempering process (including quenching). The change in thickness was observed to be about ±0.02 mm for both panels. It was also generally observed that a low-iron glass expanded more during heat treatment in comparison to a normal float glass. This suggests that the length of glass and difference in iron content contributed to the change. However, a more detailed study will be performed in future to establish numerical conclusions from test data.

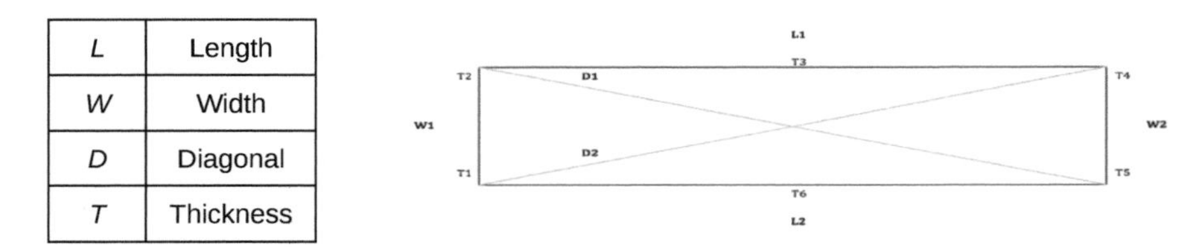

L	Length
W	Width
D	Diagonal
T	Thickness

Figure 2-1 Glass fabrication tolerance vital checks.

In order to comply with particular project-specific specifications on fabrication tolerance, some glass processors will polish one or both edges of a panel. If this process is completed before the heat treatment, it does not impact the structural integrity of glass. However, if performed after heat treatment, the zone of pre-stress within the glass is reduced – this impact on performance is not usually accounted for in structural design. In another condition, if an oversized glass panel is installed on site, it may reduce the movement tolerance allowed at the top of the capture frame, thus increasing the possibility of breakage due to expansion of the glass over life period or movement of the substructure, which may clash with the glass.

2.2 Interlayer overflow

It was observed during inspection that a number of glass panels had interlayers that were not confined to within the arris but rather overflowed beyond the edges. For open edges, such as the top edge of a glass balustrade, this is a significant visual problem. Furthermore, it has structural implications. The overflow of film on the bonding edge leaves reduced bite area for structural silicon and therefore breaches the structural design requirements as it may cause the silicon joint to overstress due to a reduce bite area than originally designed for.

Figure 2-2 Glass fin delamination on arris (left). Edge showing SG debonding from glass arris (right).

2.3 Delamination

Delamination is a frequently identified flaw in laminated glass. If it occurs in the central area of the glass, it is usually only a visual problem. With increasingly popular laminated inserts (titanium and stainless steel) for connecting glass panels, delamination near insert fittings is problematic, both structurally and visually. Most specifications focus on the visual aspects of glass and its components but there is minimal guidance on structural implications in current leading industry codes. Below are observed causes for delamination in laminated glass with preventative measures proposed where applicable:

- Tin/Air side- Better performance of lamination is achieved with the air side bonded to the PVB interlayer, however for SGP, it is found that the results are better when the interlayer faces the tin side [3]
- Water quality – for best results, the glass conductivity must be lower than 5 microsiemens when water is used for cleaning glass for best lamination results.
- Moisture in interlayer – Interlayers must always be stored in a controlled environment. The start of the roll should be discharged and not used [3].
- Autoclave cycle – Reducing the duration of autoclave cycle results in partial curing of the interlayer.
- Clean surfaces – presence of dust or other elements can cause issues for long term bonding. Clean room conditions improve the overall quality and integrity.

- Flatness of glass panes – roller waves and edge dip in thermally treated glass causes variations (peak and troughs). Excessive differences in thicknesses cause additional stresses in the interlayer, thus increasing the potential for delamination.

Figure 2-3 Delamination around insert fitting (left). Glass section with air ingress (right).

Inherent delamination in the fabrication process will grow eventually. The rate and size of growth depends on the location of delamination. If delamination occurs at the edge of a panel, it will grow at a greater rate due to the exposure to air and humidity. In most instances, delamination is only considered a visual problem and a unit is typically replaced if it is visible in the central areas. Edge of panel delamination is not usually treated with similar attention – greater importance should be afforded since structural composite behaviour is lost locally.

2.4 Coating on glass edge

Figure 2-4 Glass being checked for edge shift (left). Solution for facade glass with excessive shift (right).

With increasing use of high performance coatings on glass globally, better awareness of their application and limitations is paramount. It has been observed that if there is a Low-E coating present on the edge of the glass, as shown in figure 2-4, it poses a risk of in-adhesion to silicon due to the presence of Silver metal (Ag) in the coating. Over time, the silver (Ag) oxidises and de-bonds to the base glass pane. It is proposed that the edges are always covered with an applicator tape during the coating process (if it is applied after the tempering process), or the coating should be stripped off using standard industrial tools, such as a polish wheel grinder, with varying hardness depending on coating type.

2.5 Scratches and chips

Most European projects are based on *'Guideline to Assess the visible Quality of Glass*'[2] which currently assesses scratches visually and does not determine the depth/extent of the scratch. Scratches up to 30 micron deep are usually visible to the naked eye. Stresses in glass can rapidly increase if the edge is cracked or chipped. Such flaws are usually visually assessed and are not examined in detail for structural impact. The tensile strength of glass depends significantly on mechanical flaws on the surface [4]. Such flaws may not always necessarily visible to the naked eye.

Guidance, such as (*Hadamar*, 2009), specify the acceptable length of scratches but there is limited information on the acceptable depth of scratches. Although a number of studies have been undertaken in the past by numerous researchers, there is a missing link to industry codes for this criteria. Measuring scratches is also a difficult task without the use of complex instruments. The current visual guidance [2] identifies 3 zones in a glass pane – Main, Edge and Rebate – and gives allowable criteria based on number of scratches. In the Rebate, there is no limit on the length of scratch and other similar defects. Visually, this area is covered in a fixing frame but, structurally, it is important that the edge area is free of defects.

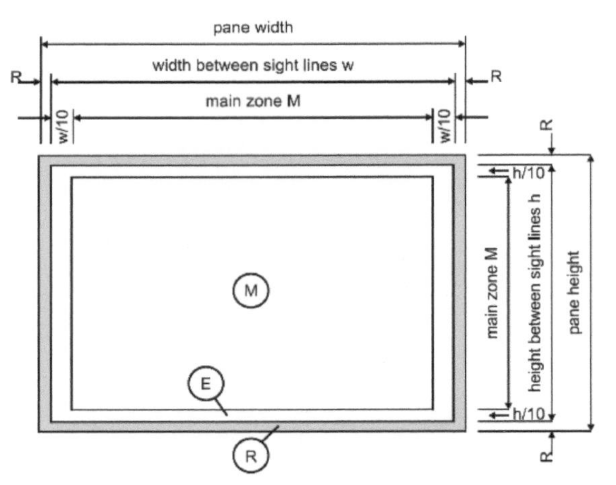

Zone	The following are allowable per unit:	
R	External shallow damage to the edge or conchoidal fractures which do not affect the glass strength and which do not project beyond the width of the edge seal.	
	Internal conchoidal fractures without loose shards, which are filled by the sealant.	
	Unlimited spots or patches of residue or scratches.	
E	**Inclusions, bubbles, spots, stains, etc.:** Pane area ≤ 1 m²: Pane area > 1 m²:	 max. 4 cases, each < 3 mm ∅ max. 1 cases, each < 3 mm ∅ per meter of perimeter
	Residues (spots) in the gas-filled cavity: Pane area ≤ 1 m²: Pane area > 1 m²:	 max. 4 cases, each < 3 mm ∅ max. 1 cases, each < 3 mm ∅ per meter of perimeter
	Residues (patches) in the gas-filled cavity:	max. 1 case ≤ 3 cm²
	Scratches: total of individual lengths:	max. 90 mm - Individual length: max. 30 mm
	Hair-line scratches: not allowed in higher concentration	
M	**Inclusions, bubbles, spots, stains, etc.:** Pane area ≤ 1 m²: 1 m² < Pane area ≤ 2 m²: Pane area > 2 m²:	 max. 2 cases, each < 2 mm ∅ max. 3 cases, each < 2 mm ∅ max. 5 cases, each < 2 mm ∅
	Scratches: Total of individual lengths:	max. 45 mm - Individual length: max. 15 mm
	Hair-line scratches: not allowed in higher concentration	

Figure 2-5 Extract from 'Guideline to assess the visible Quality of Glass' (Hadamar, 2009) [2].

The guidance is primarily for infill panels, where the glass sits in a frame and there is no load transfer through the glass itself. In the case of a structural glass, this need to be reconsidered in more detail and this missing link to the industry code must be established. The central surfaces of glass panes generally contain fewer deep surface flaws than the edges, which can be subject to higher stresses in structural applications. Edges typically contain more relatively severe flaws due to mechanical edge working. This explains why the glass strength on the edge is lower than the strength in the centre of a glass pane [4].

Table 2-1 Test samples description.

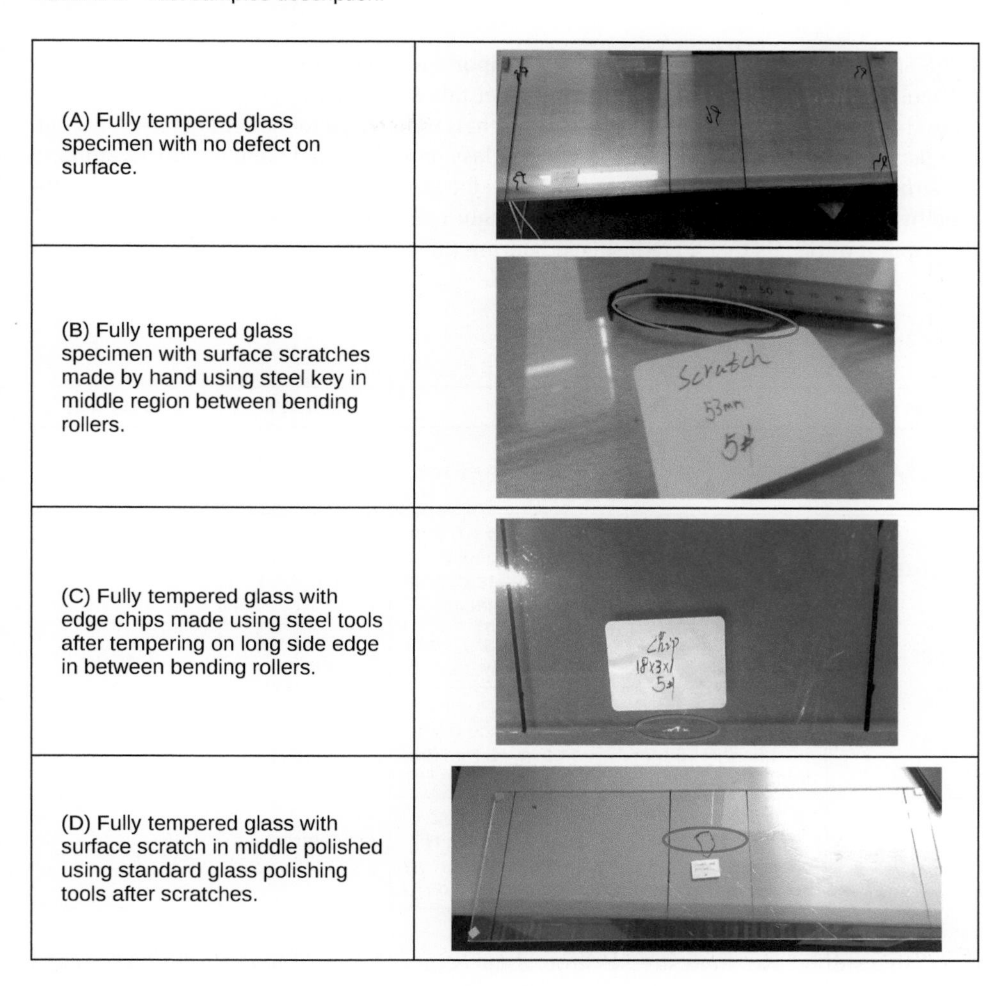

Description	Photo
(A) Fully tempered glass specimen with no defect on surface.	
(B) Fully tempered glass specimen with surface scratches made by hand using steel key in middle region between bending rollers.	
(C) Fully tempered glass with edge chips made using steel tools after tempering on long side edge in between bending rollers.	
(D) Fully tempered glass with surface scratch in middle polished using standard glass polishing tools after scratches.	

For a project related purpose, 4-point bending tests were carried out on 5 samples of each 4 types on 12 mm fully tempered glass as listed in table 2-1. The test results showed a reduction in strength due to such flaws, since the spectrum of samples was quite small (5 panels of each type) due to being project specific, this cannot be deemed definite and in future we endeavor to establish a result with higher number of specimen. A similar study conducted by Datsiou [1] presented a reduction in strength of 62-79 % in annealed glass when tested to artificial ageing method including sand abrasion and scratch induced by indenter. However, from the above test with limited samples, it can be said that the edge defects have contributed towards the strength of glass.

2.6 Edge shift

Edge shift of laminated glass panel is an important aspect for jumbo size panels. It is difficult to precisely control the length edge at this size as outlined in section 2.1. If the edge shift is significantly large, the panel weight is transferred to the setting block through one pane only. If this occurs in an insulated glass unit, the panel weight induces stress in the structural sealant/polysuphide around the edge seal. The edge shift data has been gathered by quality control inspections by the author over the past 2 years. Table 2-2 shows description of panels that were inspected during this period.

Table 2-2 Inspection panels description.

Glass panel	Description	Glass type	Size (*W* x *L*)
GL 01A	6 mm low iron heat strengthened (HS) 1.52 mm PVB + 6 mm low iron HS glass	Laminated Glass	2.4 m x 4 m
GL 01B	6 mm low iron HS +1.52 mm PVB + 6 mm low iron HS with double silver low e #412mm Air gap + 8 mm low iron tempered heat soaked(HST), border frit	Insulated glass unit (IGU)	2.4 m x 4 m
GL02	*Cold bend glass with 250 m radius:* 12 mm low iron HS 3.04 SGP +12 mm low iron HS	Laminated Glass	2.7 m x 15.6 m
GL03	8 mm low iron HS +1.52 mm SGP +8mm low iron HS with High Performance (HP) coating #4+16 mm Argon cavity+8 mm low iron HS+1.52 mm SGP +8 mm low iron HS	Insulated glass unit	2.0 m x 6.0 m
GL04	8 mm low iron HS with dot frit pattern#2 + 1.52 mm SGP +8mm low iron HS with HP coating #4 +16mm Argon cavity +8 mm low iron HS+ 1.52 mm SGP +8 mm low iron HS	Insulated glass unit	2.0 m x 6.0 m

Excessive shift causes high stresses around the edge silicon seal in case of an IGU and a reduced load bearing area in case of a laminated glass. This can be rectified on site, to an extent, by adding shims below the shorter pane or by injecting mortar resin in order to form a uniform support. Some glass processors over polish one pane of laminate to match the other pane, even after the glass is heat treated. This poses issues as the pre-stress layer is reduced – such may go unnoticed unless full time observation is performed.

A comparison between GL01A and GL01B as shown in Figure 2-6 suggests that when a laminated glass is used within an IGU, there is an increase in overall glass edge shift. It was seen that when edge shift is less than 0.5 mm, laminated glass had 12 % more panels compared to that of an IGU, however this trend changed as greater edge shift intervals are observed, in which case, number of panels with IGU edge shifts were greater.

Table 2-3 Glass edge shifts inspection record.

Edge shift	Number of panels and Glass types				
	GL01A	GL01B	GL02	GL03	GL04
Less than 0.5 mm	435	332	20	38	109
0.5 mm to 1.0 mm	307	396	12	55	210
1.0 mm to 1.5 mm	107	144	3	18	69
1.5 mm to 2.0 mm	28	33	-	6	32
More than 2.0 mm	18	15	-	2	16
Total Panels ∑	895	920	35	119	418

For edge shifts greater than 2.0 mm the result again shows lower % of IGU panels compared to laminated glass panels due to the fact that some panels were fixed by re-doing secondary seals, compared to heat treated laminated glass that in principle could not be altered once the interlayer is cured after the lamination process. The acceptable edge shift limit on panels was 1.5 mm, panels above this limit were recorded as quarantine and reviewed based on the severity of defect and panel locations.

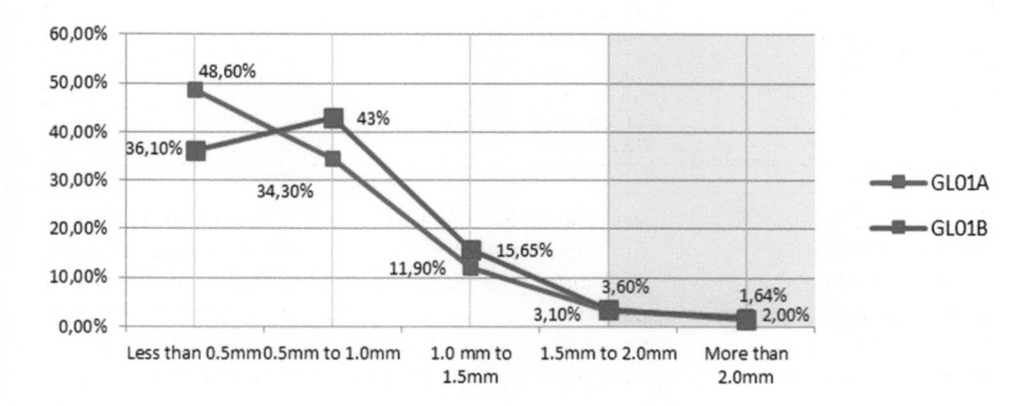

Figure 2-6 Edge shift comparison between GL01A and GL01B.

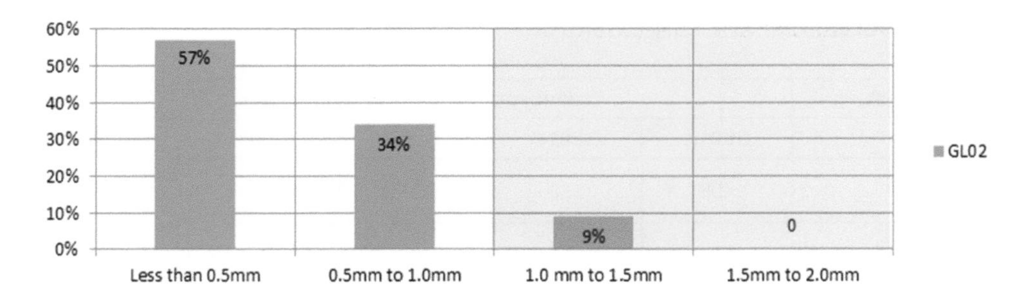

Figure 2-7 Edge shift data for GL02.

For most panels, a maximum edge shift of 1.5 mm for IGU was accepted, whereas for GL02 (laminated glass) due to design constraints, a max edge shift limit of 1.0 mm was allowed for in the specification. Since the panels were cold bent/warm bent into shape during lamination process, it was more challenging to achieve such strict requirement however as it can be seen from Figure 2-7 that almost 90 % total panels were produced to an acceptable limit and thus it can be concluded that with careful measures during production, strict fabrication tolerances can also be achieved.

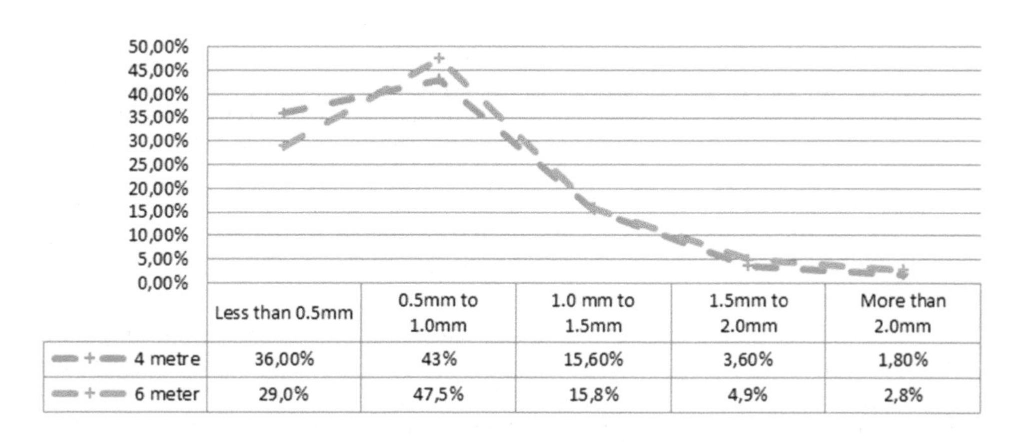

	Less than 0.5mm	0.5mm to 1.0mm	1.0 mm to 1.5mm	1.5mm to 2.0mm	More than 2.0mm
4 metre	36,00%	43%	15,60%	3,60%	1,80%
6 meter	29,0%	47,5%	15,8%	4,9%	2,8%

Figure 2-8 Edge shift comparison between 4 meter and 6 meter panel.

Figure 2-8 shows a comparison between edge shift for 4 meter and 6 meter glass panels that was recorded during glass inspection audits. As expected, it can be seen that the IGU edge shift for 6 meter panels is greater than 4 meter panels for most of inspection data, however for edge shift 'less than 0.5 mm', the results appears to be different and is expected to happen due to inconsistency in workmanship during production.

3 Observations and Conclusion

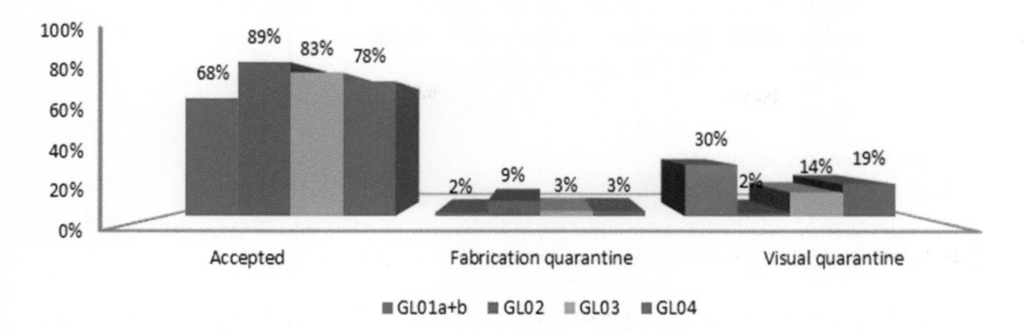

Figure 3-1 Overall inspection data.

Figure 3-1 shows overall inspection results gathered during quality inspection audits, the panels were inspected for fabrication tolerances that include but not limited to – edge shifts, dimensions, and panel bow and also for visual tolerances against set acceptability criteria as per project specification.

- GL01 had a high percentage of panels in visual quarantine/rejection primarily due to defects within high performance coating e.g. pinholes in coating and inclusions within laminate that were greater than ø 2.5 mm (acceptable criteria on project).
- GL01, 03 and 04 had very similar percentage of panels in fabrication tolerance overall and had the same inspection criteria. GL02 had a greater percentage of panels in fabrication quarantine due to more strict inspection criteria (for e.g. edge shift limit of max 1.0 mm compared to 1.5 mm for other glass types), and also since the overall spectrum of panels was lower (35 panels) compared to other panel types.

It was also observed during audits, that some basic processes such as polishing out a scratch do not have a controlled method. It is currently labour intensive and there is currently no agreed limit on how much a glass scratch can be polished. Since it is not machine controlled process, there is lack of equal pressure and consistency over the treated area. Over polishing is seen to cause distortion if done beyond a level, too less polishing may not get rid of the scratch completely. Such processes need to be standardised or have industry guidance in order to keep consistency across the industry.

For clear understanding on this subject, it is endeavored to conduct a more detailed study in future. (The study shall include test specimen, scratched to various depths and then surface polished; this shall then be reviewed visually at varying angles to establish an acceptable limit of connection between scratch depth and surface polishing.)

It can be seen that workmanship during fabrication and an emphasis on quality control plays a significant role in achieving a reliable end product. An important challenge of glass structures is the strong interaction between architecture, engineering and production techniques and facilities. Therefore, for the success of such projects it is essential that all parties are involved in the design process already at an early stage.

4 Acknowledgements

The realization of this paper would not have been possible without glass industry figures. The author would like to thank Tianjin North Glass for their support.

5 References

[1] Datsiou, C.: Evaluation of Artificial Ageing Methods for Glass, 2016, pp 09-10.

[2] Hadamar, 2009: Guideline to assess the visible quality of glass in Buildings.

[3] TROSIFOL Technical Manual, 2012: Processing of PVB film, pp 90-130.

[4] BS EN 16612 Glass in building – Determination of the load resistance of glass panes by calculation and testing.

Adhesive joints for bonded point fixings in façades and glass structures

Christiane Kothe[1], Michael Kothe[1], Jan Wünsch[1], Bernhard Weller[1]

1 Technische Universität Dresden, Institute of Building Construction, George-Bähr-Straße 1, 01069 Dresden, Germany

The wide acceptance of glass as a building material encourages many requests for transparent, almost dematerialized appearing building envelopes. The building material glass is not only used as space enclosing element in modern architecture, rather it is found increasingly in structural applications like fall protection, safety glazing and walkable or overhead glazing. Those require glass structures, which are involved in the load transfer. State-of-the-art integration of glass elements in buildings are point and linear fixings which are usually designed mechanically. However, glass is a brittle material which makes mechanical fixings unfavourable especially in regard to the occurring high stress concentrations near the necessary boreholes. Hence, an increased risk of glass breakage results. A better adapted solution are load-bearing adhesive connections which are still used successfully in civil engineering in some glass applications as well as in concrete, steel, wood or plastic and track construction. However, the technology of adhesive bonding through bonded point fixings is not common due to the high administrative and experimental effort for obtaining an individual technical approval (ZiE). Adhesively bonded point fixings and fittings represent an advantageous alternative to conventional bearing connections. In addition to cost savings, the absence of boreholes especially causes constructive facilitation. Generally, a more homogeneous stress distribution in the load introduction area arises by using bonded point fixings. Further benefits of bonded fixings are the reduction of mechanical connection elements, associated with higher quality aesthetics and an increase of creative freedom in façade design.

Keywords: glass construction, glass structures, point fixings, façade, adhesive bonding

1 Introduction

Adhesively bonded point fixings and fittings are an advantageous alternative to common bearing-type connections usually used in building and glass construction. By eliminating the mechanical glass processing, costs can be saved during the production process and facilitation of the construction can be achieved. In adhesively bonded point fixings, in contrast to mechanically attached systems, a more homogenous stress distribution occurs in the load application area.

The development of principle solutions and methods of using the adhesive bonding technology for glass facades and glass structures is the focus research activities. This is both demonstrating the feasibility, as well as increasing the general acceptance of adhesive

Engineered Transparency 2016. Glass in Architecture and Structural Engineering. First Edition.
Edited by Jens Schneider, Bernhard Weller.

bonding in glass construction. The main focus is on the theoretical and experimental investigation of punctual bonding of glass and metal fittings and the transferability of acquired laboratory-scale insights into actual product and process innovations.

2 Testing and selection of adhesives

A variety of adhesives of different groups (polyurethanes, acrylates, methacrylates, epoxy resins) are conceivable for the application of bonded point fixings in glass construction. Here, both the mechanical and chemical-physical characteristics as well as the optical appearance of the bonded joint are of particular interest. Important parameters, limiting the applicability of the adhesives, are the operating temperature range, the strength of the adhesive and the adhesion performance to glass and metal, the transparency as well as processing-relevant factors such as the viscosity or the processing time. In addition, the resistance of the adhesives to external environmental stresses influenced the field of application of the bonded joints.

Based on extensive experimental studies on various adhesive systems, epoxy resin adhesives seem to be best suited for structural bonding of point fixing systems [1,2]. A variety of commercially available epoxy resin adhesives is known and has different areas of application, like automotive industry, mechanical engineering as well as construction engineering. To fulfill the required property spectrum to its full extent, also modified adhesives are into consideration besides commercially available epoxy resin adhesives. These are adjusted in their chemical composition by the addition of fillers and additives to the necessary requirements.

These studies showed that it is possible to affect both the thermomechanical properties of the adhesive and the mechanical strength of the adhesive bond by the addition of suitable fillers and additives [3-5].

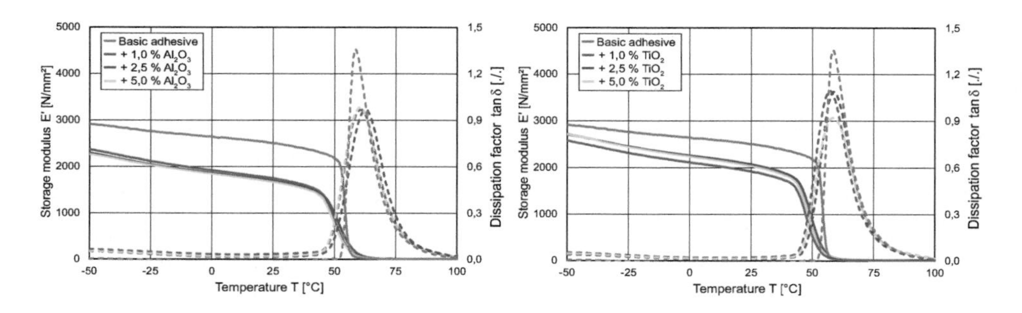

Figure 2-1 DMA thermograms of various epoxy resin adhesives containing different amounts of aluminium oxide (left) and titanium dioxide (right) as fillers.

The diagram in figure 2-1 exemplifies the influence of the addition of aluminium oxide (left) or titanium dioxide (right) on the thermomechanical properties of the adhesives. The addition of fillers results in a reduction in strength and a greater flexibility of the adhesive as shown in the curve of the storage modulus *E'*. Simultaneously, the location of the glass transition (peak position of the loss factor tan δ) is not adversely affected. This results in a greater operating temperature range.

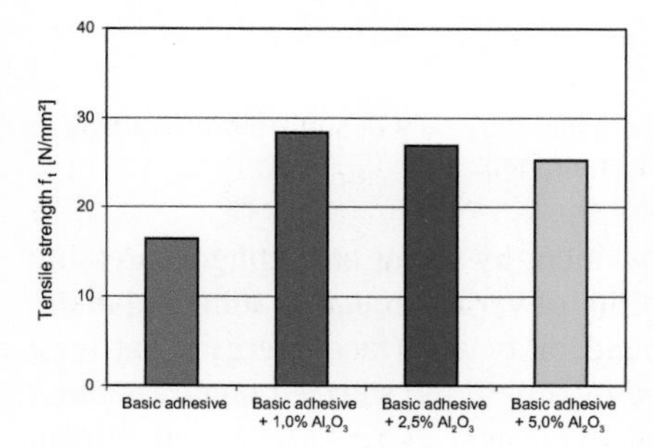

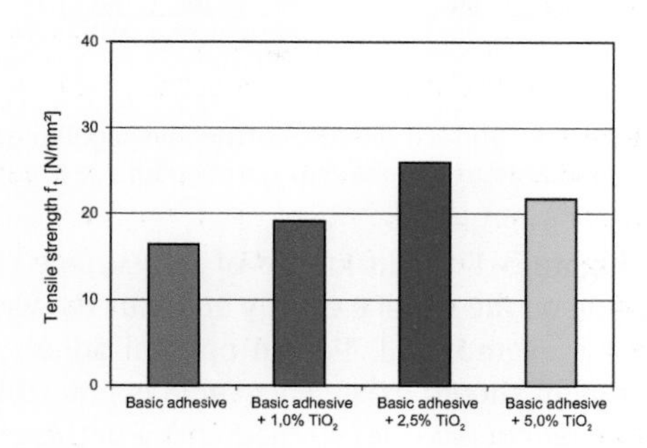

Figure 2-2 Characteristic tensile strengths of various epoxy resin adhesives with different amounts of aluminium oxide (left) and titanium dioxide (right).

The graphs in figure 2-2 show the characteristic tensile strengths from the uniaxial tensile test of the basic adhesive (grey) and the influence on the strength of the adhesive bond by addition of aluminium oxide (left diagram) and titanium dioxide (right diagram). The incorporation of aluminium oxide leads to a significant increase in the characteristic tensile strength regardless of the content of the filler. The maximum increase in strength is achieved at the slightest filler content of 1 % by weight. This is associated with an increase in strength of the adhesive bond from 16.5 N/mm² up to 28.5 N/mm² and represents an increase of 75 %. The addition of titanium dioxide also leads to a significant increase in strength, which is reflected in the characteristic tensile strengths. Here, the maximum is reached at a filler content of 2.5 % by weight, accompanied by an increase in the strength from 16.5 N/mm² to 26.0 N/mm².

3 Surface pretreatment

The surfaces of the adherend materials can be modified by specific cleaning and surface pretreatment processes. Thereby the adhesive performance is improved significantly [6]. Strong material and adhesive-specific effects of the individual surface pretreatments are found here. Optimal conditions for the adhesive process can be obtained only with a previous determination of the appropriate combination of adherend, adhesive and surface preparation.

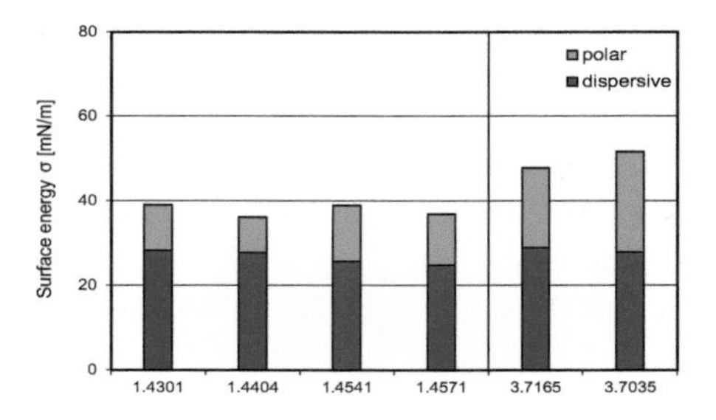

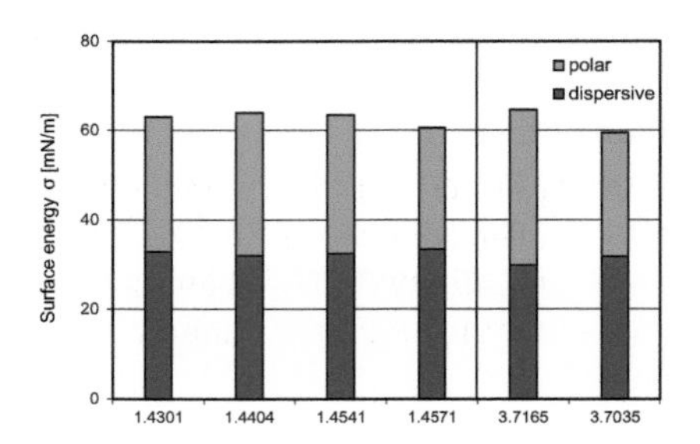

Figure 3-1 Surface energies of the investigated metallic adherends without surface pretreatment (left) and after treatment using atmospheric pressure plasma (right).

In Figure 3-1 the influence of the surface pretreatment by using atmospheric pressure plasma on the surface energy and thus the wettability of various metallic adherend materials is exemplified. For an optimal adhesive bondability a surface energy of at least 60 mN/m should exist. This value is not achieved by any of the selected metals without pretreatment (see the left diagram). For this reason, in the next step a surface pretreatment was carried out with these materials. The treatment with the atmospheric pressure plasma was chosen. Using this method, a very good wettability of the metal surfaces could be achieved. The surface energy could be increased to more than 60 mN/m (see the right diagram). For all further studies it was recommended to pretreat the metal joining part by atmospheric pressure plasma. In addition, a special primer for epoxy resins has to be used to improve adhesion [7,8].

Other aspects in the process of surface preparation are deactivation reactions which occur on the modified surfaces subsequently. The pretreated surfaces have an increased chemical reactivity. In dependence on the surface properties and the environmental conditions different deactivation reactions can be effected. Here, among others, the depositions of moisture as well as of inorganic and organic contaminants have to be mentioned. Therefore, a fast processing is required to preserve the effectivity of a previous surface preparation.

The temporal and spatial separation of the process chain is currently not advanced in the industry because of inappropriate technologies. Rather, in most industries the integration of surface modification is operated in the production processes. However, often, different methods are required for an optimal pre-treatment of the adherend surfaces, resulting in an increase of time and costs. A separation of the operations is technologically not recommended, instead, the manufacturer have to act with the shortest possible periods between surface modification and processing. That leads to small production volumes which, in addition, have to be scheduled strictly with the customers. To increase the quality and efficiency of the industrial production of bonded glass constructions, the aim should be a temporal and spatial separation of the individual production steps. Therefore, a long-term stability of the surface modifications or a suitable surface protection must be guaranteed.

To investigate the method of the atmospheric pressure plasma (APP) with regard to the mentioned aspects, the various stainless steels were pretreated and then stored under laboratory conditions at 23 °C and 35 % relative humidity or under vacuum up to two weeks. At defined intervals, samples were taken out and their wetting behaviour was analysed.

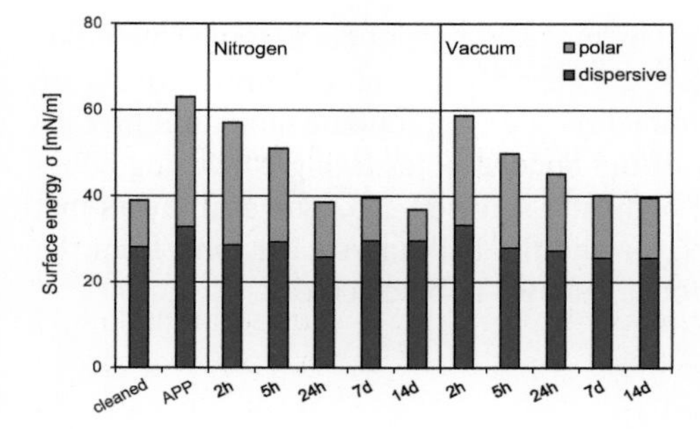

Figure 3-2 Surface energies of the adherends (material 1.4301 treated with atmospheric pressure plasma) after storage at 23 °C and 50 % relative humidity or under vacuum.

The surface energies of the stored samples decreased as a function of storage time. That occurred as well under laboratory conditions (figure 3-2, left) as under vacuum (figure 32, right). It should be pointed out that the reduction of the surface energy were similar after two hours under both storage conditions. This is due to the fact that the contact angle measurements were carried out under atmospheric conditions. Thus, the surfaces stored under vacuum could also react with the ambient air and moisture for a short period of time. Such a time-dependent effect is no longer observed at the other storage periods. It has been found that the decrease in surface energy was significantly lower for those samples which were stored under vacuum. This applies in particular for the polar fraction of the surface energy, which is especially important for the bonding process [9].

4 Development of adhesively bonded point fixings

As metallic adherend materials for punctual fixing systems, especially stainless, austenitic steels are used considering their building-authority approval. In addition to the material selection, the performance of the adhesive bonding also depends on the geometry of the bonded joint. Therefore, test results of adhesive joints with constant layer thickness were supplemented with studies of metallic adherends whose surfaces were curved (see figure 4-1) [10].

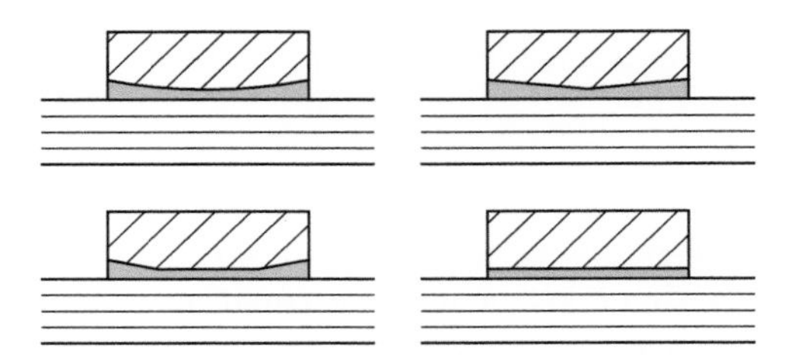

Figure 4-1 Representation of different investigated geometries of adhesive joints (exaggerated illustration of the bond line).

The durability of adhesive joints is determined by the resistance of the involved materials to various environmental influences and stresses. Especially migrating harmful media are critical factors. But the migration can be avoided through appropriate structural precautions. Figure 4-2 shows the isometric draw of the bonded point fixing consisting adhesively bonded plate, ball-bearing mounting bolts and screwed cap. The seal layers that are supposed to protect both the adhesive layer and the ball-and-socket joint from the influence of harmful media and contamination are shown in black colour.

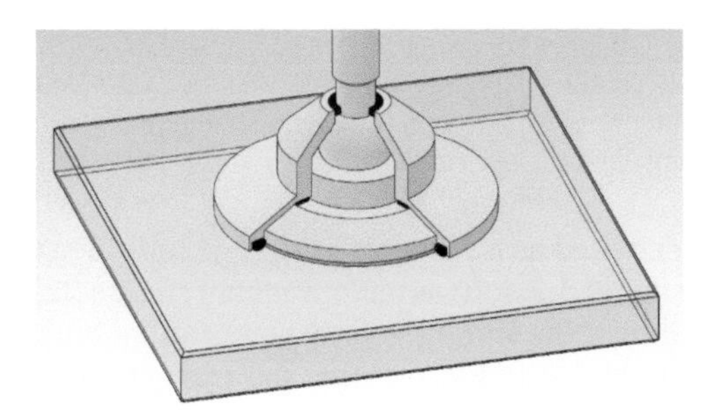

Figure 4-2 Isometry of adhesively bonded fitting with partially cut cap.

Figure 4-3 shows the profile of various types of sealing in the transition from glass to metallic holders. In principle, different circular shaped sealing profiles and sealants, which are arranged around the adhesive layer, are contemplable. In addition to rectangular and circular cross-sections also profiled cross-sections are possible. A possible combination of secondary sealing profile and primary protective sealant is also shown in figure 43 (bottom right).

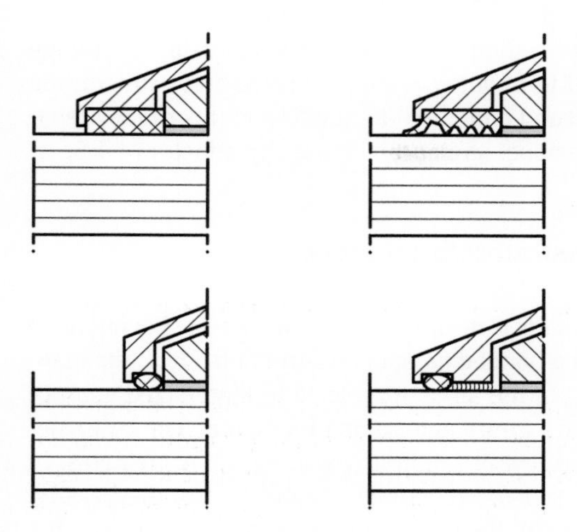

Figure 4-3 Sealing types for constructive protection of the adhesive layer against harmful media.

From a constructive point of view, circular cross-sections are the simplest variant. The availability and variety of materials are given here as these O-rings have many applications in the industrial sector. The sealing effect is produced by selective compression of the cross-section. To achieve a permanent seal of the bonded joint, the sealing materials must withstand the harmful media [11].

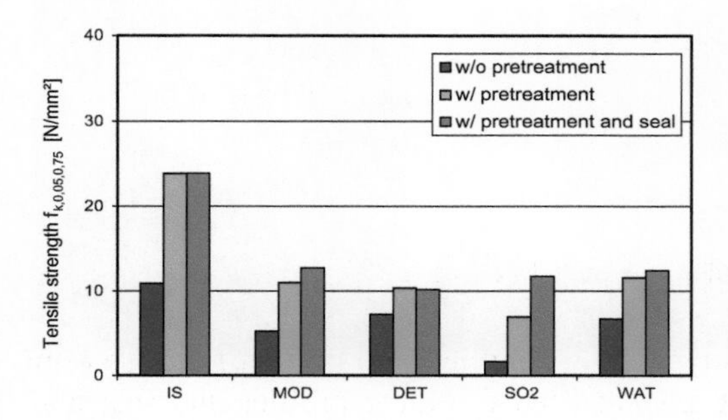

Figure 4-4 Comparison of tensile strengths of the preferred epoxy resin adhesive with and without surface treatment of adherends as well as after aging.

The diagram in figure 4-4 shows the effect of surface pretreatment on the initial strength of the adhesive bond in the uniaxial tensile test [12,13]. By the choice of optimal methods and parameters for the surface pretreatment an increase in strength by a factor of two can be achieved. At the same time, the influence of the described structural protective precautions for the residual tensile strengths was tested after artificial ageing (MOD - moderate detergent storage, DET – detergent storage, SO2 - sulphur dioxide storage, WAT –

water immersion test). The influence of the sealing can be seen especially during storage in sulphur dioxide atmosphere. Compared to bonding with no surface pretreatment, the residual strength increases by a factor of 6 and compared to bonding with surface preparation but without additional sealing the residual strength increases by a factor of 1.5.

5 Component pattern and demonstrators/prototypes

The implementation and the transfer of obtained results of small-scale test specimens to component pattern were carried out first at a bonded parapet construction. Here, the manual manufacturing process was adapted from the laboratory scale to large-sized components. This process is the starting point for a future automatic production. The construction of the load frame was made from the aspect of a simple assembly of the elements. In addition, the necessary measurement and sensor technology should be installed easily. The figure 5-1 shows the experimental setup for testing a point-fixed parapet construction under the stress of a rapid impact (pendulum test). The detail image shows the location of the measurement technique for strain measure.

Figure 5-1 Experimental setup for testing a point-fixed parapet construction by pendulum impact test (left) and position of the measurement technology (right).

By transferring the bonding process in a semi-automated process, additional test with component pattern on façade cutouts (figure 5-2) and model façades (figure 5-3) could be carried out. Figure 5-2 shows the experimental test setup for exposing a façade section with a uniform distributed load, which simulated wind pressure and wind suction. Also the construction of these test frames focused on a simple assembly of the test façade elements and the integration of measurement and sensor technology.

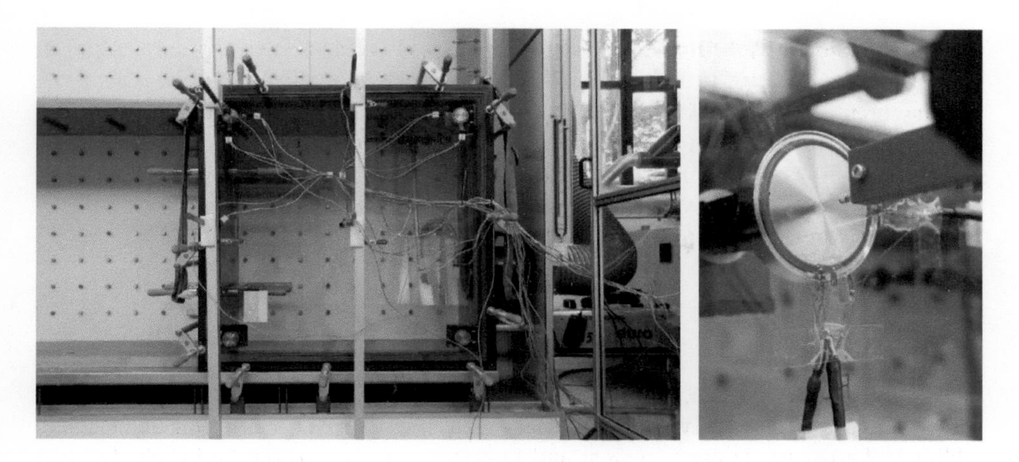

Figure 5-2 Experimental setup for testing a facade cutout by exposure to wind pressure and wind suction (left) and detailed illustration of used measurement technology (right).

The model façade with adhesively bonded punctual mounting systems (figure 5-3) consists of three by four panes of glass with dimensions of 1.46 m x 1.15 m, which are supported by four circular point fixings with the diameter of 40 mm.

Figure 5-3 Pattern a rigid facade structure with adhesively bonded punctual mounting systems.

6 Acknowledgements

The research project "GLASSKONNEX Transfer" was funded by the Federal Ministry of Education and Research (BMBF). The authors thank glass factor Ingenieure GmbH, Dresden, GWT-TUD GmbH, Dresden, MBM Metallbau Dresden GmbH, Dresden and Thiele Glas Werk GmbH, Wermsdorf, which supported as commercial and industrial project partners the research project with know-how, equipment and personnel.

7 References

[1] Weller, B.; Wünsch, J.: Transparente Klebstoffe für Glas-Metall-Verbindungen. In: Glasbau 2013, Weller, B., Tasche, S. (Hrsg.), Berlin: Ernst und Sohn, 2013, 169-183.

[2] Weller, B.; Kothe, C.; Wünsch, J.: Determination of Curing for Transparent Epoxy Resin Adhesives. In: Proceedings of Glass Performance Days 2011. Tampere, 2011, 671-676.

[3] Kothe, M.; Hempfling, S.; Weller, B.: Epoxy Resin Adhesives – Influence of Additives and Fillers to the Mechanical Properties and the Ageing Stability. In: Tagungsband Glass Performance Days 2015. Tampere. 2015. Page 301-304.

[4] Kothe, M.; Kothe, C.; Weller, B.: Epoxy Resin Adhesives for Point Fixings – A New Approach. In: Tagungsband; in-adhesives; München 2015. Seite 49-55.

[5] Kothe, M.; Kothe, C.; Weller, B.: Epoxy resin adhesives for structural purposes – a new approach. In: Tagungsband; engineered transparency; Düsseldorf 2014. Seite 383-392.

[6] Kothe, C.: Oberflächenvorbehandlung von Fügeteilen zur Optimierung adhäsiver Verbindungen im Konstruktiven Glasbau. Dissertation, Technische Universität Dresden, 2013.

[7] Weller, B.; Kothe, C.: Oberflächenvorbehandlung von Fügeteilen zur Optimierung von Klebeverbindungen. In: Glasbau 2012; Herausgegeben von B. Weller und S. Tasche; Berlin: Ernst und Sohn, 2012. Page 195-205.

[8] Kothe C.; Tasche S.; Weller B.: Optimierung adhäsiver Verbindungen im Konstruktiven Glasbau durch Atmosphärendruckplasma. In: Tagungsbeitrag; 11. Workshop Arbeitskreis Atmosphärendruckplasma; Erfurt 2012.

[9] Kothe, C.; Weller, B.; Wünsch, J.: The stability of surface pretreatments on different stainless steels. In: Tagungsband; in-adhesives; München 2015. Page 31-36.

[10] Rudolph, L.: Geklebte Punkthalter im Konstruktiven Glasbau. Diplomarbeit, Technische Universität Dresden, 2016.

[11] Kothe, C.; Rudolph, L.; Weller, B.; Wünsch, J.: Investigations on the aging resistance of sealing materials for the protection of bonded point fixings. In: Tagungsband Glass Performance Days 2015. Tampere. 2015. Page 278-280.

[12] Kothe, C.; Weller, B.: Effects of surface pretreatments on adhesive properties of glass and metals. In: Tagungsband Challenging Glass 4 & COST Action TU0905 Final Conference. Lausanne 2014. Page 353-359.

[13] Weller, B.; Kothe, C.: Investigation of surface modification methods to improve adhesive joints in glass construction. In: Tagungsband; Glass Performance Days 2011; Tampere 2011. Page 677-680.

Architecture | Shape | Structure

Teresa Daniela Lovascio[1]

1 Architectural Engineering at Technical University of Bari, Collaboration with V&A, Bari, Italy, d.lovascio@gmail.com

The fragile behaviour of the glass and its random resistance characteristics have led to the introduction of general planning principles, which are based on the following concepts: hierarchy of resistances, redundancy, robustness, fail safe approach. This kind of planning is defined as "rottura protetta" (protected breakage), during the planning, the possibility of a breakage of glass load-bearing components is accepted, as long as this data does not com-promise the safety of the entire structure towards the users. The application of hierarchical concepts (confering to each element a load bearing capacity), robustness (discretizing structural elements), and redundancy (using laminated sheets), allows to obtain that kind of missing ductility from a structural point of view. When it comes to plan glass structures, it is very important to check that the structure is able to redistribute the loads while considering possible alternative paths for the stress, accepting the spontaneous and/or accidental breakage of certain elements or parts of them. this study offers a contribution towards deeper focus and analysis for future studies for the use of qualitative methods, which aun to manage, in an optimal way, the flows of the internal loads of the structure, wether in case of planning or in case of cracks analysis. These models would permit to anticipate future behaviours of the structure under certain conditions , and according to these data, they would optimize the resources of the material through the choice of Shapes of Architecture.

Keywords: load path method, hierarchy of resistances, redundancy, robustness, fail safe approach, breakage, behaviour, critical transfer stations, width, shape, localization, development, over-view

1 The load path method

1.1 The choice of the model

In different cases, the choice of the model is neither easy nor immediate because of the contrast between the empirical model (based on experience) and the rational model (based on numerical analysis). The use of the Strut and Tie Model (STM) is proposed as instrument of analysis, while the Load Path Method (LPM) is proposed as a method to retrace geometrical profiles of the STM on the physical reality of the structural organism to be projected. This choice comes from the certainty that a model able to prefigure the general and detailed behaviour of a structural system is not a utopia, so that correlations between shape and structure are immediately evident. That is to say a model that discloses the results of the calculation by just revealing how to recognize – in the architectural shapes and pathologies – the frame of the structure, through an anamnesis able to lead till to current configuration.

Engineered Transparency 2016. Glass in Architecture and Structural Engineering. First Edition.
Edited by Jens Schneider, Bernhard Weller.

During the structural system design, the choice of the model is very important, because if it were underrated, it could generate misleading results. Therefore, it is necessary to choose a model able to link the structural shape to the behavioural characteristics in order to obtain an efficient structural system.

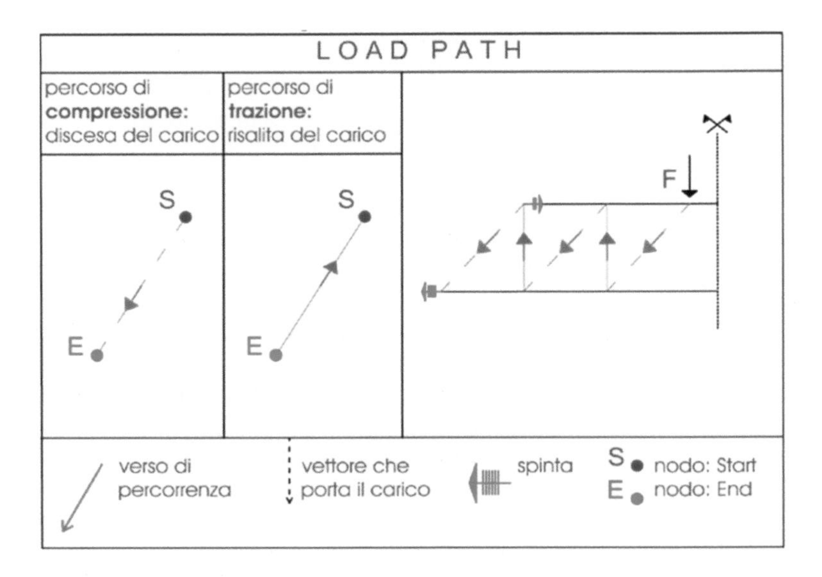

Figure 1-1 Layout LPM.

Another fundamental instrument is the ability to summarize the model linked to the structure, so that resistance tests can be carried out as immediate as possible in order to avoid long and complex processes in which one could lose the mastery of the project. Nowadays, the choices of different materials, the time and economy variables strongly influence the design evaluations. The behaviour of the glass structures strictly depends on the quality of the project, on the realization of the detailing, on the choice of the treatments and technological systems of connection and restrictions. The efficiency and versatility of an analysis structural model should be evaluated also according to its aptitude to simulate local behaviours (hubs). Actually, it is not always possible to achieve satisfactory results only through the use of theoretical models, to which it is necessary link numerical analysis.

1.2 The model

The STM model was created just to represent the complex static work of a reinforced concrete that is subjected to bending and cut. It was introduced by Ritter, 1899, who was inspired by the idea of Hennebique, idea developed in the following years by Morsch who extended the topic to torsional stresses too. Nowadays, the STM model is recognized by one of the most accredited international regulations. The Load Path Method - used as an investigation tool of the structural behaviour during the phase prior to the STM design - was introduced by Schlaich and Schafer during a Workshop carried out in Taiwan, 1996, and later mainly developed by A. Vitone, C. Vitone and F. Palmisano. Eurocode 2 latest issue recognizes the LPM as a tool for the development of suitable Strut and Tie Models for the plastic analysis of concrete structure. The goal of this study is to spread this model to the glass structure functioning, in order to understand the paths covered by loads, which generate crackings if limit resistance and duality 'shape-structure" are exceeded. These are important details for the outcome of the architectural design.

1.3 Foundation of the method

The load path represents a line on which a strength, or an element of a strength, moves through the structure from the application point (starting station) to the arrival point. At each deviation hub, the itinerant load F, must give a push H to the remaining part of the structure and receive an equal and opposite action, so that the balance is respected. The element of the strength F, carried by vector N, constantly remains along its own path within the structure. Therefore, the push H to the structure must be orthogonal to the itinerant load F. In a natural reference system, given by vertical loads and horizontal pushes, the structure is covered by descending flows of compression (dotted line) and returning traction flows (continuous lines). The internal hyperstatic structure entails a certain number of balanced paths of the structure. Among the countless possibilities in which the loads can exploit the available paths in an hyperstatic structure, the choice is influenced by the energetic evaluation, according to which the load covers the cheapest path in terms of waste of energy. A minimal settling of the structure corresponds to this configuration, that is to say the minimal amount of energy of load position that should be converted in strain energy.

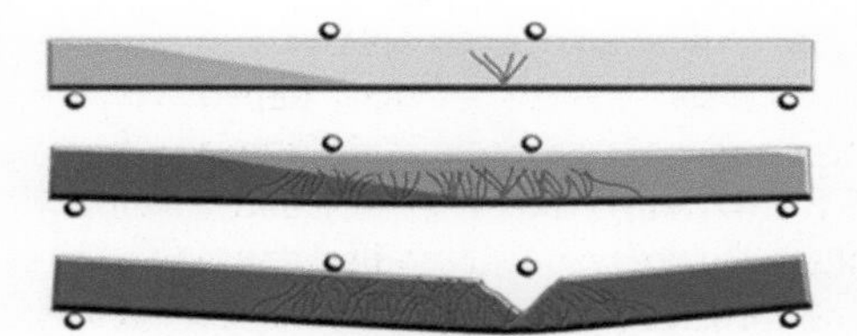

Figure 1-2 LPM Beam.

2 Fracture Mechanisms

The breaking behaviour of fragile materials, such as glass, is substantially different from the ductile material one and cannot be studied using the traditional methods of the plasticity theory. The fragile materials are sensitive to the presence of flaws and the resistance varies significantly in relation to the sample sizes. The glass breakage is due to the concentration of stresses at the top of the damage on the surface - caused by scratches, cracks or internal defects. Because of the low value of the fracture toughness and the lack of plastic behaviours, small and shallow damages, – developed during a normal handling – are enough to cause concentrations of critical stresses. Critical Transfer Stations: areas of propagation of the high tensional-density crack. The path of the load meets the crack and – in an attempt to get around the obstacle – thickens at the pinnacle of the crack, where tension peaks will occur due to the choice of compatible ways with the minimum energy investment principle. The fracture mechanics studies the conditions in which the fracture – occurred at the top of the damage – propagates causing a complete separation of the solid.

2.1 Crack morphological analysis

The visual observation of various aspects of crack morphology depends on different factors, such as:

- Width
- Shape
- Localization
- Development
- Overview
- Cracking width

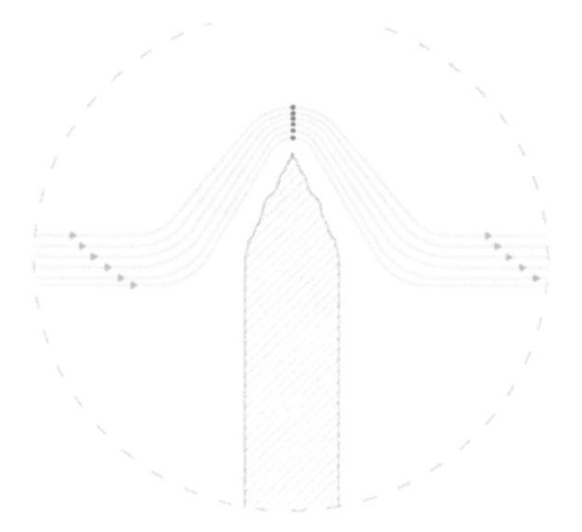

Figure 2-1 Critical Transfer Stations

2.2 Cracking width

The cracking width is linked to the potential position energy that must become energy necessary to the vector to bear the load and get around the cracking. It is possible to notice the variation in the flow intensity in the regions of continuity and discontinuity. The extremity regions are also called discontinuity regions as the retrocession path has got a much more limited development compared to the continuity regions, where there are favourable paths as they are much farer. The presence of a cracking within an ideally homogenous and isotropic body with linear elastic behaviour, entails the interruption of the load paths in correspondence with discontinuity regions. Therefore, the vector bearing the load, finding an interruption along its path, has to get around the obstacle, considering alternative paths that cause a lesser waste of energy. In the image, it is represented the simulation of the transition from non-cracked model to the cracked one of a general element subjected to traction, exploiting the application of the Effect Superposition Principle. It follows that the width is given by the reduction that the structural element experiences.

- Very wide cracks: with non-high-resistance materials;
- Very thin cracks: with high-resistance materials – although representing a difficult phenomenon;
- Short paths: the load flows usually thickens at the tip;
- Long paths: the flow change its inclination before the crack.

The load flow, in both cases, thickens at the tip, creating overvoltage that generate tears. Regarding the time interval, in which the cracking becomes bigger, it is possible to decide if the phenomenon is increasing (easy breakage) or not.

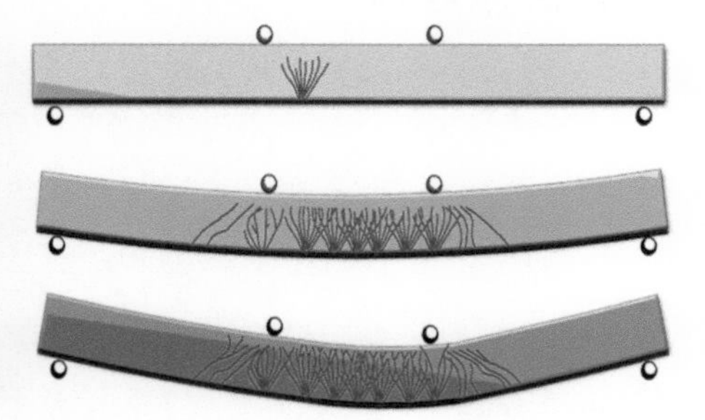

Figure 2-2 Fracture behaviour.

2.3 Shape of the cracking

The study of the cracking shape is very complex because it depends on the material characteristics and on the possible internal or surface defects. In this paper we will try to study the morphology of the cracking by connecting the sample to an ideal material, using very small elements of a solid not too heavy.

Given a flat solid, it is possible to identify, for each point, minimal and maximal tensions, connecting points with the same value we get an isostatic curve. If these curves refer to the principal maximal tension, they create a family of curves called isostatic traction curves. Whereas if the isostatic curves refer the principal minimal tension, they create a family of curves called isostatic compression curves. The breakage occurs when the value of the principal maximal tension, in the point of the solid, exceeds the breakage load of the material. In that point a crack is created with a normal behaviour. If we considered a glass beam fitted in the extremity and subjected to the action of a load homogenously distributed, it would be necessary to compare the values of normal and tangential tensions: Substituting the corresponding values, we get the size relationship in which both the portion of the tangential tension or the portion of normal tension predominate. If we considered the case in which the normal tension predominates, we would obtain $\sigma_{max} < \tau_{max}$ the maximal value of this would occur in the joint. Here the corresponding cracking frame: the breakage will start from the tensest part, that is to say in correspondence to the internal fibres. The cracking starts with a 90° performance because of the predominance of normal tensions, then it starts to bend (45°) because of the presence of tangential tension, and he continues towards the part compressed cutting it at 90°. Once the breakage has been caused, the joint sections are less and less able to resist when bending, therefore the loads change their path, aiming towards the centre of the beams where he resisting moment is at the maximum point. It causes a new breakage in correspondence to the tense part of transverse section in order to propagate towards the compressed part with vertical performance, considering the absence of tangential tensions. If the tangential tensions prevail over the normal ones, the breakage will start from the middle step, being inclined at 45° because of the tangential tensions and then start to straighten out in correspondence to the external fibres. The same would occur in case of a shelf structure.

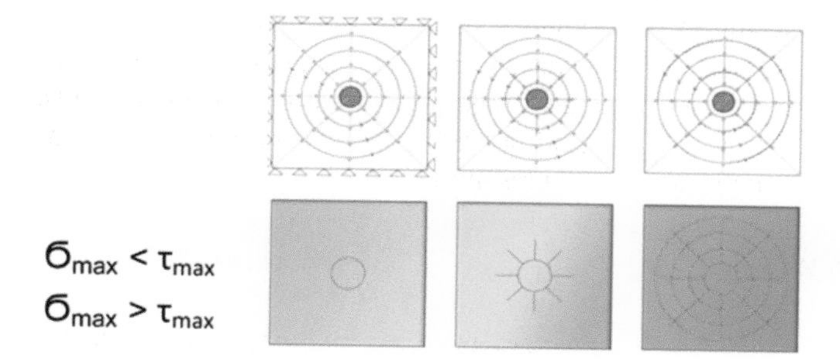

Figure 2-3 Shape of cracking.

2.4 Cracking localization

The first step in the investigation of the glass element breakage is the observation of the cracking shape and localization. During the design and building management phase, it is necessary to assess the importance of the safety for the inhabitants, avoiding the separation of the fragments and guaranteeing bevelled borders. From the sample breakage it is possible to understand the qualitative performance of the cracking and its breakage cause. Crack order: it is identified in the presence of big defects, regions with high-tension concentration that show the possibility of damage. Breakage model: it offers information about the breakage causes and the tensions that generate it. The number of crackings and the fragments extension are referred to the kind of glass used, to surface tensions during the breakage and the energy transferred by the actions that cause the breakage. The cracking localization on the glass surface confirms the information about the origin of the crack caused by the impact of a weight. One of the most representing examples about the importance of cracking localization concerns the connecting regions. The different kinds of restriction of a sheet dramatically influence the possible breakage configurations.

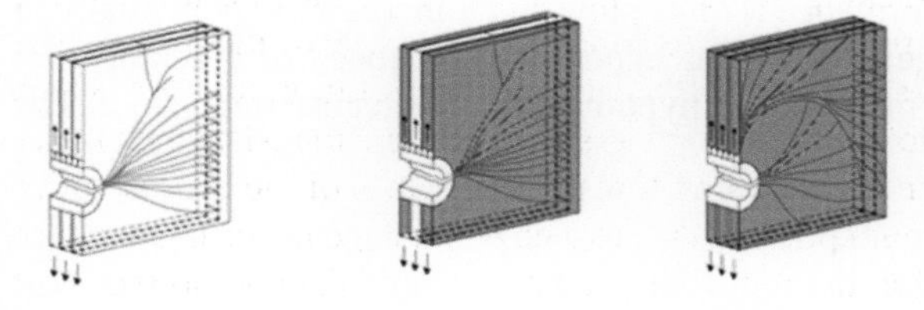

Figure 2-4 Cracking Localization.

Let us consider the case of a laminated glass sheet with point joints. In this case i twill be necessary to pierce the sheet in order to arrange the joints, connected by a polymer or resin sheath. The resin layer adhering to the wall of the pierce and bolt will be compressed and the peak of tensions will occur at the bottom of the pierce. In the third image, we could notice the first resin crackings, subjected only to compression stresses. The cracking continues touching the central glass panel and, the, the internal and external panels. The PVB traction resistance evaluate post cracking abilities. Focusing on the study of cracking localization, here are some frames of representing cases.

2.5 Cracking evolution

The danger of a cracking study depends on its development during the time, that is to say its extension, its changing shape, its propagation and possible paths intersections. It is possible to give a frame of the evolution with the following performances:

Straight line: it accepts different slopes and indicates that the cracking evolution proceeds with a speed directly proportional to the passing of the time – average danger. The linear performance allows possible suppositions about the development.

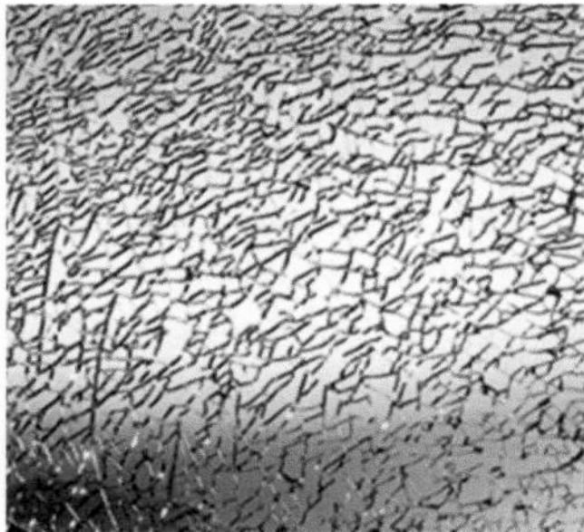

Figure 2-5 Cracking Evolution.

Branch of parabola: the initial performance is quite slow and it increases exponentially in a short space of time – high danger. Branch of hyperbola: the speed of crackings decreases with the passing of time. There are two hypothesis: 1. The cause runs out; 2. The structure finds a new equilibrium.

2.6 Overview

In conclusion, given the complex cracking models of a sheet of glass, it is useful to follow the described steps during the analysis phase. These steps simplify the problem as they set out the right guidelines, and, at the same time, they combine information about width, shape, localization and development, therefore a general cracking study is obtained. By using the load path method to evaluate the mechanism of cracked-sheet layers subjected to punching and beams under bending moments for 4 points, it was possible to define the structural behaviour, identifying the load paths before the cracking phase, during the cracking and after it, with the creation of new internal paths. The danger of a cracking study depends on its development during the time, that is to say its extension, its changing shape, its propagation and possible paths intersections.

3 Load and push paths draw the structure

The drawing of the load path along its polygons is the outline of a certain trajectory or a given geometry.

In architecture the structural shape is not just the resultant of the load path, but it is the resultant of the system consisting of the combination between the load path and its pushes. In this decomposition of load paths and push paths, we could learn and understand the interpretation of structures in architecture. We could identify two element of reference: the node and the straight section. In the first one there are the pushes that affect the structure with other paths; while, the secondo ne defines the layout. It is important to distinguish theoretically the load and push paths. Descending and re-turning load paths give another important distinction, regarding the gravitational aspect. This is fundamental for the right interpretation of the drawing in structural architecture. Often, the push path is crucial for the complete definition of architectural and structural shape and for its building technique to be adopted.

3.1 The load path without deviations

A column represents an easy example of structure. The vertical load F, applied to the superior extremity, must reach the equal and opposite load – applied to the inferior extremity – in order to gain equilibrium. Let us consider the linear layout, shown in the image, where node S (Start) and node E (End) are defined. The vertical section S-E, along which the load F is taken by vector N, defines the path.

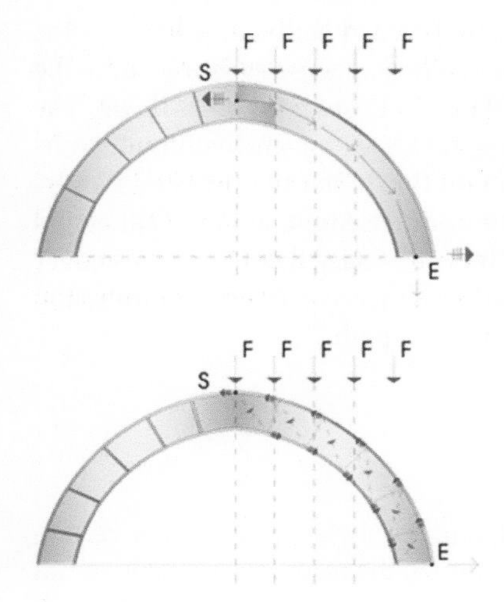

Figure 3-1 LPM.

3.2 The load path with deviations

In architecture the load paths must be conceived, in designing phase, with deviations. An example of a path with deviation is given by truss. Before tracing load and push paths, it is necessary to identify a load balanced system applied to the truss. The vertical load F is the one transferred by the roof and its own weight, which is divided into two parts: there is one starting station S and there are two end stations E. These two loads, after deviating to node S, separate and start to descend along two different slopes. The load path F suffers another deviation, becoming again vertical in order to go into the imposts.

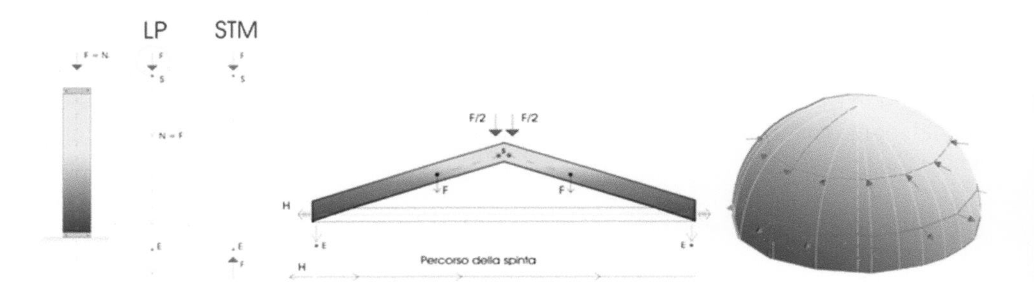

Figure 3-2 LPM.

The vector must give a push H in order to deviate and join the oblique slope path. This push H, for balance reason, must be horizontal. Moreover the asymmetry of the system guarantees the equilibrium of the push H, this push will have equal intensity in the symmetrical node. Any kind of structure can be connected to deviated path with vertical paths and horizontal pushes. The beam is a structural element whose geometry is traced by the common path of each push, along the two longitudinal parallel, and horizontal flows. The pushes are the ones given by vectors in the deviating nodes, along their oblique paths consisting of flows of compression and returning traction flows. In order to obtain a verified equilibrium, it is necessary that the pushes on the flow assemble in pairs, equal and opposite – this happens if the wall poles are parallel. It is necessary that the pushes cover the flows in the same direction, so that one push – by advancing – can trace a compression path and the other one – by descending – can trace a traction path.

4 Submission of contributions

These models would permit to anticipate future behaviours of the structure under certain conditions , and according to these data, they would optimize the resources of the material through the choice of Shapes of Architecture.

5 References

[1] Mezzina, M., Raffaele, D., Vitone, A.: Teoria e pratica delle costruzioni in cement armato, CittàStudio, Torino, 2007.

[2] Haldimann, M., Luible, A., Overend, M.: Structural Use of Glass, International Association for Bridge and Structural Engineering – ETH Zürich, Zürich, 2008.

[3] Carpinteri, A.: Meccanismi dei materiali e della frattura, Structural Engineering – Pitagora, Bologna, 1992.

[4] Tattoni, S., Cossu, G.P., Fenu, L.: L'impiego strutturale del vetro, coursework material, Università degli Studi di Cagliari, 2011.

[5] Cagnacci, E., Orlando, M., Spinelli, P.: Il vetro come materiale strutturale, Edizioni Polistampa, Firenze, 2010.

[6] Schittich, C., Stab, G., Balkow, D., Schuler, M., Sobek, W.: Atlante del vetro, UTET, 1999.

[7] Anderson, J.C., Leaver, K.D.K., Alexander, J.M., Rawlings, R.D.: Scienza dei materiali, Sansoni, Firenze, 1980.

Non-invasive measurement of gas fill for insulating glass

Miikkael Niemi[1], Mauri Saksala[2], Fana-Maria Immonen[3]

1 Sparklike Oy, Hermannin rantatie 12 A 21, 00580 Helsinki, Finland, miikkael.niemi@sparklike.com

2 Sparklike Oy, Hermannin rantatie 12 A 21, 00580 Helsinki, Finland, mauri.saksala@sparklike.com

3 Sparklike Oy, Hermannin rantatie 12 A 21, 00580 Helsinki, Finland, fana.immonen@sparklike.com

A common insulating enhancement in modern insulating glass (IG) unit is filling it with argon or krypton gas. The insulating gas is applied to the cavity either during the manufacturing process at gas press station, or manually to the ready-made units. The challenges are confirming the correct filling degree and ensuring that the initial gas concentration will remain inside the insulating glass unit (IGU). A gas escape could occur from improper sealing of the IGU, and this needs to be tested prior to shipping the IG to customers. Product liability for the insulating glass and window suppliers can last several years after the initial delivery of the product. Thus, they are looking ways to increase the security of the gas fill. Sparklike is the developer of the world´s first non-invasive gas analyser for insulating glass. The company´s Gasglass product line has become the de facto world standard for gas fill measurement of IGUs. The products are sold worldwide and are in daily use by the world´s leading insulated glass manufacturers, testing laboratories and window processors. Furthermore, to meet tightening industry standards and keep pace with the fast developments of state-of-the-art glass manufacturing, Sparklike has also developed a new laser based technology for high-performance gas fill measuring. Alongside the new technology, this article discusses the advantages of gas fill for thermal performance and the latest developments in gas fill measurement.

Keywords: gas fill measurement, non-invasive argon measurement, quality measurement on site, laser technology, argon analyser, argon detection

1 Introduction – high performance glazing

1.1 Certified energy-efficiency

Today´s architectural design favors the use of glass both as a day-lighting and a structural element in modern construction. Different types of windows increase the attractiveness of commercial and residential buildings, improve day-lighting and add to the comfort of life and work in offices and homes. At the same time glass windows are one of the most sensitive building components when it comes to assuring energy-efficiency in construction. Energy flows in and out of buildings – depending on their location and prevailing weather conditions – has been a traditional focus of attention for building owners and their suppliers particularly in view of increasing energy costs (VanBronkhorst, Persily & Emmerich 1995).

Engineered Transparency 2016. Glass in Architecture and Structural Engineering. First Edition.
Edited by Jens Schneider, Bernhard Weller.

The window industry has responded to the energy-efficiency requirement by developing different types of insulated glass structures. Multiple layer glazing, double or triple pane solutions and insulated glass units (IGUs) utilize advanced insulation materials ranging from glass types, frames, sealants and cavity fills to maximize the thermal performance and minimize the U-factor – energy penetration through the glass. Double and triple glazing, gas fill and specific coatings have contributed immensely to the energy-efficiency of windows and become the best solution and state-of-the-art choice for high performance IGUs.

1.2 Views on gas fill

Glass manufacturers are looking to maximize its resistance to heat conductivity, the so called R-factor, and thus minimize the outflow of energy through the window. To date this is best done by noble gas between the glass layers of an IGU. A vacuum solution between the glass panes is seldom recommended in areas of great temperature differences (variations of 35 degrees or more). Air fill is an option but the use of noble gases provides superior resistance to heat conductivity in comparison with air. While air has a thermal conductivity of 0.024, argon with 0.016 is only 67 % of that, and krypton at 0.0088, is just half the conductivity of argon (Thermal conductivities).

The favoured gas is, however, argon because it is clearly the most effective noble gas to use. Argon is an economic choice since krypton is significantly more expensive and xenon even more so. Compared to air fill it has been estimated that the raw material cost of well-sealed argon-filled IGUs goes up by only 1 %. But regardless of the choice of fill, the gas content of an IGU is generally considered to be adequate when exceeding 90 %. Different markets have different standards for gas content, but the European standard is EN 1279 (EN 1279-3: 2002) and the US standard ASTM (Standard test method for determining argon concentration in sealed insulating glass units using gas chromatography). Most of the time the gas content needs to exceed 90 % with a margin, but some regions have approved lower standards that match the quality performance of their local IGU-manufacturing closer. The challenge for an IGU supplier is to be able to verify this figure by reliable measurement as solid evidence of consistent product quality. (EN 1279-3: 2002; Standard test method for determining argon concentration in sealed insulating glass units using gas chromatography.)

1.3 Insulating glass performance – reason for quality control

The performance of coated, multi-layer and gas-filled insulated glass elements (IGU) increases strongly with the simultaneous application of all these elements. The coating provides protection against excess heat inflow from the outside from solar radiation while the IGUs provide protection against energy-loss from heat convection from the inside through the windows. The efficiency improvement in the U-factor is some 16 % with the use of argon gas and as much as 27 % in the case of krypton as documented (PPG Glass Education Center). It is thus easy to see why the IGUs, and more specifically gas fill, have a strong position in promoting the energy-economy of buildings. It should also be underlined in this context that reports by the EU commission show that nearly one third of all energy consumption of a modern urban society is associated with buildings, which puts the significance of energy-savings into perspective.

For IGU producers, the key high quality and energy-efficient solutions lie in proper processing of advanced glass products, such as tempered, laminated and coated glass and skillful sealing of the gas-filled IGUs through high manufacturing quality. Even if everything is done adhering to the best practices, the escape of gas fill from an IGU of some 1 % per year is considered normal and acceptable due to the pressure difference between outside conditions and the gas-filled cavity of the IGU. The requirement for manufacturing quality naturally includes assuring that the gas-fill is successful and adequate. If the processing quality is not on a high-enough level, the gas escape could be much more than 1 % per year. In the end that may lead to quality problems in the glass, and at worst, this could cause the collapsing of the entire glass structure. (M. Niemi, personal communication, May 7, 2015.)

1.4 Adequate measurement

The difficulty for IGU manufacturers has been ensuring that the performance quality of the product is adequate and the inert gas-fill up to standards. Traditionally insulation gas fill was tested by taking random samples off the production line and drilling holes in the glass to measure gas fill and thus the insulation performance of the IG-unit (Argon gas fill and insulated glass units). This was a functioning, but costly method that involved breaking the glass and destroying the expensive product either on the processing line as part of manufacturing quality control or doing the same to an already installed IGU on the field. In other words, there has been no method for measuring the gas fill without tampering with the product. Certainly no method has been available to do this on the production line as part of manufacturing on-line quality control. Thus, leaving out the possibility to gain gas content information from every piece produced. Finally, this impairs the possibility to efficiently control the quality of the production line. (M. Saksala, personal communication, May 7, 2015.)

2 Sparklike – the enabler of fast quality sampling

2.1 Applied research

For a long time IGU manufacturers had to be content with the fact that measuring the extent of gas-fill in the IGU was “difficult” and costly. Therefore, producers and quality control specialists were eagerly looking for a solution to the challenge of increasing the quantity of tested samples and to increase the reliability of the results without breaking the IGUs. A unique method was developed through cooperation between university researchers, industrialists and entrepreneurs to address the issue. These efforts resulted in the commercialization of a method for non-invasive measurement of gas fill in insulated glass. The owner of this technology, with several patents and some pending for the new technology, is the Finnish company Sparklike Oy, established in 2000. Sparklike is the only supplier on the global market today that can deliver the technology and instruments to perform this task.

2.2 Innovative entrepreneurs

The partnering entrepreneurs behind Sparklike, Niklas Törnkvist and Mats Therman, started to pursue their development idea in the year 2000. Earlier contacts with the Chemistry Department of the University of Helsinki had inspired the idea of looking into the possibilities of applying spectrometry and plasma emission spectroscopes as measuring and analysing tools to the measurement of gas fill in insulated glass windows. The first product, a semi-portable analyser, was developed in cooperation between university researchers and the entrepreneurs as early as 2001. It solved the "difficult" testing problem.

Sparklike Oy was set up to commercialize the idea – take the product to market, introduce it to processors of insulated glass and manufacture it. The same technology was applied in 2006, when Gasglass Handheld was introduced to the market. The new product was portable, and the team had been able to in-crease the accuracy even further.

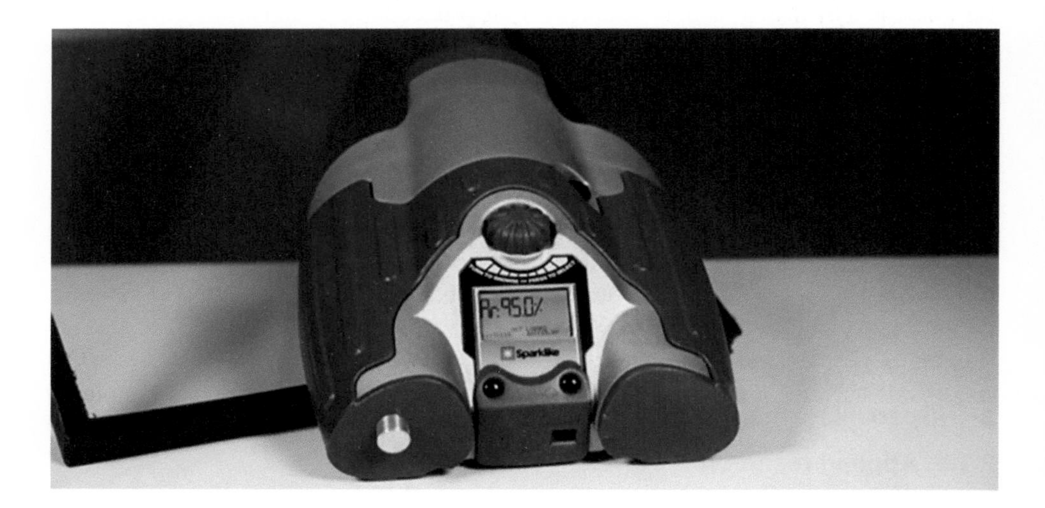

Figure 2-1 Gasglass Handheld.

The response from IGU-manufacturers, testing laboratories and window processors – the natural Sparklike target categories – was encouraging. The technology was solid, application fast and accurate. As a tool for quality control, the new technology stood out as both unique and the only one of its kind al-lowing fast and accurate non-invasive sampling of the IGU as targeted. (M. Niemi, personal communication, May 7, 2015.)

2.3 Spark emission

The function of the Gasglass Handheld device is based on igniting a high-voltage spark emission into the IGU. This creates a colour spectrum that pro-vides information on the oxygen and gas content within the IGU and displays the result in the form of a colour spectrum and a numerical reading, a percent-age of gas content. This reading is compared with the prevailing standard set for a given market, usually over 90 % gas content. This method is fast and ac-curate and is nowadays used widely on double glazed units as measuring can be done from monolithic side of the IG unit. The non-invasive method can be repeated as many times as the quality system or the end-user requires. The limitation of the spark technology is that it requires at least one monolithic side on the IG unit.

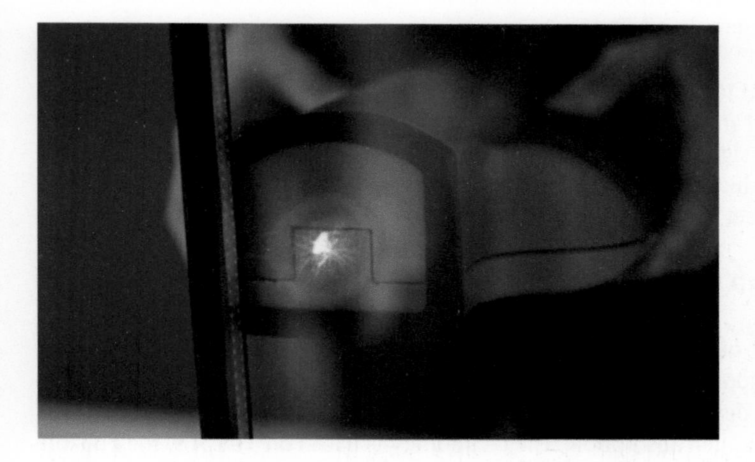

Figure 2-2 Non-invasive device measuring ionized gas emission spectrum.

With more than 1,500 devices delivered to over 40 countries, Sparklike Oy had made a measurable and credible breakthrough with its novel niche technology. What used to be a difficult "measurement problem" was now solved by applying the Gasglass Handheld instrument.

3 Gasglass laser – taking gas measurement further

3.1 Laser reflection

While the spectrum analysis technology applied in earlier Gasglass instruments solved the issue of measuring the gas content in a unique non-invasive way, it had some limitations once the industry continued to develop. New types of high-performance glass products were introduced and new coating solutions brought substantial savings in energy inflow from solar radiation.

The original high-voltage spark technology has hold out well on the market but the increasing use of products with multiple Low-E coatings or thick spacer cavities, such as many triple-glazed IGUs, raised the need for new, additional measuring capabilities. The challenge was to measure through coatings and lamination, as well as triple glazing with single measurement. Additionally, what was needed was the possibility to measure a full range of argon from 0 % to 100 %, as well as including the product as part of the manufacturing process, preferably on-line.

The glass industry and IGU manufacturers in particular had high hopes for new developments in IGU gas-fill measurement and the Sparklike product development team paid close attention to the signals from the market. With the new advanced product developments in view, Sparklike placed specified demands on the criteria for the new product that was needed to meet the measuring challenge.

3.2 Breakthrough technology

In 2011 Sparklike set out for a new technology breakthrough in initiating development work based on laser application. That resulted in the new product – Gasglass Laser. The company was able to develop a laser system to meet the market demand that focuses on the measurement of oxygen absorption. The technology is based on TDLAS, for Tunable Diode Laser Absorption Spectroscopy, in the 760 nm range. The method is indirect in measuring the oxygen absorption in the IGU and deducing that the rest of the contents consist of gas – argon, krypton, xenon – whichever is applicable.

The measuring technology is based on oxygen absorption. A laser beam from a laser diode is focused into the IG and the reflections from the IG surfaces are reflected back and detected. The device is operated from one side of the glass. The measuring time is 15 – 35 seconds, depending of the IGU structure. The maximum IG thickness for this application is 50 mm and the minimum pane thinness is 2 mm.

The new laser based technology is capable to measure both non-invasively, and through coatings and laminations. Furthermore, the new technology allows testing of complicated structures, such as energy efficient triple glazed units. This brings the level of quality assurance of these high performing glazing units at pair with advanced product expectations.

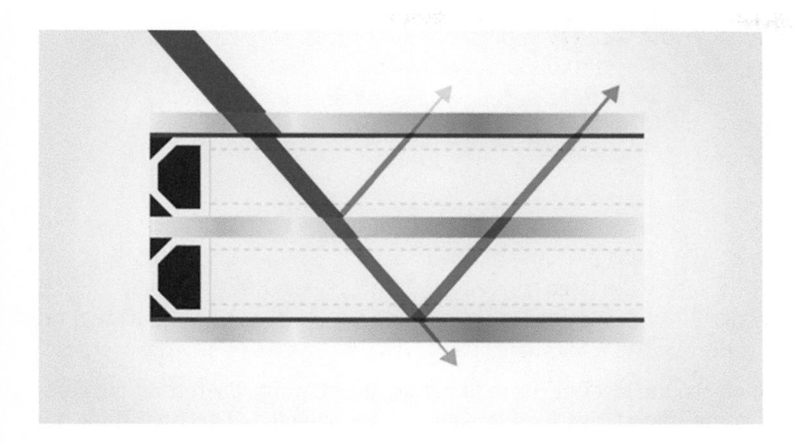

Figure 3-1 Laser measures coated triple glazed IGU.

3.3 Oxygen absorbance

Oxygen, like other gases, has certain absorption lines (wavelengths). Gasglass Laser utilizes a tunable and narrow-line width (monochromatic) diode laser. Adjusting the diode laser operation current laser frequency can be turned over an absorption line of interest and measure the variations in transmitted intensity. Techniques for laser modulation are extensively used to improve the performance of the TDLAS-system. A sinusoidal component is added to the diode laser operation current resulting in a sinusoidal wavelength and amplitude modulation of the laser output. Interaction with a wavelength dependent and non-linear transmission (absorption lineshape) results in a periodic, but non-sinusoidal transmission signal that consists of the modulation frequency itself as well as its harmonic overtones. The amplitude of the harmonic component is proportional to the absorbance, i.e. amount of oxygen. The reported reading is displayed as the noble gas content percentage. Additionally, this new method permits operator to measure of the thicknesses of glass and spacer cavities as a side product. This information is received with a high accuracy of ±50 micromillimeters.

3.4 On-line quality control

The Gasglass Laser consists of three parts: i) a main unit that includes the power source electronics and does the calculations in the background, ii) a measuring head including the more sensitive electronics and optics, and iii) a screen for user interface. The system has the capability of saving and exporting results as well as identifying data.

Figure 3-2 Gasglass Laser.

In order to make an analysis, the laser needs to penetrate the coating. Therefore, the coating transmission sets some boundaries for the signal to be detected. The final limit depends on the full structure, but as guideline transmissions above 40 % on 760 nm wavelength are measurable.

Gasglass Laser permits permanent installation on the insulated glass manufacturing line or measuring station to yield a reading for gas content as part of the manufacturing process. The system is controlled by PC through Ethernet protocol and it is capable of receiving and returning data within production system.

Several tests prove the need for analysing each produced unit. A good example is the test report conducted by ÜV Rheinland Nederland B.V. concerning the comparison of argon gas concentration in insulation glass units measured with gas chromatography and Gasglass Laser. These tests were performed in order to determine the argon gas concentration in IGUs. The reason was to compare the results from measurements done with Gasglass Laser equipment and gas chromatography.

3.5 Gas fill measurement for future

The Gasglass Laser technology was launched on the market in Spring 2015. Sparklike intends to follow up its current products Gasglass Handheld and Gasglass Laser with new product developments in line with market needs and the development of glass, coating and IGU technology. The instruments are contract-manufactured by a long-time Sparklike partner in Estonia in a cost-efficient environment. Before customer delivery each instrument is quality-controlled and calibrated in Finland by Sparklike Oy specialists. The Sparklike quality commitment "Sense of certainty" stands on solid footing. (M. Niemi, personal communication, May 7, 2015.)

In a very short period Sparklike and its originators have been able to establish themselves as the world industry leader in their specialty and make their non-invasive gas measurement technology the de facto industry standard. They have penetrated the world-market with niche technology in a typical application where market demand and technology development join hands. Sparklike service and maintenance centres in Finland and the US support all customers globally and they stand by to assist the daily users of Sparklike products, IG manufacturers, testing laboratories and window manufacturers, for both calibration and trouble-shooting purposes. Since, like any other valued equipment, Gasglass devices require maintenance to keep their accuracy and reliability at highest level. In order for the argon analyser to reach its full potential and delivering correct readings, Sparklike accredited specialists recommend the devices to be calibrated annually with Sparklike's Service Solution. (M. Niemi, personal communication, May 7, 2015.)

4 Conclusion

The new non-invasive testing methods developed by Sparklike Oy, makes it possible to measure even the most complex insulating glass structures. In other words, the latest advancements in measuring technology solve the issues attached to traditional, invasive methods, which were and still are being used to test insulating glass gas concentration. Since, the new laser based technology is capable to measure both non-invasively, and through coatings and laminations, the new technology allows testing of complicated structures, such as energy efficient triple glazed units. This brings the level of quality assurance of these high performing glazing units at pair with advanced product expectations. As the laser technology measures non-invasively the oxygen level, it can measure the percentage of any gas.

5 References

[1] Andrés, O.A.: About the freedom of free forms. In: IASS 2009. Shell and Spatial Structures, Domingo, A., Lázaro, C. (eds.), Valencia: Edititorial de la UPV, 2009, 236-237.

[2] MacKnight, W.J; Earnest, T.R.: The structure and properties of ionomers. In: Journal of Polymer Science – Macromolecular Reviews. Vol. 16, 1981, pp 41-122.

[3] VanBronkhorst, D.; Persily, A.; and Emmerich, S. (1995) Energy Impacts of Air Leakage in US Office Buildings. Proceedings of the 16th AIVC Conference – Implementing the Results of Ventilation Research. September 19 – 22, 1995: Palm Springs.

[4] Weller, B.; Härth, K.; Tasche, S.; Unnewehr, S.: DETAIL Practice Glass in Building. Principles, Applications, Examples. Basel: Birkhäuser, 2009.

[5] Thermal conductivities. National Physical Laboratory. http://www.kayelaby.npl.co.uk/general_physics/2_3/2_3_7.html, retrieved 19.7.2016.

[6] Standard test method for determining argon concentration in sealed insulating glass units using gas chromatography. ASTM International. http://www.astm.org/cgi-bin/resolver.cgi?E2269-14, retrieved 19.7.2016.

[7] EN 1279-3: 2002 Glass in building – Insulating glass units – Part 3: Long term test methods and requirements for gas leakage rate and gas concentration tolerances. CEN, 2002.

[8] PPG Glass Education Center. What is low-E glass? < http://educationcenter.ppg.com/glasstopics/how_lowe_works.aspx>, retrieved 19.7.2016.

[9] Argon gas fill and insulated glass units. RDH Technical Bulletin No. 001. < http://rdh.com/wp-content/uploads/2015/09/TB1-Argon-Gas-Fill-IGUs1.pdf>, retrieved 19.7.2016.

Glued windows and timber-glass façades – performance of a silicone joint between glass and different types of wood

Felix Nicklisch[1], Johannes Giese-Hinz[1], Bernhard Weller[1]

1 Technische Universität Dresden, Institute of Building Construction, George-Bähr-Straße 1, 01069 Dresden, Germany, felix.nicklisch@tu-dresden.de

This study assesses the aging stability of a silicone adhesive bond between timber and glass which is of great significance for novel sustainable façade and window applications. It focuses on seven different wood substrates ranging from solid soft and hard woods to wood-based products and modified solid woods. Microscopic imaging is used to highlight differences in the wood surface texture. Small-scale adhesively bonded specimens, which are composed of a wooden and a glass piece are exposed to different aging scenarios, which relate to the impacts typically encountered in structural glazing façades and glued windows such as exposure to low or high temperatures, cleaning solution as well as high or low humidity. The residual strengths and the failure patterns are evaluated according to ETAG and a specific guideline for glued windows. The results reveal a significant influence of the type of wood on the adhesive strength. Only a few materials are suitable if the surfaces are only cleaned but not pretreated. Primer application appears to be promising since an adequate strength and failure pattern is observed also on surfaces which are originally considered unsuitable.

Keywords: adhesive joint, timber, glass, aging stability, glued windows, ETAG, SEM

1 Introduction

Wooden constructions are on the rise again – encouraged by a strong public and economic trend towards sustainable and resource efficient buildings. Spurred by this growing interest, novel design principles and material assemblies in architecture and the building industry evolve. Within this scope it is possible to intensify the use of renewable materials in transparent building envelopes by substituting members of the façade substructure that are typically made from steel or aluminum by timber materials. Glued windows and timber-glass composite façades are innovative examples (Figure 1-1) of this approach.

Glued windows (Figure 1-1a) feature a circumferential bond between the glass pane and the sash profile. This enables an in-plane loading of the glass whose capacity is not used to its full potential in conventional windows as it is solely applied as an infill panel. The composite action of the glued window enables large-size windows to be designed with slender frames. [1] Additionally, the overlapping outer pane of the insulating glass unit (IGU) protects the wooden frame from direct weathering. Transparency and design possibilities of such glued windows expand the scope of application far beyond the scope of traditional windows.

Engineered Transparency 2016. Glass in Architecture and Structural Engineering. First Edition.
Edited by Jens Schneider, Bernhard Weller.

Timber-glass composite façades are also based on the concept of bonding the glazing onto a wooden substructure. The first concepts are developed by Hamm [2]. Niedermaier [3] and Edl [4] further enhanced the system. Today, a façade system (Figure 1-1b) is available on the market that has their ideas implemented. This façade comprises adapter frames made of birch plywood which are glued to the interior side of the IGU. It enables prefabrication of the composite element in the shop as well as an easy removal and exchange of damaged glass panes. The façade is comparable to conventional structural sealant glazing systems since silicone adhesives are applied and the design approach is similar.

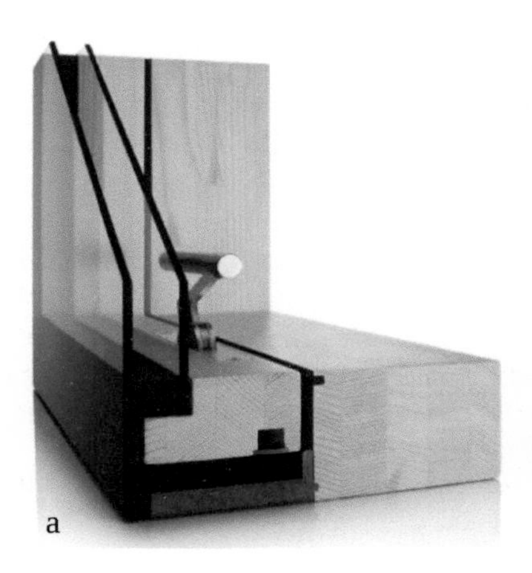

Figure 1-1 Examples of novel façade and window applications using timber as an essential part: (a) Glued window system walchfenster04 (courtesy of Walch GmbH), (b) Wooden adapter frame of the timber-glass-composite façade system UNIGLAS | FAÇADE glued on the backside of an insulating glass unit (courtesy of OTTOCHEMIE).

The building envelope is exposed to several environmental impacts such as temperature, humidity, solar radiation, water and corrosive atmosphere. All of them could affect the adhesive material, the interface between the joint and the bonded surface and the substrate material itself. It is therefore essential to consider the influence of environmental impacts on the adhesive performance. Aging tests on timber-glass joints have already been performed by Schober et. al. [5], [6] and Nicklisch [7]. Those studies basically aim for the assessment of various adhesives covering a broad range of material characteristics, but only a few wood species are involved. This study contributes to the described field of applied research by the evaluation of a broader spectrum of timber substrates which might be of interest for glued windows or timber-glass composites. All test are performed in combination with annealed glass and a silicone adhesive which is expected to be suitable because of its excellent performance in combination with glass and other substrates.

The work is part of an ongoing joint research project which is specifically devoted to glued louver windows comprising novel timber-aluminum profiles. The described experimental analysis contains excerpts from Giese-Hinz [8].

2 Materials

2.1 Adhesive

Since this evaluation especially focuses on the performance of different timber materials, only a single adhesive is part of this study. The structural sealant OTTOCOLL® S81 is used throughout the whole test series. The neutral curing two-component silicone is suitable because of its excellent weathering, aging and UV-resistance. It is often used for direct glazing, sealing of windows and glued windows. It is compatible with conventional butyl-based primary edge-sealing materials and many secondary sealants based on polyurethane, polysulfide or silicone.

2.2 Wood and wood products

Seven different wood substrates (Figure 2-1) – four solid woods and three further processed wood-based materials – are selected for this study. Pine wood (*Pinus sylvestris L.*) is a standard product in timber construction and often used for window manufacturing. However, it requires a protection against blue stain, soft rot and insects. The further choice bases on the intention to use mainly locally sourced raw material. Thus, tropical wood is not involved. Instead, alternatives such as larch (*Larix*) and oak (*Quercus*), which are also very durable, made it onto the shortlist. Beech (*Fagus sylvatica)* is the most common hardwood in Europe and therefore also chosen for the aging tests.

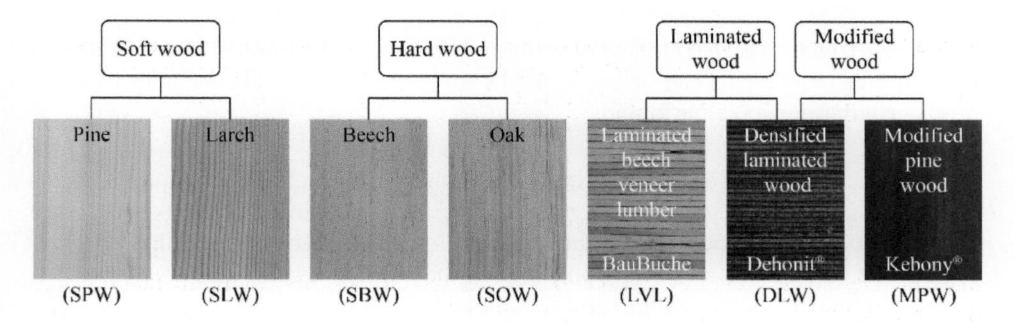

Figure 2-1 Overview and classification of the evaluated wood adherends.

Further processing of raw wood allows the enhancement of specific characteristics. Laminating multiple layers of thin veneers lead to a more uniform and stronger material. The examined BauBuche is such a laminated veneer lumber made from beech wood (*Fagus sylvatica*). The layers are glued by phenolic resin. Boards and beams are available. They provide a high surface quality which also suits window or façade profiles. Dehonit® is also a laminated but additionally densified wood. It consists of beech veneers (*Fagus sylvatica*) which are joined together by thermosetting synthetic resins under pressure and heat. The material withstands high mechanical loading as well as low and high temperatures. All veneers of BauBuche run in the same direction. The veneers of Dehonit® are glued crosswise.

Highly durable materials which are comparable to tropical wood can also be derived by chemical modification of sustainable softwood. The modified pine wood Kebony® represents this group of wood materials in the test series. It is produced by impregnation in a bio-based liquid and subsequent heat-curing. This process permanently modifies the cell walls giving the final product hard wood characteristics and a dark brown color.

The adhesion largely depends on the surface quality of the joined material. Therefore, the timber samples were examined with a scanning electron microscope (SEM) and an automated digital microscope in order to detect distinct features of each type of wood or wood-based material. Figure 2-2 shows the microscopic images of a section for those types of wood which differ the most. Clear differences are identified between pine, oak and modified pine wood. The surface images of pine, larch, beech and laminated beech veneer lumber (BauBuche) are very similar. A layering sequence is only visible for the densified laminated wood (Dehonit®).

Fiber orientation and loose particles on the surface can be clearly identified from all images (Figure 2-2, top). At higher magnification, cut open tracheids become visible on the surface. Linear grooves, which are assumed to be large vessel elements running parallel to grain, are a particular feature of the oak surface. The surface of the modified pine wood is characterized by a higher amount of visible tracheids and less loose particles compared to solid pine wood. This can be explained by more stable cell walls leading to a more precise cut during wood machining.

A linear scan performed across the samples using the digital microscope (Figure 2-2, bottom) visualizes the differences in the texture. The vessel elements on the oak sample are approx. 200 μm wide and up to 75 μm deep. We measured even higher values in other parts of the oak surface, while the roughness of the other types of wood is very similar. Changes in the topography range from -25 μm to +25 μm.

2.3 Glass

The glass component is annealed float glass with a nominal thickness of 8 mm. All four cut edges are ground. The edges are well defined and marginally chamfered (chamfer ≤ 0.1 mm). Gluing is always done onto the air-side of the glass, which is determined by means of UV-inspection. The glass surface is cleaned using a solvent. No further surface treatments were executed prior to bonding.

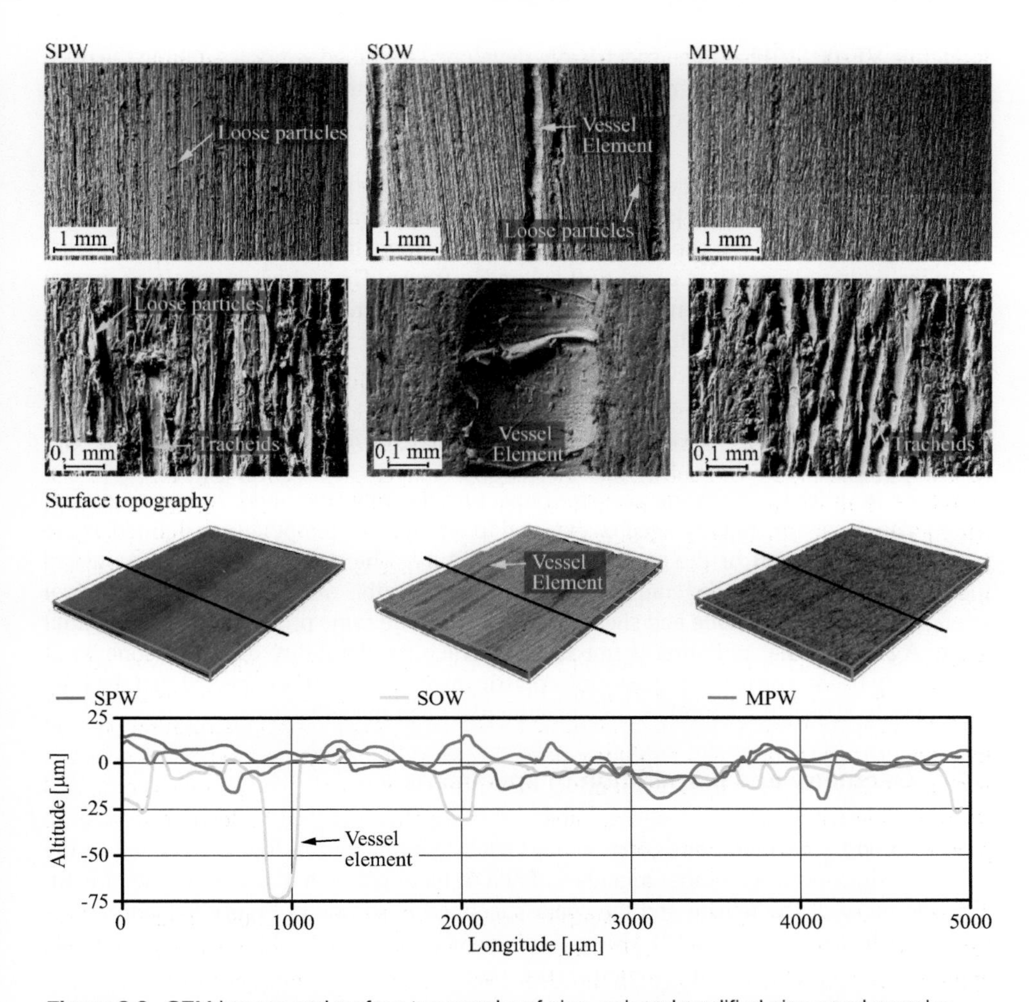

Figure 2-2 SEM images and surface topography of pine, oak and modified pine wood samples.

3 Methods

3.1 Programme

The durability of bonded joints in glass constructions and façades, which are made from structural silicone, is commonly assessed according to the European Technical Approval Guideline ETAG 002-1 [9]. It defines verification methods related to the safety in use of adhesively bonded façade elements. However, the guideline applies only to typical adherends encountered in curtain walls made from glass, aluminum and/or steel. The ift-guideline VE-08 [10] has been specifically developed for glued windows, taking the principles of ETAG 002-1 [9] and providing modified characterization methods for typical window material combinations.

According to the ift-guideline [10], the test procedure further depends on the expected exposure and loading of the joint which is basically defined by its position in the façade or the window frame. We choose a common setting where the backside of the outer pane of an insulating glass unit is glued to the wooden frame. The guideline specifies several tests comprising shear and tensile tests at different temperatures as well as the determination of the bond strength after UV-exposure, after immersion in cleaning solution and after storage in dry or humid climate.

The presented study bases on these specifications. First, the initial value of the mechanical strength (tension and shear) is determined at room temperature. The values serve as a reference in order to assess the possible impact on the adhesive bond. In a second step, the specimens are artificially aged or exposed to high or low temperatures defined as the upper and lower limit of the application temperature. The various scenarios are tested individually and neither in combination nor one after the other. Finally, the residual strength is derived in tensile and shear tests following the same procedure as in the initial stage. Aging by solar radiation is not assessed since the durability of the silicone bond has already been approved for glass and a plastic material. [11] It is expected that the UV-light mainly affects the boundary between the glass and the adhesive.

3.2 Geometry and preparation of specimens

Two different specimen configurations are used to evaluate the aging behavior of the adhesive joints between timber and glass. Each of them relate to a different type of load. The geometry of the tensile and the shear specimen is shown in Figure 3-1 and corresponds to the experiments in [7]. However, the veneer orientation of the layered materials is modified to agree with an assumed arrangement of the profiles in a window.

The geometry of the specimen loaded in tension mode is derived from the test pieces used to assess the mechanical resistance of structural sealant joints according to ETAG 002-1 [9] or ift-guideline VE-08 [8], respectively. The area of the glued joint measures 50 x 12 mm. The thickness is 3 mm. The lower piece of the specimen was rotated 90° with respect to the upper piece in order to provide the space for the brackets of the test fixture (Figure 3-1). All solid wood pieces were aligned in such a way that the fibers run parallel to the longitudinal direction of the joint. In contrast, the veneers of the layered materials stand perpendicular to the longitudinal direction of the joint.

The joint dimensions of the single lap shear specimens corresponded to those of the tension mode. The glass part measured 50 x 50 mm and was 8 mm thick. The timber piece had a size of 50 x 60 x 18 mm. The glass and timber component overlap by 12 mm. It was glued in a way that the veneers were perpendicular to the bond line plane and the fiber direction of the solid wood was parallel to the loading direction.

All timber pieces were stored for more than two weeks in a constant climate of +20 °C and 65 % relative humidity prior to bonding. The moisture content (table 3-1) was measured by the oven dry method according to EN 13183-1 [12] on a random sample of each wood specimen. The pine wood of which the specimens were sawn reached a moisture content of 11.5 %. Beech wood reaches 12.7 % – the highest moisture content of all samples – while the lowest value for solid woods was measured on the oak piece. The wood samples which are further processed contain less moisture due to absorption of substances used for modification or gluing. The moisture contents correspond roughly to reference values from handbooks or manufacturer specifications where applicable.

The timber surfaces were thoroughly cleaned using compressed air but not treated any further beyond their delivered state. The glass, however, was wiped clean with a lint-free cloth and isopropyl alcohol. All individual pieces of each specimen and spacers were aligned and clamped on a fixture plate. Gluing was done using side-by-side cartridges with a mixer nozzle. The bonded specimens cured for a minimum of 28 days under the constant conditions previously described.

3.3 Test procedure

3.3.1 General specifications – testing at different temperatures

The test configurations for both load types are shown in Figure 3-1. The electromechanical testing system (Instron 5880) comprises a temperature chamber. The samples to be tested at low or high temperature were preconditioned at test conditions for 24 ± 4 hours in advance. All other variations were performed under standard laboratory conditions (approx. +23 °C and 50 % rH). The study comprises seven different material combinations as well as two different types of loading (tension and shear). The standard test lot involves five specimens. Taking into account the initial testing and five different aging scenarios, the total amount adds up to 420 individual tests.

Two identical brackets hold the timber and the glass part of the specimen in the tensile tests. The load was raised to F = 10 N to 15 N at the beginning, to diminish unwanted slip in the entire force measurement chain and to ensure a firm hold of the specimen in the loading device. At this point, the strain signals were set to zero. The shear device comprises two halves sliding parallel against each other. The load is applied in the timber and at the glass edge near the joint. Additional cleats fix the far ends of the specimen to prevent rotation. In this case, the preload was F = 30 N. A video extensometer measured the relative displacement of two contrasting marker points on the specimen. Both tests run with a constant strain rate of 1 mm/min until either the bond or the joined materials failed.

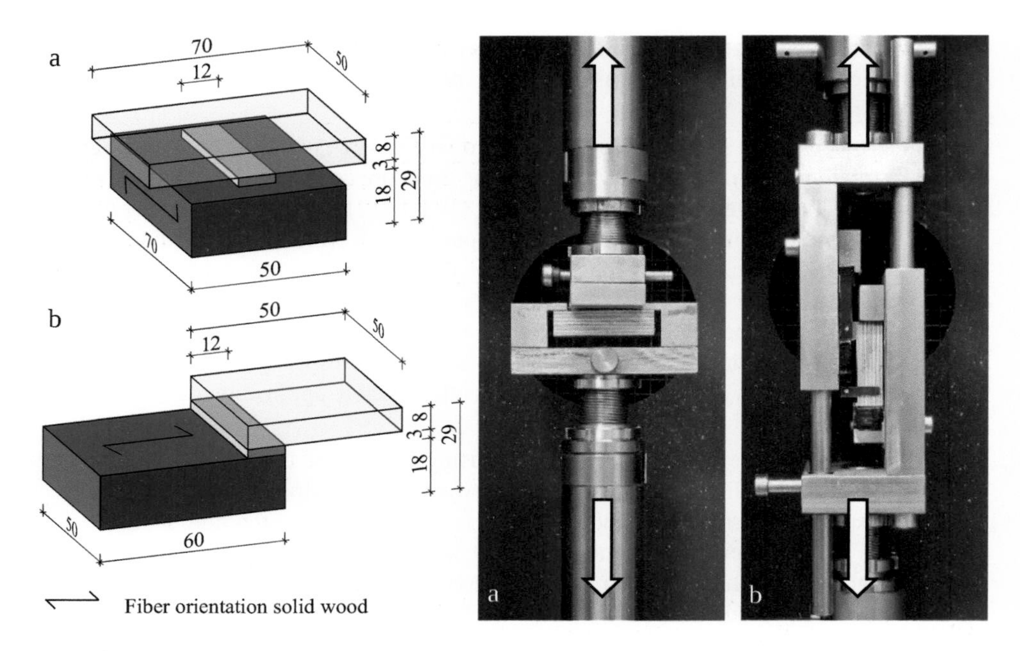

Figure 3-1 Test specimens and loading device: (a) tension and (b) shear.

3.3.2 Immersion in cleaning agent

A set of specimens was immersed for 21 days in a water bath containing 1 % solution of cleaning agent (Pril Original, pH-value 5.2 – 5.8.). The specimens were completely covered by the cleaning solution. The containers were stored in a heating cabinet at +45 °C for the whole test period. After the immersion follows a 7-day re-drying phase at standard laboratory conditions. The residual strength is assessed immediately after the drying.

3.3.3 Wetting or drying at room temperature

The impact of high or low moisture contents on the adhesion was examined by storing the samples in a climate chamber (Feutron KPK 600). Wetting and drying is performed independently in each case on five specimens per material variation. The first set was stored at humid conditions (+20 °C | 90 % rH), the second also at room temperature but in combination with a low relative humidity (+20 °C | 30 % rH). Once the corresponding moisture content in the wood was reached, the specimens were tested immediately to avoid a change caused by a longer exposure to the ambient conditions in the laboratory. The adaption period in the chamber was around 14 days. Moisture contents measured on the specimens are shown in table 3-1.

Table 3-1 Moisture content of selected wood pieces at the time of bonding and after storage at wet and dry conditions.

Storage conditions		SPW	SLW	SBW	SOW	LVL	DLW	MPW
Initial	+20 °C | 65 % rH	11.5 %	12.2 %	12.7 %	11.0 %	9.6 %	8.7 %	4.8 %
Wet	+20 °C | 90 % rH	16.9 %	18.0 %	18.3 %	16.8 %	16.8 %	10.9 %	7.3 %
Dry	+20 °C | 30 % rH	6.7 %	5.7 %	6.0 %	6.7 %	5.3 %	5.4 %	2.9 %

4 Results and discussion

4.1 Initial strength

The bond strengths for both types of loading are shown in Figure 4-1a – specified as nominal stresses at break (force divided by the bond area). The mean values of the tensile and shear stress are displayed separately for each type of wood. The percentage of cohesive failure can be read below the related column in the chart. Full cohesive failure (Figure 4-1b) is the desired failure pattern for silicone joints. Adhesive failure or near-surface cohesive failure (Figure 4-1c) is often related to poor adhesive strength on the affected surface. Hence, the guidelines [9], [10] limit the allowable percentage of adhesive failure to 10 % as average over all specimens and to 25 % on an individual specimen.

The highest initial tensile strength of the joint between glass and timber was measured for the specimens made from pine (SPW). The mean stress is approximately 1.11 N/mm^2. All specimens failed cohesively in the joint (Figure 4-1b). Similar values were achieved with beech (SBW), BauBuche (LVL) and Kebony® (MPW) where the tensile specimens failed around 1.10 N/mm^2, also with a fully cohesive failure of the glue. Thus, those types of wood have no influence on the ultimate strength. In contrast, the load-bearing capacity of the joints is less, if the adhesion to the timber surface is not sufficient. This is clearly visible for the specimens made from oak (SOW), larch (SLW) and to a minor extend also for those made from densified laminated wood (DLW). Extensive loss of adhesion occurred on the oak wood surfaces (Figure 4-1c) – only 11 % of the joint area fails cohesively – leading to a significantly lower tensile strength. The degree of adhesive failure observed on larch specimens also exceeds the permissible values, however, the average strength is not affected that much. The minor occurrence of adhesive failure on the densified laminated wood Dehonit® is considered negligible.

The clear differences between the individual wood species derived from the tensile tests are not as obvious when tested in shear mode. Nearly all specimens failed fully cohesively, while the shear strength ranged on a similar level around 1.00 N/mm^2. Only 2 % of the joint area on the oak surface exhibited a loss of adhesion.

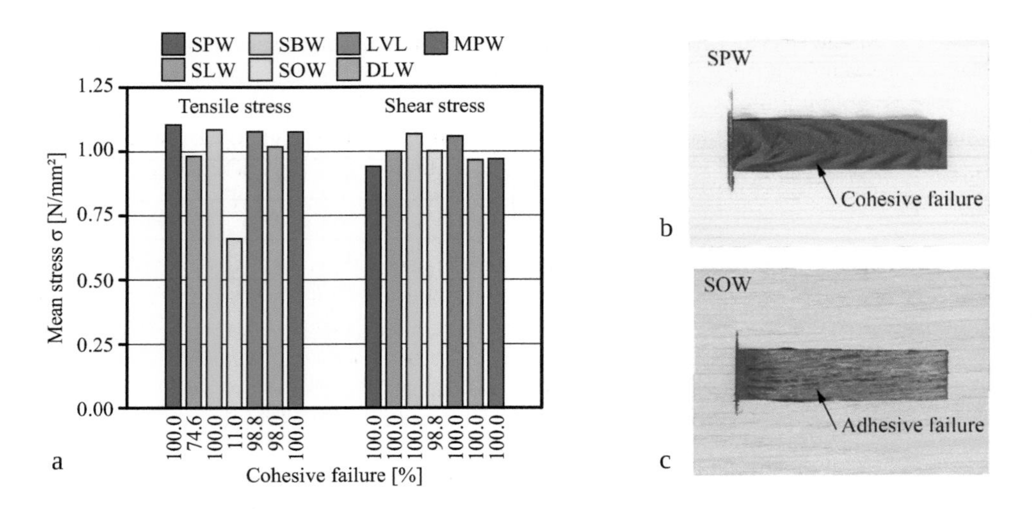

Figure 4-1 Tensile and shear test results at room temperature: (a) mean failure stress for each type of wood, typical failure pattern of tensile specimens made from (b) pine and (c) oak.

4.2 Residual strength after aging

The influence of different temperature, cleaning agent and moisture on the adhesive bond between wood and glass is assessed by five different aging scenarios. Figure 4-2 displays the residual tensile strength in relation to the values before aging. The tensile tests are considered more meaningful and thus used to evaluate the impact of aging on the strength. The results of the shear test correspond to those of the tensile test, but are not shown here. The dashed horizontal lines indicate the maximum allowable loss of strength according to ETAG [9] (25 %) and to the ift-guideline [10] (50 %).

The results reveal a temperature dependent material behavior of the adhesive joint. Compared to the initial values, the tensile strength increases at a test temperature of -20 °C and decreases for the majority of tested materials at +80 °C. This is a well-known characteristic of adhesives. The failure patterns correspond basically to the observations during the initial testing. Loss of adhesion occurs mainly on oak and larch wood surfaces. The most significant impact of temperatures was measured on Dehonit® (DLW) at +80 °C. Here, the ultimate strength decreases by slightly more than 25 %. Nevertheless, extreme temperatures are not considered a highly critical impact.

The greatest loss of strength relates to the immersion in cleaning agent. Although the residual strength did not fall below the allowable values of the ift-guideline, a severe loss of adhesion is observed on many specimens after this aging scenario. A sufficient adhesive strength is only achieved on solid pine wood and on solid beech wood where the joints still failed cohesively by almost 100 %. The specimens made from densified laminated wood and glass even failed without any mechanical loading. The timber part expanded strongly after being put into the water bath (Figure 4-3a). Additionally, certain substances might be released from the modified wood which negatively affect the adhesive. This is assumed due to an intense discoloration of the solution and the unique failure pattern (Figure 4-3c). As a consequence, the bond line is partially damaged and already fails while the specimens are still in the bath.

A high moisture content in the wood components basically leads only to a moderate reduction of the ultimate tensile strength. Compared to the initial tests, larch and densified laminated wood exhibit a significantly larger amount of adhesive failure. Drying of the bonded specimens generally improves the joint performance. The adhesive strength increased for all material combinations. The oak-glass specimen even reached similar values to the other types of wood and exhibit nearly no adhesive failure.

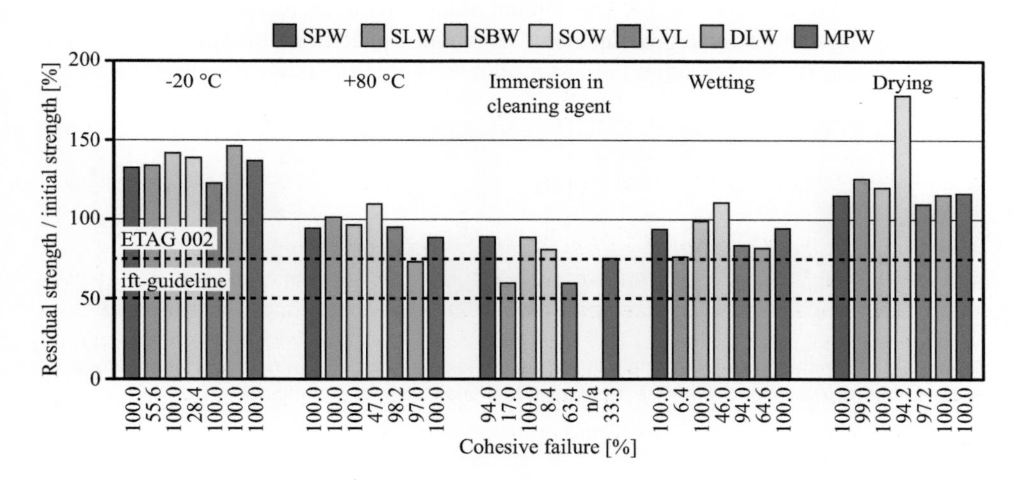

Figure 4-2 Mean breaking stress in percent based on the tensile stress at room temperature.

Figure 4-3 Specimens made from densified laminated wood (a) before and (b) after immersion in cleaning solution, (c) moisture expansion and material incompatibility leading to a premature failure.

5 Improvement by further surface preparation

As a consequence of the poor behavior of some timber-glass bonds in the aging tests, the influence of additional surface preparations is assessed on larch wood and on oak wood which exhibited a high percentage of adhesive failure. Wood tends to accumulate oil, fat and other contaminations on its surface. This might have affected the adhesion quality.

Three different surface preparations are assessed. First, the surface is just freshly planed to obtain a surface which contains less contaminations. Second, two different primers (OTTO Primer 1105, OTTO Primer 1215) are applied to enhance the surface quality. Thus, three differently prepared surfaces are assessed in this extended test series. The effect of the additional preparation becomes clearly visible by SEM imaging. Figure 5-1 shows the oak surface in the different states: before preparation, after re-planing and after re-planing and primer application. The amount of loose particles on the surface is reduced by planing, but the distinctive vessel elements remain apparent on the surface. The film-forming primer further smoothes the surface but does not fully cover the vessel elements.

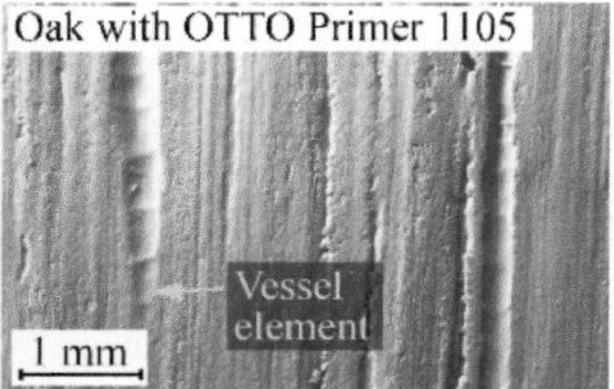

Figure 5-1 Preparation steps exemplarily shown on oak surfaces, first without any preparations, second after re-planing and third after re-planing and application of a primer.

The gluing was done within an hour after pretreatment to avoid a re-contamination of the surfaces. The specimens cured under the same conditions as the previous set of specimens. Testing was done after a 28-day curing period at room temperature.

Figure 5-2 shows the results of the tensile test compared to the less prepared specimens used in the initial test set. Except from the application of primer 1105 on solid larch wood, all surface preparations lead to a significant increase of the tensile strength as well as to an improved adhesion. It is possible to reach the same strength as the pine-wood-glass-joints whose average tensile strength at room temperature is indicated by the dashed line in blue color. The best results derive from a fresh planing and the application of primer 1215. This preparation lead to nearly full cohesive failure of the joint for both materials. Figure 5-2 shows the SEM images of the failed surface of an oak specimen without and with preparation.

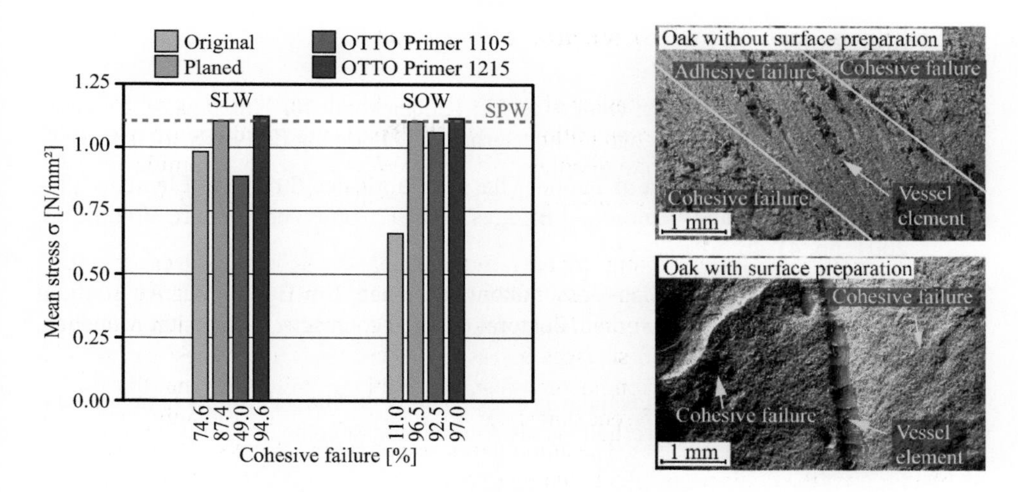

Figure 5-2 Tensile test results after further preparation and SEM images of the failed oak surface.

6 Conclusion

This study assesses the aging stability of a silicone adhesive bond between seven timber materials and glass. It is concluded that the joint strength and durability strongly depends on the chosen type of wood while neither the pure adhesive nor the adhesion to glass exhibits any degradation. High temperatures, wet conditions or a high moisture content in the timber adherend typically lead to a loss of strength and the increase of adhesive failure on the timber surface. An improved adhesive strength results from the storage in dry conditions. However, only the adhesive bonds to solid pine and solid beech wood show residual strengths and failure patterns which are in the permissible range. Loss of adhesion was never observed on the glass-side throughout the study. The severe adhesive failure, which occurred on oak and on larch already in the unaged state, can be eliminated by further surface treatments such as re-planing shortly before the bonding and by the use of a specific primer. The two examined laminated woods are not considered suitable since distinct swelling of the wood-based materials damaged the joint before it could be tested. Material incompatibilities are observed for the densified laminated wood after exposure to cleaning solution. Further research needs to prove that the improvement achieved by the pretreatment still remains after the specimens have aged.

7 Acknowledgements

The study is part of the research project “Bonded Timber-Alloy Louver Windows” funded through the Central Innovation Programme for SMEs (ZIM) by the German Federal Ministry for Economic affairs and Energy (BMWi). The authors would like to thank the adhesive manufacturer Ottochemie and the window manufacturer Eurolam for their support.

8 References

[1] Lieb, K.; Schober, K. P.; Uehlinger, U.: Klebetechnik für Holzfenster. In: Glaswelt, Sonderheft Glaskleben im Fensterbau, 61. Jahrgang (2009), S. 10-13.

[2] Hamm J., "Development of Timber-Glass Prefabricated Structural Elements", Innovative Wooden Structures and Bridges. IABSE Conference Report, Volume 85, 2001, pp. 41-46.

[3] Niedermaier P., Holz-Glas-Verbundkonstruktionen. Ein Beitrag zur Aussteifung von filigranen Holztragwerken, doctoral thesis, Technische Universität München, 2005.

[4] Edl T., Entwicklung von wandartig verklebten Holz-Glas-Verbundelementen und Beurteilungen des Tragverhaltens als Aussteifungsscheibe, doctoral thesis, Technische Universität Wien, 2008.

[5] Schober, K. P. et al.: Grundlagen zur Entwicklung einer neuen Holzfenstergeneration. Endbericht 1. Projektjahr. Wien: Holzforschung Austria, 2006.

[6] Schober, K. P. et al.: Grundlagen zur Entwicklung einer neuen Holzfenstergeneration. Endbericht 2. Projektjahr. Wien: Holzforschung Austria, 2007.

[7] Nicklisch, F.: Ein Beitrag zum Einsatz von höherfesten Klebstoffen bei Holz-Glas-Verbundelementen [Application of high-modulus adhesives in load-bearing timber-glass-composite elements]. Doctoral thesis. Technische Universität Dresden, 2016.

[8] Giese-Hinz, J.: Klebverbindungen zwischen Holz und Glas – Haftverhalten und Oberflächenbeschaffenheit bei verschiedenen Holzarten [Timber-glass adhesive joints – adhesion and surface condition of various types of wood]. Diploma Thesis (unpublished). Technische Universität Dresden, 2016.

[9] ETAG 002-1: Guideline for European technical approval for structural sealant glazing kits (ETAG) – Part 1: Supported and unsupported systems. European Organisation for Technical Approvals (EOTA), Brussels: 2012.

[10] ift-Richtlinie VE-08/3 Beurteilungsgrundlage für geklebte Verglasungssystem, ift Rosenheim, 2014.

[11] Test report no. 507 37730 R1 dated 17.03.2009. Characterization of a bonded system (Ottocoll S81, annealed glass, PVC) according to ift-guideline VE-08/1. Issued by ift Rosenheim GmbH. Online: http://www.otto-chemie.de/!product-i18n-download/197/e6bc2eb1cfb4859bae8183fbbbfd8906 [19.07.2016].

[12] DIN EN 13183-1: moisture content of a piece of sawn timber – Part 1: Determination by oven dry method. Deutsches Institut für Normung e.V. Berlin: Beuth, 2002.

How to handle futures glass design by coatings (Let's look 10 years ahead)

Paul Bastianen[1]

1 Vindico Applied Technologies BV, Netherlands; Columnist Intelligent Glass Solutions (IGS), Glaswelt, editor Dutch magazine Glas in Beeld, Damloper 69, 4902 CE Oosterhout

At this moment, the glass industry is not quite ready for changes in construction like 3D printing and robotizing. First the glass quality itself (certainly in Europe) is suffering by bringing aluminium dioxide into compliance with EU norms to optimize the production cost (lower temperature in the oven and save on the use of expensive raw materials). This means glass surfaces are no longer resistant against acids and alkaline (aluminium increases the resistant Al_2O_3 > 0.7 %, EU norm 0.1 - 0.2 %). Such a phenomenon is also observed in greenhouses. There is a huge condensation (up to 25 liter / m² per day) on the inside of the roof glass. The water is collected in the gutters and recycling for use. For that reason the corroded glass by osmosis process (every year after the harvest) is cleaned from inside with fluorine acids (solution of max 7 %). This is not allow anymore for use for a window cleaning. In the following diagram you see the result of a study and measurement of cleaned glass against the polluted (blue line) and even corroded glass (red line) and how it influences the light transmission in greenhouses (light transmission which is important for the PV modules, is between the wave length of 380 and 1050 nm). The light transmission can be decreased up to 50 %, which translates to power losses up to 25 %.

Keywords: sodium migration, osmoses, glass aging, protection, maintenance

1 Sodium migration

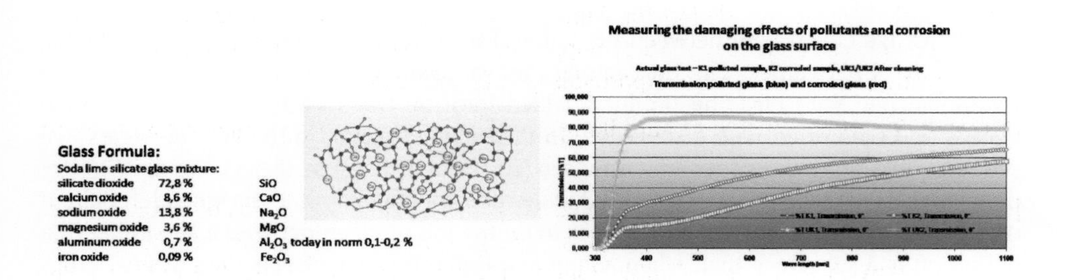

Figure 1-1 Float glass formulation (left). Light transmission measurement in greenhouse cleaned glass, polluted glass and corroded glass (right).

To prevent such losses due to this factor, glass needs to be provided with a coating. The coating can create a barrier for sodium migration. However, in this article we don't treat

Engineered Transparency 2016. Glass in Architecture and Structural Engineering. First Edition.
Edited by Jens Schneider, Bernhard Weller.

this issue in full details. This is available in the papers presented at the Solar show Valencia 2010: "Potential Induced Degradation of solar cells and panels and Characterization of Multi crystalline Silicon Modules with System Bias Voltage Applied in Damp Heat." How to overcome these complex issues that affects power production in the PV projects? The next section deals with some options.

2 Cleaning systems

There are two main cleaning systems in the market used for cleaning windows as well as for PV modules.

- In common one can clean by: Spraying, dipping or manually.
- Using systems: pressure and temperature. Rinsing with water and sewage treatments.
- Cleaning products can be Liquid or powder (alkaline, acid or neutral). Use of smell additives or collors in the cleaners for a certain purpose.

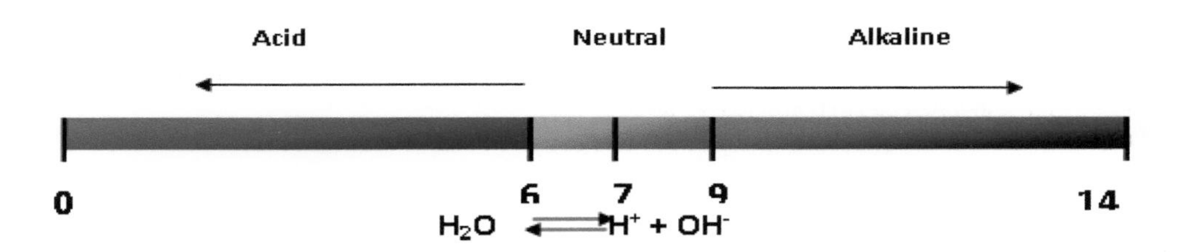

Figure 2-1 pH-Range.

pH values we called acid between pH value 0-6, neutral between 6-9 pH and alkaline between 9-14 pH values. The first among the two systems used in glass cleaning is the traditional one with cleaning products added into the water (warm), using a window washer and then dried with a squeegee. In the second system, in the last few years, we see more and more use of reverse osmoses (RO) water, because there is no need to dry the glass surface with the squeegee. RO water doesn't contain minerals when left behind on the surface. Cleaners contain tensides to do the job of cleaning, which according to a definition is a substance that, when added to a liquid (water), affects the physical properties of the surface, e.g. increasing its wetting properties or formulating of emulsified liquids.

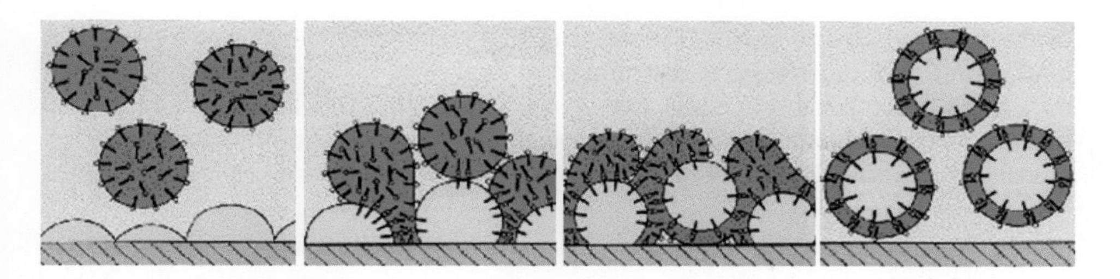

Figure 2-2 Working of tensides.

The tensides in the cleaner depose the pollution on the surface and soak it up. With rinsing water afterwards, the pollution can be taken off from the surface. Most of the time the tensides cannot be remove completely and the surface gets another surface energy; a hydrophobic behaviour occurs [hydrophobic means water (hydro) repellant (phobic)]. Water is repelled by the surface and tends to form small droplets that easily run off the glass rather than wetting the glass through making a thin layer of water, which is called hydrophilic.

Glass itself has a hydrophilic surface energy, which means a contact angle of about 25°. We called hydrophobic surface when the contact angle is above the 100° (lotus effect, the behaviour of the lotus flower with a contact angle of above 200°) Hydrophilic surfaces has a contact angle of less than 10°.

Figure 2-3 Difference between hydrophilic (left) and hydrophobic (middle) surface energy, explanation contact angle measurement.

The tensides which is left on the surface provide the polluted surface fast cleaning and most of the time for example with rain it takes more than a month that the surface gets its own surface energy.

Cleaning glass with cleaners has a risk, because most of the cleaners contain acid. When the acid is too strong it could damage the other materials around the glass, like nature stone, brickwork, aluminium etc. On the other hand the cement and concrete cause damaging the glass, because the cement and concrete water in combination with the CO_2 in the air makes a silica acid, which bonds very well to the glass surface and cannot be removed simply by cleaning.

In the past the window cleaners were using fluorinated acid, but this is forbidden in most countries in Europe, because it proved to cause a cancer of the bones. That is the reason, the reverse osmoses water has been adopted for glass cleaning, and the glass industry is using this for many years especially in the production of secondary glass products, such as DGU (double glazed units), laminated glass or tempered glass.

But there is a difference in use in the industry of reversed osmoses water. One is used during the production process and another is used for the maintenance of the façades and the cleaning of PV modules. The use of RO Water for PV modules cleaning is dangerous for the surfaces.

3 Reverse osmoses water (ROW)

Cleaning with the reverse osmoses water is based on demineralised water, which means that all minerals and iron etc are extracted out of the water. This can be measured by the conductivity. The high contents of minerals and the iron in the water have a high conductivity. The drinking tap water, depending on the country and place, can have conductivity between 500 - 700 µS (microsiemens is a measurement of conductivity). The reserve osmoses water is below the 30 µS. The washing machines used in the industry are using water with lower than the 10 µS.

The working mechanism of cleaning with reverse osmoses water is based on ionic activity. The water contains less ions concentration whereas the pollution and dirt contain high concentration ions. The ions move from high to low concentration. Therefore, they get dissolve in to RO water, so the surface gets striped off from pollution. Because this water contains less minerals, the surface doesn't have to dry afterwards, there is no risk of leaving minerals behind which caused spots on the surface. To study the effect of the reverse osmoses water, the equipment is fixed on the roof of the building, or mobile on a truck.

Figure 3-1 Equipment to make reverse osmoses water.

There is a greater variety of construction materials used around the glass in the architectural designs we see today than there was previously in the 80’s and 90’s, back then there were more ‘‘glass-glass’’ solutions used in facades. In future the diversity of materials will be even greater. Concrete and cement water from the wall gives off an acid which badly affects glass. Not only must the glass be saved in the future with high performance coatings to increase the durability of the material, but also certifications like BREEAM, LEED™ etc will pose more demands on lowering maintenance costs.

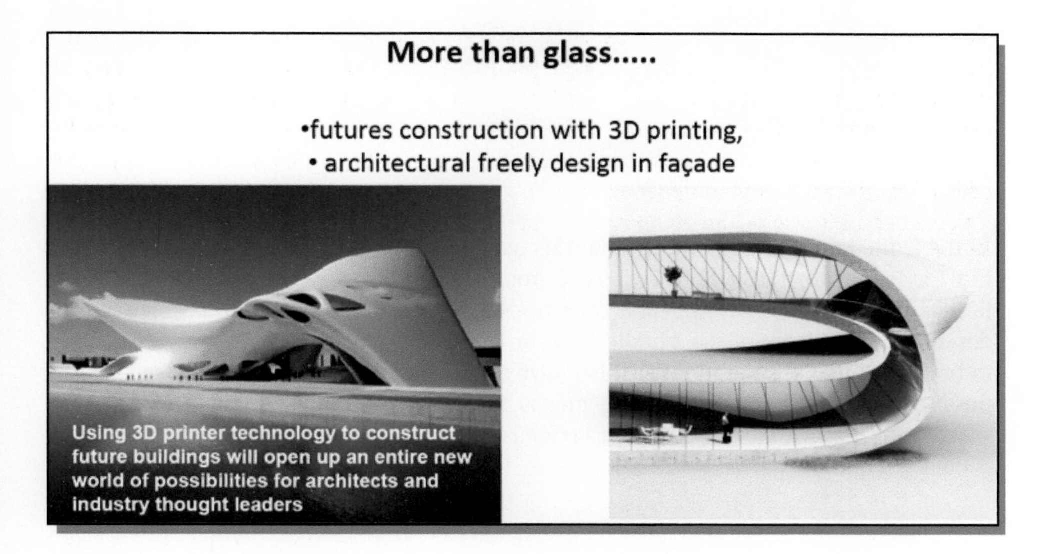

Figure 3-2 Futures 3D printed constructions.

Mistakes are also made in different environments and different climates. For example in cities like Dubai or futures Cairo skyscrapers are built with commonly used building materials (Cairo are planned by Arabic sheiks to build city like Dubai, maybe they can learn from the mistakes the made in existing cities) from Western Europe that are totally inappropriate for the climatic conditions in the Middle East. Before the BurjKhalifa (828 m) was built, it was well known that stainless steel needs a certain oxygen grade which is not supplied at this great height. The embarrassing result was that less than three years after construction of the world's tallest building, the stainless steel had begun to corrode and rust. The combination of the characteristics of the chosen material, its composition (316L) and its finish (brushed grid 400), the local atmospheric circumstances and the type and intensity of aggravate pollution a high level of maintenance and extra attention is required. To avoid corrosion and to bring down the cleaning and maintenance effort, protection of the material is required. The inspected installed items need serious attention in less than 2 months to avoid irreparable marks.

Figure 3-3 Research at the BurjKhalifa.

In the United States and in other countries outside of Europe facades were not meant to last for longer than 10 years. Here in Europe we consider more than 60 years, this is largely because our environmental conditions are different and more agreeable than the desert climate or that of the Middle East, or Texas. Not only do desert conditions have to be taken into consideration, we also have to consider the influence of the sea (like in the Netherlands or England) which means that during certain periods of time in the season, façades will be exposed to condensation. Even today, with increased use of triple glazed units.

The largest source of glass damage or complaints

- coating changes (LowE, solar controlled glass)
- scratches on tempered glass(after tempering process glass weak)
- compatibility different materials
- condensation outside IGU (triple glass U-value < 1 W/m².K)

Figure 3-4 Largest source of glass damage or complains.

U-values < 1.0 W/m²K, there is still a risk of osmosis (sodium and calcium ions migrate to the condensate droplets) to buildings. This process is known to corrode glass in showers and with condensation it will have a similar effect on glass in the façade. It is directly for this reason glass has to be protected with coatings that are bonded to the glass. Glass companies are continually developing anti-condensation coatings, which mean that not only will the surface energy change but also there will be a change in the emissivity of the glass in order to prevent condensation occurring in the first place. Other construction materials suffer more from condensation than glass does and are more likely to use different coatings, e.g. a sacrificial coating like powder coated and anodized aluminium.

4 Conclusions and lessons

It is clear that the Investors are looking only for the profitability of investment in construction, façades, PV or thermal solar and not to saving the planet by better environment. But the total maintenance cost can be extremely high when you don't consider what can go wrong with the used materials, equipment, electric systems, life time of an expensive converter etc.

There is no precise calculation how to clean the polluted surfaces which have a huge influence on internally office climate and performance in the workplace, the power gain and this is indicated in the long term study by FAGO group Eindhoven and the Bern Academy in Switzerland, so be aware of the problem and make the decision to invest small money in a combination coatings comparing the complete investment of the total façade and system.

5 References

[1] A In a long time studyoftheacademyof Bern (Langzeiterfahrungen in der Photovoltaik-Systemtechnik, Heinrich Häberlin, Berner Fachhochschule, Technik und Informatik (BFH-TI) Fachbereich Elektro- und Kommunikationstechnik, Photovoltaik-Labor J lcoweg 1, CH-3400 Burgdorf / SCHWEIZ Tel. 034 426 68 11, Fax 034 426 68 13, Internet: www.pvtest.ch, e-Mail: heinrich.haeberlin@bfh.ch).

[2] benchmarkMartifer Solar Portugal.

[3] Potential Induced Degradation of solar cells and panels by J. Berghold, O. Frank, H. Hoehne, S. Pingel, B. Richardson*, M. Winkler SOLON SE, Am Studio 16, 12489 Berlin, Germany * SOLON CORPORATION, 6950 S. Country Club Rd, Tucson, Arizona 85756.

[4] Characterization of Multicrystalline Silicon Modules with System Bias Voltage Applied in Damp Heat by Peter Hacke, Michael Kempe, Kent Terwilliger, Steve Glick, Nathan Call, Steve Johnston, Sarah Kurtz Ian Bennett1, Mario Kloos1 National Renewable Energy Laboratory, 1617 Cole Blvd. Golden, CO 80401 USA 1 Energy Research Centre of the Netherlands, 1755 ZG Petten, NL.

[5] Inspection & Recommendation Report Stainless Steel „Bull Nose" of Burj Dubai version V4.0.

[6] Arabian Aluminum, Mr. pHd John Zerafa, Tel.: +971 50-8751364. Performance improvement techniques for PV strings in Qatar: Results of first year of outdoor exposure. Diego Martinez-Plazaa*, Amir Abdallaha, Benjamin W. Figgisa, TalhaMirzab. Qatar Environment and Energy Research Institute, P.O. Box 5825, Doha, Qatar. GreenGulfInc, P.O. Box 210290, Doha, Qatar.

Adhesion of a two-part structural acrylate to metal and glass surfaces

Vlad A. Silvestru[1], Oliver Englhardt[2]

1 Institute of Building Construction, Graz University of Technology, Lessingstraße 25/III, 8010 Graz, Austria, silvestru@tugraz.at

2 Institute of Building Construction, Graz University of Technology, Lessingstraße 25/III, 8010 Graz, Austria, englhardt@tugraz.at

The research discussed in this paper concentrates on a two-component acrylate adhesive which is considered to be suitable for linear connections between glass and metal substrates. The influence of different surface conditions of stainless steel and aluminum on the adhesion behavior of the mentioned acrylate adhesive to these substrates was investigated in double-lap-shear tests. Additional double-lap-shear tests on connections between glass and stainless steel were performed. The results of the roughness measurement, the failure patterns as well as the obtained stress-strain-curves are presented and discussed. Furthermore, the results of the experimental tests are compared to values determined with a finite element simulation using a multi-linear material model for the acrylate adhesive. The performed investigations are part of a larger research project on glass-metal façade elements with composite load-bearing behavior.

Keywords: hybrid glass-metal structure, linear adhesive bonding, acrylate, stainless steel, aluminum, double-lap-shear test

1 Introduction

Over the last years adhesives gained an increasing role in the design of structural glass connections. Numerous research projects on the properties and the applications of novel adhesive materials have been completed. For silicone adhesives, which show lower strength, higher flexibility and are often used in facades, a large amount of results and experience is available. For stiffer adhesives, like acrylates or epoxies, the applications are more experimental and only limited knowledge about their resistance to different environmental conditions and their durability exists. The two component acrylate adhesive investigated in this paper is considered to be suitable for linear connections between glass and metal substrates, where shear loads have to be transferred between the two adhesively bonded materials. Its predecessor products have been used by Wellershoff [1] for activating square glass panes as shear panels by circumferentially bonding them to a hinged steel frame as well as by Netusil [2] for hybrid steel-glass beams.

One of the main demands to adhesive connections is to avoid a failure by debonding from one of the substrates. To ensure this request suitable surface conditions have to be chosen and recommended surface treatments have to be applied. The manufacturing of surfaces

Engineered Transparency 2016. Glass in Architecture and Structural Engineering. First Edition.
Edited by Jens Schneider, Bernhard Weller.

and the available methodologies for measuring these surfaces are described by Whitehouse [3]. Methods of improving adhesive joints in glass structures by surface treatments have been investigated by Kothe [4]. For the research presented in this paper different stainless steel and aluminum surfaces as well as glass surfaces were investigated mainly in double-lap-shear tests in order to determine the failure type and the performance of the mentioned acrylate adhesive. The performed investigations are part of a larger research project on glass-metal façade elements with composite load-bearing behavior presented in Silvestru et al. [5].

2 Methods and material properties

The performed investigations include double-lap-shear tests on adhesive metal-metal connections, double-lap-shear tests as well as tensile tests on adhesive metal-glass connections and finite element simulations of the metal-glass tests. All the used test specimens were manufactured and conditioned for 28 days at 23 °C and 50 % relative humidity. The tests were performed displacement controlled with a loading speed of 2 mm/min. The different test specimen geometries and test setups are described in the next subsections.

For the metal-metal connections substrates made of stainless steel no. 1.4404 and of aluminum no. 3.3206 were used. The stainless steel surfaces were untreated and grinded with different grain sizes, while the aluminum surfaces were untreated, powder-coated and anodized. To characterize the roughness of these surfaces the profile roughness parameter R_a was measured with a stylus instrument. For the metal-glass connections the metal substrates were made of stainless steel no. 1.4404 with a P300 grinded surface, while the glass substrate was made of fully tempered glass without any coatings. The main mechanical properties from the technical data sheet of the used two component acrylate adhesive are summed up in Table 2-1.

Table 2-1 Properties of the used two component acrylate adhesive.

Density [kg/l]	Open time [min]	Fixture time [min]	Young's modulus * [N/mm²]	Poisson's ratio [-]	Tensile strength * [N/mm²]	Elongation at break * [%]
1.19	~ 10	~ 25	~ 250	~ 0.47	~ 10.0	~ 200
* Determined according to ISO 37 [6]						

For the performed shear tests a double-lap test setup was chosen because the symmetrical arrangement allows a pure shear loading without undesired bending moments. Previous experimental programs on adhesives (e.g. Overend et al. [7] and Belis et al. [8]) with single-lap-shear tests showed that the rotation occurring due to different axes of the loading and the support is difficult to measure.

3 Double-lap-shear tests on metal-metal connections

A total of 32 test specimens were tested in double-lap-shear. Table 3-1 shows the different surfaces used for the stainless steel and the aluminum substrates as well as the test specimen numbers for each surface.

Table 3-1 Test specimen nomenclature for the different investigated stainless steel and aluminum surfaces.

Stainless steel no. 1.4404					Aluminum no. 3.3206		
untreated	**grinded P50**	**grinded P80**	**grinded P150**	**grinded P300**	**untreated**	**powder-coated**	**anodized**
TS 01-03	TS 04-06	TS 07-09	TS 10-12	TS 13-17	TS 21-23	TS 24-26	TS 27-29
				TS 18-20			TS 30-32

For all the surface types an adhesive joint geometry with the dimensions of 50 x 12 x 3 mm (geometry 1) was tested based on the H-sample proposed in ETAG 002 [9] for structural silicones. Only the thickness was reduced from 12 mm to 3 mm according to the recommendations of the adhesive manufacturer. For the P300 grinded stainless steel surface and the anodized aluminum surface a second adhesive joint geometry with the dimensions of 50 x 60 x 3 mm (geometry 2) was tested in order to analyze the performance of a five times larger, more laminar adhesion area.

3.1 Roughness measurement

The roughness of the metal surfaces was determined by scanning them with a stylus instrument (Mitutoyo SJ-201, see Figure 3-1 left) over a predefined profile length with a predefined speed. Based on the expected roughness, a profile length of 5 x 2.5 mm and a speed of 0.5 mm/sec were used.

The mean values and the deviation range of the measured roughness parameter R_a are plotted in the diagram in Figure 3-1 right. For the stainless steel surfaces a good agreement with the used grain size can be observed in the tendency of the measured values. Considering the value of 3 µm, which is the maximal recommended value by silicone adhesive producers for substrate surface roughness (R_a), together with the ranges of the measured values, it can be concluded that a grinding of the stainless steel surfaces with P150 or higher should be used. For the used acrylate no such recommendation is available. The aluminum surfaces show roughness values in the range of the finer stainless steel surfaces or lower.

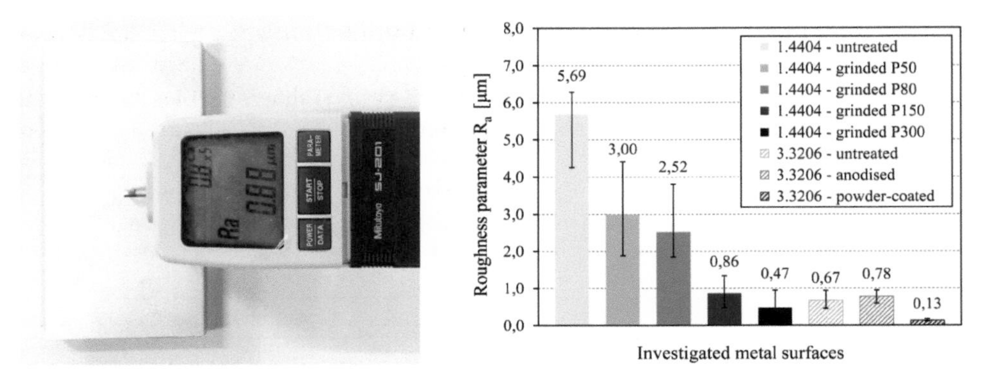

Figure 3-1 Procedure used for the roughness measurement (left); mean values of the profile roughness parameters R_a (µm) and deviation range for the different investigated stainless steel and aluminum surfaces (right).

3.2 Manufacturing of the test specimens

The double-lap-shear test specimens for the metal-metal connections consist of two lateral metal sheets and one middle metal sheet which are adhesively bonded together by the acrylate adhesive. The test specimen geometry as well as the test setup with the load application by clamping jaws, the support conditions with fixing bars and the measurement instrumentation (one displacement transducer for each of the two adhesive joints) is illustrated in Figure 3-2 left.

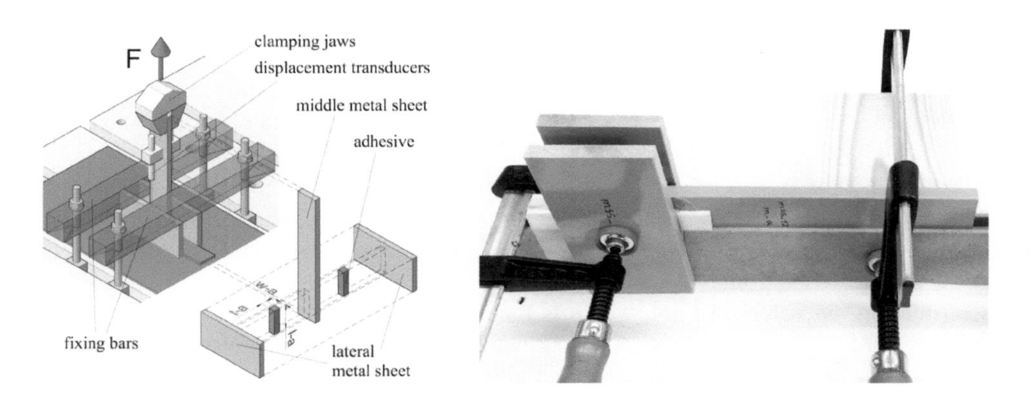

Figure 3-2 Geometry of the double-lap-shear test specimens for metal-metal connections (left); manufacturing procedure for the test specimens for metal-metal connections (right).

The manufacturing of the specimens was performed in the laboratories of the adhesive producer. Firstly, the adhesion area of the metal surfaces was wiped over with a special cleaner to remove dust and residual grease. Secondly, a primer was applied on the surfaces to increase the adhesive strength. Afterwards, PTFE blocks were fixed with tape to the middle metal sheet to determine the joint thickness, an adhesive stripe was applied with a static mixer in the resulting area and the lateral metal sheets were pressed on the stripe with screw clamps (see Figure 3-2 right). Finally, the joint was tooled with a spatula and the tapes and the PTFE blocks were removed.

Table 3-2 Dimensions of the adhesive joints in the double-lap-shear test specimens for the metal-metal connections.

Dimension [mm]	Geometry 1		Geometry 2	
	target value	real value *	target value	real value *
length l-a	50.0	48.9	50.0	50.8
width w-a	12.0	13.3	60.0	58.3
thickness t-a	3.0	-	3.0	-

* the specified real values are mean values of the measured dimensions

Since the dimensions of the adhesive bite have an influence on the mechanical behavior of the joint, the real values were measured for all specimens and are given in Table 3-2 along with the target values.

3.3 Typical failure patterns

Typical failure patterns for both investigated geometries of the metal-metal connections are illustrated in Figure 3-3.

Figure 3-3 Typical failure patterns of the double-lap-shear test specimens for metal-metal connections with joint geometry 1 (left) and joint geometry 2 (right).

For geometry 1 both adhesive and cohesive failure was observed (see Figure 3-3 left) depending on the surface condition. The stainless steel surfaces showed a cohesive or mixed failure for grindings with P150 or P300, while for the other conditions adhesive failure occurred. In the case of aluminum only the anodized surfaces showed mixed failures, while for the other surfaces adhesive failure was noticed. For the specimens with geometry 2 the failure was mainly cohesive as shown in Figure 3-3 right. In subsection 3.4 it can be observed, that the failure occurred always at higher load levels after plasticizing of the adhesive.

3.4 Mechanical results

The results of the double-lap-shear tests on metal-metal connections are plotted in the form of engineering stress vs. shear displacement curves in Figure 3-4 separately for joint geometry 1 (diagram left) and joint geometry 2 (diagram right). The occurring shear stresses τ were calculated by Equation (3.1) according to ETAG 002 [9], where F is the recorded force and $w_{a,i}$ and $l_{a,i}$ are the width and the length of the adhesive bite for the two sides of the double-lap-shear test specimens.

$$\tau = \frac{F}{w_{a,1} \cdot l_{a,1} + w_{a,2} \cdot l_{a,2}} \tag{3.1}$$

The elongations were determined as shear displacements d (%) by Equation (3.2) according to ETAG 002 [9], where Δl is the extension and $t_{a,i}$ are the thicknesses of the adhesive for the two sides.

$$d = tan^{-1}\,\gamma = \frac{\Delta l}{(t_{a,1} + t_{a,2})/2} \tag{3.2}$$

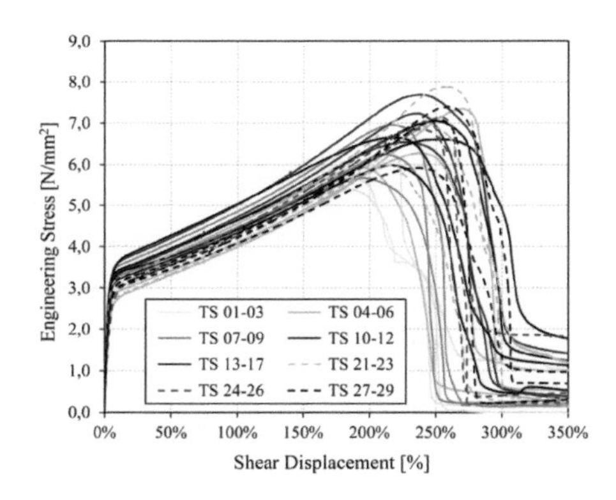

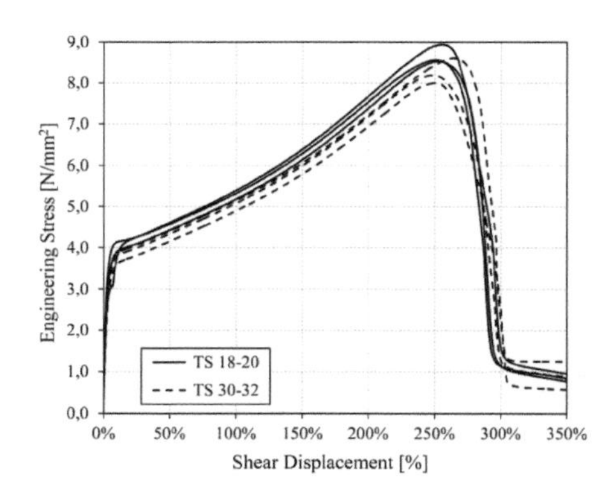

Figure 3-4 Engineering stresses plotted vs. shear displacements for the double-lap-shear test specimens for metal-metal connections with joint geometry 1 (left) and joint geometry 2 (right).

A good agreement of the results for the different test specimens with the same joint geometry can be observed especially for geometry 2. For geometry 1 the stiffness is in good agreement, while for the stresses after plasticizing as well as for the failure stresses and the corresponding elongations a variation of the results can be observed. The differences of the stresses in the plastic range can be explained by the small variation of the joint dimensions, which has an influence on the effect of stress peaks. The different failure stresses and corresponding elongations are determined by the different occurring failure types. Table 3-3 provides mean values for stresses in a relevant strain range (up to 25 %) as well as failure stress and corresponding maximal load and elongation for the two investigated joint geometries.

Table 3-3 Mean values of the shear stress at different shear displacements (strains), maximal load and corresponding elongation for the adhesive metal-metal connections.

Specimen type	Stress at 5 % strain [N/mm²]	Stress at 15 % strain [N/mm²]	Stress at 25 % strain [N/mm²]	Failure stress [N/mm²]	Maximal load [kN]	Elongation at max. load [mm]
small	2.12	3.18	3.33	6.64	8.60	7.1
large	1.67	3.38	3.83	8.47	50.13	7.9

4 Tension and shear tests on metal-glass connections

Five test specimens were tested in tension and another five in double-lap-shear for metal-glass connections. For the metal substrate stainless steel with a P300 grinded surface was used. The dimensions of the adhesive joint were 100 x 20 x 3 mm. The dimensions of the glass panes were chosen based on availability (300 x 200 x 10 mm), while the dimensions of the metal elements were chosen based on reasons concerning manufacturing and test setup.

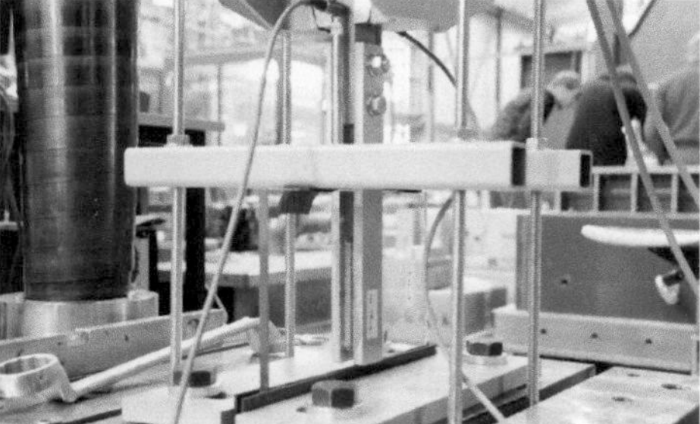

Figure 4-1 Experimental setup for the tension test specimens (left) and the double-lap-shear test specimens (right) for metal-glass connections.

Figure 4-1 illustrates the test setups for tensile loading (left) and for shear loading (right) of the adhesive joint. Similar to the metal-metal connections, the load was applied with clamping jaws and the glass pane was fixed with steel bars. Since glass is a brittle material, the contact between the glass pane and steel components was avoided by using EPDM rubber layers between them. Displacement transducers were used to measure the relative movement between the metal sheets and the glass pane.

4.1 Manufacturing of the test specimens

The test specimens were manufactured in the laboratory at Graz University of Technology with a similar procedure as the metal-metal connections. The only difference is that for the specimens tested under tension the self-weight of the metal elements was used instead of fixing with screw clamps. The real values of the adhesive bite dimensions were measured for all specimens and are given in Table 4-1 along with the target values.

Table 4-1 Dimensions of the adhesive joints in the metal-glass connections.

Dimension [mm]	target value	real value *	
		Tension test specimens	Shear test specimens
length l-a	100.0	100.1	100.8
width w-a	20.0	18.8	20.0
thickness t-a	3.0	3.1	3.1
* the specified real values are mean values of the measured dimensions			

4.2 Typical failure patterns

Typical failure patterns for both the tension and the shear test specimens for metal-glass connections are illustrated in Figure 4-2.

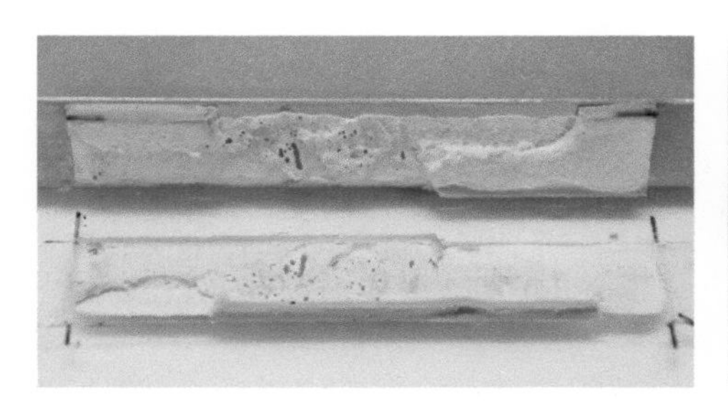

Figure 4-2 Typical failure patterns of the tension test specimens (left) and the double-lap-shear test specimens (right) for metal-glass connections.

For the specimens tested under tension only cohesive failures as the one shown in Figure 4-2 left were noticed. For the specimens tested in shear generally cohesive failure as the one shown in Figure 4-2 right occurred, but in particular cases also a mixed failure was noticed. A partial debonding from the glass surface was observed.

4.3 Mechanical test results

For the metal-glass connections, the results of the tension tests are plotted in the form of engineering stress vs. engineering strain curves (see Figure 4-3 left) and the results of the double-lap-shear tests in the form of engineering stress vs. shear displacement curves (see Figure 4-3 right). A good agreement of the test results for both loading types can be observed between the different test specimens regarding stiffness. In the case of the tensile loading also the maximal stresses are in good agreement, while for the shear loading a variation can be noticed. These differences can be explained by the occurrence of mixed (cohesive and adhesive) failure of some of the adhesive joints loaded in shear.

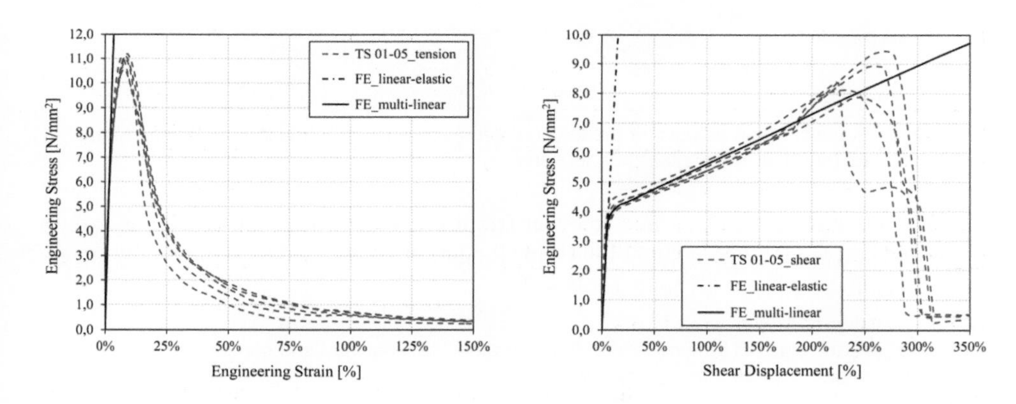

Figure 4-3 Engineering stresses plotted vs. engineering strains for the tension test specimens (left) and vs. shear displacements for the double-lap-shear test specimens (right) for metal-glass connections.

Additional to the curves from the mechanical tests, results from finite element simulations with a linear-elastic and a multi-linear (bilinear) material model for the adhesive are represented in the two diagrams. For the linear-elastic model, the Young's modulus from Table 2-1 is used. For the multi-linear model a yielding stress (7.0 N/mm^2) is provided above which the stiffness of the material becomes significantly lower. The glass and the stainless steel are modelled as linear-elastic. For all the materials 20-node solid elements were used, for the glass and the stainless steel type C3D20 and for the adhesive type C3D20H (Abaqus). The contact between the adhesive and the substrate materials was modelled with tie constraints, excluding an eventual adhesive failure. For both setups only one quarter of the system was modelled by applying symmetry boundary conditions in order to reduce the computation time (see Figure 4-4). The linear-elastic models show a good agreement with the experimental results only for small strains were the adhesive behaves almost linear. For higher strains the stiffness of the system is overestimated. The multi-linear model shows a good agreement also for higher strains, especially for the double-lap shear setup.

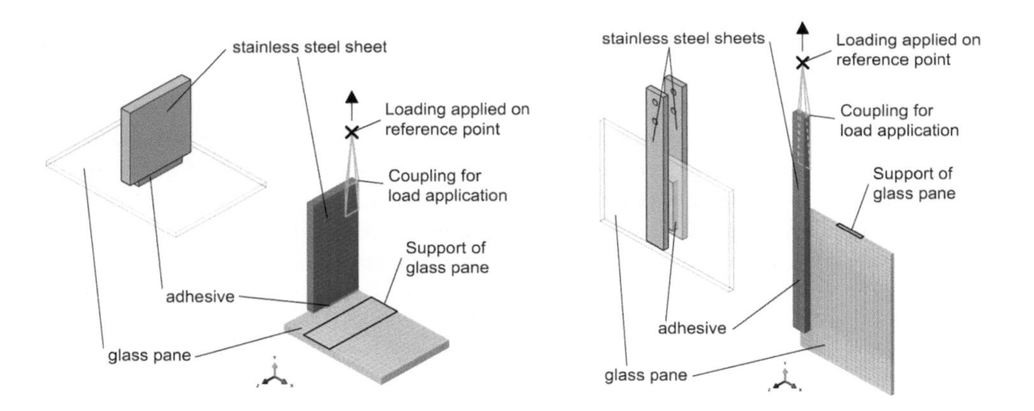

Figure 4-4 Finite element models for the tension test specimens (left) and the double-lap-shear test specimens (right) for metal-glass connections.

The mean values for stresses in a relevant strain range (up to 25 %) as well as failure stress and corresponding maximal load and elongation for the two experimentally investigated loading types are given in Table 4-2.

Table 4-2 Mean values of the stress at different strains, maximal load and corresponding elongation for the adhesive metal-glass connections.

Specimen type	Stress at 5 % strain [N/mm^2]	Stress at 15 % strain [N/mm^2]	Stress at 25 % strain [N/mm^2]	Failure stress [N/mm^2]	Maximal load [kN]	Elongation at max. load [mm]
tension	9.89	7.67	3.73	10.97	20.71	0.25
shear	3.44	4.26	4.42	8.55	34.52	7.55

5 Conclusions and outlook

The presented investigations aimed to provide a solid knowledge base on the behavior of the selected two-part structural acrylate adhesive in connections to different materials and surface conditions. Since the planned application for the adhesive is metal-glass elements with composite structural behavior, the transfer of shear loads was of special interest and therefore mainly double-lap-shear test setups were chosen. The results on connections under shear loading showed that no adhesive failure occurs while the material is in the elastic area. After plasticizing of the adhesive, the stiffness of the connection decreases and above shear displacements of around 200 % an adhesive failure may occur on the metal substrates. A proper surface condition (grinded with P150 or P300 for the stainless steel and anodized for the aluminum) results in a mixed or a cohesive failure. On the glass surfaces cohesive or mixed failure was noticed. In the case of the metal-glass connections under tension only cohesive failure was noticed and the stiffness was almost constant until the maximum engineering stress was reached.

The obtained results for the considered environmental boundary conditions prove the suitability of the selected adhesive for metal-glass connections where a higher stiffness and strength than the ones reachable with silicones are required. Further investigations are necessary for the influence of low and high temperatures as well as of moisture on the behavior of the adhesive connections, especially when an application in the façade area is aimed for. Another aspect to analyze is the loading rate dependency of the acrylate behavior, especially regarding optimization of material models for finite element simulation.

6 Acknowledgements

This research was carried out at Graz University of Technology as part of the FFG research project no. 838561. The authors would like to acknowledge the collaboration with Waagner-Biro Stahlbau AG and the Laboratory for Structural Engineering at Graz University of Technology on this research project as well as the financial support from the Austrian Research Promotion Agency (FFG) and from Waagner-Biro Stahlbau AG for this project. Further, the authors acknowledge the support of Sika Services AG in Widen (CH) and Sika Österreich GmbH in providing the adhesive and in manufacturing the specimens.

7 References

[1] Wellershoff, F.: Nutzung der Verglasung zur Aussteifung von Gebäudehüllen. PhD thesis: RWTH Aachen, 2006.

[2] Netusil, M.: Hybrid steel-glass beams. PhD thesis: CTU Prague, 2011.

[3] Whitehouse, D.: Surfaces and Their Measurement. Elsevier, 2004.

[4] Kothe, Ch.: Surface modification methods for improving adhesive joints in glass structures. PhD thesis: TU Dresden, 2013.

[5] Silvestru, V.A.; Kolany. G.H.E.; Englhardt, O.: Glass Building Skins – Presentation of the research project and intermediary findings. In: Proceedings of Advanced Building Skins, Graz, 2015, pp 118-126.

[6] ISO 37: Rubber, vulcanized or thermoplastic – Determination of tensile stress-strain properties, 2011.

[7] Overend, M.; Jin, Q.; Watson, J.: The selection and performance of adhesives for a steel-glass connection. In: International Journal of Adhesion and Adhesives, Vol. 31 2011, 587-597.

[8] Belis, J.; Van Hulle, A.; Out, B.; Bos, F.; Callewert D.; Poulis H.: Broad screening of adhesives for glass-metal bonds. In: Proceedings of GPD Finland 2011, Tampere, 2011, pp 286-289.

[9] ETAG 002: Guideline for European Technical Approval for Structural Sealant Glazing Systems (SSGS), 2001.

Glass resists – Temporary coatings for sheet and hollow glass

Michael Gross[1]

1 KIWO – Kissel + Wolf GmbH, In den Ziegelwiesen 6, 69168 Wiesloch, Germany,
michael.gross@kiwo.de

What is a resist? Using a resist, a glass surface can be masked with a design, either partially or all-over, prior to further finishing, processing or transportation, thus protecting them from outside influences. The resists are medium to highly viscous emulsions and depending on the product, can either be screen printed or applied by spraying, roller coating or dipping.

Keywords: resists, coatings, surface protection, surface masking, surface processing, protection

1 Resists in general

1.1 For which applications are resists used for?

Resists & Coatings include products for partial etching and sandblasting, decorative sputtering and surface protection. These include heat-curable etching resists for use in acidic and alkaline etching and plating baths for sharp contoured screen-printing of glass or mono- and polycrystalline solar wafers; or peel and water-resistant, screen printable masks with good resistance to aqueous cleaning processes, as they are commonly used in the pre- and post-cleaning stage prior to glass processing. But also sandblasting-resistant masks which are elastic and easily removable again as well as UV-reactive sandblasting photoresists for squeegee or spray application with attainable contours fineness up to 75 microns (with suitable fine-grained sand blasting agents) are examples of interesting product developments. With a higher coating thickness, the resist is also suitable for deep sandblasting to represent multi-dimensional effects.

1.2 Application examples

1.2.1 Sandblasting

A common method for producing matt surfaces is by sandblasting. Also in this type of surface treatment, it is possible to mask surfaces such as glass or metal with a printable resist, in order to protect them from the sandblast corundum. Imagine, you want to sandblast a design on a sheet of glass. The easiest way to do that is using a stencil which usually consists of a plotted self-adhesive film. But if you want to produce more than one piece with the same motif, a new stencil is required for each piece that you wish to produce. With the possibility to use a printable resist, you can speed up your production dramatically and this saves you a lot of money. Removing a printable sandblast resist is

Engineered Transparency 2016. Glass in Architecture and Structural Engineering. First Edition.
Edited by Jens Schneider, Bernhard Weller.

very easy. The resist is peelable like a film or you can wash it off with high pressure water or special chemicals. This is a further advantage compared to self-adhesive plotted film.

Figure 1-1 Sandblasted glass.

1.2.2 Decorative sputter coating of architectural glass

Sputtering, also called magnetron sputter coating, is the removal of precious metal atoms, such as platinum or gold, in a vacuum chamber by energetic ion bombardment, to coat a substrate with the sputtered metal particles. This coating has mainly functional reasons and is common for heat reflection at architectural glass – well known as mirrored glass on skyscraper facades in financial metropoles. The same application method is used for decorative sputter coating of architectural glass. Only that for the decorative sputter coating the metallic deposition is applied partially and not over the entire surface. The areas which should not to be coated, are masked with a resist. The resist is usually applied by means of screen-printing. After the sputter coating process, the resist and the metallic coating above can be removed with a special cleaning chemical either manually or in an automatic washing unit.

Figure 1-2 Sputter coating.

We are very proud to be able to offer our new development – an inkjet compatible sputter resist for flat glass application. The resist is compatible with large format digital printers for the flat glass industry. This opens up many new opportunities. Glass facades are mainly decorated with a continuous pattern. Every sheet of glass has another design. Masking by means of screen-printing would mean that for each sheet of glass a new screen-printing stencil is required. This is not the case with inkjet printing. A further opportunity is that inkjet is able to serve all sizes of float glass. The biggest screen-printing machine for flat glass has a maximum printing size of 3.25 to 8 meters. Digital printers for flat glass can print 3.25 to 18 meters.

1.2.3 Liquid protective film

Once a glass or metal surface has been processed, it can be protected from dirt and mechanical stress, with a "liquid" protective film. After drying, the surfaces are temporally protected against various chemicals and mechanical stress. Is the protection no longer needed, the resist can be removed as film, without any residue on the substrate. Depending on the product, the resist can be applied by screen-printing, spraying, roller, curtain or dip coating.

Figure 1-3 Liquid protective film.

Canopy for the tram and bus stop "Krefeld Ostwall"

Markus Kramer[1]

1 IB KRAMER Structural Engineering FEM-Analysis, Nierenhofer Str. 68A, D-45257 Essen, info@ib-kramer.de.

The „Ostwall", one of Krefeld's most important public traffic intersections, is getting modernized completely for the moment. The highlight of the project is the new canopy for the tram and bus stop. It was decided to realize a steel-glass structure, to get most possible transparency. The structure has a length of 125 m and a width of 12 m. The main structure has 9 column brackets, that are bearing a space truss. These elements are made of steel. For the roofing, partially curved glass elements are used, that are bonded to a stainless steel frame by structural silicone. The slope of the roofing reaches to the gutter in the middle axis of the building. For each side, only one glass element is used, to guarantee a good flow of rain water. The glass elements have a developed length of 5.2 m and a width of 2.2 m. At the front ends of the structure, spherically curved glass elements are used. The glass roof is accessible for cleaning and inspection work. To get a smooth view of the bottom surface, the glass is bonded under the steel frame, so the silicone bonding is used to bear the self weight of the glass elements. The pre-fabricated elements are fixed to the main structure at the bottom girder of the framework directly, and at the top girder by a suspension bar. To get a minimized cross section of the stainless steel bars, they are stabilized by the glass elements to prevent them from torsional-flexural buckling.

Keywords: silicone bonding, curved glass, FEA, structural design, glass as bracing element

1 Introduction

Krefeld's most important public traffic intersection, the "Ostwall" is going to be changed and modernized for the moment. The architectural highlight should be the canopy for the tram and bus stop "Ostwall / Rheinstrasse". Even Krefeld is well known for its textile industry, the city council decided to realize a steel-glass structure, after a long political discussion.

The architectural design was done by Stefan Schmitz Architekten, Cologne. A 125 m long and 12 m width structure with 9 pairs of pillars, bearing the main structure, made of a 3D-Vierendeel-girder was designed. The surface of the canopy, drafted as a hanging element, should be made of glass, with a smooth and curved shape, sloping to the center line, where the gutter was placed. Along the outer edge, a metal frame was designed to take an illumination made of LED-stripes. Additionally the suspensions for the contact wire of the tram should be fixed to the structure of the canopy (Fig. 1-1).

Engineered Transparency 2016. Glass in Architecture and Structural Engineering. First Edition.
Edited by Jens Schneider, Bernhard Weller.

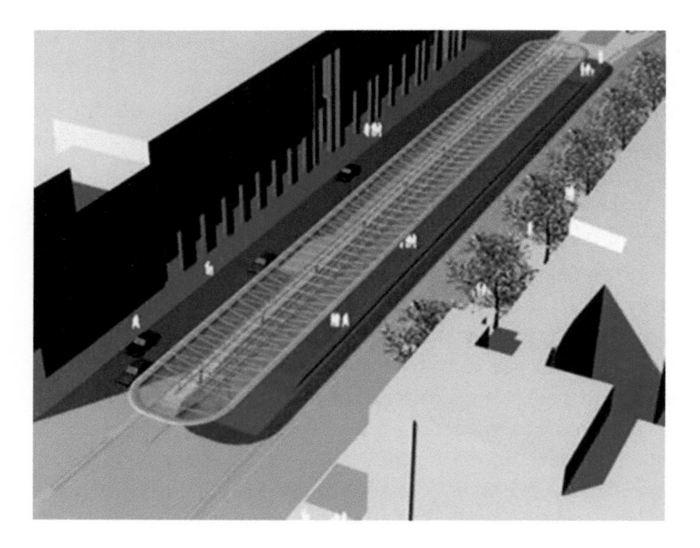

Figure 1-1 Architectural design, ©Stefan Schmitz Architekten.

2 Load bearing system

2.1 Main steel structure

For the main steel structure, 9 pairs of columns, bearing a 3D-Vierendeel-girder was chosen. The columns had a slope in lateral direction. To brace the structure in lateral and longitudinal direction, the columns were restrained to concrete fundaments. The 3D-Vierendeel-girder on top of the columns was designed as a Gerber-girder with 7 elements. The hinged connections were done with sockets, also designed, to take the dilatation due to thermal loading (Fig. 2-1).

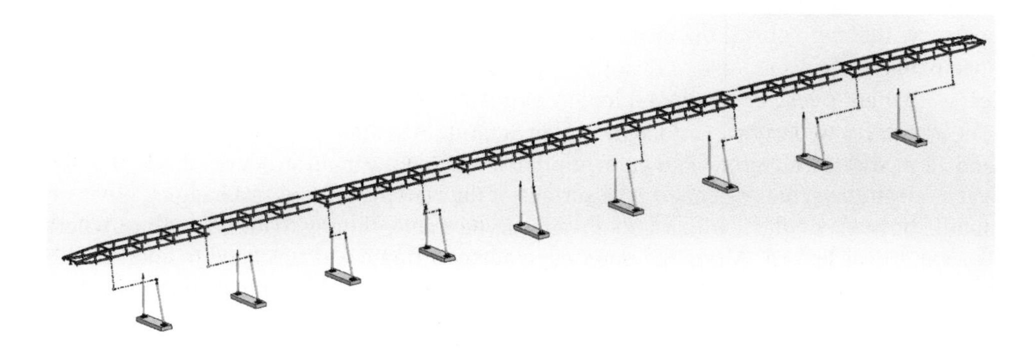

Figure 2-1 Load bearing system of main steel structure.

2.2 Steel-glass elements

For the glass surface two types of hybrid steel-glass elements with structural silicone bonding were used. The 104 regular glass elements in the middle had a developed length of 5290 mm and a width of 2215 mm. The shape was planar at a length of 3420 mm, cylindrically curved at a developed length of 1677 mm, and planar at the end at a length of 168 mm. To "close" the structure at the front ends, four spherically curved glass elements were used, building up a semi-circle in the top view (Fig. 2-2).

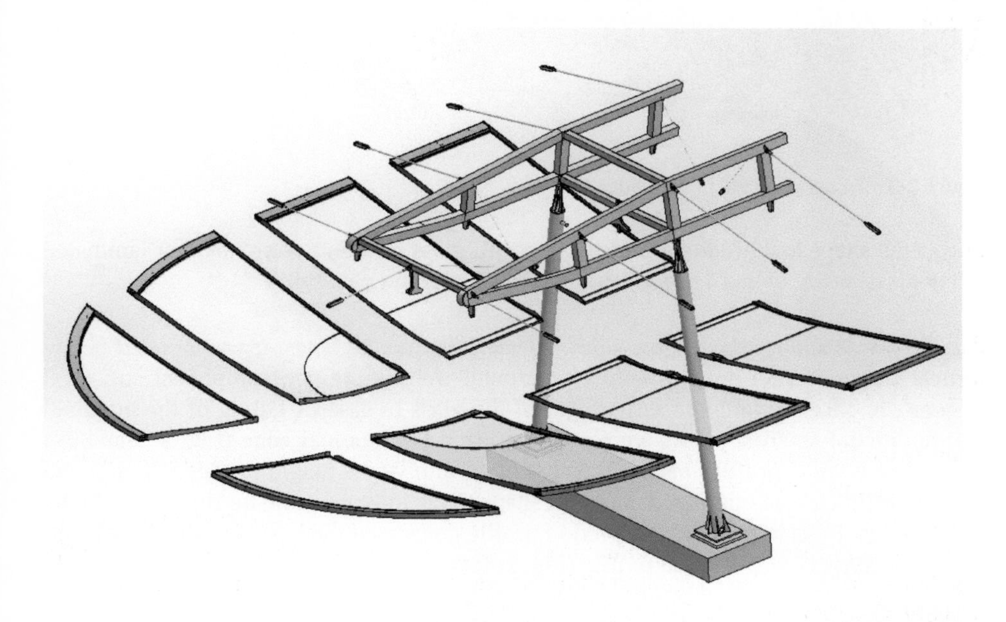

Figure 2-2 Regular and front end glass elements.

The elements had a slope of 11° at the upper planar area. It was decided to use only one glass element over the developed length to guarantee a proper water flow with no gaps perpendicular to the flow direction. The use of semi-curved elements was reasonable to optical and statical aspects: The curved part of the glass helped to get a smooth and dynamical bottom view of the canopy, and from the statical point of view, it stiffened the glass element, so that the quite wide upper planar area had a quasi-four-sided support. Especially the last aspect was the reason why it was possible to realize the required pattern of 2.25 m for the glass elements, and to avoid a support of the short edge at the bottom of the glass element, that would have restrained a proper water flow to the gutter. At the bottom edge the glass was designed with a step, to give the ability to fix a drip plate (Fig. 2-3).

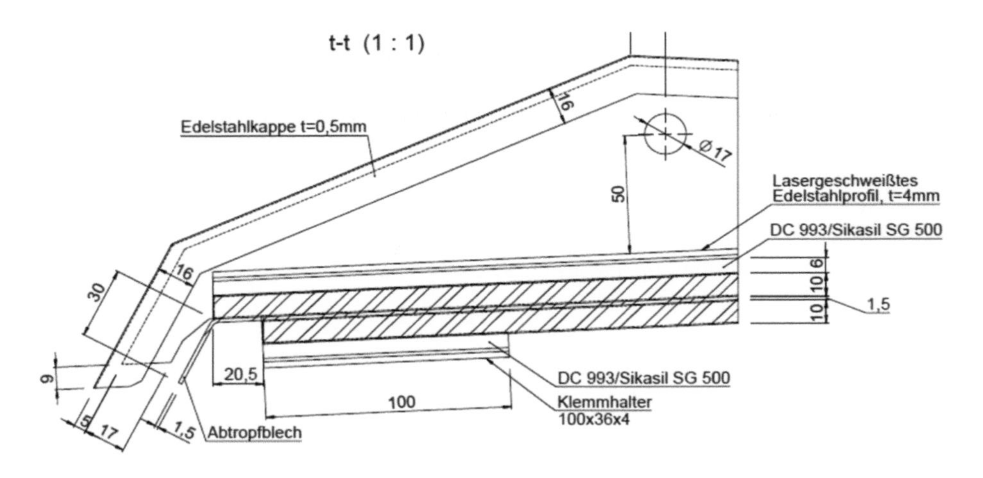

Figure 2-3 Drip plate at the bottom edge.

Due to the shape and production restrictions, all glass elements were made of laminated safety glass, made of 2x10 mm float glass, PVB interlayer 1.52 mm.

The glass was supported at three sides. At the longitudinal sides, it was bonded below vertical stainless steel profiles with a horizontal web by the two-component structural silicone DC 993. To take the self-weight of the glass in case of failing of the bonding, four additional small supports were added at each longitudinal edge (Fig. 2-3 and 2-4 a),b)). The vertical profiles had a thickness of 10 mm, and a height between 75 mm and 111 mm following an optimized, laser cut shape to get most possible transparency. At the bottom edge a rectangular hollow section profile connected the lateral profiles, and supported the glass by a U-section profile (Fig. 2-4c)).

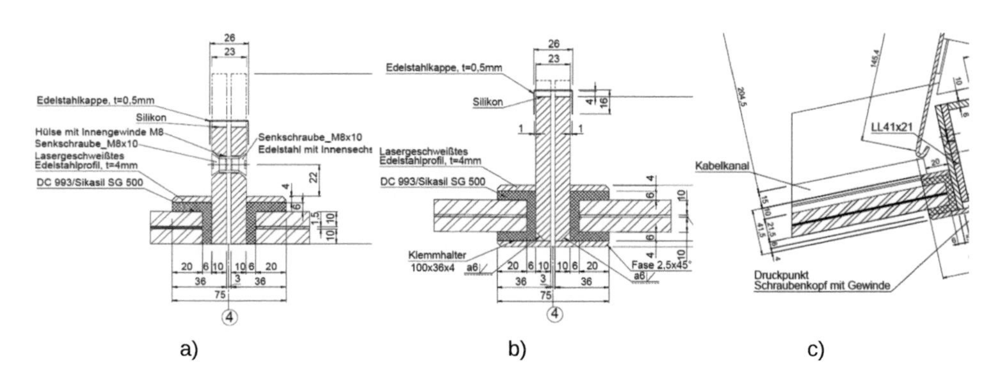

Figure 2-4 a), b), and c) – Cross section of stainless steel frame profiles.

The three stainless steel profiles and the semi-curved glass built a hybrid steel-glass element that was put to the main structure by two suspension rods, fixed to the upper boom of the 3D-Vierendeel girder, and directly to the lower boom by a gab (Fig. 2-5 and 2-6).

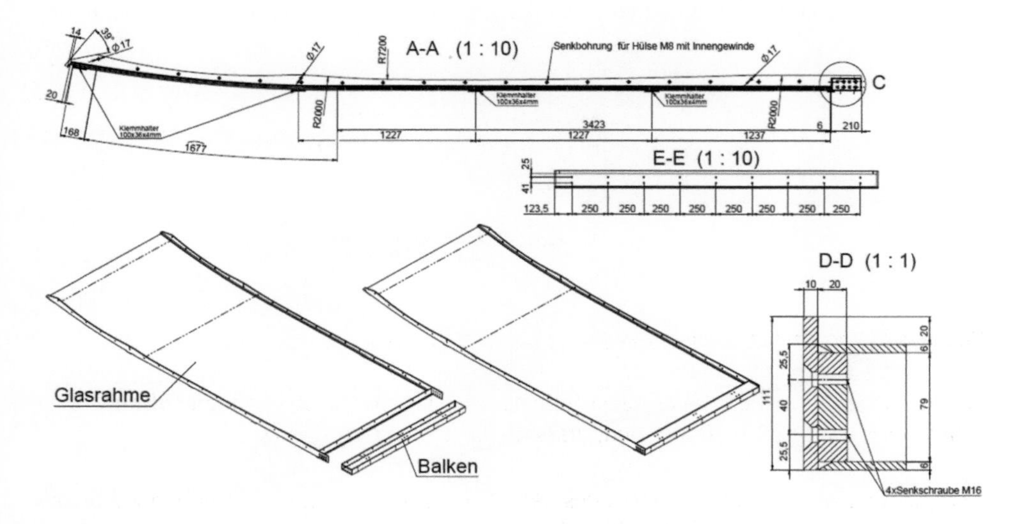

Figure 2-5 Steel-glass frame.

Normally the statical system of the suspended steel-glass frame would cause severe stability problems to the lateral steel profiles, because the normal force is quite high, due to the small slope of the forward suspension rod (Fig. 2-6).

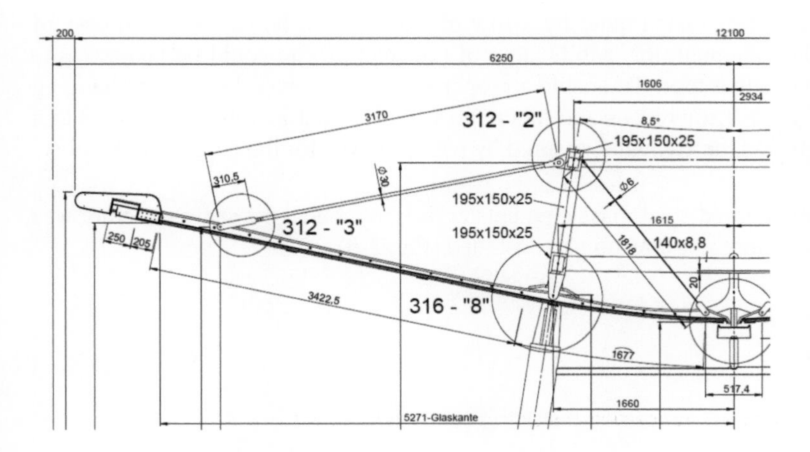

Figure 2-6 Cross section of installed regular steel-glass element.

But by creating a hybrid load-bearing element, bonding the steel and glass elements together, it was possible to avoid any problem of lateral torsional buckling. The glass elements acted as a bracing element, preventing the very slender steel profiles from lateral spreading, and the bonding was sufficient to generate an elastic rotating support. So the steel profiles could be designed very slender and small to get most possible transparency (Fig. 2-7).

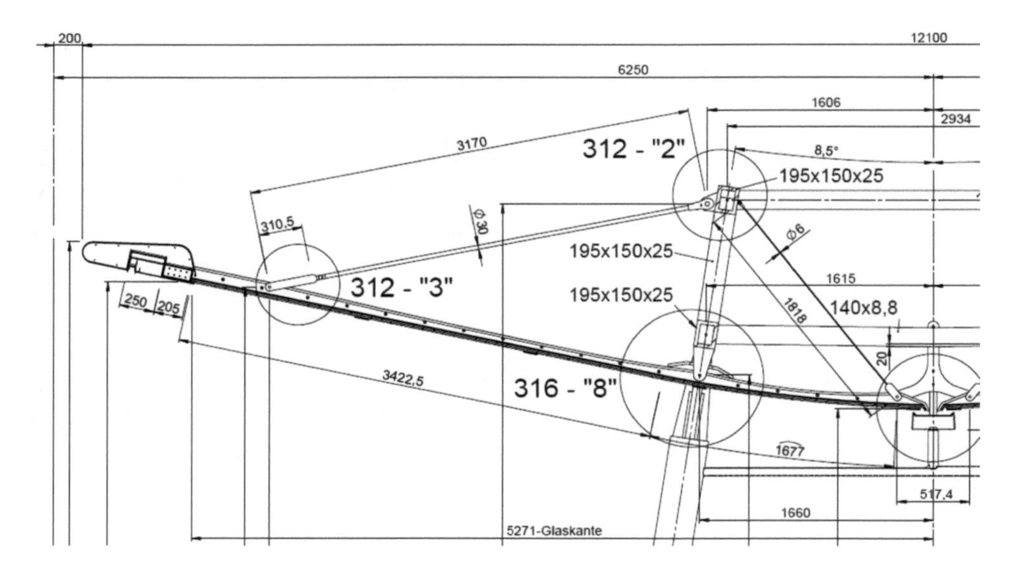

Figure 2-7 Bottom view finished canopy.

The frames were fixed to each other by a row of countersunk bolts, so that in case of failing of one glass element, the stabilization of the steel profile could be taken by the steel-glass elements next to it. To enable a proper installation, even in case of changing an element, and to take tolerances and dilatation from thermal loading, a gap of 3 mm was designed between the elements, assured by using sleeves for the countersunk bolts.

For the spherically curved steel-glass elements at the front ends, the same structure was chosen principally, but there was no gap designed (Fig. 2-8).

Figure 2-8 Front end.

The metal frame around the outer edge is made of a load bearing, welded U-section profile, containing the LED-stripes, and covered by an aluminum cap. To avoid dismounting the LED-lights and the aluminum cap in case of changing one of the steel-glass elements, the U-section profile was designed to bridge the gap of a missing glass element. Also the supports for the contact wire were fixed to the surrounding frame (Fig. 2-9).

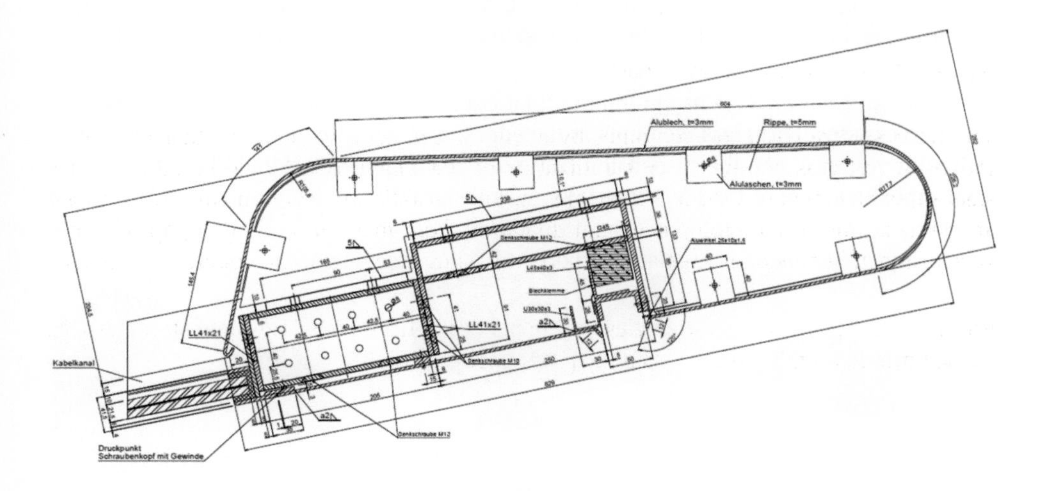

Figure 2-9 Surrounding frame with aluminum cap and LED.

2.3 Fabrication

The columns, the seven parts of the 3D-Vierendeel girder and all steel-glass elements were pre-fabricated. So all structural bonding were done in factory under defined temperature and defined humidity conditions.

3 Proofs

3.1 Analytical proofs

The main steel structure was proofed according to DIN 18800 for steel structures in combination with the German national technical approval "Allgemeine bauaufsichtliche Zulassung Nr.: Z-30.3-6 Erzeugnisse, Verbindungsmittel und Bauteile aus nichtrostenden Stählen", which is the basic to proof stainless steel elements in Germany. This approval refers to DIN 18800, so this code was chosen to proof the complete steel structure instead of using DIN EN 1993.

Loads and load combinations were assumed according to DIN 1055-100. In detail dead weight, wind loads, and snow loads were used to analyze the main structure with a standard 3D-framework software.

For the steel-glass elements a complete 3D-finite element model was generated and analyzed with ANSYS 14.5. In addition to the loads assumed for the calculation of the main steel structure, a man load according to GS BAU 18 in context to DIN 4426 was set to the structure, because the glass elements had to be entered by cleaning service staff.

The glass and the profiles at the longitudinal edge were generated with shell elements, the stainless steel and glass elements at the edges, and the silicone bonding were generated with volumes, and the suspension rods with beam elements (Fig. 3-1). For the glass no composite action of the PVB-interlayer was assumed.

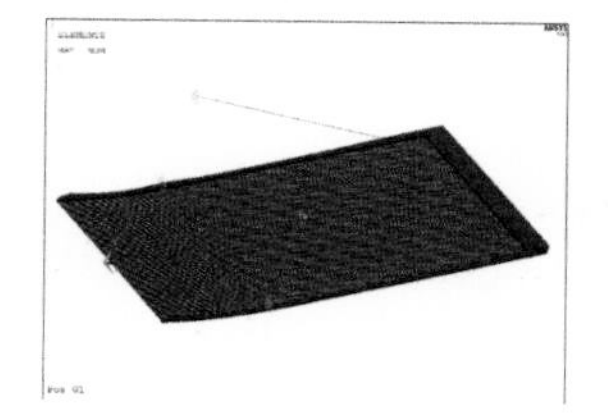
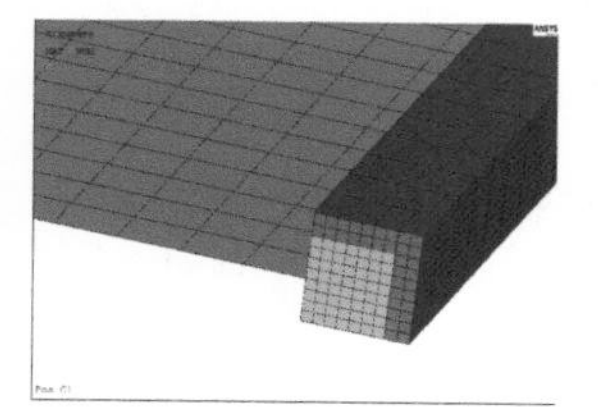
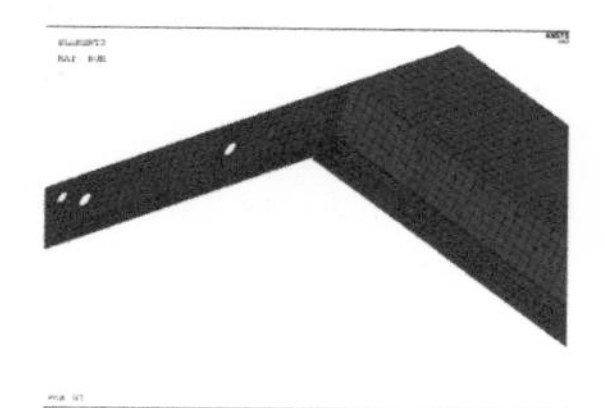

Figure 3-1 Complete model (left). Glass and silicone (centre). Stainless steel (right).

Because the expected strain of the silicone bonding is very small, and is even not allowed to be high, due to the needed elastic supporting effect for stiffening the slight steel profiles, a linear elastic material law is defined to generate the silicone in the FEA-model, instead of using a hyper-elastic material law. To get the Young's modulus for the silicone, five small specimen with a length of 300 mm and original dimensions of the silicone bonding, and the stainless steel profile, were tested at the Labor für Stahl- und Leichtmetallbau, Munich. The glass panel was restrained, and the steel profile was loaded by a single load. At six points, the displacements were measured. To verify the Young's modulus of the silicone, a FEA-model of the specimen was generated, and the load-displacement-curve of the FEA was compared with the curves from the tests. A sufficient conformity was obtained with a Young's Modulus of 1.2 N/mm² (Fig. 3-2).

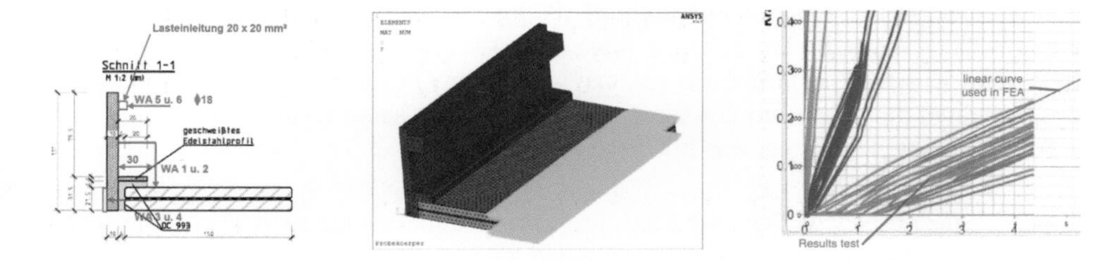

Figure 3-2 Specimen (left). FEA model (centre). Comparison load-displacement-curves (right).

The verified Young's modulus was applied to the complete FEA-model. To proof stability of the slight steel profiles, a pre-deformation following the ideal lateral-torsional-buckling figure, was put to the FEA-model, and the total deformation and stresses for all materials were calculated for the loads given above. The stainless steel elements were proofed according to DIN 18800 in combination with the national technical approval for stainless steel. Needed results from the FEA were the von-Mises-stress.

To proof the glass the first principle stresses were decisive. Proofs were done according to DIN 18008 "Glas im Bauwesen". For the characteristic resistance of the curved float glass, values according to BF Bulletin 009/2011 "Guideline for thermally-curved glass in the building industry" were used, with $f_k = 40$ N/mm².

Directional stresses were used to proof the silicone bonding according to ETA-01/0005 "DC 993 and DC 895", and ETA-03/0038 "Sikasil SG 500". The minimum allowable design stress was used for the proofs, because the type of silicone bonding should be open before contracting.

For the spherically curved elements only a shell model of the glass was generated, because the loading to the steel elements was lower than at the regular elements, due to the smaller width of influence.

3.2 Proofs based on tests

Proofs based on tests at an original mockup were needed to check the post breakage behavior of the glass elements, and to check safety to prevent from falling thru the glass element in case of cleaning staff falling onto the glass.

For both tests three original steel-glass-frames had been produced and installed at a test structure in original position.

First the test to simulate a falling person according to GS BAU 18 was done. Therefor a bag filled with 50 kg of glass beads was dropped from a height of 1.2 m onto the pre-damaged glass unit at two different positions (Only the upper glass of the laminated glass unit was destroyed). After this a weight of 100 kg was put to the place, the bag laid, at an area of 200 x 200 mm. The test was fine, when the glass elements resisted the loading more than 15 min. This target was reached without any problem. Even the lower pane of the laminated glass unit did not show any damage. The same result was getting for the second element, so this test was fulfilled.

To check post breakage behavior, both pre-damaged glass units were used. The second layer was destroyed, too, and then a simulated snow load of 0.5 kN/m^2 was put onto the glass. Both units were able to take the load more than 24 h (Fig. 3-3). To check the effect of the bonding, at one unit the silicone sealant was cut completely. Even this additionally damaged element did not fail.

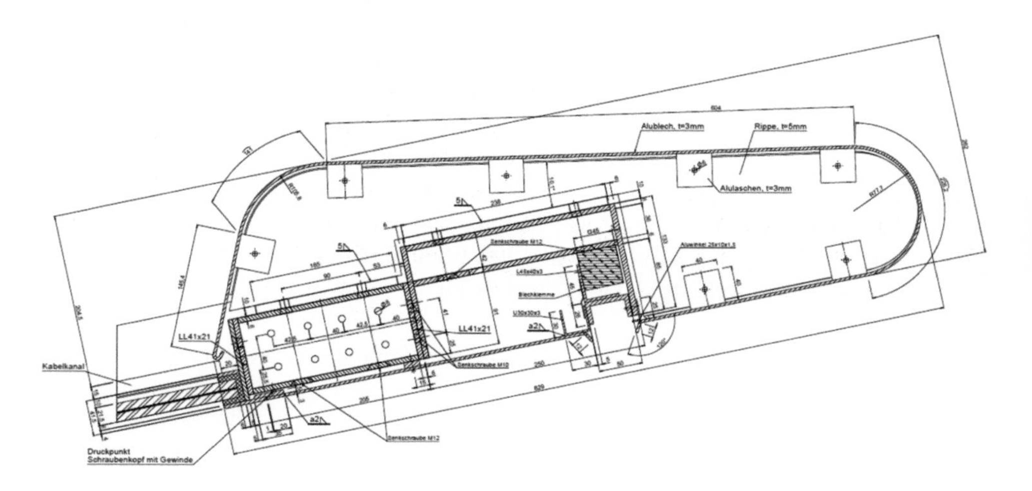

Figure 3-3 Test for post breakage behavior.

The very positive post breakage behavior is reasonable for the cylindrically curved part of the glass. Normally a two, or three side supported flat glass pane would not had have any chance to confirm this test, having a width of more than 2.2 m. But due to the high stiffness of the curved part of the glass element, the plane part of the glass element shows a behavior similar to a four side supported pane, so the curved part of the glass is acting like an additional supporting for the large plane area of the element.

4 Conclusion

For the canopy of the tram and bus stop "Krefeld Ostwall", a high transparent steel-glass structure could be realized. The consequent use of glass as a stiffening and load bearing element helped, to minimize the steel elements of the steel-glass frames, because stability failure could be avoided. Also the silicone bonding, connecting the stainless steel and the glass elements was suitable for this kind of fixing, because it was elastically enough for the brittle material glass, but stiff enough to generate an elastic supporting for the steel profiles to prevent them from lateral torsional buckling.

All in all, the realized structure is not so far away from the first architectural design (Fig. 4-1).

Figure 4-1 Realized project.

Resilience, Damage Tolerance & Risk Analysis of a Structure Comprising Structural Glass

Peter Lenk[1], *Dániel Honfi*[2]

1 Arup, London, UK, Peter.Lenk@arup.com

2 SP Technical Research Institute of Sweden, Gothenburg, Sweden

With increased popularity of structural glass applications, questions about structural safety and risk associated with such applications to the general public shall be examined in detail. Often these types of applications are used in public or semi-public spaces thus they receive increased attention and the consequences of failure or underperformance are crucial. Further, a comprehensive, standardised methodology of designing glass structurally is still lacking. In this paper we will review resilient concepts currently adopted by engineers designing with structural glass to ensure that structural reliability and robustness criteria are met. This is especially challenging, since a well-established probabilistic basis, that is (at least partly) available for traditional construction materials, has not yet been developed for structural glass. To ensure the consideration of a system and life-cycle perspective, operational objectives, damage sensitivity and damage tolerance for possible load exposures during the design lifetime will be outlined.

Keywords: resilience, risk assessment, structural glass

1 Introduction

Failure of structural glazing poses potential risks to building occupants, therefore it is expected that the failure of such applications, or more general the risks associated with their failure, should be below an accepted level. This raises several questions concerning e.g. methods of risk assessment, acceptance of risk and possibilities for mitigation.

2 Review current risk assessment methods

2.1 Definition of risk

While progress and development in any (technological) field is undoubtable linked to risk and individuals willing to take risk are often seen as pioneers, only those who evaluated and grasped all potential outcomes of theirs actions fulfilled their duty to the society. Proceeding in activities with limited understanding could be acceptable only if appropriate risk management techniques are adopted. In principle this means that comprehensive assessment of all eventualities was carried out and outcome of such analysis is presented to key stakeholders. This applies for design and construction of structures made of glass as well. Achieving larger spans and heights or more slender elements, application of novel

Engineered Transparency 2016. Glass in Architecture and Structural Engineering. First Edition.
Edited by Jens Schneider, Bernhard Weller.

type of connections and materials requires taking risks that needs to be assessed and evaluated carefully, which glass designers, often even intuitively, do. However, scientific methods and guidance are available to support their work.

In general the risk is characterized as a combination of probability of an event and its consequence, see e.g. ISO/IEC Guide 73 [1]. According to Faber and Stewart [2] technical risk is typically defined as the expected consequences associated with a given activity. Mathematically risk R can be expressed as the product of the probability P that an event (associated with the given activity) will occur and the consequences C given the event occurs. Activities usually involve more than one events, thus the risk associated with a given activity R_A relating to all possible events n_E which may follow as a result of the activity may be expressed as:

$$R = \sum_i^{n_E} P_{E_i} C_{E_i} \tag{2.1}$$

where E_i denotes event i. Risks could be expressed in economical or sociological terms. In structural engineering both are important while obviously human life is of key consideration. Therefore it is essential to have a common basis for comparison/aggregation of different type of risks. General risk assessment methods are given in ISO 31000:2009, where principles and guidelines are outlined applicable to every human activity. Thus its recommendations could be seen as a common ground and applicable to assess risk associated with the design of structural glass applications.

Risks are not measurable only in terms of safety to public; capital cost, disruption in business and reputation of business etc. should also be considered. Thus just by simply comparing surface strength of a plane glass, as calculated according to various codes around the world, it is clear that same glass design may be difficult to justify if relocated geographically. Perception of safety, in structural glass, is clearly subjective. This subjectivity is even reflected in the words people use expressing their perceptions of risk Kent [3]. To overcome this subjectivity of risks, standardized procedures of risk assessment have been developed in various fields. Some risk assessment methods used in structural glass design are presented in the following subsections.

2.2 Ciria C 632

In designing with glass and particular in structural glass safety is an important topic for discussion. Detailed glass risk assessment is usually a requirement of Construction Design and Management (CDM) Regulations [4], but it can inform clients and guide contractors who wish to offer alternative design. CIRIA C 632 – Guidance on glazing at height 2006 [5] is a generally followed document to complete qualitative risk assessment in form of risk matrices (Figure 2-1); however, in its appendix need for more sophisticated probabilistic methods is recognized.

5. Extremely High (almost certain, p=0.93)	5	10	15	20	25
4. High (probable, p=0.75)	4	8	12	16	20
3. Medium (very possible, p=0.5)	3	6	9	12	15
2. Low (possible, p=0.3)	2	4	6	8	10
1. Insignificant (Very unlikely, p= 0.07)	1	2	3	4	5
Likehood / Consequence	1. Insignificant (Minior injury not requerig first aid)	2. Low (Minor injury requering first aid)	3. Medium (Injury causing absence from work for three or more days)	4. High (Serious injury, fractures, disablement)	5. Extremely High (Death, disablement)

Figure 2-1 Qualitative risk matrix as per recommendations of CIRIA C 632.

2.3 LDSA probabilistic method

Example of probability based risk assessment was presented by the London District Surveyors' Association in [6]. The example shows the calculation of the risk of fatality of single toughened vertical glass sheet due to the nickel sulfite inclusion only however this methodology can be extended to any other potential tread and application:

$$F = F_f x P_0 x P_c x P_d x P_p x P_f \tag{2.2}$$

Total frequency of fatality F is simple product of partial probabilities: F_f – frequency of potential tread, P_0 – probability that glass will fall outwards, P_c – probability that glass fragments will fail in chunks (risk intensity- wet blanket, dislocation of entire element or it's significant portion), P_d – probability that protection screen, canopy (if present) will not break up glass chunks or give partial protection, P_p – probability that area under falling glass fragments is occupied by people (risk mitigation, moving entrance no go areas etc.), P_f – probability of a fatality (consequence severity) if person is struck by glass fragments. Above methodology is simple however detail methodology guidance to each partial probability needs to be agreed on.

2.4 NEN 2608+C1:2012

Latest Dutch code NEN 2608+C1:2012 in appendix D offers a risk assessment methodology. The method is commonly known as the Fine/Kinney method, since it is based on early work of Fine 1971 [7], which then was later revised by Kinney 1976 [8]. Fine proposed to calculate risk due to a hazard by evaluating the potential consequences of an accident, the exposure or the frequency of occurrence of the hazard-event that could lead to the accident and the probability that it will result in the accident with the above defined consequence. It shall be noted that it is not clear from the work if rating has been arbitrarily set based on experience or some deeper analysis was considered. The weight of probability of exposure is equal to consequence, while probability rating resemble some correlation with percentage of estimative probability by Kent [3] summarized in Table 1. In this work risk levels are correlated to urgency of action required to mitigate the risk. While as per our opinion key value of this work is in justification of effectiveness of proposed risk mitigation. In NEN risk is also associated with level of damage of the structural element.

3 Acceptable and tolerable risk limits

According to Health and Safety Executive in UK (HSE) all risks to public shall be as low as reasonably possible (ALARP) [9]. This method is also widely accepted in civil engineering applications, which fall outside the scope of codified design. Distinctions between individual risk (specific to a person, location and time) and Collective risk (an aggregate of all events being considered irrespective of who is affected). As per HSE's Reducing risk Protecting People criteria for individual risks, i.e. the probability of being killed in an activity, are $1/10^6$ as generally acceptable risk and $1/10^4$ as tolerable risk, however if risk is in between those two boundaries, it has to be demonstrated that it is ALARP. Just for completeness according to the EN 1990:2002, failure probability of $1/7.2 \times 10^5$ /50 yr, i.e. reliability index $\beta = 3.8$ ($1/1.44 \times 10^6$ /1yr, $\beta = 4.7$) was specified as criteria for the consequence class 2 in the reliability calculation. These target reliabilities are calibrated with consideration of the possible consequences of failure, the associated costs, and the level of efforts and procedures necessary to reduce the risk of failure and damage.

According to ISO 2394, the fundamental principle of the marginal lifesaving costs should be applied for the regulation of life safety. This principle ensures a certain level of safety for the occupants of the structure. The safety level is determined so that the costs associated with saving additional lives i.e. applications of additional safety measures is higher than the corresponding marginal lifesaving costs. Marginal lifesaving costs, and thus minimum target reliabilities, can be based on the Life Quality Index (LQI), see more details in Nathwani et al. [10].

If there is no risk of loss of human lives associated with structural failures, the target failure probabilities can be determined based on pure economic optimization. Target reliabilities given in the JCSS Probabilistic Model Code [11] are presented as a function of consequences of failure and the relative costs of safety measure. These targets are indicative values for supporting economic optimization and might not be acceptable concerning life safety.

4 Direct and indirect risks with glass applications

As mentioned earlier, quantitative assessment of risks involves the quantification of probability of occurrences of (exposure) events and the quantification of consequences. Concerning exposures, for permanent and environmental actions on structures well developed probabilistic models can be found in the literature, e.g. JCSS [11], based on historical data. However, for structural glass applications, failure is often caused by accidental situations, for which it is more difficult to define reliable probabilistic distribution functions (Vrouwenvelder et al. [12]). A possibility is to look at glass failure statistics and analyze the failure occurrences.

Possible failures of the retention system are outlined in Table 1. Similarly, list of possible glass breakage probabilities with proposed frequencies of occurrence are presented in Table 2. The values in these tables can be seen as risks associated to failure of single elements, i.e. the direct risks of some particular events. This is an important measure to characterize a system's vulnerability to a certain exposure.

However, to completely avoid glass failure is (theoretically) impossible and (practically) prohibitively expensive. Therefore it is important to ensure that if glass breaks 1) the failure propagation should be limited and 2) the failure mechanism should be safe. The consideration of these aspects requires a system perspective and could be assessed by investigating the indirect risk, which in turn indicate the robustness of the system in consideration. Calculation of indirect risks requires the conditional probability of glass breakage leading to indirect consequences, i.e. partial or complete collapse of the entire structure. Furthermore, the quantification of consequences is need in terms of e.g. extent of subsequent system failure, number of injuries and fatalities, cover-up costs, losses due to business interruption etc.

Table 4-1 Implied frequency and potential retention system failures based on previous incidents.

Potential retention system failures based on previous incidents	Implied frequency
Inadequate sealant bite or bond area	1 in 50 years
Carrier frame screws fail to engage	1 in 50 years
Sealant deterioration	1 in 25 years
Inadequate edge capture of pressure cap	1 in 25 years
Glazing product defect e.g. igu edge seal failure	1 in 20 years
Substructure failure	1 in 100 years
Substructure connection failure	1 in 100 years

Table 4-2 Implied frequency and potential glass breakage causes based on previous incidents.

Potential glass breakage causes	Heat Treated Glass	Annealed Glass
Inclusions e.g. NiS	1 inclusion per 130 tonnes would generate ~ 1 per 500 tonnes per yr	Negligible
Variable glass processing e.g. uneven tempering	1 in 100 years	Negligible
Incipient damage during installation and gales or thermal stress	Negligible, if damage occurs during installation likely to break there and then, not years later	Say 3 incidents in 10 years as edges more prone to crushing
Cradle impact	1 in 10 years	1 in 5 years
Vandalism: catapults, air rifles, bricks applicable up to 12m	1 in 20 years	1 in 20 years
Stones flying up from adjacent rail / road traffic or roadwork's up to 12 m only	1 in 50 years	1 in 25 years
Bird strike or birds dropping objects	1 in 50 years	1 in 20 years as less impact resistant
Hailstorm	1 in 100 years	1 in 50 years
Impact from inside e.g. moving furniture	1 in 50 years	1 in 20 years as less impact resistant
Impact from timbers, roofing pebbles, falling glass etc. from adjacent buildings during gales	1 in 50 years	1 in 20 years as less impact resistant
Structural movement in the building causing glass to metal edge contact and squeezing of the glass	1 in 50 years	1 in 50 years
Thermally induced failure in base build	1 in 50 years	1 in 50 years
Thermal from inside through unforeseen action e.g. tenant fits black out screens, black leather sofas up against glass	1 in 50 years	1 in 50 years
Overzealous maintenance: e.g. jet wash or scrapers especially during resealing	1 in 50 years	1 in 20 years as less impact resistant
Inadequate glass design	1 in 100 years	1 in 100 years
Statistical variability in glass strength	Negligible	Negligible

The robustness of the system I_R can be measured by a robustness index defined as the ratio of direct risks to total risks after Baker [13]:

$$I_R = R_{dir} / (R_{dir} + R_{ind}) \tag{4.1}$$

Where R_{dir} represents the direct risks and R_{ind} is the indirect risks. This formulation is, however; just a crude approximation, as pointed out by Faber [14], since the consequences are decoupled from the specific failure scenarios. Nevertheless (some kind of) assessment of robustness is extremely useful for structural glass applications, due to the brittle nature of the glass. Robustness gives information about how much damage the system can tolerate without collapse.

However, since design of ultimately robust structures would be uneconomical, the discourse about the safety and reliability of engineering systems has been shifted towards resilience lately, as reported by e.g. Ayyub [15]. Resilience is a broad concept and emphasizes the temporal dimension of dealing with hazards, typically including phases of anticipation, resistance, absorption and restoration. A resilient system is able to "bounce back" quickly or even "bounce forward" i.e. built back better.

Therefore, an advisable strategy for designing structural glass is not simply increase robustness and build glass "fortessess", but to find a balance of providing alternative load path and sacrificing plies/elements/subsystems in a way that functioning of the building should be disrupted as little as possible. This might include availability of spare glass planes and/or connection elements, solutions where repair/replacement does not require business shut down etc.

5 Conclusions

Large glass surfaces without secondary supports together with the complex forms and shapes of modern architecture raise several challenges in regard to the structural design. This paper discussed some of these important issues and proposes methods to deal with risks concerning structural glass failure in a rational way. List of possible glass breakage probabilities with proposed frequency of occurrence was presented and concept of robustness of system discussed. In a subsequent publication the authors intend to provide illustrative examples based on probabilistic risk assessment to highlight the usefulness and limitations of the methods presented in this paper.

6 References

[1] ISO/IEC Guide 73: 2002, Risk management – Vocabulary – Guidelines for use in standards.

[2] Faber, M.H.; Stewart M.G.: Risk assessment for civil engineering facilities: critical overview and discussion. Reliability Engineering and System Safety. Vol. 80, 2003, pp 173-184.

[3] Kent, S.: Words of Estimative Probability, 1964.

[4] CDM 2015 , Managing health and safety in construction – Construction (Design and Management) Regulations 2015. Guidance on Regulations, ISBN: 9780717666263, HSE, London, UK, 2015. Online available at: http://www.hse.gov.uk/pubns/priced/l153.pdf.

[5] CIRIA C 632 Guidance on glazing at high 2006.

[6] London District Surveyors Association, Risk assessment – Management of health and safety at work regulations 1992. Specimen risk assessment and staff instructions, ISBN 095165182X, London, UK 1994.

[7] Fine, W.T.: Mathematical evaluations for controlling hazards, 1971.

[8] Kinney, G.F.: Practical Risk Analysis for Safety Management. Naval Weapons Center, NTIS report number NWC-TP-5865, 1976.

[9] HSE: Reducing risk Protecting People, London, UK, 2001.

[10] Nathwani, J.S.; Lind N.; Pandey M. Affordable Safety by Choice: The Life Quality Method. University of Waterloo, Waterloo, 1997.

[11] JCSS. Probabilistic Model Code, Joint Committee of Structural Safety, ISBN 978-3-909386-79-6, 2001, Online availiable at: www.jcss.byg.dtu.dk/Publications/Probabilistic_Model_Code.aspx.

[12] Vrouwenvelder, A.; Stieffel, U.; Harding, G.: Eurocode 1, Part 1.7 Accidental Actions, Background Document, First Draft, January 2005.

[13] Baker, J.W.; Schubert, M.; Faber M.H.: On the assessment of robustness. Structural Safety Vol. 30, 2008, pp 253-267.

[14] Faber, M.H.: Codified Risk Informed Decision Making for Structures. Symposium on Reliability of Engineering System, SRES'2015, Hangzhou, China, Oct. 15-17, 2015.

[15] Ayyub, B.M.: Systems Resilience for Multihazard Environments: Definition, Metrics, and Valuation for Decision Making. Risk Analysis. Vol. 34(2), 2014, pp 340-55.

Education City Doha – Glazed cable net system for tram stop stations

Felix Schmitt[1], Dr. Martin Bechtold[2], Ralf Dinort[3]

1 Josef Gartner GmbH, Beethovenstraße 5, 97080 Würzburg, Felix.Schmitt@josef-gartner.de

2 Fatzer AG, Salmsacherstraße 9, CH-8590 Romanshorn, Martin.Bechtold@fatzer.com

3 if_group, Am Dachsberg 3, 78479 Reichenau-Waldsiedlung, dinort@if-group.de

There is a People Mover System in progress for the New Education City in Doha. For 17 tram stops of this system, Grimshaw-Architects did the design together with SBP. The canopy's of the stops consist of two cable-nets which are fixed onto surrounding cables and pre-stressed together with connecting pins at all crossing points and reverse anchored by 6 masts fixed by tension rods. The project includes a structural ropes product novelty: the stainless steel full locked coil rope was used over the conventional stainless steel spiral strand rope. The striking technical and architectural properties provide the ultimate touch for this high profile project and made this new product a key aspect in the bidding phase. The benefits of the locked coil rope construction such as the smooth surface, the maximized contact to clamps and the maximized cross sectional area are combined with those of the stainless steel material with its shiny appearance and its corrosion resistance. The glass panels are triangular to allow for flat surfaces in this double curved geometry. The glass panels are fixed at the cable connection points by clamping their corners. The glass built-up is a mixture of several requirements and contains a hydrophobic coating on the top, two ceramic frit surfaces to reach a 3D-Rosette pattern and an acid etch surface on the inside. The technical and visual function of the building was confirmed by building Performance and Visual Mock-Up's.

Keywords: tram stop, cable net, stainless steel FLC ropes, glazed roof, hydrophobic coating

1 Introduction

1.1 Project objectives

In the west of Doha, the capital city of the state Qatar arises a new campus for the Education City. The city contains in 14 km^2 a lot of new buildings for various Universities, Academies as well as mixed and single sex residential buildings and more. This campus area will be served by an 11.5 km tram network by Siemens Avenio equipped with supercapacitor energy storage for wireless operation. The batteries will be charged during the night in the tram-depot station and during every stop in the tram stations.

The tram stop were designed by the world renowned and well known pioneers of high-tech architecture and transport projects "Grimshaw Architects". The network comprises one depot and 24 stops, of which 17 are open air stops with the cable net design. The architects looked for a design which is in context with the historic style and constructions in the gulf region, fulfill all requirements for a tram stop and will also be recognized as a

Engineered Transparency 2016. Glass in Architecture and Structural Engineering. First Edition.
Edited by Jens Schneider, Bernhard Weller.

landmark from a distance. This results in a tent like building which is fixed by cables. Grimshaw Architects developed the Construction Documentation together with a team of different companies in which “Schlaich Bergermann and Partner” acts as structural engineer and “Front” as façade consultant. In April 2015 “Permasteelisa Gartner Qatar” signed the contract for recalculation, delivery and installation of the 17 canopies of tram stops including 34 waiting rooms.

Figure 1-1 Rendering by Grimshaw-Architects.

2 Structure of the canopies

2.1 Structural system

Three tram stop configurations of the People Mover System have been developed by Grimshaw Architects in cooperation with Schlaich Bergermann & Partners. Type 1 accommodates two-way tram traffic with dual platforms and waiting rooms at each side. It is of standard system length of 40.8 m and width of 19.5 m. Type 2 and Type 3 are similar to type 1, but with a width of 13.8 m. The difference between Type 2 and 3 is that type 3 has only a one-way traffic, but is prepared for a two-way service.

The final execution differs only in a few items, which have been agreed between Gartner and the design team during construction design progress. These differences are in particular the orientation of the spreading elements and the type of the edge cables. The orientation of the spreader elements has been changed from vertical to perpendicular to the glass panels in order to avoid misalignments between the cable and the glazing joints transitions. The second change has been a novelty in cable structures. The first time full locked coil ropes in Stainless Steel have been applied in a cable structure. In this project FLCs Ø 45 mm, stainless steel 1.4436, have been used for the edge cables.

Out of the three configurations only type 1 will be discussed, since this type is the most common and is used in the largest station. 8 of totally 17 stations consist of type 1. Its overall dimensions are 40.8 x 19.5 m and its total height is approximately 8.4 m. The structural system consists of 4 corner columns and two intermediate columns which divide the long side in half. The corner columns are tied back by two pairs of tie rods in a V-shaped arrangement. Each V-pair consists of one rod connected to the column's head and another one connected to the column at a level of +4.50 m above ground. The two rods are attached to the same support bracket. The columns are supported on a spherical bearing. The intermediate columns are tied back only in their slope line by a pair of upper and lower tension rods. The columns support the roof's cable net structure.

The roof is composed of two curved cable nets, one convex and one concave, which are kept apart by struts respectively by tension rods in the outer region, since approximately 2m from the edge both nets are penetrating each other. The convex shaped net carries the glass via 2 parallel cables OSS Ø 12 mm with a distance of 45 mm. The other nets use single cables OSS Ø 16 mm except 24 mm for the lower cable of the cable truss between the intermediate columns. The cables forming the net are attached to the FLC Ø 45 mm edge cable by clamps. The edge cables themselves are connected to the columns opposite the tie back joints and transfer the lateral loads directly into these rods. The deflected load results in a compression force in the column. The following figures show the general dimensions in section and plan view. The section also shows the final orientation of the spreader elements perpendicular to the glass panels.

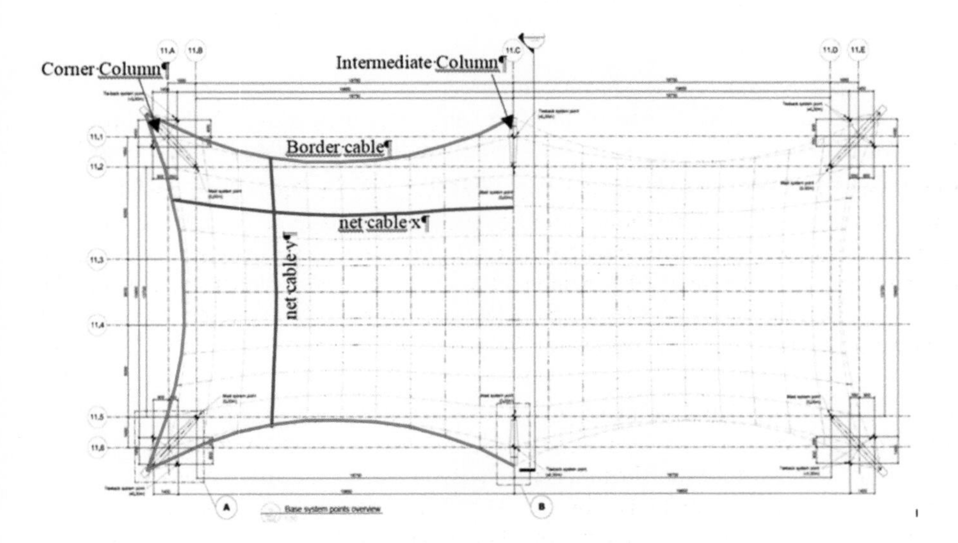

Figure 2-1 Plan View Type 1.

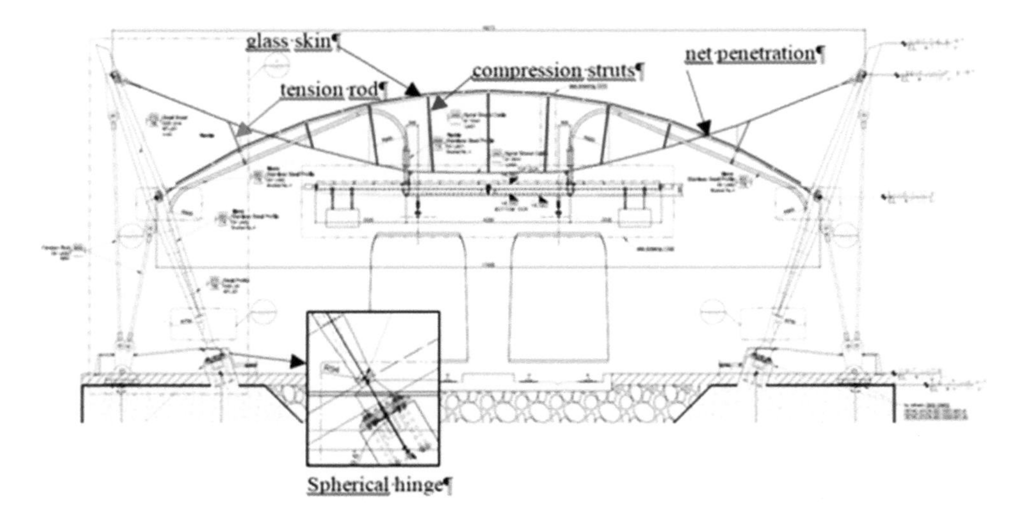

Figure 2-2 Cross section Type 1.

2.2 Load bearing principle

The People Mover Canopy is a lightweight structure composed of tension and compression elements. The stiffness of the structure is gained by the system's pretension. The pretension is achieved pushing both cable nets apart from each other using compression struts and tension rods between the nets. The pretension of the cables forming the net is transferred into the edge cables resulting in high cable force herein. The edge cables carry the force to the columns and outer tie rods. Columns and tie rods provide a pair of tension-compression elements and lead the cable forces corresponding to their decomposition of forces into the foundation.

External loads are carried by one of the two cable nets depending on their direction. Downward forces are allocated to the concave bottom net and upward forces accordingly to the concave shaped upper net, which is permanently loaded by the glass weight.

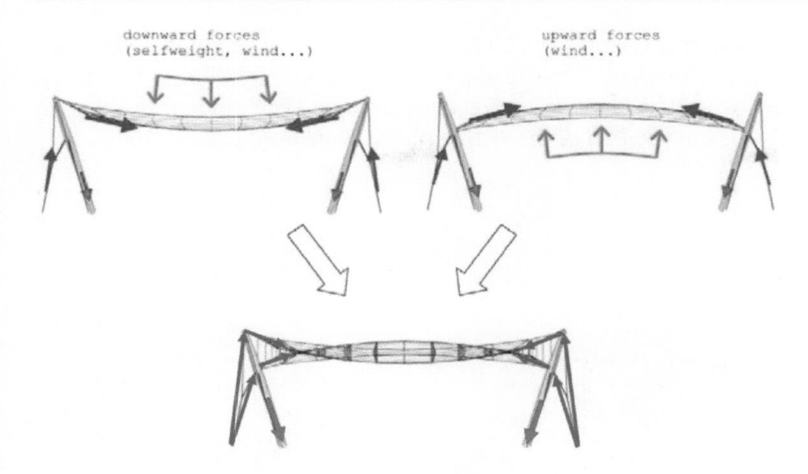

Figure 2-3 Load Bearing System [1].

Since cables are pure tension elements, the value of pretension must be chosen at a high enough level, that the cables will never completely lose their pretension, i.e. that the cables never will turn slack under external loads. This pretension method can be compared to a pre-stressed concrete construction. Due to their pretension both cable nets are also able to take "compression" forces. Another issue governed by the level of pretension is the limitation of deflections. In case of the People Mover there have been two criteria. The two edge cables at the front sides and the middle cable truss are supporting the OCR-frame (overhead charging rail-frame) since these are the stiffest axis. The deflection over the length of this frame is limited to ±10 mm for the 1 year return period of the governing wind load. The second criterion has been the vertical deflection under the wind load with 50 years return period. The cable net and glazing details shall allow for these deflections. The maximum upward deflection of the cable net has been specified with 351 mm and the maximum downward deflection with 230 mm.

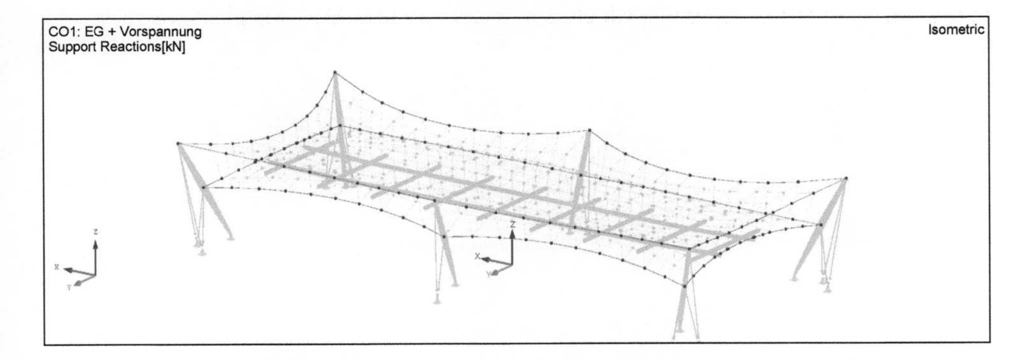

Figure 2-4 RSTAB model isometric view.

The structural calculation was performed with RSTAB/RFEM, a 3D nonlinear Structural Frame & Truss Analysis Software. The calculation was performed to 3rd order theory, taking element strain as well as large deflections into account. The influence of the wet silicone joints as well as the yielding of the setting blocks have been examined.

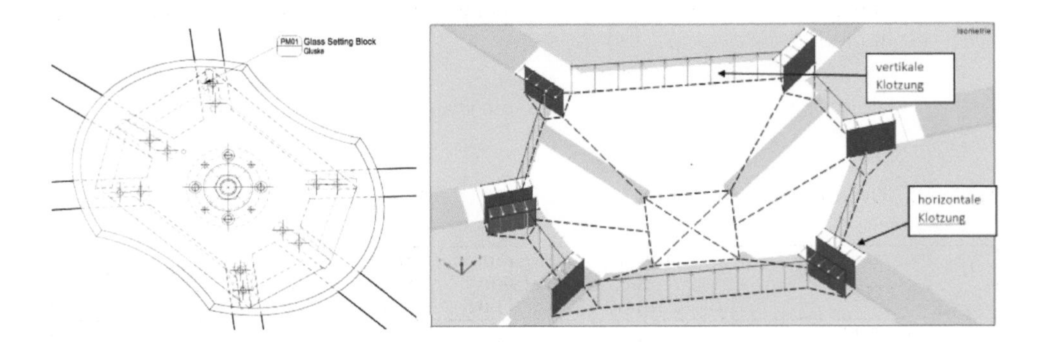

Figure 2-5 Glazing Support with Setting Blocks (left); FE Model of Glazing Support (right).

Because it has not been possible yet to calculate the whole model for all load combinations with such a precise mapping of gasket properties in an FE program a small reference model has been developed to calibrate the stiffness properties of substitute bracing elements used in RSTAB.

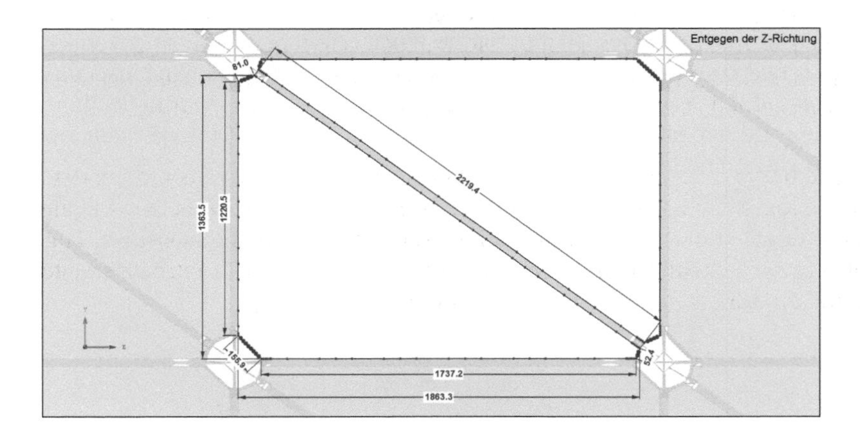

Figure 2-6 Reference Glass Panels within a Cable Net Mesh.

3 A rope novelty

3.1 Stainless steel cable assemblies

The tram stops are located close to the coast of the Persian Gulf. The hot desert climate with temperatures of up to 45°C and a high dew point create a severe corrosive environment. For this reason all stainless steel components are of grade 1.4436 or 1.4462 instead of the usual grade "A4" (1.4401).

An overview of the cable assemblies used in the structure are shown in table 3-1.

Table 3-1 Overview of the cable assemblies.

Rope Grade 1.4436	Diameter [mm]	Minimum Breaking Load [kN]	Stiffness [MN]	No.	Sockets Grade 1.4462
Stainless Steel Full Locked Coil Rope	45	1'740	177	204	HYEND Open Spelter Socket
Stainless Steel Spiral Strand Rope	24	489	45.5	17	HYEND Open Swaged Socket
	16	216	20.1	439	
	12	127	11.4	1014	
				34	HYEND Open Spelter Socket

3.2 State of the art in stainless steel cable assemblies

The reason to use stainless steel cable assemblies over galvanized cable assemblies in the first place is their unrivaled visual appearance. For that even quite significant disadvantages are taken on board as in comparison to a galvanized rope with the same diameter the breaking loads as well as the stiffness are lower whereas the price is higher.

The standard and by far most popular rope for glass facades and glass roofs are stainless steel spiral strand ropes up to 36 mm in diameter. These can be socketed with the swaging method which is efficient in manufacturing and the sockets have small dimensions. This method involves a socket with a cylindrical part with a bore. The rope is introduced into that bore and the cylindrical part is subsequently swaged onto the rope.

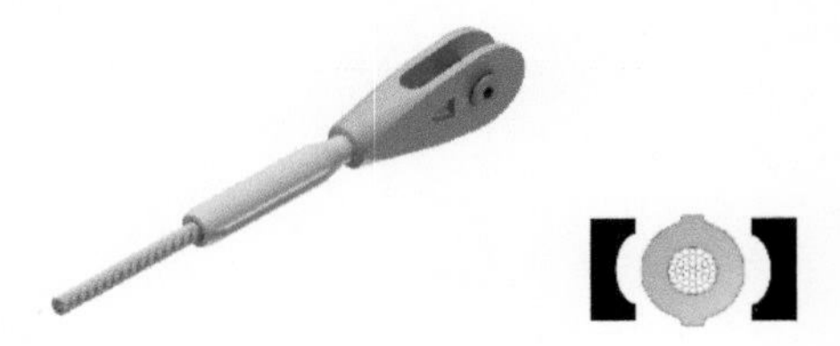

Figure 3-1 HYEND Open Swaged Socket and icon for the swaging method.

3.3 Larger diameter cable assemblies

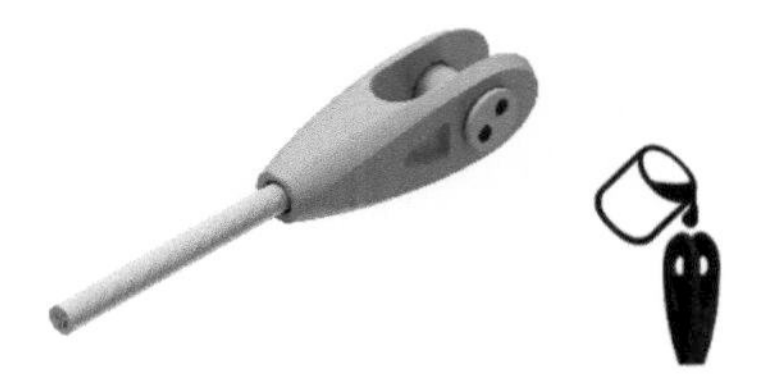

Figure 3-2 HYEND Open Spelter Socket and icon for the spelter method.

A rope diameter that goes beyond 36 mm is a game changer as a different type of socketing method is required. Swaging is no longer viable as the load transfer efficiency reduces. This very roughly can be explained by the degreasing ratio of circumference over cross sectional area for the rope. The spelter socketing method needs to be used.

This method involves a socket with a conical inside. The rope is opened into a brush which is placed inside the conical part of the socket. This space is then filled with a socketing medium. Due to the necessity of having to house the rope brush the sockets involved are bigger than for the swaging method. Also the manufacturing process takes longer. However this method transfers the full breaking load of the rope whereas the swaging method a 10 % reduction in breaking load efficiency applies.

Hot metal socketing as well as resin socketing are frequently used with galvanized ropes. The hot metal socketing uses a molten zinc alloy and requires heating of the conical part of the socket and the rope brush up to approximately 320 °C. The resin socketing uses a polyester resin which means the socket and the rope do not require heating. During the curing the chemical reaction can raise the temperature up to 50 °C.

Cable supported structures with diameters greater than 36 mm often switch to galvanized ropes and mostly the full locked coil rope is used over the spiral strand rope as it has a greater stiffness, a greater breaking load, a closed and therefore less exposed surface and is better suited for clamping. The closed and therefore less exposed surface is important for the corrosion protection of galvanized ropes.

Figure 3-3 Spiral Strand Rope and Full Locked Coil Rope.

The better clamping has its reason in the smoother surface. As more rope surface is in contact with the groove of the clamp the clamping force that is applied via bolts can be significantly higher (e.g. by 66 %). This obviously provides the possibility to design clamps smaller which has a direct visual impact on a facade or roof that is designed for transparency and lightness.

The galvanizing is dissatisfactory in high profile structures as it compromises the visual appearance on rope and socket. Therefore on some rare occasion's large diameter stainless steel spiral strand ropes have been produced up to 50 mm. In this case however the advantages of the full locked coil rope are missing.

3.4 A new development serves the need for this high profile project

The discerning reader may already see where this is leading to as the obvious missing product in this line of thought is the stainless steel full locked coil cable assembly. However some of the advantages that count for the galvanized full locked coil rope such as the greater breaking load, the closed and therefore less exposed surface do not apply or are of no significance for the stainless steel version. Firstly the breaking load for the stainless steel locked coil rope is slightly lower than that of the equivalent spiral strand. Secondly the corrosion protection is addressed by the material itself and the corrosion protection is not a sacrificial coating and therefore the exposed area and the openness of the rope is not an issue.

For the visual appearance however the openness of the rope plays an important role as it means more or less room and space for collection of dirt and impurities. The smooth surface of the full locked coil rope is also easily cleaned which means the structure can be kept in a good shape. Of course the advantage of the more efficient clamp design also fully applies.

The benefits of the locked coil rope construction such as the smooth surface, the maximized contact with clamps and the maximized cross sectional area are combined with those of the stainless steel material with its shiny appearance and its corrosion resistance.

Figure 3-4 Stainless steel full locked coil rope and HYEND Open Spelter Socket.

The socketing method with polyester resin was chosen as the socketing with hot metal would lead to a tempering color and the risk of embrittlement of the stainless steel in contact with the hot zinc alloy. Whilst this was an obvious choice more significant developments had to be done to produce the z-shaped wires and to define their quality levels as well as to find the optimum rope construction in terms of combination of wire.

4 Glass

The whole canopy is closed with glasses which are designed to fulfill a combination of requirements. An important function is to be readily identifiable in the landscape as a tram stop. At day the viewer perceives the color and curvature of the roof. Viewing from within the individual fittings, details of the filigree cable construction and waiting rooms are to ascertain. To allow the different functions of the glasses there are a lot of components in the build-up of the laminated glass:

Top: 8 mm fully tempered – Clear glass with an hydrophobic coating on top side
1.52 mm SGP Interlayer

Middle: 12 mm fully tempered – Grey tinted glass with ceramic frit on #3
1.52 mm SGP Interlayer

Bottom: 8 mm fully tempered – Low iron glass with ceramic frit on #5 and acid etch on #6

- Structural: The triangle glasses are supported at the three trimmed corners with flexible clamps to allow the movements of the cable-structure. The thicknesses of the three layers of glasses are calculated for the dimensions and different load- combinations.
- Dust and sand: To allow for easier cleaning with water hose, there is a hydrophobic coating on top side of the glass. The effect is also known as “Lotus-effect”.

- Water tightness: To have water tight roof all joints are closed with wet grey silicone on site.
- Sun shading: The visual light transmission is reduced with the combination of grey tinted glass and ceramic frit; whereas the acid etched bottom surface diffuses the transmitted daylight.
- Aesthetics for daylight under the canopy: There is a special developed ceramic frit pattern (Rosette pattern) at two different positions (No. 3 & 5) of different pattern, which gives by sun at each viewing location another visual appearance. The best scale of the pattern was selected at a Visual Mock-Up on site.
- Aesthetics at night-time: To maximize the night-time light reflection off the underside of the canopy to the platform, the white ceramic frit pattern of 70 % was selected. Furthermore, the acid etch treatment minimizes the glare from the up-lights. The selection of the glass-built-up causes a soft glow on the upper surface, which becomes a part of the night-time identity of the tram stops.

5 Mock-Up

All the performances and properties are tested with a Performance Mock-Up (PMU) and a Visual Mock-Up (VMU) for the canopy as well as for the Waiting Rooms. Whereas the tests took place in the testing area of Josef-Gartner in Gundelfingen, the two VMU where built in Doha to have the real light conditions. After the approval of all components the fabrication and delivery of the material could start. The installation of the first tram stop started at mid December 2015 and the completions of all 17 stations are expected in the year 2017.

Figure 5-1 VMU with different frit pattern (left). Installed cables (right).

Figure 5-2 Installation step.

6 References

[1] Gregory Haley, Grimshaw, 1874C-00GE-RT-0100-4 – Architectural Design Report.

[2] Robert Hellyer, SBP, 1874C-03SG-RT-0100-5 – Structural design report.

[3] Nathanael Anton, IF-group, Master Thesis: Einflüsse der Glaseindeckung auf die Verformungen und Schnittgrößen einer Seilnetzkonstruktion".

[4] Wacker Ingenieure ,1874C-03SG-RT-0300-0 – WIND TUNNEL REPORT, Wacker Ingenieur.

[5] Structural Design Report, Josef Gartner / IF-group, Canopy Type 1, 1874C-03SG-RT-0300-00.

Author Index

Keyword Index

T

U

V

W